Wood and Biofiber International Conference (WOBIC 2017)

IOP Conference Series: Materials Science and Engineering
Volume 368

Selangor, Malaysia
21 - 23 November 2017

Editors:

Mohammad Jawaid
Norkhairunnisa Mazlan

ISBN: 978-1-5108-6483-2
ISSN: 1757-8981

Printed from e-media with permission by:

Curran Associates, Inc.
57 Morehouse Lane
Red Hook, NY 12571

Some format issues inherent in the e-media version may also appear in this print version.

Printed by Curran Associates, Inc. (2018)

For permission requests, please contact the Institute of Physics
at the address below.

Institute of Physics
Dirac House, Temple Back
Bristol BS1 6BE UK

Phone: 44 1 17 929 7481
Fax: 44 1 17 920 0979

techtracking@iop.org

Additional copies of this publication are available from:

Curran Associates, Inc.
57 Morehouse Lane
Red Hook, NY 12571 USA
Phone: 845-758-0400
Fax: 845-758-2633
Email: curran@proceedings.com
Web: www.proceedings.com

Table of contents

Volume 368

The Wood and Biofiber International Conference (WOBIC 2017)

21–23 November 2017, Selangor, Malaysia

Accepted papers received: 30 May 2018
Published online: 13 June 2018

Preface

Papers

IOP Conf. Series: Materials Science and Engineering **368** (2018) 011001 doi:10.1088/1757-899X/368/1/011001

Preface

Present IOP Conference Series: Materials Science and Engineering covers selected papers in the field of Material Science and Engineering presented during The Wood and Biofiber International Conference (WOBIC 2017), 21-23 November 2017, Hotel Bangi-Putrajaya, Selangor, Malaysia. All papers selected for IOP Conference Series: Materials Science and Engineering after peer review process.

The IOP Proceeding will give a broad updated survey and information of the major innovation in the field of natural fibres, nano cellulose, polymer composites and nanocomposites for industrial applications and provide a reference material for future research in polymer-based composites and nanocomposites.

Present Proceeding covers papers written by eminent scholar and leading experts in the field of Material Science and Engineering. WOBIC-IOP Conference Series covers topics such as Characteristics of cellulose and cellulose nanofiber from oil palm mesocarp fibres, Material selection criteria and green technology for automotive industry, carbon fiber reinforced composite, bioethanol production from oil palm frond, biocomposite reinforced with sugarcane bagasse nanofiber, green nanocomposites from Sugar palm, and oil palm, treatment of palm oil mill effluent, date palm fibres based composites, pulp blending between oil palm empty fruit bunch and citronella leaf fibers, Kenaf, oil palm, Coir, sugar palm based biocomposites and hybrid composites, cellulose nanocrystals, bio-nanocomposite films, Recycling of waste rubber as fillers; carrageenan, starch, seaweed/calcium carbonate composite films, Peronema canescens and Acacia hybrid, paper incorporated with nanocellulose, Agarwood volatile compounds, oil palm veneer, water-borne coating, Biochemical methane, starch-based films, Ionic liquid, pulp and paper manufacturing from oil palm and bamboo, Activated carbons from oil palm shell, 2D woven kenaf, rubberwood, Plywood, bleached bamboo paper, and laminated kenaf woven fabric composites for below-knee prosthesis socket application.
The IOP conference series will fill the gap in the published literature or work on natural fibre, polymer composites and nanocomposites, its characterization and applications in different areas and provide a reference material for future research in this field. This IOP series is written by renowned experts from China, UK, Iran, Malaysia, Sudan, Japan, and Nigeria. We are highly grateful to all the authors who contributed in IOP Series and make it valuable contributions in Material Science and Engineering. We are also grateful to IOP Publishing, UK, supporting team especially Steph Gill and Anete Ashton for helping us to finalize this IOP Conference Series.

Guest Editors,
Mohammad Jawaid, Universiti Putra Malaysia, Serdang, Malaysia
Norkhairunnisa Mazlan, Universiti Putra Malaysia, Serdang, Malaysia

Peer review statement

All papers published in this volume of *IOP Conference Series: Materials Science and Engineering* have been peer reviewed through processes administered by the proceedings Editors. Reviews were conducted by expert referees to the professional and scientific standards expected of a proceedings journal published by IOP Publishing.

IOP Conf. Series: Materials Science and Engineering **368** (2018) 012001 doi:10.1088/1757-899X/368/1/012001

Characteristics of cellulose from oil palm mesocarp fibres extracted by multi-step pretreatment methods

L N Megashah[1], H Ariffin[1, 2]*, M R Zakaria[1] and Y Ando[3]

[1] Department of Bioprocess Technology, Faculty of Biotechnology and Biomolecular Sciences, Universiti Putra Malaysia, 43400 Serdang, Selangor, Malaysia.
[2] Laboratory of Biopolymer and Derivatives, Institute of Tropical Forestry and Forest Products (INTROP), Universiti Putra Malaysia, 43400 UPM Serdang, Selangor, Malaysia.
[3] Eco-Town Collaborative R&D Center for the Environment and Recycling, Kyushu Institute of Technology, 2-4 Hibikino, Wakamatsu-ku, Kitakyushu, Fukuoka 808-0196, Japan.

*Corresponding author: hidayah@upm.edu.my

Abstract. Characteristics of cellulose isolated from oil palm mesocarp fiber (OPMF) using multi-step pretreatment methods were determined in this study. It was postulated that the succession of pretreatment steps for cellulose extraction from lignocellulosic biomass may affect the cellulose properties. Pretreatment steps involved in cellulose extraction were superheated steam (SHS) treatment, enzymatic hydrolysis and alkaline treatment. OPMF was subjected to SHS treatment to open up the fibres wall structure followed by enzymatic hydrolysis using xylanase for effectiveness enhancement the alkaline treatment afterwards. In order to obtain bleached cellulose, peracetic acid, a totally chlorine free (TCF) chemical was used. Cellulose obtained from these multi-step pretreatment methods had purity of 87.5 %, with crystallinity of 69.2 % and thermally stable up to 360 °C.

1. Introduction

Plant based materials provides potential applications such as in biofuel and biosugar production. The unique properties of plant material are mainly due to its carbohydrate composition which could be converted into value-added products. There have been many reports on the conversion of these lignocellulosic materials such as sugarcane bagasse [1-2], corn stover [3], rice straw pulp [4], raw cotton linter [5], sawdust, rice husk and wheat straw [6]; into value-added products. In Malaysia, oil palm biomass have attracted interest from many researchers [7–9]. There are several types of oil palm biomass such as oil palm fronds (OPF), oil palm empty fruit bunch (OPEFB), oil palm trunks (OPT), oil palm shells (OPS) and oil palm mesocarp fibers (OPMF).

IOP Conf. Series: Materials Science and Engineering **368** (2018) 012001 doi:10.1088/1757-899X/368/1/012001

Cellulose is a linear polymer of anhydroglucopyranose units (AGU) which is linked through ether bonds, while hemicellulose has mixed monomeric unit of C5 and C6. Commonly, hemicellulose contains more than one sugar type and known as amorphous polysaccharide. For lignin, it has a complex crosslinking polymer with phenylpropane unit [10]. Cellulose was first discovered by a French chemist named Anselme Payen in 1837 who presented a carbohydrate in various plant material yielded with fibrous substance after purification with acid ammonia treatment and the extraction with water, alcohol or ether. Generally, isolation of cellulose from lignocellulosic materials requires harsh conditions considering the complex plant structure. Until today, various methods for isolation of cellulose, such as alkaline treatment, bisulfite/sulfate process, biological treatment, and enzymatic treatment as well as the combination with mechanical treatment have been tested. It was shown that different process resulted in diverse in physical and chemical properties of cellulose. Table 1 shows examples of pretreatment methods used for cellulose extraction from plant biomass with their advantages and shortcoming.

Table 1. The advantages and shortcoming from different methods of cellulose extraction.

Method for cellulose extraction	Biomass	Advantages	Shortcoming	References
Hydrogen peroxide and acetic acid combination	Rice straw, pine wood & oak wood	Simple methods	Low purity of cellulose obtained	[11]
Pulping, bleaching (50% hydrogen peroxide) and purifying process (5% potassium hydroxide)	Oil palm frond	Increase the brightness of bleached pulp	High energy consumption and high alkaline charge for pulping process	[12]
Mechanical commutation, steam washing, sulphuric acid treatment, organosolvent extraction and alkaline hydrogen peroxide process.	Sweet sorghum bagasse	Economical and eco-friendly processing	High cost and required many chemical replacement	[13]
Oxygen, hydrogen peroxide and sodium chlorite treatment. Followed by enzymatic saccharification.	Mixed hardwood chips	Attributed to removal of hemicellulose and lignin	Required chlorinated chemicals	[14]
Dewaxed with (2:1, v/v) mixture of toluene and ethanol, alkali treated and acid hydrolysis (2x). Followed by sodium chlorite bleaching.	Arecanut husk fibres	Hemicellulose removal	Many chemical replacement and required chlorinated chemicals	[15]

Recent studies of these pretreatment and delignification process have been investigated by considering the treatment conditions which give less impact on the environment. In this study, multi-step pretreatment methods consisted of enzyme, superheated steam, alkali and chlorine-free bleaching have been used for cellulose isolation from oil palm mesocarp fibre (OPMF). The aim of using combined pretreatment was to obtain highly purified cellulose. Characteristics of the cellulose obtained was studied by using several analytical tools such as by using Wide-angle X-ray Diffraction (WAXD), Scanning Electron Microscopy (SEM), Fourier Transform Infrared Spectroscopy (FTIR) and Thermogravimetric Analyzer (TGA) for determining the cellulose crystallinity, morphology, functional groups and thermal stability.

2. Experimental

2.1 Materials
OPMF was collected from Seri Ulu Langat Palm Oil Mill, Dengkil Selangor, Malaysia. The samples were thoroughly washed with tap water and sun dried until free from moisture to avoid fungal growth in samples. The dry samples were kept in sealed plastic bag and stored at room temperature prior to further treatment. All chemicals used in pretreatment were of laboratory grade. Commercial xylanase enzyme (pulpzyme HC 2500) was supplied by Novozymes A/S (Denmark).

2.2 Multi-step treatment (Superheated steam, Enzymatic hydrolysis, Alkaline treatment & Bleaching)
The superheated steam pretreatment was conducted using a lab scale superheated steam oven (QF-5200C, Naomoto Corporation, Osaka, Japan). In brief, the conditions used were according to the previous study [16]. The temperature and retention time was 260 °C for 30 min. About 100g of SHS treated fibers was soaked with sodium phosphate buffer (pH5.5) of xylanase which has an activity of 2500 AXU/g in standard condition. Hydrolysis using xylanase was conducted at liquid to solid content of 5:1 (w/w) at temperature of 55 °C with agitation speed of 160 rpm for 24 hours. Alkaline treatment was conducted using 10 % (w/v) sodium hydroxide (NaOH) and heated at 90 °C for 60 min. The ratio of solid to liquor fixed at 1:10 (w:w). Finally, bleaching was conducted using peracetic acid (PA) solution. PA solution was developed by mixing acetic acid and hydrogen peroxide at equal volume ratio (5:5) [11] with addition of catalyst using 98 % sulphuric acid (H_2SO_4) at 1.1 % (w/v). The samples were vigorously stirred for 24 hours at 70 °C.

Solids obtained after multi-step pretreatments were washed with distilled water until the pH became neutral. Characteristics of samples obtained from each stage of treatment were determined.

2.3 Characterization of untreated and treated OPMF

2.3.1 Chemical composition analysis. Chemical composition analysis of fibres was conducted based on Technical of Pulp & Paper Industry (TAPPI) Standard Test Method. Composition included the testing of moisture content (T264 om-88), extractives (T204 cm-97), lignin content (T222 om-88), cellulose and hemicellulose content (TAPPI useful method 249-75) [17].

2.3.2 Thermo gravimetric Analysis (TGA). Thermal stability of fibres was determined by conducting thermogravimetric (TG) analysis on a Thermo gravimetric/ Differential Thermal Analyzer (TG/DTA) (model: EXSTAR6000, Hitachi High-Tech Science Corporation, Tokyo, Japan). The samples were heated from 40 °C to 550 °C at heating rate of 10 °C/min under nitrogen flow of 100 ml/min.

2.3.3 Scanning Electron Microscopy (SEM) analysis. The morphology of sample surface was examined using a scanning electron microscopy (SEM) model JCM-6000PLUS NeoScope Benchtop SEM (JEOL Ltd., Japan). Prior to observation, the samples were coated with Pt for 30 s by using Hitachi Ion Sputter E-1030. The view was conducted at several magnifications: 500X, 2000X and 8000 X with beam voltage of 15 kV.

2.3.4 Wide Angle X-ray Diffraction (WAXD) analysis. Crystallinity of samples was determined using a Rigaku MiniFlex 600 bench top X-ray Diffractometer operated at 40 kV and 15 mA and the measurement condition was in 2θ range of 5 to 50° with scanning speed at 3.6 L/min (about 40 minutes). The crystallinity index (CrI) was calculated by the subtraction of total intensity (I_{002}) at 2θ=23 and the amorphous scatter (I_{am}) at 2θ=18 by divided with I_{002} and represent as percent of crystallinity [18].

2.3.5 Fourier Transform Infrared Spectroscopy (FTIR) analysis. The functional groups of fibers were determined using Nicolet iZ10 FT-IR spectrometer in standard analyses using transmission or ATR technique. The spectral resolution was scanned in the range of 700 to 4000 cm^{-1}.

3. Results and Discussion

In this study, cellulose extraction from OPMF was conducted in multiple steps with the combination of superheated steam pretreatment, enzymatic hydrolysis, alkaline treatment and peracetic acid (PA) bleaching.

3.1 Effect of chemical composition in each stage of multistep treatment

Chemical composition of fibre samples is shown in Table 2. The data is presented as the mass fraction (%) of each component of cellulose, hemicellulose, lignin and extractives presence in untreated and in each stage of treated samples. Untreated OPMF contained 28.2 %, 32.7 % and 32.4 % cellulose, hemicellulose and lignin, respectively. Extractives in raw OPMF were about 6.5 %. Extractives may indicate the presence of oil and wax in fibers which are available at small amount in biomass. The fibers were then subjected to multistep pretreatment with the aim to remove the non-cellulosic component by increasing the cellulose fraction in fibres. The purpose of using combined treatments was to break down the bondings between non-cellulosic components like hemicellulose and lignin, without degrading the cellulose.

It is seen that hemicellulose content was decreased from 32.7 % to 20.2 % after SHS treatment. The composition of lignin increased after SHS treatment due to condition in which lignin depolymerized while in condensation at high temperature and limited the removal of lignin [19]. It has been reported that lignin isolates from pretreated biomass such as using superheated steam exhibited better enzyme binding on fiber surface compared to directly use enzyme on untreated biomass [20]. Therefore, subsequent treatment using enzymatic hydrolysis would enhance the removal of recalcitrant hemicellulose and at the same time extract the cellulose. The xylanase helped to loosen the link of xylan or known as recalcitrant hemicellulose in between cellulose and lignin [21-22]. As a result, the hemicellulose greatly reduces within treatment to 12.7 %. The use of xylanase also potentially accelerates the delignification process towards fibers. By this reason, the fibres subject to delignification using alkaline chemical which results in swelling the fibres and partly removal of lignin. The removal of some lignin cause the loose of network binding for fibre bundles [23]. The extensive lignin removal proposed at final stage of bleaching using PA. PA bleaching was taken place to remove remaining lignin component by reaction of oxidation which the lignin depolymerized at temperature of 70 °C and consequently more dissolving of lignin. Thus, the fraction of cellulose revealed and take place the highest percentage of cellulose content which gives the purity of cellulose from OPMF about 87.5 %.

Table 2. Chemical composition (%) for lignocellulose component of OPMF in each multistep treatment.

Multi-step stage	Cellulose (%)	Hemicellulose (%)	Lignin (%)	Extractives (%)
Raw OPMF	28.2 ± 0.8	32.7 ± 4.8	32.4 ± 4.0	6.5± 0.1
SHS pretreatment	27.3 ± 0.0	20.2 ± 0.0	42.0 ± 0.3	6.0 ± 0.9
Enzymatic hydrolysis	33.6 ± 0.1	12.7 ± 0.0	48.3 ± 0.1	5.41 ± 0.7
Alkaline treatment	39.9 ± 1.2	11.4 ± 1.0	37.6 ± 0.2	2.6 ± 0.0
PA bleaching	87.5 ± 1.2	12.5 ± 1.1	0.0 ± 0.0	0.0 ± 0.0

*Note: Results are representative of three replicate analysis

3.2 Morphological changes at each stage of pretreatment

In case of morphology, the effect of pretreatment on fiber surface can be clearly evaluated using a standard resolution of SEM micrograph in Figure 1. The physical structure of the raw OPMF can be seen clearly, i.e. the silica bodies were highly attached in pores. SHS treatment was found to remove the silica deposits, as being reported by [9]. Further treatment by xylanase which removed some amount of hemicellulose resulted in smaller size fibrils. The subsequent alkaline treatment contributed to ultimate removal of silica bodies, which opened the pores. In this stage, the size of fibres reduced to 10 micrometer due to separation of fiber matrix. PA bleaching which removed lignin eventually disrupted the structure of the lignocellulose, leaving small size microfibrils consisted of mainly cellulose (Table 2).

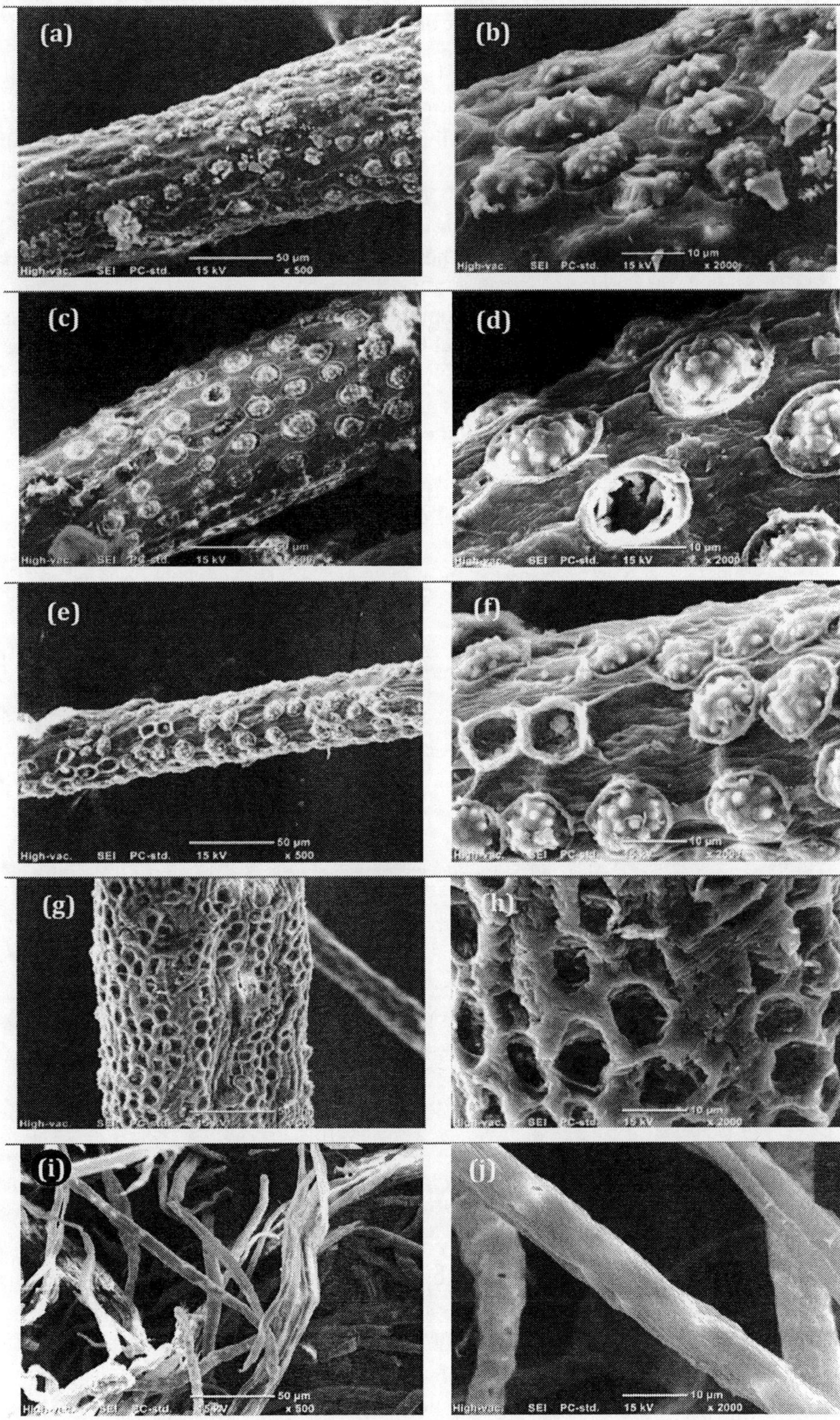

Figure 1. Morphology of OPMF treated from each pretreatment step: (a) raw OPMF at 500x; (b) raw OPMF at 2000x; (c) superheated steam treated at 500x; (d) superheated steam treated at 2000x; (e) enzymatic treated at 500x; (f) enzymatic treated at 2000x; (g) alkaline treated at 500x; (h) alkaline treated at 2000x; (i) PA bleaching at 500x; (j) PA bleaching at 2000x.

3.3 *Functional groups of untreated and treated OPMF*

The changes of functional groups in sample from each treatment stage were characterized by using ATR-FTIR. As shown in Figure 2, the OPMF fibres had a strong and broad hydrogen bond of O-H stretching at 3348 cm^{-1} region which indicated as cellulose group with water absorption [12,24]. The C-H stretch vibration was notably low at 2840-2970 cm^{-1}. The peak intensity at around 1700-1730 cm^{-1} indicates as the presence of C=O stretch which apparent to hemicellulose or lignin position [8,24–26]. There was no peak corresponded to the hemicellulose and lignin components after alkaline treatment and PA bleaching, which indicated almost complete removal of the components. The disappearance of the peak in PA-bleached sample corresponds to mainly cellulose. As discussed earlier, the rises of cellulose indicates the removal of non-cellulosic component. The aromatic skeletal of C=C at 1613 cm^{-1} corresponded to existing of lignin which can be observed clearly for raw OPMF fibers [8,26]. The following region around 1429 cm^{-1}, 1463 cm^{-1}, 1509 cm^{-1} and the rest were assigned to CH_3, C=O stretch and C=C stretch or aromatic skeletal of lignin. Whereas, the peak intensity at around 900-1200 cm^{-1} apparent to be strong as fibres are treated which corresponds to C-O stretch, C-O-C stretch and C-H bending which can be defined as cellulose and hemicellulose [12].

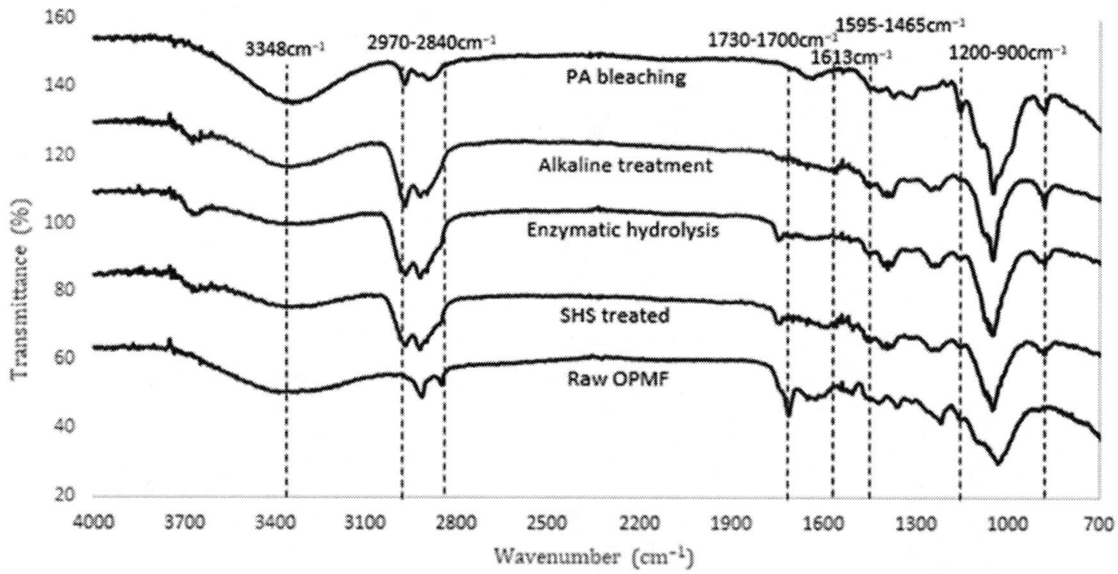

Figure 2. Band characteristic of FTIR spectra after each stage multistep treatment.

3.4 *Thermal stability of untreated and treated OPMF*

In correlation with other analysis, the prediction of cellulose, lignin and hemicellulose composition can be analyzed using a TGA. Thermal stability and decomposition peak of the cellulose fiber are presented Figure 3. As shown in the figure, weight loss within temperature range of 50 to 100 °C can be attributed to water content [10, 27]. It is interesting to note that the thermal stability of the OPMF treated samples varied according to the treatment step, which could be related with the chemical composition of the samples. Based on TG thermogram, raw OPMF showed multi-step degradation, which could be attributed by cellulose, hemicellulose and lignin degradation. [28] reported that hemicellulose is degraded at lower temperature due to its amorphous structure with low molecular mass.

Subsequent pretreatments of SHS followed by enzyme hydrolysis resulted in fibres with higher thermal stability with T_{max} (from DTG thermogram) at around 350 °C. On the other hand, despite of further hemicellulose removal by alkaline treatment, thermal stability of alkaline-treated sample was reduced. This could be contributed by reduced lignin composition and cellulose degradation at high concentration of alkali. TG thermogram showed complete fibre degradation for fibre obtained after PA treatment, indicating that the fibre obtained after the last step of treatment consisted of mainly cellulose.

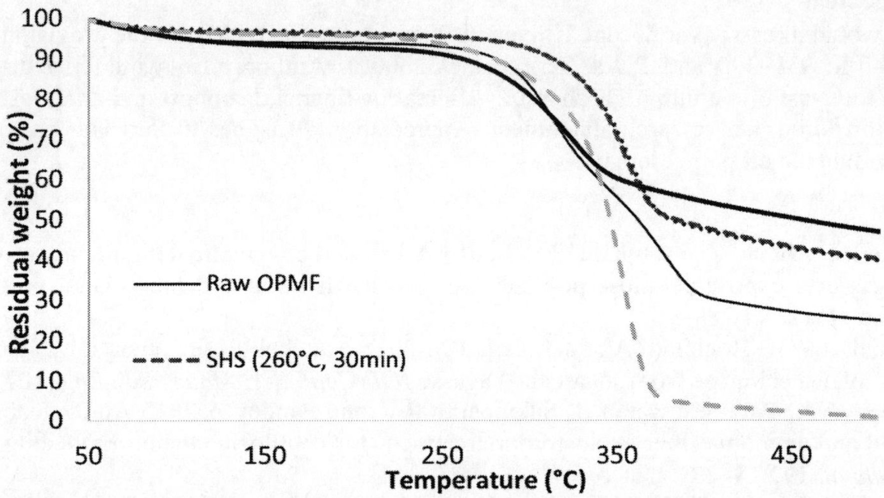

Figure 3. TGA thermogram of OPMF samples obtained from multi-step pretreatment.

Crystallinity of untreated and multi-step pretreated OPMF are conducted by WAXD as shown in Figure 4. Untreated OPMF showed broader peak which attributed to amorphous cellulose. After multi-step pretreatment, sharp peaks at 16° and 23° was observed indicating the crystalline fraction of OPMF cellulose. Crystallinity of the OPMF fibre increased from 34.3 % (before pretreatment) to 69.2% (after multi-step pretreatment). The increase of crystallinty is mainly due to the removal of amorphous structure of hemicellulose and lignin [29]. As shown in diffraction pattern, the strongest intensity was found in treated fibers which eventually affected the crystallinity of cellulose.

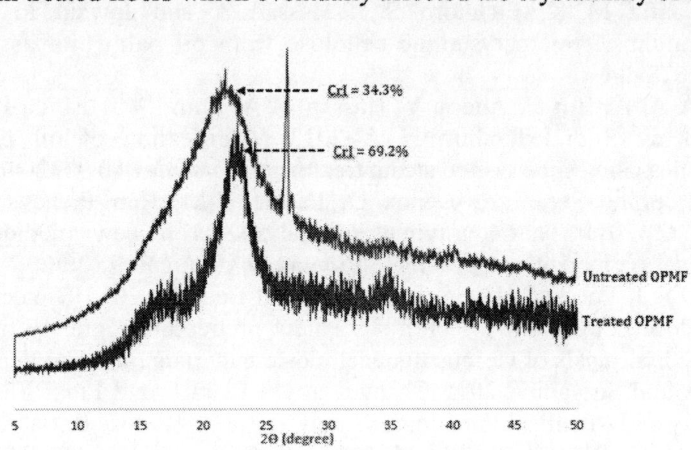

Figure 4. WAXD analysis of untreated and multi-step pretreated OPMF.

4. Conclusions

Results from this study exhibited that cellulose extraction could be achieved by combination of multiple-step pretreatment methods. Each step of pretreatment affecting differently on the fiber, resulted in fibers with diverse characteristics. It was interesting to note that thermal stability was greatly affected after each pretreatment stage, which is mainly due to the chemical composition of the fiber after pretreatment which resulting increment in cellulose purity from 28.2 % to 87.5 %. Produced cellulose obtained after pretreatment had a crystallinity of 69 %, suggesting the suitability of the sample for further applications such as in biocomposite or nanocellulose production.

Acknowledgement

The authors would like to acknowledge Universiti Putra Malaysia (UPM) for the provision of research grant (Putra-IPS: 9519300 and Putra Berimpak 9523000). Authors also would like to extend the appreciation to Kyushu Institute of Technology, Japan for financial support and analyses provided to the first author during her research attachment. Appreciation also goes to Seri Ulu Langat Palm Oil Mill for providing the oil palm biomass.

References

[1] Asgher M, Ahmad Z and Iqbal H M N 2013 Alkali and enzymatic delignification of sugarcane bagasse to expose cellulose polymers for saccharification and bio-ethanol production *Ind. Crops Prod.* **44** 488–95

[2] Wulandari W T, Rochliadi A and Arcana I M 2016 Nanocellulose prepared by acid hydrolysis of isolated cellulose from sugarcane bagasse *IOP Conf. Ser. Mater. Sci. Eng.* **107** 12045

[3] Mathew A K, Parameshwaran B, Sukumaran R K and Pandey A 2016 An evaluation of dilute acid and ammonia fiber explosion pretreatment for cellulosic ethanol production *Bioresour. Technol.* **199** 13–20

[4] Hasanjanzadeh H, Hedjazi S, Ashori A, Mahdavi S and Yousefi H 2014 Effects of hemicellulose pre-extraction and cellulose nanofiber on the properties of rice straw pulp *Int. J. Biol. Macromol.* **68** 198–204

[5] Morais J P S, Rosa M D F, De Souza Filho M D S M, Nascimento L D, Do Nascimento D M and Cassales A R 2013 Extraction and characterization of nanocellulose structures from raw cotton linter *Carbohydr. Polym.* **91** 229–35

[6] Sindhu R, Binod P and Pandey A 2016 Bioresource Technology Biological pretreatment of lignocellulosic biomass – An overview *Bioresour. Technol.* **199** 76–82

[7] Iberahim N I, Jahim J M, Harun S, Nor M T M and Hassan O 2013 Sodium Hydroxide Pretreatment and Enzymatic Hydrolysis of Oil Palm Mesocarp Fiber *Int. J. Chem. Eng. Apl.* **4** 101–5

[8] Mohamad Haafiz M K, Eichhorn S J, Hassan A and Jawaid M 2013 Isolation and characterization of microcrystalline cellulose from oil palm biomass residue *Carbohydr. Polym.* **93** 628–34

[9] Nordin N I A A, Ariffin H, Andou Y, Hassan M A, Shirai Y, Nishida H, Yunus W M Z W, Karuppuchamy S and Ibrahim N A 2013 Modification of oil palm mesocarp fiber characteristics using superheated steam treatment *Molecules* **18** 9132–46

[10] Carrier M, Loppinet-Serani A, Denux D, Lasnier J-M, Ham-Pichavant F, Cansell F and Aymonier C 2011 Thermogravimetric analysis as a new method to determine the lignocellulosic composition of biomass *Biomass and Bioenergy* **35** 298–307

[11] Wi S G, Cho E J, Lee D S, Lee S J, Lee Y J and Bae H J 2015 Biotechnology for Biofuels Lignocellulose conversion for biofuel : a new pretreatment greatly improves downstream biocatalytic hydrolysis of various lignocellulosic materials *Biotechnol. Biofuels* 1–11

[12] Syamani F A and Suryani A 2015 Changes in Oil Palm Frond Fiber Morphology , Cellulose Crystallinity and Chemical Functional Groups during Cellulose Extraction Phases **7** 105–14

[13] Ganesh Kumar C, Pradeep Kumar M, Gupta S, Shyam Sunder M, Veera Mohana Rao K, Jagadeesh B, Swapna V and Kamal A 2015 Isolation and Characterization of Cellulose from

Sweet Sorghum Bagasse *Sugar Tech* **17** 395–403

[14] Park J, Shin H, Yoo S, Zoppe J O and Park S 2015 Delignification of lignocellulosic biomass and its effect on subsequent enzymatic hydrolysis *BioResources* **10** 2732–43

[15] S J C C, George N and Narayanankutty S K 2016 Isolation and characterization of cellulose nanofibrils from arecanut husk fibre *Carbohydr. Polym.* **142** 158–66

[16] Warid M N M, Ariffin H, Hassan M . and Shirai Y 2016 Optimization of Superheated Steam Treatment to Improve Surface Modification of Oil Palm Biomass Fiber *BioResources* **11** 5780–96

[17] Teramoto Y, Tanaka N, Lee S and Endo T 2008 Pretreatment of Eucalyptus Wood Chips for Enzymatic Saccharification Using Combined Sulfuric Acid-Free Ethanol Cooking and Ball Milling **99** 75–85

[18] Yasim-Anuar T A T, Ariffin H, Norrrahim M N F and Hassan M A 2017 Factors Affecting Spinnability of Oil Palm Mesocarp Fiber Cellulose Solution for the Production of Microfiber *Bioresources* **12** 715–34

[19] Tian D, Chandra R P, Lee J S, Lu C and Saddler J N 2017 Biotechnology for Biofuels A comparison of various lignin - extraction methods to enhance the accessibility and ease of enzymatic hydrolysis of the cellulosic component of steam - pretreated poplar *Biotechnol. Biofuels* 1–10

[20] Zeng Y, Zhao S, Yang S and Ding S ScienceDirect Lignin plays a negative role in the biochemical process for producing lignocellulosic biofuels *Curr. Opin. Biotechnol.* **27** 38–45

[21] Roncero M B, Torres A L, Colom J F and Vidal T 2000 Effects of xylanase treatment on fibre morphology in totally chlorine free bleaching (TCF) of Eucalyptus pulp *Process Biochem.* **36** 45–50

[22] Gangwar A K, Prakash N T and Prakash R 2014 Applicability of Microbial Xylanases in Paper Pulp **9** 3733–54

[23] Abdel-Halim E S 2014 Chemical modification of cellulose extracted from sugarcane bagasse : Preparation of hydroxyethyl cellulose *Arab. J. Chem.* **7** 362–71

[24] Fahma F and Iwamoto S 2010 Isolation , preparation , and characterization of nanofibers from oil palm empty-fruit-bunch (OPEFB) 977–85

[25] Fatah I Y A, Abdul Khalil H P S, Hossain M S, Aziz A A, Davoudpour Y, Dungani R and Bhat A 2014 Exploration of a chemo-mechanical technique for the isolation of nanofibrillated cellulosic fiber from oil palm empty fruit bunch as a reinforcing agent in composites materials *Polymers (Basel).* **6** 2611–24

[26] Ren H, Zhao T, Xing J, Zhang Y, Li Z, Wang Y and Ma W 2015 Isolation and Characterization of Microcrystalline Cellulose from Distillers Grains *J. Biobased Mater. Bioenergy* **9** 515–22

[27] Nascimento P, Marim R, Carvalho G, Celso R, Cid G, Celso R and Cid G 2015 Nanocellulose Produced from Rice Hulls and its Effect on the Properties of Biodegradable Starch Film **19** 1–8

[28] Acquah G E, Via B K, Fasina O O, Adhikari S, Billor N and Eckhardt L G 2017 Chemometric modeling of thermogravimetric data for the compositional analysis of forest biomass 1–15

[29] Singh S, Cheng G, Sathitsuksanoh N, Wu D, Varanasi P, George A, Balan V, Gao X, Kumar R, Dale B E, Wyman C E and Simmons B A 2015 Comparison of different biomass pretreatment techniques and their impact on chemistry and structure **2** 1–12

Material selection criteria for natural fibre composite in automotive component: *A review*

M Noryani[1,2], S M Sapuan[1*], M T Mastura[3], M Y M Zuhri[1] and E S Zainudin[1]

[1]Department of Mechanical and Manufacturing Engineering, Faculty of Engineering, Universiti Putra Malaysia, 43400 UPM Serdang, Selangor, Malaysia
[2]Faculty of Mechanical Engineering, Universiti Teknikal Malaysia Melaka, Hang Tuah Jaya, 76100 Durian Tunggal, Melaka, Malaysia.
[3]Faculty of Engineering Technology, Universiti Teknikal Malaysia Melaka, Hang Tuah Jaya, 76100 Durian Tunggal, Melaka, Malaysia.

*Corresponding author: sapuan@upm.edu.my

Abstract. Emphasizing the green technology in manufacturing process with proper materials selection of natural fibre reinforced composite (NFRC) can save our future generation. There is multiple consideration of the criteria from the reseacher for material selection process since the database of NFRC is not completely available. Product design specification (PDS) of automotive component is studied from past researcher with the summarized of the elements considered. Literature from year 2000 until year 2016 were collected which is focused on related automotive component design. In this study, it is found that the most commonly used elements of PDS of automotive component was consider from the previous researcher were cost, density, tensile strength, Young's modulus and elongation at break. Generally, the mentioned elements are recommended to be considered by other researchers in automotive component design in material selection process.

1. Introduction

In recent years, the annual car production is increase in automotive industry by the year 2020. It will give a special advantage by emphasize renewable material and green operation in manufacturing process according to automotive perspective [1]. It's become a good effort from the industry to overcome the projections world energy consumption 1990-2040 as reported by the International Energy Outlook [2]. In addition, a good green product can be delivered to the end user and give a good environment to the future generation. Based on World Nuclear Association, the electricity energy is the higher demand increase by more than two-thirds as overall energy in between 2011-2035 that may need a nuclear power suggested by major international report on energy. To beat this global issue in future, the design engineers should properly select the right materials in manufacturing process especially in automotive component to meet the standard product requirement by the industry.

In a recent year, an engineer keeps on seeking the best materials that can produce a better product with a satisfactory performance. The main factor is to meet the product design specification (PDS) for each of the automotive component before the assembly process of the product in a whole design and manufacturing process. Good materials, low cost, lightweight and performance are the factors that contribute to the last user which is customer. Finally, final product such as a car meet the

IOP Conf. Series: Materials Science and Engineering **368** (2018) 012002 doi:10.1088/1757-899X/368/1/012002

customer satisfaction and can reflect to the customer loyalty to the product [3-4]. Material selection is one of the critical process in design and manufacturing of a product as shown in Figure 1. Ashby [5] describe the characteristic of one material can increase the precision of the product compare to a multiple and diversity of the materials characteristics. A good receipt of the materials will give a better performance on the physical, mechanical and environmental effect [6–10]. In fact, it can reduce cost and time [11–13].

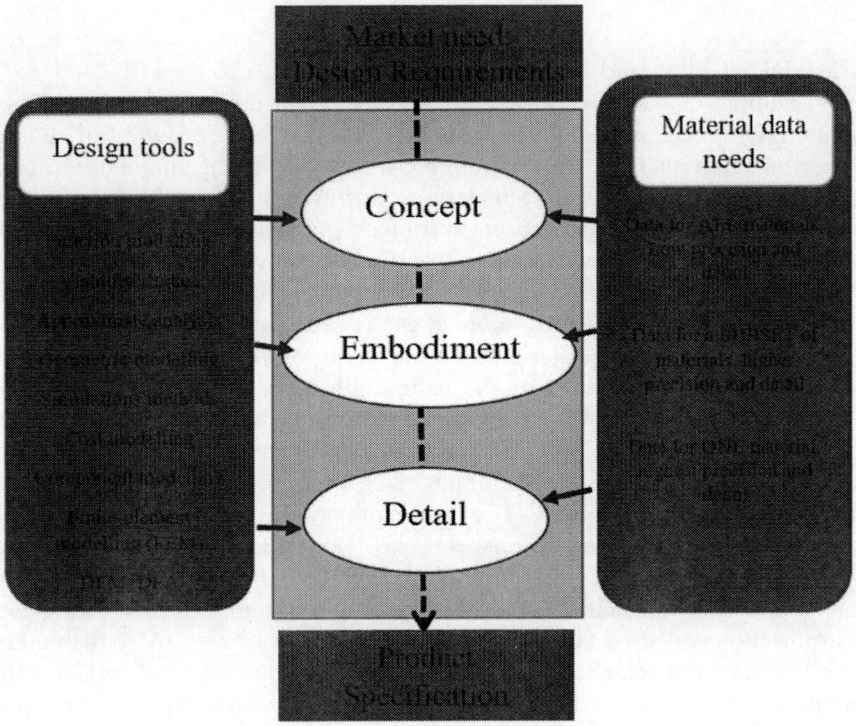

Figure 1. A Framework of Design Process [5].

Determination the best material of natural fibre composite in manufacturing process is difficult compare to the synthetic fibre because the properties of this material was distinctive. The elements of physical, mechanical, chemical and environmental properties of NFRC are not consistent and diversity [14]. These composites are custom-made at least from two combinations of materials, it causes variety performance of the properties. These materials constraints become a conflict to the manufacturing process to fulfil the design requirement of automotive component in industry.

Most of the previous researcher study on material selection of automotive component application in car manufacturing such as clutch pedal, car front hood, buggy bonnet, brake disc, bumper beam, dashboard panel, break lever, car roof and anti-roll bar [15–23] were consider multiple criteria and sub criteria based on the application of the automotive component. In materials selection process, there are a lot of methods in multiple-criteria decision-making (MCDM) was used in varied application such as Analytical Hierarchy Process, Analytical Network Process, Multi-attribute Utility Theory, Preference Selection Index, Technique of Ranking Preferences by similarity of the Ideal Solutions, Vlse Kriterijumska Optimizacija Kompromisno Resenje (VIKOR), Elimination and Choice Expressing the Reality, Simple Additive Weighting, Data Envelopment Analysis, Preference Ranking Organization Method for Enrichment Evaluations, Quality Function Deployment, Quality Function Deployment for the Environment and questionnaire [24–27]. All these method is implemented by different approach such as computer aid engineering, knowledge-based expert, fuzzy and integrated method of MCDM [28–33]. Researcher should determine the criteria to desired properties of interest application before performing the analysis. In general, as observed from prior studies, the most researchers focus on four main properties which are physical, mechanical, chemical and environmental on their study.

This review paper attempts to provide a knowledge and idea to the new researcher about the major criteria used for material selection. The criteria, types of materials, case study on different automotive component from previous study was summarized. The major concern on types of criteria with command materials of fibre and matrix also listed in this study that will be a quit acess to other researcher.

2. Design Material Selection and Requirement

By considering multiple characteristics of the composites can produce multiple performance of the materials and a proper procedure should be addressing to find the best output in automotive industry. Decision making on materials is extremely intuitive when considering single criterion problem. It can create a bias problem to other alternative materials. However, in the real world, the engineer should identify an appropriate criterion, goal, aspects, attributes and possible alternatives in MCDM. Most of MCDM used similar systematic evaluation steps which involving (1) determination the relevance criteria and feasible alternatives, (2) numerical measurement of the criteria and evaluation of the alternatives and (3) determining a ranking score of each alternatives [34].

Identification on factors and criteria at the beginning of selection process can perform a better judgment on the potential capabilities of the natural fibre composites [35]. A lot of features and properties should be considered in difference application. Decision maker need to identify the criteria or factors based on the application and PDS is used as a guide to the decision maker. In this study, the literature from year 2000 until 2016 was summarized and the criteria that influence the selection process is discussed as shown in Table 1. The most main criteria consider by the decision maker were physical, mechanical and environmental aspect [23], [29], [36-37]. Some researchers used these properties to select the concept design for a particular application such as selection of bumper system, brake disk, break lever and wheelchair [36], [38–40]. Consideration the criteria for selection process is one of the instrument on concurrent engineering [33]. From this review, more than 50 types of criteria were considered for material selection focusing in automotive component from 22 articles. We found the most commonly used criteria of PDS are cost, density, tensile strength, Young's modulus and elongation at break. The description of top 10 criteria was explained in Table 2.

Besides physical, mechanical and environmental properties have been identified as major contributing factors for material selection by referring on Table 1, other factors such as chemical, maintenance, thermal and acoustic insulating properties are considered by minority of the researchers.

2.1 Product Design Specification (PDS)

Each of automotive component have a specific product design requirement. This PDS describe how a design is made, what it is intended to do and how far it complies with the requirement. The requirement is formal documented to be satisfied by the material, design, product and service generally. To avoid any automotive component failure during the product design testing, the engineer should optimize follow the PDS. For example, Davoodi et al. [54] consider design parameters such as thickness, bumper beam curvature and strengthening ribs to fulfil the desired PDS to produce vigorous bumper beam. Other than market investigation, conceptual design, detail design and manufacture, PDS is one of the main factors that should be consider in design stages as mentioned by Sapuan [15].

Table 1. Selection requirement for automotive component design.

Description	Design requirement	Application	Materials	Author
Material selection of natural fibre	Density Elongation at break Micro-fibril angle Fibres' length Moisture content Cellulose Hemicellulose Lignin Young's modulus Bio-degradability Raw cost Tensile strength Availability Production rate	Anti-roll bar (ARB)	Sugar palm Kenaf Oil palm Sisal Jute Hemp Flax Pineapple Coir	[23]
Material selection of matrix (thermoplastic)	Density Young's modulus Fracture toughness Elongation at break Tensile strength Impact strength Chemical resistance Water absorption CO2 footprint Raw cost Thermal conductivity Recycle	Anti-roll bar (ARB)	PP PS HDPE LDPE TPU	[41]

Material selection of matrix (thermoplastic)	Tensile strength Young modulus Elongation at break Impact strength Coefficient of thermal expansion Process melting temperature Raw material cost Density Water absorption Heat deflection temperature	Parking Brake Lever	PP LDPE HDPE Nylon 6	[29]
Material selection of natural fibre	Strength Stiffness Density Raw material cost	Parking Break Lever	Kenaf Oil palm Flax Hemp Jute Ramie Coir Sisal Cotton Bagasse Pineapple Banana Sugar palm	[21]
Material selection	Stiffness Density Cost Water absorption Availability	Car front hood	Kenaf Jute Sisal Flax Hemp	[16]

IOP Conf. Series: Materials Science and Engineering **368** (2018) 012002 doi:10.1088/1757-899X/368/1/012002

Material selection for buggy bonnet	Density Young's modulus Specific Young's modulus Tensile strength Specific tensile strength Ultimate elongation at break	Buggy bonnet	Flax Hemp Jute Abaca Sisal Coir E-glass	[17]
Material Selection	Density Thermal conductivity Coefficient of thermal expansion Glass transition temperature Acoustic insulation properties Elastic modulus Fracture toughness Elongation to break Yield strength Impact strength Curing temperature Curing pressure Curing time Resistance of chemicals Level of hydrophobic nature Weather resistance Service temperature Sunlight and UV resistance Wettability Cost	Automotive	Biocomposites	[42]

IOP Conf. Series: Materials Science and Engineering **368** (2018) 012002 doi:10.1088/1757-899X/368/1/012002

Material selection	Dent resistance (yield strength, panel thickness, panel stiffness) Ease of manufacturing Noise, vibration, harshness (NVH) Density Material cost, manufacturing cost Temp. performance Crashworthiness Durability Bending stiffness Torsional stiffness	Body in White panel: Roof	Steel-BH Steel-DP Steel-HSLA Steel-martensite Aluminium-5xxx Aluminium-6xxx Magnesium Titanium GFRP HDPE	[22]
Material selection	Safety Size Maintenance Performance Installation Materials Weight Environment Process, Cost Disposal	Bumper beam for passenger	Glass+Thermoplastic Kenaf/Glass+Epoxy	[43] [44]
Material selection	Young modulus Tensile strength Density	Dashboard panel	NFRPC	[20]

IOP Conf. Series: Materials Science and Engineering **368** (2018) 012002 doi:10.1088/1757-899X/368/1/012002

Material Selection	Density Permissible slide speed Permissible pressure Elastic modulus Tensile strength Thermal expansion coefficient Thermal conductivity Hardness Wear resistance Corrosion resistance Heat-treatability Manufacturability Recycle cost Material cost Processing cost Disposal Reusability Energy consumption Polluting	Automotive Engineering		[45]
Concept design selection	Environment Size Weight Materials Standard Patents Safety Cost Performance Maintenance Manufacturing Shape	Automotive bumper beam	Glass+Epoxy Carbon+Epoxy Carbon+10%PP Glass+40%PP Glass+30%Polyester Glass+60%Vinylester	[46] [37]

IOP Conf. Series: Materials Science and Engineering **368** (2018) 012002 doi:10.1088/1757-899X/368/1/012002

Material selection	Performance (elastic modulus, density) Cost Aesthetics Comfort	Interior motorcar panel	Polypropylene +hemp 40% Polypropylene +flax 40% Wood Cork	[47]
Material selection for polymeric composite	Energy absorption Impact toughness Performance Flexural strength Flexural modulus Cost Density Service conditions Corrosion resistance Water absorption Manufacturing process Shape Environment consideration Recycling Disposal Availability	Automotive bumper beam	Glass fibre epoxy Carbon fibre epoxy Carbon fibre-reinforced polypropykene (10%) Glass fibre-reinforced polyprolyene (40%) Glass fibre-reinforced polyester (30%) Glass fibre vinylester SMC (60%)	[19]
Design Processing Materials	Strength Stiffness Density Maximum service temperature Durability with water Toxicity Price Energy and co2 footprint Safe for disposal	Eco-aware lightweight automotive friction materials	Kenaf Jute Ramie Asbestos	[12]

IOP Conf. Series: Materials Science and Engineering **368** (2018) 012002 doi:10.1088/1757-899X/368/1/012002

Material selection of natural fibre composite	Density Tensile strength Young's modulus Biodegradability Toxic level Material cost Manufacturing cost	Automotive component	Hemp+PP Kenaf+PP Jute+PP Sisal+PP Flax+PP PP Coir+PP EFB+PP	[48]
Material selection	Compressive strength Friction coefficient Wear resistance Thermal conductivity Specific gravity Cost	Automotive brake disc/rotor system	Cast iron (GCI) Aluminium alloy (based metal matrix composite) Titanium alloy Ceramics Composites	[18]
Material selection	Density Porosity Microstructural analysis Hardness Compressive Strength Compressive Strain Compressive Load	Automotive Break Pad	Coconut fibre reinforced composite	[50]
Material selection	High strength to weight ratio Easy manufacturing Cost Longer life High oxidation and Corrosion resistances Higher creep and fatigue resistances	Spar of Human Powered Aircraft	Super Alloys Hastelloy S, X Haynes HR-120, 160, Haynes 230, 282, 625 Inconel 600, 601 CMSX-4	[51]

IOP Conf. Series: Materials Science and Engineering **368** (2018) 012002 doi:10.1088/1757-899X/368/1/012002

Material selection for thermal conductor	Density Compressive stress Ultimate tensile stress Spring back index Bend force index Static load index Hardness Yield stress Elastic modulus Thermal diffusivity Thermal conductivity Cost	Cylindrical thermal conductor sheet	Copper-2-beryllium Copper-cobalt-beryllium Electrolytic tough-pitch, copper, soft Electrolytic tough-pitch, copper, hard Wrought aluminium alloy Wrought austenitic stainless steel Commercial bronze, soft Carbon steel	[52]
Material selection	Failure strength Fracture toughness Density Price Fragmentability	Flywheel	300M 2024-T3 7050-T73651 Ti-6Al-4V E glass-epoxy S glass-epoxy Carbon-epoxy Kevlar29-epoxy Kevlar49-epoxy Boron-epoxy	[53]
Material selection	Strength Modulus Manufacture Cost	Clutch pedal	Polymeric based composite	[15]

IOP Conf. Series: Materials Science and Engineering **368** (2018) 012002 doi:10.1088/1757-899X/368/1/012002

Table 2. The criteria of automotive component [23].

Criteria of PDS	Description
Density	The degree of consistency measures by the quantity of mass per unit volume
Tensile strength	The resistance of a material to breaking under tension
Young's modulus	The measurement of elasticity, ratio of the stress to strain
Thermal conductivity	The rate at which heat passes through on material, the amount of heat that flows per unit time through a unit area with a temperature gradient of one degree per unit distance
Cost	The price of raw materials and manufacturing
Elongation at break	The capability of a material to resist changes of shape without crack formation
Fracture toughness	The ability of a material containing a crack to resist
Water absorption	The fluid uptake on the material
Stiffness	The ability of an object to resist deformation in response to an applied force
Easy of manufacturing	The process of converting raw materials, component or part into finished goods

2.2 *Materials*

In this review, most of the reference used natural fibre reinforced composite as the material for selection process. The combination of natural fibre and the matrix will enhance the properties of the composite for example a hybrid materials from sugar palm fibre and fibre glass in fabrication of boat can increase the tensile and impact properties of the composites [55]. In different study by Ishak et al. [56] also show the composites reinforced with kenaf bast fibres produce better score on tensile, flexural and impact test compare with kenaf core fibre composites with different level of fibre loading. The advantages of the composite are low density and high specific strength and stiffness [57–59]. Moreover, the fibre from the composite is renewable resources which need less energy production and can reduce CO_2 [60]. The most preferable concern used this material was the cost which it's cheaper compare to synthetic fibre [11] . Joshi et al. [61] compare the natural fibre and glass fibre with comparative life cycle assessment and found that the natural fibre composite achieve higher on key driver of their relative environmental performance. However, the composite has wider variability of properties because of the diversity combination from the fibre and matrix itself. On the other hand, high moisture absorption of the composite will result swelling [62], [63]. The low durability problem of the composite can be improved with the chemical treatment such as two different alkali pre-treatment treated palm fibre reinforces mortars can increase the durability and other mechanical properties studied by Ozerkan et al. [64]. The salt-fog environment condition also can be considered as a practical approach for enhancing the durability of natural flax fibre with external basalt layers [65]. Different consideration on geometry parameter by Davoodi et al. [54] such as cross section, thickness, added ribs and fixing method optimizations can improve the impact properties in epoxy composite bumper beam. From this review on material selection, most of the researchers used NFRC on their studies. The top five natural fibre was jute, hemp, flax, kenaf and coir. It is because these natural fibres are easy to handle and process during the experimental work. Furthermore, the properties of these natural fibres are available. In this study, we found that polypropylene (PP), high density polyethylene (HDPE) and low density polyethylene (LDPE) are the command matrix used as the reinforced material to the fibre due to high recyclable of these thermoplastic nature compared to thermosetting such as epoxies and silicones. The application by using natural fibre for automotive component is a good transformation in automotive industry. Besides, based on Table 1 there is still metal based material used such as steel, aluminium and alloy for critical automotive component like turbine blade, thermal conductor and brake disc.

3. Discussion

Material selection process depends on the application and decision-making tools that may vary the result accordingly. As shown in Mansor et al.'s [21] study, kenaf fiber was selected as the most suitable natural fiber for lever break with regards to its design specifications. While, Mastura et al. [23] obtained sugar palm as the most suitable natural fibre in their study for the automotive anti roll bar using different method of decision making. Different study also considers different requirement such as environmental aspect, customer voice and design and manufacturing guideline. The selection process on natural fibre and matrix is done separately by the authors like Mansor et al. [40] and Mastura et al. [41]. This kind of implementation can be applied by others researcher or product design engineering that interested on single materials at the beginning of product design requirement on manufacturing process. In addition, the finding can contribute to a satisfactory performance on the final composite. Different practice done by Farag [47] and Ahmad Ali [48] which selection process is done on the composite which focus on certain type of matrix for example PP is selected as the matrix for the composite on automotive component application studied by Ahmad Ali et al. [48].

In other application on buggy bonnet, Furtado et al. [17] shown the jute fibre composite have higher damping behaviour compare to glass fibre. The jute fibre is a realistic choice in automotive application where the attenuation of vibration and noise is desirable. Kenaf was identified as an appropriate NFC to achieve the transportation weight reduction on application of car front hood by using VIKOR [16]. A study done by Mustafa et al. [12] also found kenaf have a potential as an alternative sources of friction materials not only consider the design requirement but impact on environment and human health is count into the study.

The concept of different application will have different criteria also practical in other industry such as in medication, the medical device consider availability, design flexibility, cost per unit, performance properties, regulatory compliance, bio-compatibility, aesthetics and usability, manufacturing efficiency, sterilization and cleaning and sustainability for choosing the right materials for the medical device design [66].

4. Conclusion

Numerous types of criteria were considered from the researcher in year 2000 until 2016 on different automotive component application for material selection. From over 50 types of criteria was summarized, we can conclude the most criteria used for material selection to satisfy the PDS in automotive industry were cost, density, tensile strength, Young's modulus and elongation at break. Furthermore, the availability, ability and flexibility of these natural fibre such as jute, hemp, flax, kenaf and coir were the reason they become the favourite natural fibre used in automotive component. In addition, the command filler to the natural fibre was PP, HDPE and LDPE. From this study, the finding can be used to other researcher to start their study on automotive component design in material selection process generally.

Acknowledgements

The authors would like to thank Universiti Putra Malaysia for the opportunity doing this study as well as Universiti Teknikal Malaysia Melaka for providing the scholarship award to the principal author in this project.

References

[1] Nunes B and Bennett D 2009 Green technologies in the automotive industry *International Association for Management of Technology Management of Green Technology Industry Perspective* **1** 7-8

[2] Conti J, Holtberg P, Diefenderfer J, LaRose A, Turnure J T, and Westfall L 2016 *International Energy Outlook 2016 Eith Projections to 2040.* Washington: U. S. Energy Information Administration

[3] Das D, Bhattacharya S, and Sarkar B 2016 Decision-based design-driven material selection: A normative-prescriptive approach for simultaneous selection of material and geometric variables in gear design *Mater. Design* **92** 787–793

[4] Al-Oqla F M, Sapuan S M, Anwer T, Jawaid M, and Hoque M E 2015 Natural fiber reinforced conductive polymer composites as functional materials : A review *Synth. Met.* **206** 42–54

[5] Ashby M F 2005 *Materials Selection in Mechanical Design*, Third. Oxford: Elsevier Butterworth Heinemann

[6] Węcławski B T, Fan M, and Hui D 2014 Compressive behaviour of natural fibre composite. *Compos. Part B-Eng* **67** 183-191

[7] Fairuz A M, Sapuan S M, Zainudin E S, and Jaafar C N A 2016 Effect of filler loading on mechanical properties of pultruded kenaf fibre reinforced vinyl ester composites *J. Mech. Eng. Sci.***10** 1931–1942

[8] Ramesh M, Palanikumar K, and Reddy K H 2012 Mechanical Property Evaluation of Sisal Jute-Glass Fiber Reinforced Polyester Composites *Compos. Part B-Eng* **48** 1–9

[9] Razali N, Salit S, Jawaid M, Ishak M R, and Lazim Y 2015 A Study on Chemical Composition, Physical, Tensile, Morphological, and Thermal Properties of Roselle Fibre:Effect of Fibre Maturity *BioResources* **10** 1803–1824

[10] Yahaya R., Sapuan S M, Jawaid M, Leman Z, and Zainudin E S 2015 Effect of layering sequence and chemical treatment on the mechanical properties of woven kenaf-aramid hybrid laminated composites *Mater. Design* **67** 173–179

[11] Monteiro S N, Lopes F P D, Ferreira A S, and Nascimento D C O 2009 Natural-Fiber Polymer-Matrix Composites: Cheaper, Tougher, and Environmentally Friendly *JOM* **61** 17–22

[12] Abdollah M F B, Shuhimi F F, Ismail N, Amiruddin H, and Umehara N 2015 Selection and verification of kenaf fibres as an alternative friction material using Weighted Decision Matrix method *Mater. Design* **67** 577–582

[13] Tezara C, Siregar J P, Lim H Y, Fauzi F A, Yazdi M H, Moey L K, and Lim J W 2016 Factors that affect the mechanical properties of kenaf fiber reinforced polymer: A review. *J Mech Eng Sci* **10** 2159-2175

[14] AL-Oqla F M, Sapuan S M, Ishak M R, and Nuraini A A 2015 A Model for Evaluating and Determining the Most Appropriate Polymer Matrix Type for Natural Fiber Composites *Int J Polym Anal Ch* **20** 191–205

[15] Sapuan S M 2005 Concurrent design and manufacturing process of automotive composite components *Assembly Autom* **25** 146–152

[16] Ishak N M, Malingam S D, and Mansor M R 2016 Selection of natural fibre reinforced composites using fuzzy VIKOR for car front hood *Int J Mater Prod Tec* **53** 267–285

[17] Furtado S C R, Araújo A L, Silva A, Alves C, and Ribeiro A M R 2014 Natural fibre-reinforced composite parts for automotive applications *I J Auto C* **1** 18–38

[18] Maleque M A, Dyuti S, and Rahman M M 2010 Material Selection Method in Design of Automotive Brake Disc *World Congress on Engineering* 3–7

[19] Hambali A, Sapuan S M, Rahim A. S, Ismail N, and Nukman Y 2011 Concurrent Decisions on Design Concept and Material Using Analytical Hierarchy Process at the Conceptual Design Stage *Concurrent Eng-Res A* **19** 111–121

[20] Sapuan S M, Kho J Y, Zainudin E. S, Leman Z, Ahmed Ali B A and Hambali A 2011 Materials selection for natural fiber reinforced polymer composites using analytical hierarchy process *Indian J Eng Mater S* **18** 255–267

[21] Mansor M R, Sapuan S M, Zainudin E S, Nuraini A A, and Hambali A 2013 Hybrid natural and glass fibers reinforced polymer composites material selection using Analytical Hierarchy Process for automotive brake lever design *Mater. Design* **51** 484–492

[22] Mayyas A, Shen Q, Mayyas A, Shan D, Qattawi A, and Omar M 2011 Using quality function deployment and analytical hierarchy process for material selection of body-in-white *Mater. Design* **32** 2771-2782

[23] Mastura M T, Sapuan S M, Mansor M R, and Nuraini A A 2017 Environmentally conscious hybrid bio-composite material selection for automotive anti-roll bar. *Int J Adv Manuf Technol* **89** 2203-2219

[24] Tzeng G H and Huang J J 2011 Multiple Attribute Decision Making Methods and Applications *CRC Press*

[25] Mardani A, Jusoh A, Nor K, Khalifah Z, and Valipour A 2014 Multiple criteria decision-making techniques and their applications – a review of the literature from 2000 to 2014 *Econ. Res. Ekon. Istraz.* **28** 516–571

[26] Jahan A, K Edwards L, and Bahraminasab M 2016 *Multi-criteria Decision Analysis For Supporting the Selection of Engineering Materials in Product Design*, Second. Oxford: Elsevier Butterworth Heinemann

[27] Marini C D, Fatchurrohman N, Azhari A, and Suraya S 2016 Product Development using QFD, MCDM and the Combination of these Two Methods *IOP Conf. Ser. Mater. Sci. Eng.* **114** 12089

[28] Ahmed Ali B A, Sapuan S M, E Zainudin. S, and Othman M 2013 Java based expert system for selection of natural fibre composite materials *J. Food, Agric. Environ.* **11** 1871–1877

[29] Mansor M R, Sapuan S M, Zainudin E S, Nuraini A A, and Hambali A 2014 Application of Integrated AHP-TOPSIS Method in Hybrid Natural Fiber Composites Materials Selection for Automotive Parking Brake Lever Component *Aust. J. Basic & Appl. Sci.* **8** 431–439

[30] Mousavi-Nasab S H and Sotoudeh-Anvari A 2017 A comprehensive MCDM-based approach using TOPSIS, COPRAS and DEA as an auxiliary tool for material selection problems *Mater. Design* **121** 237-253.

[31] Rao R V 2013 *Decsion Making in the Manufacturing Environment using Graph Theory and Fuzzy Multiple attribute Decision Making Methods*, 2nd ed. Springer

[32] Sapuan S M 2001 A knowledge-based system for materials selection in mechanical engineering design *Mater. Design* **22** 687–695

[33] Sapuan S M and Mansor M R 2014 Concurrent engineering approach in the development of composite products : A review *Mater. Design* **58** 161–167

[34] Yang S S, Nasr N, Ong S K, and Nee A Y C 2015 Designing automotive products for remanufacturing from material selection perspective *J. Clean. Prod.*

[35] Al-Oqla F M, Salit M S, Ishak M R, and Aziz N A 2015 Selecting natural fibers for bio-based materials with conflicting criteria *Am. J. Appl. Sci.* **12** 64–71

[36] Davoodi M M, Sapuan S M, Ahmad D, Aidy A, Khalina A, and Jonoobi M 2011 Concept selection of car bumper beam with developed hybrid bio-composite material *Mater. Design* **32** 4857–4865

[37] Hambali A, Sapuan S M, Ismail N, and Nukman Y 2010 Material selection of polymeric composite automotive bumper beam using analytical hierarchy process *J Cent South Univ T* **17** 244–256

[38] Sapuan S M, Maleque M A, Hameedullah M, Suddin M N, and Ismail N 2005 A note on the conceptual design of polymeric composite automotive bumper system *J Mater Process Tech* **159** 145–151

[39] Fatchurrohman N, Sulaiman S, Sapuan S M, M Ariffin K A, and Baharuddin B T H T 2013 New Technique for Conceptual Selection of Manufacturing Process and Material, Case Study: Metal Matrix Composite Component *Adv Mat Res* **789** 82–86

[40] Mansor M R, S Sapuan M, Zainudin E S, A Nuraini A, and Hambali A 2014 Conceptual design of kenaf fiber polymer composite automotive parking brake lever using integrated TRIZ – Morphological Chart – Analytic Hierarchy Process method *Mater. Design* **54** 473–482

[41] Mastura M T, Sapuan S M, Mansor M R, and Nuraini A A 2017 Materials selection of thermoplastic matrices for ‘ green ’ natural fibre composites for automotive anti - roll bar with particular emphasis on the environment *Int J Pr Eng Man-Gt* **5** 111–119

[42] Al-Oqla F M, Sapuan S M, Ishak M R, and Nuraini A A 2015 A Model for Evaluating and Determining the Most Appropriate Polymer Matrix Type for Natural Fiber Composites *Int J Polym Anal Ch* **20** 191–205

[43] Davoodi M M, Sapuan S M, Aidy A, Osman N A A, Oshkour A A, and Abas W A B W 2012 Development process of new bumper beam for passenger car : A review *Mater. Design* **40** 304–313

[44] Davoodi M M, Sapuan S M, Ahmad D, Aidy A, Khalina A, and Jonoobi M 2012 Effect of polybutylene terephthalate (PBT) on impact property improvement of hybrid kenaf / glass epoxy composite *Mater Lett* **67** 5-7
[45] Liu H, You J, Zhen L, and Fan X 2014 A novel hybrid multiple criteria decision making model for material selection with target-based criteria *Mater. Design* **60** 380–390
[46] Hambali A, Sapuan S M, Ismail N, and Nukman Y 2009 Composite manufacturing process selection using analytical hierarchy process *IJMME* **4** 49–61
[47] Farag M M 2008 Quantitative methods of materials substitution: Application to automotive components *Mater. Design* **29** 374–380
[48] Ali B A, Sapuan S M, Zainudin E S and Othman M 2015 Implementation of the expert decision system for environmental assessment in composite materials selection for automotive components *J Clean Prod 107* 557-567
[49] Maleque M A and Dyuti S 2010 Materials selection of a bicycle frame using cost per unit property and digital methods *IJMME* **5** 95–100
[50] Maleque M A, Atiqah A, Talib R J, and Zahurin H 2012 New natural fibre reinforced aluminium composite for automotive brake pad *IJMME* **7** 166–170
[51] Kasaei A, A Abedian, and Milani A S 2014 An application of Quality Function Deployment method in engineering materials selection *Mater. Design* **55** 912–920
[52] Jahan A, Ismail Y, Shuib S, Norfazidah D, and Edwards K L 2011 An aggregation technique for optimal decision-making in materials selection *Mater. Design* **32** 4918–4924
[53] Jee D and Kang K 2000 A method for optimal material selection aided with decision making theory *Mater. Design* **21** 199–206
[54] Davoodi M M, Sapuan S M, Ali A, Ahmad D, and Khalina A 2010 Thermoplastic impact property improvement in hybrid natural fibre epoxy composite bumper beam *IOP Conf. Ser. Mater. Sci. Eng.* **11** 012013
[55] Misri S, Z Leman, Sapuan S M, and Ishak M R 2010 Mechanical properties and fabrication of small boat using woven glass/sugar palm fibres reinforced unsaturated polyester hybrid composite *IOP Conf. Ser. Mater. Sci. Eng.* **11** 1–13
[56] Ishak M R, Leman Z, Sapuan S M, Edeerozey A M M, and Othman I S 2010 Mechanical properties of kenaf bast and core fibre reinforced unsaturated polyester composites *IOP Conf. Ser. Mater. Sci. Eng.* **11**
[57] Li M, Wang Z, Liu Q, Wang S, Gu Y, Li Y, and Zhang Z 2017 Carbon nanotube film/epoxy composites with high strength and toughness *Polym Composite* **38** 588-596
[58] Maya M G, George S C, Jose T, Sreekala M S, and Thomas S 2017 Mechanical properties of short sisal fibre reinforced phenol formaldehyde eco-friendly composites *PFRR* **8** 27–42
[59] Sharma S and Dwivedi S P 2017 Effects of waste eggshells and SiC addition on specific strength and thermal expansion of hybrid green metal matrix composite *J Hazard Mater* **333** 1–9
[60] Ghoushji M J, Eshkoor R A, Zulkifli R, Sulong A B, Abdullah S, and Azhari C H 2017 Energy Absorption Capability of Axially Compressed Woven Natural Ramie/Green Epoxy Square Composite Tubes *J Reinf Plast Comp* **36** 1028–1037
[61] Joshi S V, Drzal L T, Mohanty A K, and Arora S 2004 Are natural fiber composites environmentally superior to glass fiber reinforced composites? *Compos Part A-Appl S* **35** 371–376
[62] Kord B and Roohani M 2017 Water transport kinetics and thickness swelling behavior of natural fiber-reinforced HDPE/CNT nanocomposites *Compos Part B-Eng* **126** 94–99
[63] Das S 2017 Mechanical and water swelling properties of waste paper reinforced unsaturated polyester composites *Constr. Build. Mater.* **138** 469–478
[64] Ozerkan N G, Ahsan B, Mansour S, and Iyengar S R 2013 Mechanical performance and durability of treated palm fiber reinforced mortars *Int. J. Sustain. Built Environ.* **2** 131–142
[65] Fiore V, T Scalici, Calabrese L, Valenza A, and Proverbio E 2016 Effect of external basalt layers on durability behaviour of flax reinforced composites *Compos Part B-Eng* **84** 258–265
[66] Cheryl G (2016, May 17). 10 Criteria for Choosing the Right Materials for your Medical

Device Design *MDT*. Retrieved from http://www.mdtmag.com

IOP Conf. Series: Materials Science and Engineering **368** (2018) 012003 doi:10.1088/1757-899X/368/1/012003

Physical characterization of carbon fiber reinforced composite with blended epoxidized Jatropha oil (EJO) resin.

M R M Hafiezal[1,2,4], A Khalina[1,2*] , Z A Zurina[1,2], M D Azaman[3] and Z M Hanafee[5].

[1]Department of Biological and Agricultural Engineering, Faculty of Engineering, Universiti Putra Malaysia, 43400 UPM Serdang, Selangor, Malaysia.
[2]Laboratory of Biocomposite Technology, Institute of Tropical Forestry and Forest Product (INTROP), Universiti Putra Malaysia, 43400 UPM Serdang, Selangor, Malaysia.
[3]School of Manufacturing Engineering, Universiti Malaysia Perlis, 02600 Arau, Perlis, Malaysia.
[4]Faculty of Engineering Technology, Universiti Malaysia Perlis, 02100 Padang Besar, Perlis, Malaysia.
[5]Aerospace Malaysian Innovative Centre, No.3-6-01, GMI Bangi, Malaysia.

*E-mail: khalina@upm.edu.my

Abstract. Jatropha oil is an inedible oil that is mainly used for bio-fuel in cars, lorries, machineries and jets as it provides clean combustion, reduced carbon dioxide emission and additional power with unlimited renewable resource; jatropha plant. Several studies have been carried out by researchers to diversify jatropha oil's uses such as lubrification, bio-fuel, traditional medicine etc. This time, jatropha oil was being used as matrix in composite structure by epoxidizing jatropha oil (EJO) itself. This paperwork will present about the physical characterization (density, visual assessment, fiber volume fraction) of EJO blended with synthetic epoxy as resin in carbon fiber reinforced composite panel by vacuum assisted resin transfer moulding VARTM or vacuum infusion technique. It is found that by blending EJO with synthetic epoxy, their physical characteristics are almost the same and comparable except the color of blended EJO was a bit yellowish and matte. Density of composite panel with blended EJO was 4% lower than synthetic epoxy which contributed to lighter materials that could be built and the surface morphology of both are identical.

1. Introduction

Thermoset or epoxy derived from petroleum is very synonym in composite sector, is used as resin to bond fibers as one body in order to resist a certain amount of applied loading. It is thermosetting polymer which are formed starting from liquid state into solid state by an irreversible curing process. Recently bio-based composite is very popular in composite research world. They used natural fiber as reinforcement and natural thermoplastic or natural thermoset as matrix. This bio-composite attracting many attention from manufacturers and researchers as the finding are interesting and could be highly optimized by various process. More than 60% of global production of epoxy were used in coating industry, total balance of 40%were used in adhesive/assembly parts, assembly concretes, sandwich panels for aircraft and automobiles [1]. Price of petroleum is one of the main factor besides its footprints towards environment. As the price of petroleum is not stable, it is a good leap to find alternative, environmental-friendly and renewable resources for an alternative replacement of epoxy.

One of an alternating replacement of epoxy is bio-based epoxy derived from crude jatropha oil, local unwanted plant in Malaysia as its oil is inedible. For the moment, jatropha oil is used as bio-fuel in cars, lorries machineries and jets as it burns clean combustion, emitted less carbon dioxide, and provide additional power with unlimited renewable resources [2]. There were several other natural epoxidized oils such as soya oil, cottonseed oil, algae oil, karanja oil, pine oil, corn oil, palm oil and sugar can oil. Epoxidation process differed from each other depend on nature of natural oil, including types of applications from aircraft panel to food packaging products [3-4].

In this paper, an investigation was carried out on composite panel made of carbon fiber with blended Bio-based epoxy derived from crude jatropha oil. The purpose of this research work is to improve the amount of renewable materials in aircraft interior parts as more than 80% of aircraft body made from composite materials. The comparison study of physical appearance, physical properties, density, thickness and fiber volume fraction were carried out between composite panel with 100% synthetic epoxy and composite panel with 18% blended Bio-epoxy.

2. Experimental

2.1 Fabrication and Materials of Composite Panel.

Two composite panels were fabricated in this experiment. The first one was using 100 % synthetic epoxy which would be named 0 % bio-epoxy panel. The second one was using epoxy resin blended with 18% bio-epoxy by weight percentage. Synthetic epoxy brand was Epoximite 100 with slow hardener 103 with instruction ratio of 28:1. 10 layers of 2 × 2 twill weave carbon fiber 3 K 200 gsm were used for each panel. Carbon fibers and synthetic epoxy were supplied by Mechasolve Engineering, Malaysia, and Bio-epoxy was supplied by in-laboratory collaboration of Faculty of Engineering University Putra Malaysia (UPM) with Aerospave Malaysian Innovative Centre (AMIC) Malaysia. Bio-epoxy was derived from crude Jatropha Oil supplied by Bionas Malaysia Sdn. Bhd.

Dimension of composite panel fabricated by VARTM was 300 mm × 300 mm × 2 mm (width × length × thickness). The composite panel 0% Bio-epoxy was the control specimen and composite panel with 18% Bio-epoxy was the evaluated specimen.

A flat mold was used as base surface for fabricating the composite panel. Release agent was applied onto the surface of flat mold for easy release. Then, 10 layers of carbon fiber were stacked and arranged. Then peel ply, media distributor, EnkaFusion mat, and bagging plastic were organized in order as in figure 1 with one inlet for resin, and one outlet for vacuum. All these materials were secured by sealant tape surrounding them to trap air and resin from escape. Once all were settled, air was sucked out by vacuum pump until it reached around 12 mbar. Afterwards, inlet pipe was open to let the resin infused through the carbon fiber, media distributor and so on. Inlet pipe was closed once almost all resin have been wetted. The set up was let to stay around and cure at room temperature 3 to 4 hours before the vacuum pump was shut off as in figure 2. Finally after 24 hours in room temperature, the composite panel was cured and carbon fiber composite panel was retrieved from the set up.

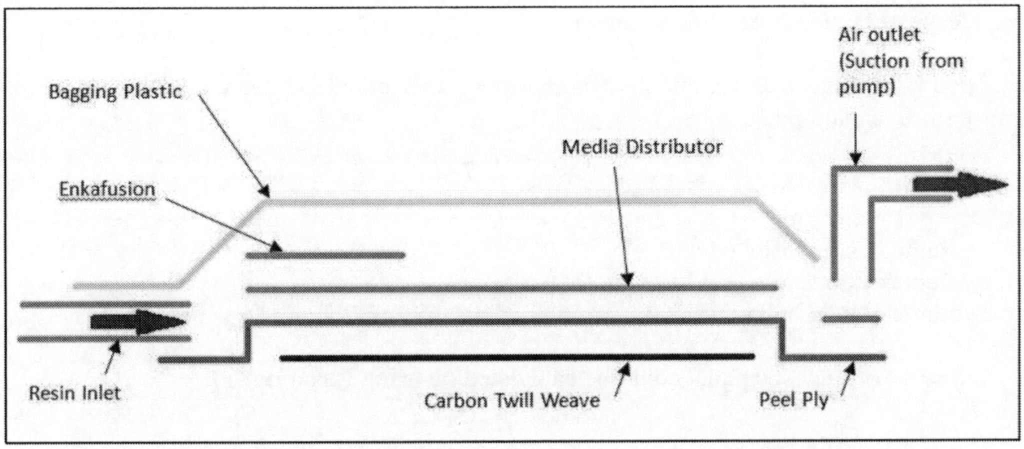

Figure 1. Set up diagram of Vacuum Assisted Resin Transfer Method (VARTM).

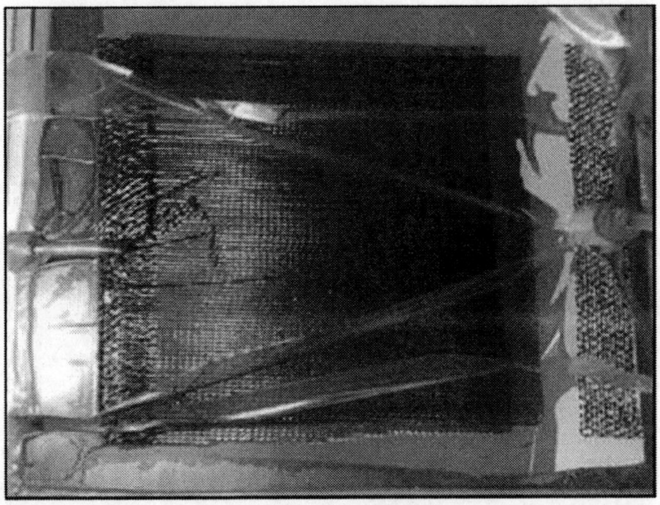

Figure 2. Layers of carbon fibers in the middle of infusion process.

2.2 *Burnt off test.*

For burnt off test, specimens were prepared by cutting them into 25 mm × 25 mm according to ASTM D3171-09. Aluminium crucibles were used to support the specimens during the burning process in an electrical muffled furnace at 550 °C for two hours. Weight of the specimens and crucibles were measured before and after burning process. Specimens' density were measured by a water displacement density calculator.

Rule of Mixture was use to compare the fiber volume fraction of specimen theoretically by using Equation (1) and experimentally measuring the weight of specimen obtained by this burnt off test.

$$Vf = \frac{Mf}{Mi} \times 100 \times \frac{\rho c}{\rho f} \quad (1)$$

Where;

Mi : initial mass of the specimen

Mf : final mass of specimen after combustion

ρc : the density of the specimen

ρf : density of in the fiber reinforcement

Theoretical constituent content of composite panel could be calculated by using following Rule of Mixture, Equation (2):

$$Vcp = Vf + Vr = 1cm^3 \tag{2}$$

Where;
Vcp : volume of composite panel
Vf : volume of fiber
Vr : volume of total resin

Density of composite panel could be calculated by using Equation (3):

$$\rho cp = \frac{Mcp}{Vcp} \tag{3}$$

Where;
ρcp : density of composite panel
Mcp : mass of composite panel
Vcp : volume of composite panel

2.3 Bio-epoxy derived from crude Jatropha Oil.
Through the process of addition, removal, stirring, and washing the mixture of crude jatropha oil with acids and amberlite, bio-epoxy derived from crude Jatropha oil [5]. In the Table 1 below, are presented some physical properties of crude Jatropha oil and its Bio-epoxy (Epoxidized Jatropha oil) that have been extracted.

Table 1. Physical properties of Jatropha oil and Bio-epoxy [5].

Physical Property	Crude Jatropha Oil (CJO)	Bio-epoxy from CJO	Synthetic Epoxy
Appearance	Liquid (clear yellow color)	Liquid (yellow color)	Liquid (Clear color)
Density (g/cm^3)	0.89304	1.298	1.1
Specific Gravity	0.8948	1.302	1.1
Dynamic Viscosity at 25 °C (cP) / (Pa.s)	46.8 cP	546 cP	650 cP
Kinematic Viscosity (mm^2/s)	52.41	588	-
Epoxy Equivalent Weight (EEW) (gr/eq.)	263.16	328.95	-
Epoxy Value (eq/100 gm)	0.38	0.30	-

3. Results and discussion

3.1 Fiber Volume Fraction.
In figure 3, an example of specimen after burning process is shown. For the specimen with 18 % Bio-epoxy, its fiber volume fraction was 42 % compare to 47 % for specimen with 0 % Bio-epoxy. That was reduction of 5% of fiber volume fraction. Theoretical values of fiber volume fraction for 18 % Bio-epoxy and 0 % Bio-epoxy were 60 % and 75 % respectively. Both show similar reduction trend in fiber volume fraction property. Despites the big gap between theoretical value and experimental value of fiber volume fraction, by achieving more than 40 % of fiber volume fraction are considered good fiber volume fraction because of different type of woven and different type of resin and filler, would differ the volume fraction of composite structure [6-11].

Figure 3. Sample in crucible after burnt off test

3.2 Density and Thickness.

For the specimen with 18 % Bio-epoxy, its density was 1.415 g/cm^3, while density of specimen with 0 % Bio-epoxy was 1.474 g/cm^3. This decrement of density shows improvement in weight of part that blended with Bio-epoxy. This weight reduction of 4% in overall composite structure could bring a significant improvement impact in aerospace application as it could improve fuel efficiency since more than 50% of airplane materials are composite structure [12-13]. With the same amount of 10 layers carbon fibers, the thickness of the composite panel with 18% Bio-epoxy is slightly thicker by 0.22 mm compare to composite panel with 0 % Bio-epoxy. This increment show blended Bio-epoxy used up more space in bonding with fiber carbon compare to synthetic epoxy. This is due to bio-epoxy is vegetable oil based, which contains many fatty acids which has small molecular chains compare to petrochemical based epoxy. The length of molecular chain contribute to weight of the cured epoxy [14]. Furthermore, fatty acids in bio-epoxy is known to be bulky and flexible thus slightly increasing the volume or thickness of the cured bio-epoxy [15]. Refer to figure 4 for result comparison of theoretical & experimental result for density and fiber volume fraction.

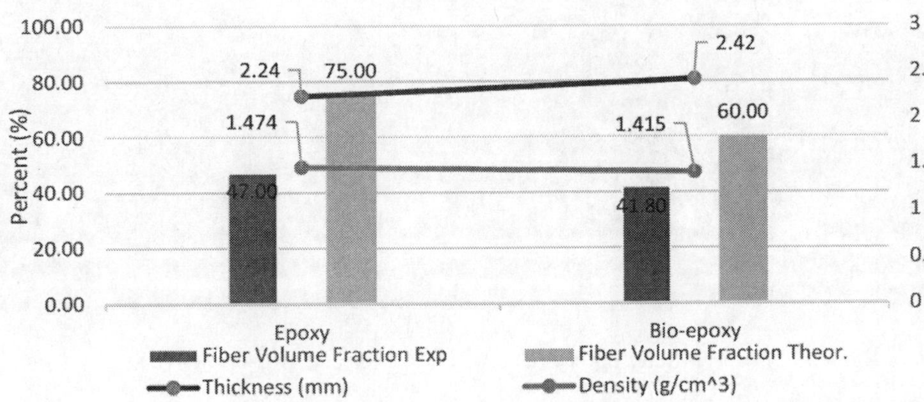

Figure 4 Thickness, Density and Fiber volume fraction comparison between composite panel with 18% Bio-epoxy and 0% Bio-epoxy.

3.3 Physical appearance.

For the specimen with 18 % Bio-epoxy, the blended resin show translucent color compare to specimen with synthetic epoxy which shows transparent color. This is due to size of molecules and crystalline/amorphous structure inside the cured epoxy. Small size molecules will allow light to pass through it compare to bigger molecules which scatters the light away [16]. As the bio-epoxy becoming thickened, it is getting opaque as in figure 6. This is due to crystalline structure is higher in bio-epoxy compare to synthetic epoxy. As shown in figure 5, we could categorize the composite to be shining and reflective finishing and matte with less reflective finishing. Both finishing has advantage and disadvantage according to application such as controlling the thermal effects by sun radiation as used in NASA devices and equipment like garments, spacesuits etc. [17].

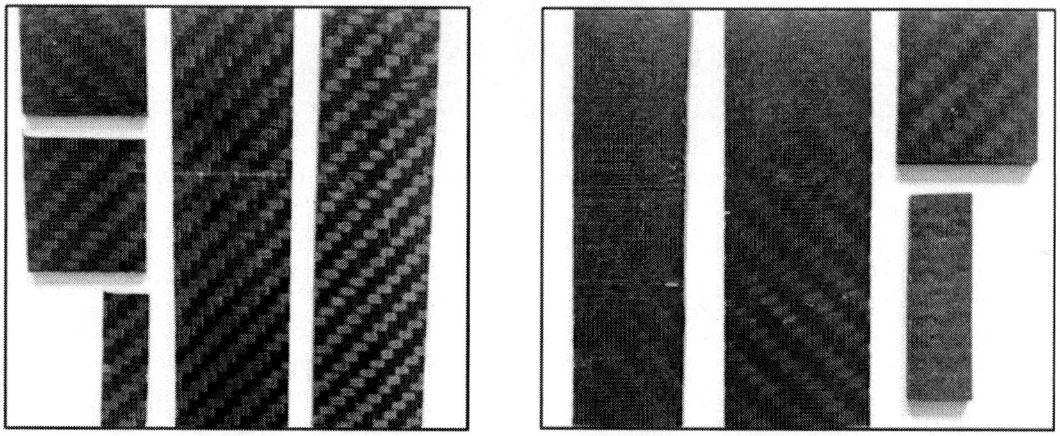

Figure 5 Specimens with 0 % Bio-epoxy or synthetic epoxy (a), while (b) specimens with 18 % blended Bio-epoxy.

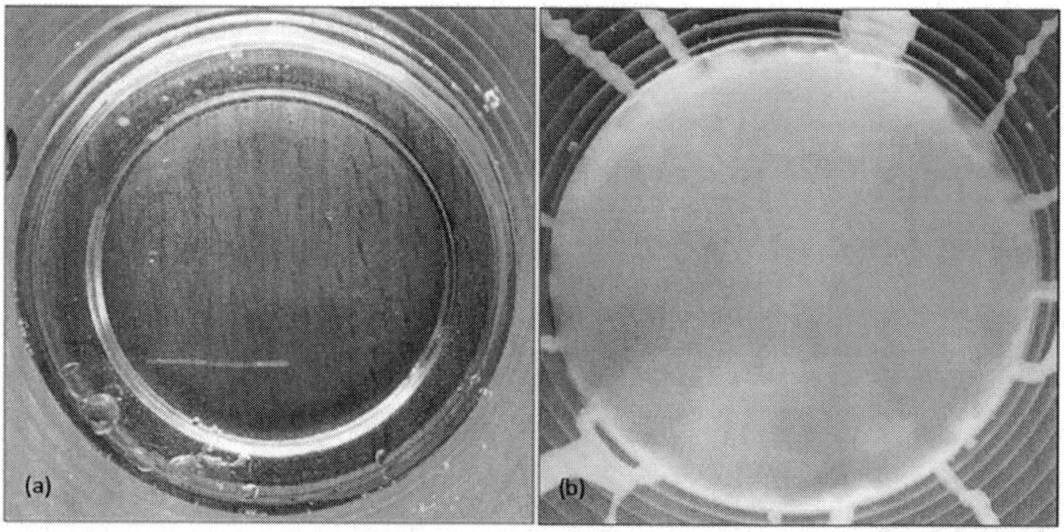

Figure 6 0 % Bio-epoxy or synthetic epoxy (a) and 18 % blended Bio-epoxy (b).

3.4 Microstructure from Scanning Electron Microscopy (SEM).

The surface characteristics of both composite panels were compared through scanning electron microscopy. Their cross sectional view of both panels presented shown in figure 7. The purpose of SEMs' images were to compare 0% bio-epoxy with 18% bio-epoxy in term of the interfacial bonding

IOP Conf. Series: Materials Science and Engineering **368** (2018) 012003 doi:10.1088/1757-899X/368/1/012003

between fibers and resins, internal cracks or defects came from fabrication process. As images between both specimens were compared, it could be seen that both specimens were pretty much similar as no significant different could be seen. The interfacial bonding between carbon fibers and resin blended with 18% bio-epoxy indicated good interfacial properties. For the defects, almost no void occurred between fibers and resin for 18% bio-epoxy. Those fractured fibers and broken resins were resulted from cutting process in obtaining the cross sectional area. Apart from that, very little cracks due to fabrication process could be seen. The slight expansion of thickness for 18% bio-epoxy did not degrade the composite structure even though there are bio-epoxy content in the resin. Normally by adding impurities, fillers or other elements in the resin, there would be increment or decrement of surface morphology (such as porosity, opening planar network), defects (cracks and debonded fibre to resin) and performance itself [18-20].

The SEM images are important as both specimen show high similarity to ensure good performance in mechanical properties. These images also proved that the vacuum infusion method were qualified to be used with this new bio-epoxy.

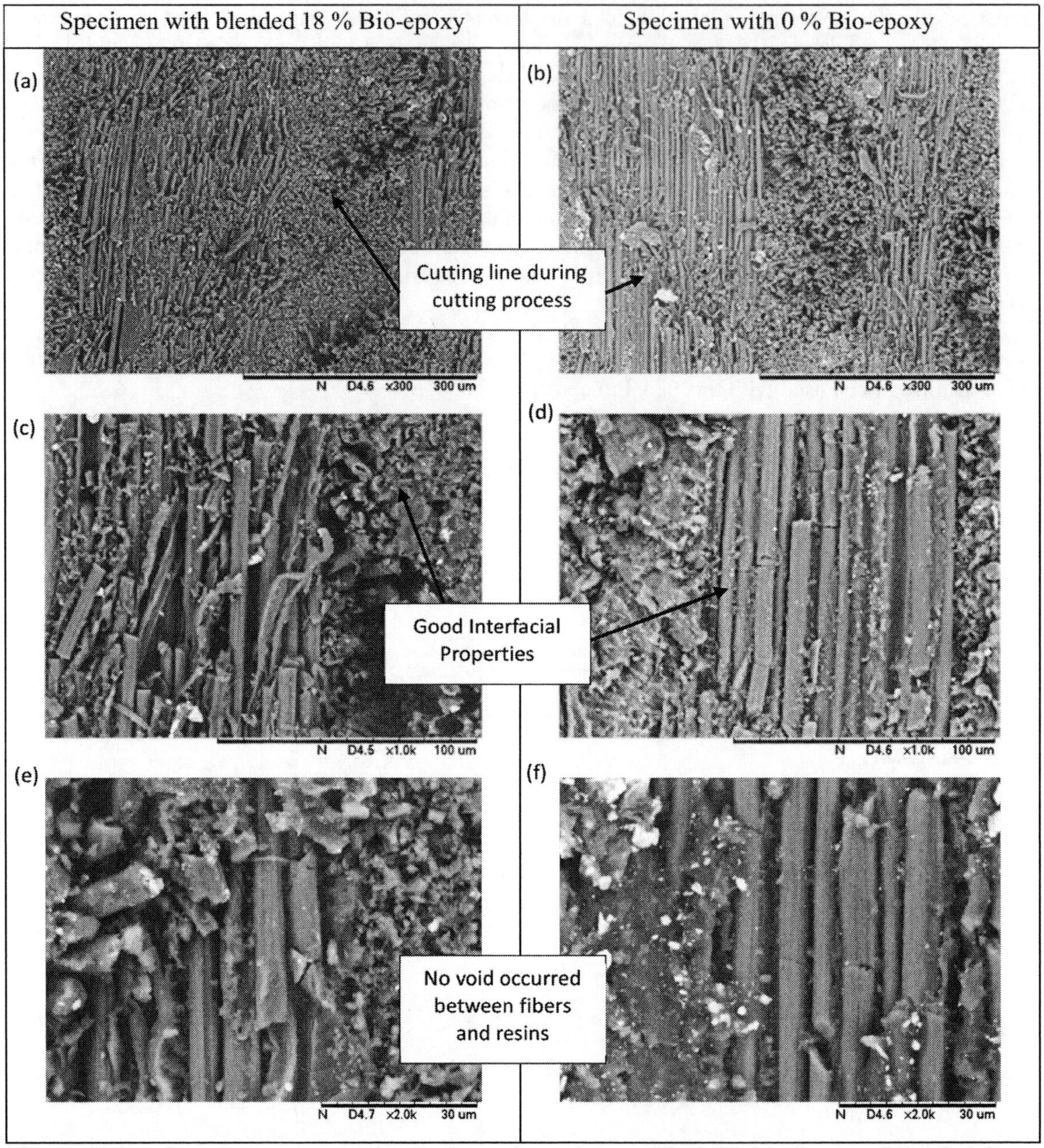

Figure 7 (a,c,e) are are cross sectional view of specimen with 0 % bio-epoxy while (b,d,f) are specimen with blended 18 % bio-epoxy. From top to bottom, image with different magnification.

4. Conclusion
From the investigation of physical characterization of carbon fiber reinforced with blended bio-epoxy derived from crude Jatropha oil, they were found that by adding 18% of bio-epoxy the density of composite panel was reduced by 4% that could lead to better fuel efficiency if it used as proper parts in aerospace application. The thickness was comparable as it differed only by 0.22mm and its fiber volume fraction also is comparable to synthetic epoxy. Both are fabricated by VARTM process and by adding bio-epoxy no sign of defects to change to other type of the fabrication process.

The appearance of composite panel with Bio-epoxy show translucent color which reflect less light compare to synthetic epoxy which is almost 100% clear. The interfacial bonding between fibers and bio-resin were similar to fibers with synthetic epoxy as seen from cross sectional view in SEM.

5. Acknowledgement
This research was initiated by collaboration between University Putra Malaysia (UPM) and Aerospace Malaysian Innovative Centre (AMIC) under grant number: 630081. Also we are thankful to Mecha Solve Engineering for providing Vacuum Infusion Machine Model 20/2 brand Vacmobiles.com.

References
[1] Auvergne R, Caillol S, David G, Boutevin B, and Pascault J P 2013 Biobased thermosetting epoxy: present and future *Chem Rev* **114** 1082-1115
[2] Paramarta A and Webster D C 2016 Bio-based high performance epoxy-anhydride thermosets for structural composites: The effect of composition variables *React Funct Polym* **105** 140-149
[3] Pawar M, Kadam A, Yemul O, Thamke V, and Kodam K 2016 Biodegradable bioepoxy resins based on epoxidized natural oil (cottonseed & algae) cured with citric and tartaric acids through solution polymerization: A renewable approach *Ind Crop Prod* **89** 434-447
[4] Tee Y B, Talib R A, Abdan K, Chin N L, Basha R K, and Yunos K F M 2015 Comparative study of chemical, mechanical, thermal, and barrier properties of poly (lactic acid) plasticized with epoxidized soybean oil and epoxidized palm oil *BioResources* **11** 1518-1540
[5] Ezzah F 2017 Production, characterization and kinetic modelling of malaysia Jatropha oil based bio-epoxy resins, unpublished *Master Thesis* University Putra Malaysia 43-62.
[6] Wong I M M, Tan C L, Azmi A I, Leong K W, Radzi M, and Nazri M 2014 Evaluations of mechanical properties and residual strength of drilled glass fiber reinforced polymer (GFRP) composites *Appl Mech Mater* **660** 270-274
[7] Tan C L, Azmi A I, and Muhammad, N 2014 Performance evaluations of carbon/glass hybrid polymer composites *Adv Mat Res* **980** 8-12
[8] Pozo Morales A, Güemes A, Fernandez-Lopez A, Carcelen Valero V, and De La Rosa Llano, S 2017 Bamboo–Polylactic Acid (PLA) Composite Material for Structural Applications *Materials* **10** 1286
[9] Fu S Y, and Lauke B 1996 Effects of fiber length and fiber orientation distributions on the tensile strength of short-fiber-reinforced polymers *Compos Sci Technol* **56** 1179-1190
[10] Anbukarasi K dan Kalaiselvam S 2015 Study of effect of fibre volume and dimension on mechanical, thermal, and water absorption behaviour of luffa reinforced epoxy composites *Mater Design* **66** 321-330
[11] Machrafi H, Lebon G, and Iorio C S 2016 Effect of volume-fraction dependent agglomeration of nanoparticles on the thermal conductivity of nanocomposites: Applications to epoxy resins, filled by SiO2, AlN and MgO nanoparticles *Compos Sci Technol* **130** 78-87
[12] Hale J "Boeing 787 from the Ground Up 06" *Aeromagazine: Quarter 4 2006* retrieved 2 January2018 <http://www.boeing.com/commercial/aeromagazine/articles/qtr_4_06/article_04_1.html >
[13] "Advancing Aerospace Materials - AIRBUS" *Airbus Newsroom* pp. 1–13 2016 retrieved 12 December 2017 <http://www.airbus.com/newsroom/news/en/2016/10/composites-silicon-carbide.html>.
[14] Bruckner R "Advance Organic Chemistry" *Academic Press* edition 2002 ISBN 0121381102 13-15.
[15] Liu K, Madbouly S A, dan Kessler M R 2015 Biorenewable thermosetting copolymer based on soybean oil and eugenol *Eur Polym J* **69** 16-28
[16] Groover M P Fourth edition 2010 Fundamental of modern manufacturing *John Wiley & Sons* ISBN 9780470467002 161-163.
[17] NASA January 1975 Reflective superinsulation materials by Midwest Research Institute *NASA CR2507* 5-13.
[18] Ibrahim B A and Kadum K M 2010 Influence of polymer blending on mechanical and thermal

properties *MAS* **4** 157
[19] Gogoi P, Boruah M, Sharma S, and Dolui S K 2015 Blends of epoxidized alkyd resins based on jatropha oil and the epoxidized oil cured with aqueous citric acid solution: A green technology approach *Acs Sustain Chem Eng* **3** 261-268
[20] Gogoi P, Boruah M, Bora C, and Dolui S K 2014 Jatropha curcas oil based alkyd/epoxy resin/expanded graphite (EG) reinforced bio-composite: Evaluation of the thermal, mechanical and flame retardancy properties *Prog Org Coat* **77** 87-93

IOP Conf. Series: Materials Science and Engineering **368** (2018) 012004 doi:10.1088/1757-899X/368/1/012004

Environmental performance of bioethanol production from oil palm frond petiole sugars in an integrated palm biomass biorefinery.

S J H M Yusof[1,2], A M Roslan[3,4*], K N Ibrahim[5], S S S Abdullah[6], M R Zakaria[3,4], M A Hassan[1,3] and Y Shirai[7]

[1]Department of Food and Process Engineering, Faculty of Engineering, Universiti Putra Malaysia, 43400 Serdang, Selangor, Malaysia
[2]School of Bioprocess Engineering, Universiti Malaysia Perlis, Kompleks Pusat Pengajian Jejawi 3, Kawasan Perindustrian Jejawi, 02600 Arau, Perlis, Malaysia
[3]Department of Bioprocess Technology, Faculty of Biotechnology and Biomolecular Sciences, Universiti Putra Malaysia, 43400 Serdang, Selangor, Malaysia
[4]Laboratory of Biopolymer and Derivatives, Institute of Tropical Forestry and Forest Products (INTROP), Universiti Putra Malaysia, 43400 Serdang, Selangor, Malaysia
[5]Section of Environmental Engineering Technology, Universiti Kuala Lumpur, Malaysian Institute of Chemical and Bioengineering Technology, Vendor City Taboh Naning, 78000 Alor Gajah, Melaka, Malaysia
[6]Section of Bioengineering Technology, Universiti Kuala Lumpur, Malaysian Institute of Chemical and Bioengineering Technology, Vendor City Taboh Naning, 78000 Alor Gajah, Melaka, Malaysia
[7]Department of Biological Functions and Engineering, Graduate School of Life Science and System Engineering, Kyushu Institute of Technology, 2-4 Hibikino, Wakamatsu, Fukuoka 808-0916, Japan

*Corresponding author: ar_muhaimin@upm.edu.my

Abstract. Generation of biofuels from renewable resources such as lignocellulosic biomass is a promising approach to reduce the sole reliability on the depleting fossil fuel. The aim of this work is to assess the potential environmental impact of bioethanol production from oil palm frond in a conceptual oil palm based biorefinery model, utilizing wet disc milling as a pretreatment method. A cradle-to-gate approach was selected, beginning with the harvesting and transportation of the frond petiole from the plantation, followed by production of oil palm frond petiole sugars via pretreatment and saccharification prior to bioethanol fermentation, and finally purification of the fermentation products to obtain anhydrous bioethanol. A life cycle assessment was performed using CML 2 baseline 2000 method (SimaPro v8.0), where ten impact categories were evaluated. It was found that the most significant environmental impact was from sugar recovery process with contribution of more than 90 %. This is mainly due to high power consumption by wet disc milling during pretreatment. Apart from that, production of enzyme and chemicals which were used during saccharification consumes high energy thus contributing major problems to the surrounding. Finding of this study helps to identify the hotspot which can be improved to establish a more energy efficient and greener system for bioethanol production from oil palm frond petiole sugars.

1. Introduction

Oil palm frond (OPF) is among the major biomass produced in the oil palm industry with generation of 54.17 and 54.24 million tonnes in 2010 and 2011, respectively [1]. It is obtained during replanting, pruning and harvesting, hence is available throughout the year. In the common practice, pruned frond is left in the plantation for nutrient recycling and soil conservation [2-3]. However, it was discovered that the basal (lower) part of OPF is rich in cellulosic materials and sugars, which are needed in the production of the biofuels and biobased chemicals [4-5]. Therefore, it is suggested that this part of OPF to be collected and serve as a raw material for biobased products whereas the remaining two third that contains high nutrients still serve as fertilizers in the plantation. With high carbohydrates and nutritive contents, OPF petiole can be converted into value added products such as biofuels, biobased chemicals, biofertilizer and animal feed [6–9]. Moreover, OPF juice was found to have high amount of free sugars with 70 % of glucose, and can be easily obtained by simple pressing method [10].

First generation bioethanol is generated using sugars and starch from feedstocks such as sugarcane and corn [11–14]. However, bioethanol production from lignocellulosic materials including crop residues, forestry and municipal waste is getting much attention these days to overcome the limitation in first generation biofuel production [15–17]. Apart from eliminating competition with food source, most of these wastes are available in large quantity and cheap. Examples of potential crop residues are wheat straw, sweet sorghum, cane bagasse, rice straw and corn stover [12,18–22]. However, the conversion of carbohydrate polymers to biofuel is a challenge due to the complex structure of lignocellulose. Pretreatments to alter its original structure is necessary prior to saccharification and fermentation [6,11-12,22].

Previously, feasibility study on bioethanol production from OPF petiole sugars was conducted within an integrated palm biomass biorefinery [6]. It was demonstrated that integration of a biorefinery to an existing palm oil mill was possible and has high potential for scaling up. Nevertheless, the environmental aspect should also be considered in promoting a sustainable biofuel production process [12]. Life cycle assessment (LCA) is a common tool to evaluate the impacts of a product system to the surrounding through its life cycle and have been extensively used to measure the environmental performance of bioethanol production [11,13,16,23–25]. Therefore, a LCA for bioethanol production from OPF petiole sugars was conducted in this study to evaluate the potential harm of the process to the environment.

2. Material and methods

2.1 Process description

2.1.1 OPF collection and transportation. Model used in this study was as reported by Abdullah et al.[6]. Bioethanol was produced from OPF petiole sugars collected from 4 neighboring oil palm mills. Approximately 57 600 tonnes of OPF petioles can be obtained from 240 000 tonnes of fresh fruit bunches (FFB) harvested every year. The harvested OPF petioles were transported to the mills which located approximately 15 km from the plantation. It was proposed that an additional cart for carrying OPF petioles to be attached to the existing truck carrying the harvested FFB [6]. The sugar production was primarily conducted at the individual mills using the surplus energy from the existing cogeneration system. Then, concentrated OPF petiole sugars from every mill was combined at a biorefinery which is located at one of the four mills. Each mill was assumed to be located 80 km radius from each other [6].

2.1.2 OPF sugar production. Cleaned petiole were milled to obtain the OPF juice which contain free sugars. It was reported that OPF juice comprise of 54 g/L, 21 g/L and 2 g/L of glucose, sucrose and fructose respectively [10]. Approximately 50 % (w/w) of OPF juice can be obtained, leaving another 50 % of OPF pressed fiber. The OPF pressed fiber was then subjected to wet disc milling (WDM) pretreatment at conditions described by [26]. 80 % of holocellulose was recovered after pretreatment, where 95 % of it was converted to glucose and xylose following saccharification.

Saccharified sugars and the OPF juice were subsequently concentrated using evaporators to remove 50 % of their water content before transmitted to the selected mill which locates the bioethanol refinery. A total of 13 900 tonnes/year/mill of sugars can be generated.

2.1.3 Bioethanol production. The concentrated sugars were sterilized prior to inoculation of *S. cerevisae* culture at 10 % (v/v). Urea was added to the fermentation media as nitrogen source. The fermentation was conducted in a reactor at 30 °C for 24 hours. Separation of solids and supernatant in the fermentation broth was subsequently performed by centrifugation. The supernatant which contains bioethanol was further purified using a continuous distillation, rectification, stripping and molecular sieves drying to obtain 98.9 % of anhydrous bioethanol. Whereas the solid residue was subjected to the waste treatment system for production of dried distiller grains with soluble (DGGS).

2.2 Life cycle assessment

LCA is a method to assess the environmental impacts of a product through its life cycle from the raw material acquisition and production, to end-use and disposal, with a generic framework provided by ISO 14040 and 14044 [16,27]. According to ISO 14040 (1997), LCA consists of 4 different phases, starting with goal and scope definition, inventory analysis, life cycle impact assessment and finally interpretation. Each step was conducted based on ISO standards as guidelines to ensure precise decision can be concluded. In this study, the characterization model CML 2 baseline 2000 v2.05 incorporated in software SimaPro v8.0 (Pre Consultants 2014) was adopted, based on the previous LCA study on bioethanol production which is almost similar to the present case [27-28].

2.2.1 Goal and scope of study . The aim of this study is to quantify the environmental impact of bioconversion of OPF petiole sugars for bioethanol production. The impacts were assessed according to all processes involved, starting from the harvesting of the OPF petiole at the plantation, transportation of petiole to the mill where it will undergo pretreatment and saccharification processes for sugar production, followed by the sugar conversion to bioethanol via fermentation and finally purification of bioethanol to obtain anhydrous bioethanol (cradle-to-gate). The evaluated impact was based on functional unit of 1 tonne of anhydrous bioethanol. Figure 1 shows the life cycle assessment system boundaries for the case studied.

2.2.2 Inventory analysis. Data for harvesting and transportation of OPF petiole were obtained from Pusat Penyelidikan Tun Razak, Pahang. While data on material input and output for bioethanol production were attained from the earlier study [6] and also calculated based on findings from the previous works [6, 9, 26, 29]. Most of emission data were generated from simulation by Superpro Designer software v9.5. Since the life cycle data for cellulase production is not available, the material and energy demands were estimated from the inventory data generated using values from previous study utilizing similar enzyme, assuming comparable enzyme production steps were adopted [30-31]. Inventory for urea, sodium and citric acid were obtained from the EcoInvent 3.1 database. Table 1 listed data use for the Life Cycle Impact Assessment.

2.2.3 Assumptions. In order to facilitate the evaluation of the impact arising from the proposed model, few assumptions were implemented. Steam required to run the bioethanol biorefinery is solely provided by the oil palm mill, through cogeneration system. Approximately 140 053 tonne/ year of steam was needed and this requirement can be fulfilled by the mill using the existing excess steam left following FFB processing [6]. Hence, impact of steam was excluded. Whereas the electricity is obtained from both the cogeneration and the national grid. This is because only 7.72 GWh of electricity can be produced by the cogeneration system based on boiler and steam efficiency of 77.4% [6]. 4.08 GWh of electricity per tonnes FFB was required for FFB processing, leaving an excess energy of 3.64 GWh of electricity. On the other hand, total electricity required for bioethanol production using WDM pretreatment method was 1 076 GWh, beyond the surplus electricity available. Therefore, an extra electricity supply of 1 072 GWh must be obtained from the external source and the value was considered in the inventory.

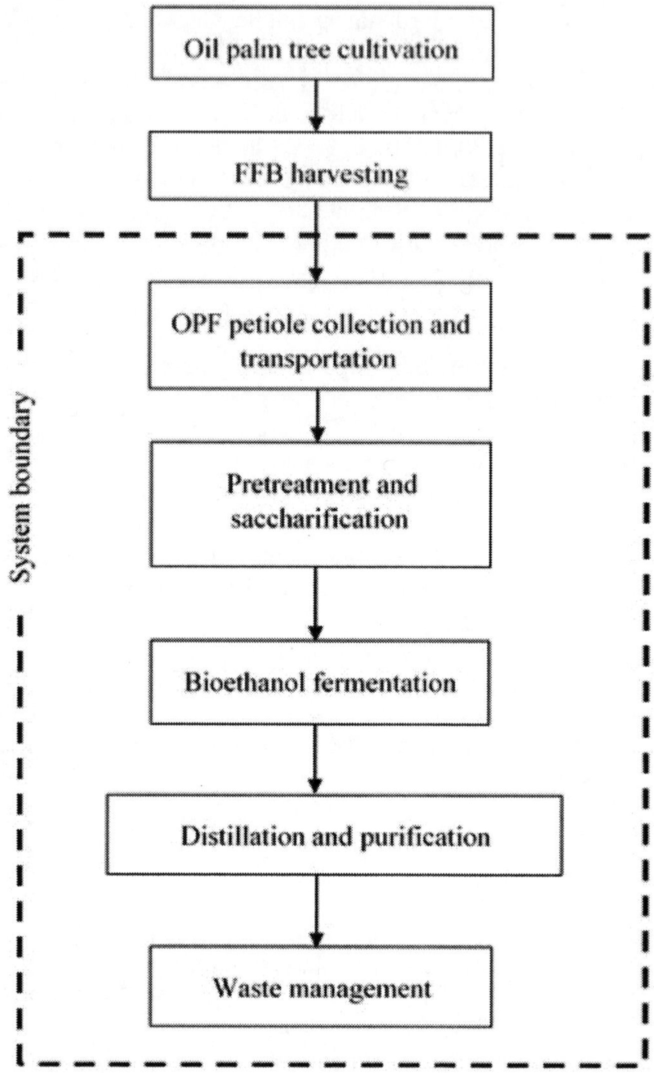

Figure 1 Life Cycle Assessment system boundaries for the bioethanol production from OPF petiole sugars.

Table 1 Inventory data for bioethanol production from OPF petiole sugars.

	Values	Reference
Transportation		
Input		
Trucks, 6 tonnes capacity, tkm/year	7 199 985	This study
Trucks, 20 tonnes capacity, tkm/year	3 324 480	This study
Bioethanol production		
Input		
OPF petiole, tonnes/year/mill	57 600	[6]
OPF pressed fiber, tonnes/year/mill	20 160	[6]
OPF petiole sugars, tonnes/year/mill	13 900	[6]

IOP Conf. Series: Materials Science and Engineering **368** (2018) 012004 doi:10.1088/1757-899X/368/1/012004

Citric acid, tonnes/year	2327	This study
Sodium citrate, tonnes/year	4688	This study
Enzyme, tonnes/year	2685	This study
Urea, tonnes/year	1286	This study
Air, tonnes/year	1 121 681	This study
Water, tonnes/year	2 707 952	This study
Electricity, MWh/year	1 072 110	This study
Output		
Anhydrous bioethanol, tonnes/year	26 700	This study
Stillage, tonnes/year	521 800	This study
DGGS, tonnes/ year	2611	[6]
Saccharified fiber residue, tonnes/year	33 600	[6]
Energy from OPF residue, MJ/year	1767	[6]
Emission		
CO_2, tonnes/year	27 138	This study
N_2, tonnes/year	861 012	This study
O_2, tonnes/year	261 387	This study
Water vapor, tonnes/year	34 560	This study
Ethanol, tonnes/year	82	This study
Waste water, tonnes/year	270 000	[6]
Water for recycle, tonnes/year	57 600	[6]
Enzyme production		
Input		
Cellulosic hydrolysate, kg	7600	[30]
Water, kg	4474	[30]
Maize steep water, kg	119	[30]
Diammonium phosphate (nutrient), kg	47	[30]
Electricity, kWh	2236	[30]
Output		
Enzyme broth, kg	11 455	[30]
CO_2, kg	879	[30]
Furfural, kg	8.2	[30]
Acetic acid (VOC), kg	2.8	[30]

2.2.4 Characterization model and impact categories. CML 2 baseline 2000 model (SimaPro v8.0) was applied to perform the Life Cycle Impact Assessment. Ten mid-point impacts was evaluated: abiotic depletion potential (ADP), acidification potential (AP), eutrophication potential (EP), global warming potential (GWP100), ozone layer depletion potential (ODP), human toxicity potential (HTP), fresh water ecotoxicity potential (FETP), marine ecotoxicity potential (METP), terrestrial ecotoxicity potential (TETP), and photochemical oxidation potential (PCOP).

3. Results and discussion

Physical or mechanical pretreatment such as ball milling and WDM aims to increase the surface area by reducing the biomass particle size, together with defibrillation and reduction in crystallinity degree [32]. This provides higher accessibility of enzymes to substrate, thus escalating the saccharification rates and yield. Previous study showed that with WDM application, 95 % of holocellulose was successfully converted into simple sugars following saccharification [26]. Nevertheless, despite of its efficiency in enzymatic hydrolysis, WDM has high energy consumption and application in a larger scale was somehow difficult [29,33-34]. It was reported that approximate energy required were 48 MJ/kg for sugarcane bagasse and 39.6 MJ/kg of sugarcane straw for 20 cycles [29]. Whereas for OPMF, 18.9 MJ/kg was consumed for 9 cycles [35].

Figure 2 depicted impact values for each category according to major process involved. The process was classified into 4 sub-processes: transportation, WDM pretreatment, saccharification and ethanol fermentation and purification. By comparing contribution of each sub-process to the environmental burden, the process which needs to be improved can be identified.

In general, life cycle impact assessment demonstrated that pretreatment step by wet disc milling had a major impact to all categories, except for EP (Figure 2d), ODP (Figure 2e) and TETP (Figure 2j) where saccharification impact dominates. WDM consumed high electricity thus requires high usage of fossil fuel. This explains its significant impact in ADP (92%). ADP (Figure 2a) which refers to the decrease of non-living natural resources (including energy resources) such as minerals or crude oil, is a very widely used impact category reflecting fossil fuel energy use [27]. Similarly, application of enzyme and chemicals (citric acid and urea) during saccharification and fermentation, involved high utilization of fossil fuel during their production and diesel consumption for transportation.

GWP (Figure 2b) is a quantification of an equivalent amount of carbon dioxide released to the surroundings. Due to high electricity consumption, wet disc milling contributed the highest impact to global warming potential (90%) as atmospheric emissions are directly related to energy and resource consumption [36]. Whereas saccharification has the second highest impact with almost 10%, due to enzyme application. Previously, enzyme production was found to be the main hotspot for carbon dioxide emission [27, 36]. This is because enzyme production consumes a substantial amount of fossil or combustion electricity for air compression which also generates a considerable amount of carbon dioxide emission in the chain [16]. Utilization of fossil fuel energy by all transport and during production of urea used in fermentation resulted in carbon dioxide emission. Additional contribution comes from carbon dioxide produced during fermentation.

AP refers to environmental damage cause by acidic gas such as sulfur dioxide (SO_2). Since fossil fuel consumption resulted in SO_2 emissions [27], thus WDM produced a significant impact in AP (Figure 2c). Similarly, high usage of fossil fuel explained the reason for saccharification, fermentation and transportation contribution in this impact category.

For EP category, it was demonstrated that the impact from saccharification and pretreatment steps were almost equal, which was approximately 50 % (Figure 2d). Eutrophication potential is associated with environmental impacts of excessively high nutrients (N and P) in the river and lakes. Consequences are like shifting in species composition, increased biological productivity such as algal bloom and subsequent oxygen consuming degradation processes [19, 27]. Significant burden from saccharification step was due to application of diammonium phosphate which serves as a nitrogen source in enzyme production, contributing to nutrients emission. High eutrophication effect was

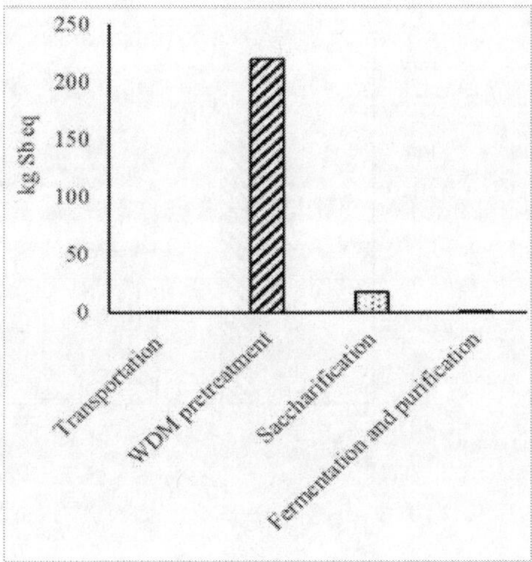
kg Sb eq
250
200
150
100
50
0
Transportation
WDM pretreatment
Saccharification
Fermentation and purification

(a)

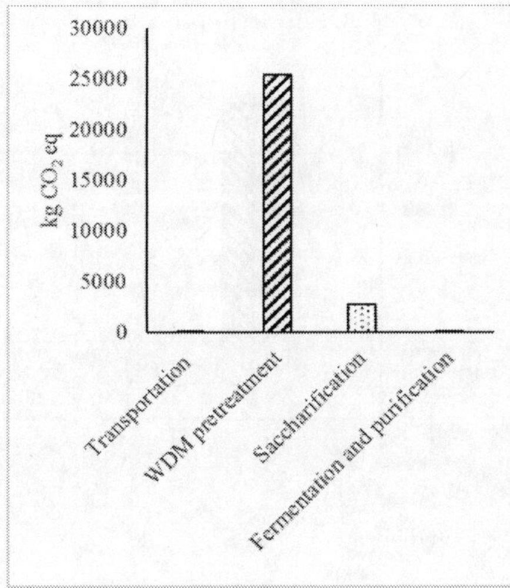
kg CO_2 eq
30000
25000
20000
15000
10000
5000
0
Transportation
WDM pretreatment
Saccharification
Fermentation and purification

(b)

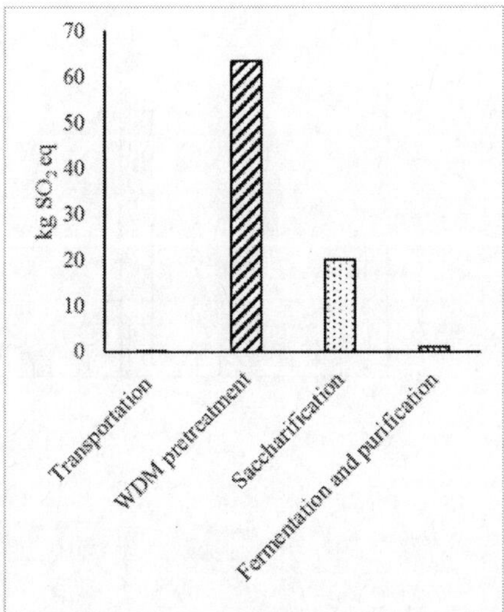
kg SO_2 eq
70
60
50
40
30
20
10
0
Transportation
WDM pretreatment
Saccharification
Fermentation and purification

(c)

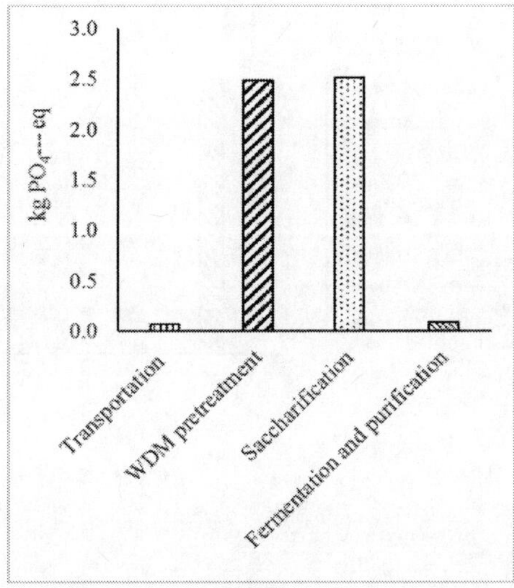
kg PO_4^{---} eq
3.0
2.5
2.0
1.5
1.0
0.5
0.0
Transportation
WDM pretreatment
Saccharification
Fermentation and purification

(d)

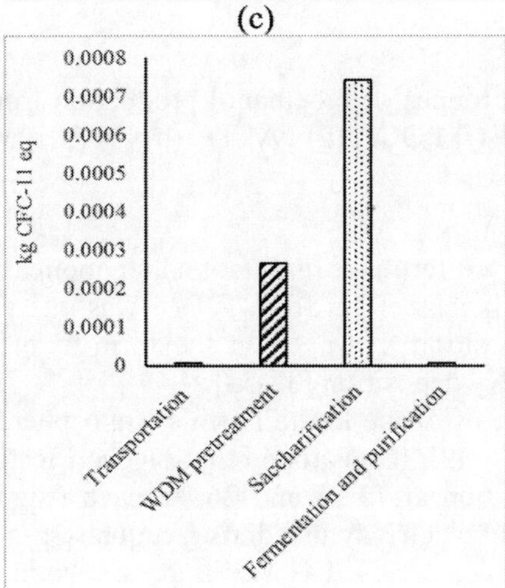
kg CFC-11 eq
0.0008
0.0007
0.0006
0.0005
0.0004
0.0003
0.0002
0.0001
0
Transportation
WDM pretreatment
Saccharification
Fermentation and purification

(e)

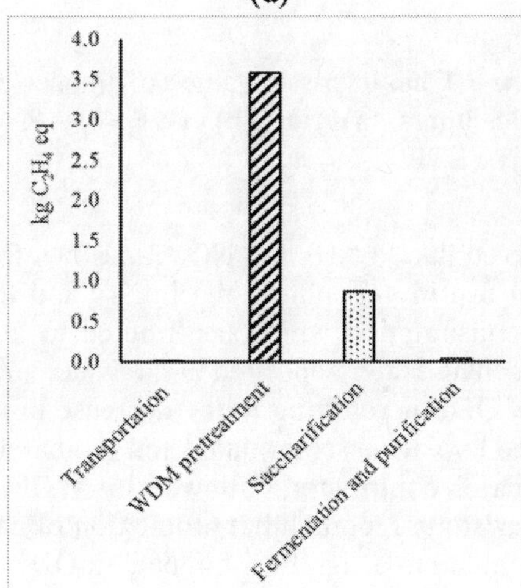
kg C_2H_4 eq
4.0
3.5
3.0
2.5
2.0
1.5
1.0
0.5
0.0
Transportation
WDM pretreatment
Saccharification
Fermentation and purification

(f)

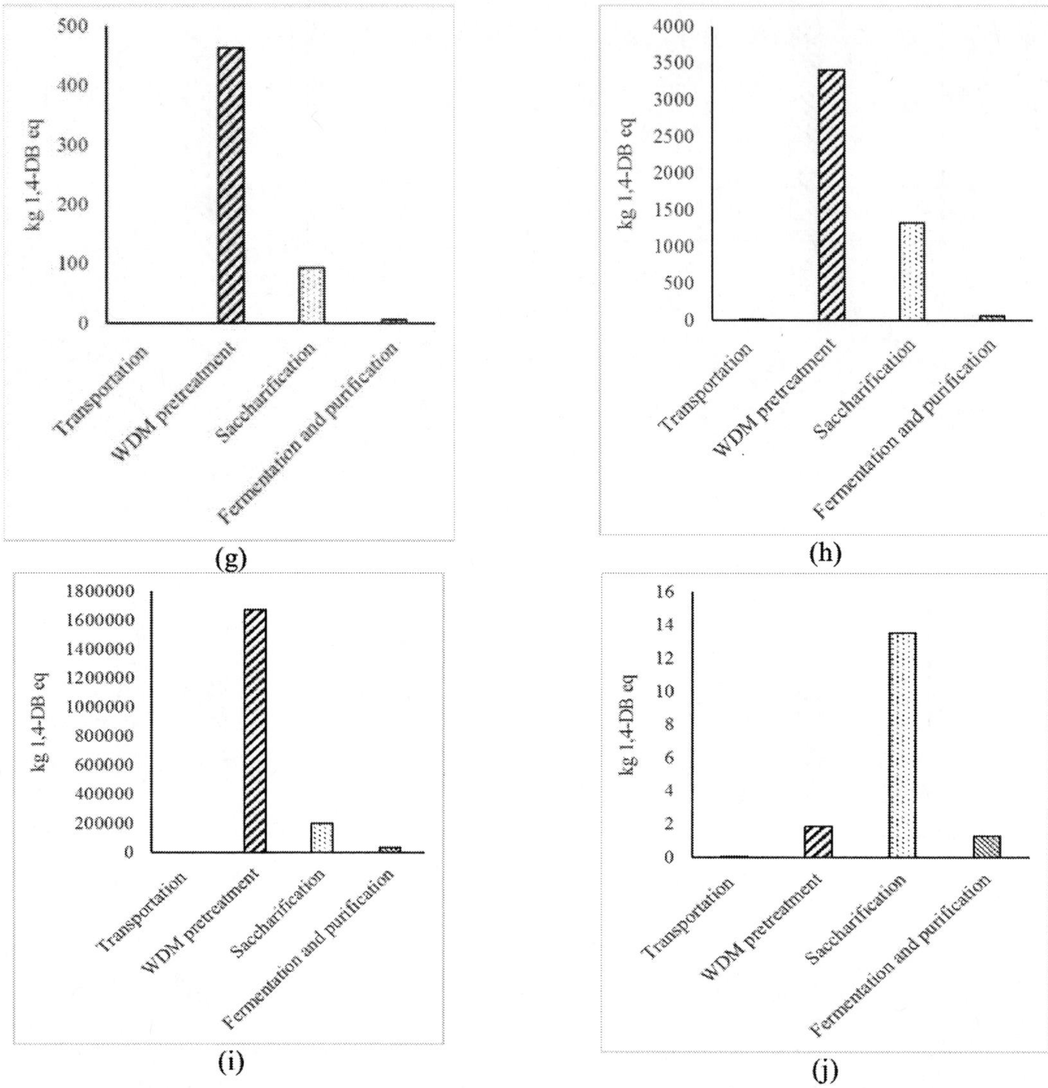

Figure 2 Characteristic values of impact categories for 1 tonnes of bioethanol production from OPF petiole sugars. (a) ADP, (b) GWP, (c) AP, (d) EP, (e) ODP, (f) POCP, (g) FAETP, (h) HTP, (i) MAETP and (j) TETP.

observed due to NH_3 and NO_x emissions from nitrogen based fertilizer production and application and diesel use in agricultural machinery and tractors [16]. Whereas effect from WDM was due to fossil fuel consumption which contributed to atmospheric deposition. Nitrogen in the form of NO_x, was released and later deposited in the water source as wet or dry deposition [37-38].

ODP is referring to the decrease in the total volume of ozone in the Earth's stratosphere and is caused by various chlorinated and bromated substances [27, 39]. It was found that saccharification was the major contributor, followed by WDM and transportation at 73 % and 26 %, each (Figure 2e). Earlier study reported that production of carbon source and electricity use during cellulase production contributed to a significant impact in ODP [40]. Halons which are used as fire suppressant and coolant in the gas pipeline distribution system was emitted during electricity generation, increasing the risk of ozone depletion [41].

Whereas for POCP, WDM has the highest impact with 80% and saccharification was the second highest with 19 % (Figure 2f). The photochemical oxidation, also referred as summer smog, is the result of reactions between NO_x and hydrocarbons or volatile organic compounds (VOC) [27]. Intensity of impact increased with increment of amount of emissions from fossil fuel consumption. Previous study showed that bioethanol conversion unit produced the highest POCP, with SO_2 as major pollutant, generating from coal combustion [19, 39].

HTP, FAETP, MAETP and TETP characterization compares a large number of chemicals that can contribute to cancer or have other negative effects on human health, fresh aquatic ecosystem health, marine aquatic ecosystem health and terrestrial ecosystem health, respectively [19]. Coal which is commonly use as fossil energy for electricity and steam production contains high composition of heavy metals such as As, Hg, Pb and Cr, thus contributes to high impact in ecotoxity potential. This clarifies the significant impact from high energy requirement process such as WDM and saccharification (Figure 2g-2j).

Based on these findings, several approaches can be suggested to reduce the environmental impact. Other pretreatment methods with higher or comparable performance could be used to replace WDM. Hydrothermal pretreatment, including liquid hot water, hot compressed water and superheated steam are among potential methods as it is more environmental friendly since it use only water and heat (chemical free) [42]. With process optimization it was demonstrated that application of hot compressed water of oil palm frond prior to enzymatic hydrolysis can generate glucose yield of 93% [43]. Furthermore, few studies on LCA of pretreatment methods in biofuel production have shown that liquid hot water pretreatment produced lower carbon dioxide or GHG emission [27,39,44]. Apart from that, with combination of WDM and hot compressed water, the amount of energy consumed was reduced up to 21 % [35]. Nonetheless, there are few drawbacks that also needs to be considered. Hydrothermal pretreatment consumed substantial amount of water which resulted in more waste water that have to be treated [44]. Moreover, the process also generated significant amount of inhibitors which needs to be removed as they could interrupt the saccharification and fermentation process [45]. Therefore, further investigation should be conducted prior to selection of the most appropriate method.

Modifications related to enzyme production and hydrolysis can helps to improve the environmental performance cause by enzyme application. Among alternatives that can be considered are to decrease the enzyme quantity, improving the enzyme production process, applying enzyme recycling and utilizing recombinant yeast which produce hydrolytic enzymes, thus eliminating the requirement of enzyme production step [46]. Other than that, the environmental performance can also be improved by using the waste stream such as COD in the stillage for energy recovery [47-48].

4. Conclusions

A cradle-to-gate LCA on the conversion of OPF petiole sugars to bioethanol reveals that pretreatment and saccharification steps were the hotspots which needs to be improved. Despite of its high efficiency in sugar production, application of WDM is somehow environmentally unfavorable. Selection of proper pretreatment method, as well as improving enzyme and chemicals utilization, are therefore crucial to reduce burden to the environment. This study could serve as a base case for comparison with other LCA work on bioethanol production from OPF petiole sugars utilizing other pretreatment technique in the future.

Acknowledgements

The authors would like to thank Science and Technology Research Partnership for Sustainable Development (SATREPS) under Japan Science and Technology Agency (JST) and Japan International Cooperation Agency (JICA) for partly funding this work. The authors also thank Universiti Putra Malaysia for technical assistance and Universiti Kuala Lumpur for the permission to use the SimaPro v8.0 (PreConsultant 2014) during this study. Appreciation also goes to Universiti Malaysia Perlis and Ministry of Higher Education, Malaysia for the provision of study leave and scholarship for the first author.

References

[1] Sabiha-Hanim S, Siti Normah M D S, Khalilah A K and Aishah B 2016 Xylooligosaccharides production from oil palm frond by Trichoderma longibrachiatum xylanase *MJAS* **20** 525–30

[2] Wan Zahari M, Abu Hassan O, Wong H K and Liang J B 2003 Utilization of oil palm frond - based diets for beef and dairy production in Malaysia *Asian Austral J Anim* **16** 625–34

[3] Zahari M A K M, Abdullah S S S, Roslan A M, Ariffin H, Shirai Y and Hassan M A 2014 Efficient utilization of oil palm frond for bio-based products and biorefinery *J Clean Prod* **65** 252–60

[4] MIA 2013 National Biomass Strategy 2020: New wealth creation for Malaysia's biomass industry

[5] Roslan A M, Zahari M A K ., Hassan M . and Yoshihito S 2014 Investigation of Oil Palm Frond Properties for Use as Biomaterials and Biofuels *JTAFS* **58** 26–9

[6] Abdullah S S S, Shirai Y, Ali A A M, Mustapha M and Hassan M A 2016 Case study: Preliminary assessment of integrated palm biomass biorefinery for bioethanol production utilizing non-food sugars from oil palm frond petiole *Energ Convers Manage* **108** 233–42

[7] Ofori-Boateng C and Lee K T 2014 Sono-assisted organosolv/H2O2 pretreatment of oil palm (Elaeis guineensis Jacq.) fronds for recovery of fermentable sugars: Optimization and severity evaluation *Fuel* **115** 170–8

[8] Lee K . and Ofori-Boateng C 2013 *Advances in Biofuels* ed R Pogaku and R H Sarbatly (Springer New York)

[9] Zahari M A K M, Ariffin H, Mokhtar M N, Salihon J, Shirai Y and Hassan A 2014 Case study for a palm biomass biorefinery utilizing renewable non-food sugars from oil palm frond for the production of poly(3-hydroxybutyrate) bioplastic *J Clean Prod*

[10] Zahari M A K M, Zakaria M R, Ariffin H, Mokhtar M N, Salihon J, Shirai Y and Hassan M A 2012 Renewable sugars from oil palm frond juice as an alternative novel fermentation feedstock for value-added products *Bioresource Technol* **110** 566–71

[11] Borrion A L, Mcmanus M C and Hammond G P 2012 Environmental life cycle assessment of lignocellulosic conversion to ethanol : A review *Renew Sust Energ Rev* **16** 4638–50

[12] Morales M, Quintero J, Conejeros R and Aroca G 2015 Life cycle assessment of lignocellulosic bioethanol : Environmental impacts and energy balance *Renew Sust Energ Rev* **42** 1349–61

[13] Ometto A R and Hauschild M Z 2009 Lifecycle assessment of fuel ethanol from sugarcane in Brazil 236–47

[14] Dias M O D S, Filho R M, Mantelatto P E, Cavalett O, Rossell C E V, Bonomi A and Leal M R L V 2015 Sugarcane processing for ethanol and sugar in Brazil *J. Environ. Dev* **15** 35–51

[15] Kumar D and Murthy G S 2011 Impact of pretreatment and downstream processing technologies on economics and energy in cellulosic ethanol production *Biotechnol Biofuels* **4** 27

[16] Borrion A L, Mcmanus M C and Hammond G P 2012 Environmental life cycle assessment of bioethanol production from wheat straw *Biomass Bioenergy* **47** 9–19

[17] Goh C S, Tan K T, Lee K T and Bhatia S 2010 Bioresource Technology Bio-ethanol from lignocellulose : Status , perspectives and challenges in Malaysia *Bioresource Technol* **101** 4834–41

[18] Canilha L, Chandel A K, Suzane dos Santos Milessi T, Antunes F A F, Luiz da Costa Freitas W, das Graças Almeida Felipe M, and da Silva S S 2012 Bioconversion of sugarcane biomass into ethanol: an overview about composition, pretreatment methods, detoxification of hydrolysates, enzymatic saccharification, and ethanol fermentation *BioMed Res Int,* 2012

[19] Wang M, Chen Y, Xia X, Li J and Liu J 2014 Bioresource Technology Energy efficiency and environmental performance of bioethanol production from sweet sorghum stem based on life cycle analysis *Bioresource Technol* **163** 74–81

[20] Chen H and Fu X 2016 Industrial technologies for bioethanol production from lignocellulosic biomass *Renew Sust Energ Rev* **57** 468–78

IOP Conf. Series: Materials Science and Engineering **368** (2018) 012004 doi:10.1088/1757-899X/368/1/012004

[21] Relvas F M, Morais A R C and Bogel-Lukasik R 2015 Kinetic modeling of hemicellulose-derived biomass hydrolysis under high pressure CO2-H2O mixture technology *J Supercrit Fluid* **99** 95–102

[22] Zabed H, Sahu J N, Boyce A N and Faruq G 2016 Fuel ethanol production from lignocellulosic biomass : An overview on feedstocks and technological approaches *Renew Sust Energ Rev* **66** 751–74

[23] Nguyen T L T and Gheewala S H 2008 Life cycle assessment of fuel ethanol from cane molasses in Thailand *Int J Life Cycle Assess* **13** 301

[24] Botero Agudelo J, Castaño Peláez H, and Naranjo Merino C 2011 Life Cycle Assessment for bioethanol produced from cassava in Colombia *PML* **6** 69-77

[25] Kunimitsu Y and Ueda T 2013 Economic and environmental effects of rice-straw bioethanol production in Vietnam *Paddy Water Environ* **11** 411-421

[26] Zahari M A K M, Abdullah S S S, Roslan A M, Ariffin H, Shirai Y and Hassan M A 2014 Efficient utilization of oil palm frond for bio-based products and biorefinery *J Clean Prod* **65** 252-260

[27] Wang L, Littlewood J and Murphy R J 2013 Environmental sustainability of bioethanol production from wheat straw in the UK *Renew Sust Energ Rev* **28** 715–25

[28] Chiew Y L and Shimada S 2013 Current state and environmental impact assessment for utilizing oil palm empty fruit bunches for fuel , fiber and fertilizer e A case study of Malaysia *Biomass Bioenergy* 1–16

[29] da Silva A S A, Inoue H, Endo T, Yano S, and Bon E P 2010 Milling pretreatment of sugarcane bagasse and straw for enzymatic hydrolysis and ethanol fermentation *Bioresource technol* **101** 7402-7409

[30] Sheehan J, Aden A, Paustian K, Brenner J, Walsh M and Nelson R 2004 Energy and Environmental Aspects of Using Corn Stover for Fuel Ethanol *J Ind Ecol* **7** 117–46

[31] Budsberg E, Rastogi M, Puettmann M, Caputo J, Balogh S, Volk T, Gustafson R and Johnson L 2012 Life-Cycle Assessment for the Production of Bioethanol from Willow Biomass Crops via Biochemical Conversion *Forest Prod J* **62** 305–13

[32] Teixeira R S S, Silva A S, Moutta R O, Ferreira-leitão V S, Barros R R O, Ferrara M A and Bon E P S 2014 Biomass pretreatment : a critical choice for biomass utilization via biotechnological routes *BMC proceedings* **8** 8–10

[33] Hideno A, Inoue H, Tsukahara K, Fujimoto S, Minowa T, Inoue S, Endo T and Sawayama S 2009 Wet disk milling pretreatment without sulfuric acid for enzymatic hydrolysis of rice straw *Bioresource Technol* **100** 2706–11

[34] Zakaria M R, Hirata S, Fujimoto S and Hassan M A 2015 Combined pretreatment with hot compressed water and wet disk milling opened up oil palm biomass structure resulting in enhanced enzymatic digestibility *Bioresource Technol* **193** 128–34

[35] Zakaria M R, Norrrahim M N F, Hirata S and Hassan M A 2015 Hydrothermal and wet disk milling pretreatment for high conversion of biosugars from oil palm mesocarp fiber *Bioresource Technol* **181** 263–9

[36] Roy P, Tokuyasu K, Orikasa T, Nakamura N and Shiina T 2012 Evaluation of the Life Cycle of Bioethanol Produced from Rice Straw *Biomass Bioenergy* **37** 188–95

[37] European Commission 2013 *Science for Environment Policy In-depth Report Nitrogen Pollution and the European Environment Implications for Air Quality Policy*

[38] Howarth R and Rielinger D M *Nitrogen from the atmosphere: Understanding and reducing a major cause of degradation in our coastal waters*

[39] Wang L, Littlewood J and Murphy R . 2014 An economic and environmental evaluation for bamboo-derived bioethanol *RSC Advances* **4** 29604–11

[40] Gilpin G S and Andrae A S G 2017 Comparative attributional life cycle assessment of European cellulase enzyme production for use in second-generation lignocellulosic bioethanol production *Int J Life Cycle Assess* **22** 1034–53

[41] Atilgan B and Azapagic A 2015 Life cycle environmental impacts of electricity from fossil

fuels in Turkey *J Clean Prod* **106** 555–64

[42] Alvira P, Tomas-Pejo E, Ballesteros M and Negro M J 2010 Pretreatment technologies for an efficient bioethanol production process based on enzymatic hydrolysis: A review *Bioresource Technol* **101** 4851–61

[43] Goh C S, Lee K T and Bhatia S 2010 Hot compressed water pretreatment of oil palm fronds to enhance glucose recovery for production of second generation bio-ethanol *Bioresource Technol*

[44] Prasad A, Sotenko M, Blenkinsopp T and Coles S R 2016 Life cycle assessment of lignocellulosic biomass pretreatment methods in biofuel production *Int J Life Cycle Assess* **21** 44–50

[45] Zakaria M R, Hirata S, Fujimoto S, Ibrahim I and Ali M 2016 Bioresource Technology Soluble inhibitors generated during hydrothermal pretreatment of oil palm mesocarp fiber suppressed the catalytic activity of Acremonium cellulase *Bioresource Technol* **200** 541–7

[46] Sebastião D, Gonçalves M S, Marques S, Fonseca C, Gírio F, Oliveira A C and Matos C T 2016 Life cycle assessment of advanced bioethanol production from pulp and paper sludge *Bioresource Technol* **208** 100–9

[47] Wilkie A C, Riedesel K J and Owens J M 2000 Stillage characterization and anaerobic treatment of ethanol stillage from conventional and cellulosic feedstocks *Biomass Bioenergy* **19** 63–102

[48] Kaparaju P, Serrano M, Belinda A, Kongjan P and Angelidaki I 2009 Bioresource Technology Bioethanol , biohydrogen and biogas production from wheat straw in a biorefinery concept *Bioresource Technol* **100** 2562–8

IOP Conf. Series: Materials Science and Engineering **368** (2018) 012005 doi:10.1088/1757-899X/368/1/012005

Evaluation of water absorption of polyvinyl alcohol-starch biocomposite reinforced with sugarcane bagasse nanofibre: Optimization using Two-Level Factorial Design

M Asem[1], W M F W Nawawi[1], D N Jimat[1]*

[1]Department of Biotechnology Engineering, Faculty of Engineering, International Islamic University Malaysia, 50728 Gombak, Selangor, Malaysia

*Email: jnoraini@iium.edu.my

Abstract. Global concern on petroleum based plastics which is non-degradable in our environment has led researchers to develop biodegradable plastics. However, biodegradable plastics have poor barrier properties because of their hydrophilic character of biopolymers. It is known that incorporation of nanocellulose extracted from plant sources to improve barrier properties of biocomposites because of its nanoscopic structure. This study aims to develop biodegradable film based from PVA/Starch and nanocellulose from sugarcane bagasse. Investigation on the effect of sugarcane bagasse nanofibre (SCB-NF) content, PVA content, starch content and water content to the water absorption property of polyvinyl alcohol (PVA)/Starch (S) composites reinforced with sugarcane bagasse nanofibre (SCB-NF) was carried out using Design Expert Version 9.0 with a two-level factorial design (2-FI). Composition of SCB-NF, PVA, Starch and water content was varied from the range of 1 to 9%, 3-8 gram, 1-4 gram, and 80 to 100mL. The nanofibre content was found to have significant effect ($p=0.0099$) on the water absorption of biocomposite film, which is in parallel to the theory of nanofillers to decrease water absorption of biocomposite. As expected, the individual and interaction effects of the compositions can have effect on the water absorption of the biocomposite because of the chemical bonds interaction that they form during the synthesis of biocomposite film.

1. Introduction

Presently, food packaging industry predominantly use petroleum based plastic materials. The most commonly used petroleum based plastics in food packaging includes low density polyethylene (LDPE), polystyrene (PS), and polyethylene terephthalate (PET). However, the problem with petroleum based plastics in food packaging is their non-degradable nature, and the limited resources [1].One of the attempts to overcome the problem of petroleum based plastic waste is by developing a biodegradable polymer. Polyvinyl alcohol (PVA) is one of the promising polymers since it is the only vinyl type polymer with biodegradable characteristics. However, the application of PVA materials is limited due to the high cost and slow degradation process especially under anaerobic condition [2]. Considering this limitation, PVA is often blended with other cheap biodegradable polymers [3].

The best known renewable resources capable of making biodegradable plastics are starch and cellulose. Starch is well known abundant raw material and relatively cheap biodegradable polymer. Previous studies have reported that blending of PVA and starch can enhance their tensile strength, elongation and toughness [4]. Furthermore, the presence of hydroxyl group in both PVA and starch makes it form a strong hydrogen bond and relatively good compatibility [5]. The disadvantages of using these PVA/starch blends for packaging is limited due to their poor water barrier properties, hydrophilic nature due to presence of large number of hydroxyl groups. Thus, several studies have proposed the incorporation of nanofillers into PVA/starch blends in order to improve water barrier property [3].

Water barrier property is important in developing film for food packaging. Nanocellulose has been used as nanodimension filler in composite material for reinforcing strength, improving biocompatibility, controlling water sorption, and barrier properties [6]. Nanocellulose is an emerging sector in the food packaging industry as nanoreinforcment for polymers matrix. It is known that nano size cellulose is one of the strategies to improve the polymer composite adhesion and compatibility [7]. Morphology of the nanocomposites, bearing tortuous mixing nanofillers with a macromolecular matrix, leads to higher barrier properties and thus lowering the permeability towards water [8]. In addition, it is known that gas or water molecules penetrate in crystalline ordered structure of nanocellulose, and their dense percolating network held together by strong inter-particle bonds idealize them as good candidate for use as barrier films [9]. This phenomenon was noted by Yoon et al. [10], where water resistance of the composites improved by 70% by addition of nanosized poly (methyl methacrylate-co-acrylamide) particles.

The structure of nanocomposites is important in having remarkable barrier properties. Properties of nanocomposite film depend on many factors. According to Kumar [11], the choice of the filler is important in determining the adhesions between the matrixes and the filler. The strength of bonding will be stronger if the filler is functionalized meaning more availability of OH bonds presence on the surface to bind to the matrix and the homogenize dispersion of the filler in the matrix. Another important factor is the strong adhesion between filler and matrix, which affect the percolation thresholds, and lessen empty spaces (voids) between the polymer matrix interfaces.

By understanding various factors that affect water absorption of nano composite film, the best approach for the optimization of various factors is a 2 level factorial design by Design Expert software. Two- level factorial is designed to study the effects of factors and their interactions simultaneously at two levels over a range of chosen factor levels. The range of loadings chosen was estimated based from previous researchers. Lani, 2014 concluded that with a 7: 3 ratio of PVA and starch is the best combination for optimum properties. Besides, Lani also found that nanocellulose from EFB fibre loading of 5% showed the best water resistance properties of the nanocomposite film. Further studies by Naguib et al. [12], showed that with 5% nanocellulose reinforcement from bagasse fiber, the composite showed the least water absorption.

In developing a new material which is biodegradable and have potential in food packaging this study aims to investigate the effect of factors that determine the water absorption property of the PVA/starch/nanocellulose biocomposite. It is hypothesized that amount of constituents of polymer matrix composites and their interaction may have effect on the final property of water absorption of biocomposite film. Using 2 level factorial design (2-FI) the factors involved are PVA content, starch content, water content (solvent), and nanocellulose loading, and their interaction are optimized that give the least water absorption of the biocomposite film. Nanocellulose is chosen as nanoreinforcment for the PVA/starch blend matrix. Nanocellulose having the characteristics of three hydroxyl groups laterally along surface of cellulose chain, and the high surface area makes it an ideal candidate to functionalize with the polymer matrix of PVA/starch matrix. Furthermore, the advancement of nanocellulose fibre derived from sugarcane bagasse is not widely exploited yet [13].

2. Experimental

2.1. Materials

Sugarcane bagasse originated from green canes (*Saccharum officinarum, Gramineae* family) was collected from night market, Taman Melati, Kuala Lumpur, Malaysia. All chemical reagents were of analytical reagent grade. Polyvinyl alcohol (MW 89,000) with degree of hydroxylation 98 % was purchased from Sigma Aldrich (Malaysia). Pure potato starch was obtained from Systerm (Malaysia). Sodium hydroxide and hydrogen peroxide were purchased from Friendemann Schmidt. Sulphuric acid was purchased from Laboratory reagent.

IOP Conf. Series: Materials Science and Engineering **368** (2018) 012005 doi:10.1088/1757-899X/368/1/012005

2.2. Extraction of nanocellulose from sugarcane bagasse

2.2.1 Alkali treatment. Sugarcane bagasse undergoes alkali treatment to remove the lignin and hemicellulose structure from the sugarcane bagasse fibres. The sugarcane bagasse was treated with an alkali solution (2%w/v NaOH) placed in water bath at 80 °C. The solid residue was then filtered and washed several times using deionized water under vacuum pump.

2.2.2. Bleaching. Following pulp disintegration process, the bleaching process comes with aqueous hydrogen peroxide solution in the ratio of 1:1 at 75 °C. The mixture was also washed and filtered several times to ensure proper washing.

2.2.3. Hydrolysis. Bleached fibres were hydrolysed with a 1% v/v H_2SO_4 solution at 80 °C.

2.2.4. Ultra sonication. Hydrolyzed sugarcane bagasse fibres were then sonicated to break down the individualized fibres by using microcentrifuge tubes filled with ice packed in a beaker. Hydrolyzed fibres was filled with distilled water, vortexed, and ultrasonicated with 80 Hertz. Following ultrasonication, the microtube was centrifuged for 20 minutes at 10,000 rpm, and kept in refrigerator at 4 °C.

2.3 Preparation of PVA/Starch/Nanocellulose biocomposites using Two factor Factorial design
Preparing the biocomposite film was done after extracting the nanocellulose from previous experiment. In order to determine the significance of the parameters, experiments were designed in determining the best parameters and amount needed for the preparation of PVA/Starch/nanocellulose.

A two factor factorial design from statistical software Design Expert version 9 (Stat-Ease, Minneapolis, USA) was used to optimize the conditions for the preparation of the composites. The ranges and levels of the factors investigated in this study are shown in Table 1. Process condition was carried out for 4 hours at 130 °C with rotor speed at 450 rpm, and the water absorption test was measured using ASTM D570. Samples of 20 mmx 20 mm in dimensions were immersed in 10mL of distilled water in room temperature for 24 hours. Before immersion, samples were dried in an oven for 50 °C for 24 hour and initially weighed. According to standard test method for water absorption of plastics ASTM D570, following immersion for 24 hour, water accumulated on the sample surface was gently wiped and samples were reweighed for the nearest 0.001. Water absorption was calculated using the following equation.

$$\text{Percentage water absorption, \%} = \frac{\text{final weight-initial weight}}{\text{final weight}} \times 100 \quad (1)$$

Four independent variables were employed by factor factorial design (FFD). The variables used were nanocellulose loading (A), PVA(B), starch(C) , and water (D). The design consists of 19 runs including 3 center points. The response measured is the water absorption.

Table 1: Values and coded levels of each variable in 2 factor factorial design.

Variables		Coded levels	
		Low level (-1)	High level (+1)
A	Nanocellulose loading (%)	1	9
B	PVA (g)	3	8
C	Starch (g)	1	4
D	Water (mL)	80	100

3. Results and discussion

The nanofibre dimension from sugarcane bagasse was confirmed by FESEM analysis (JEOL JSM 5600). Figure 1 illustrates the even surface morphology of sugarcane bagasse fibres at 100k mmagnification. From the figure, fibres are clearly seen in individualized forms covering the area. The fibres is between 20 -50 nm in diameter and several micrometer in length.

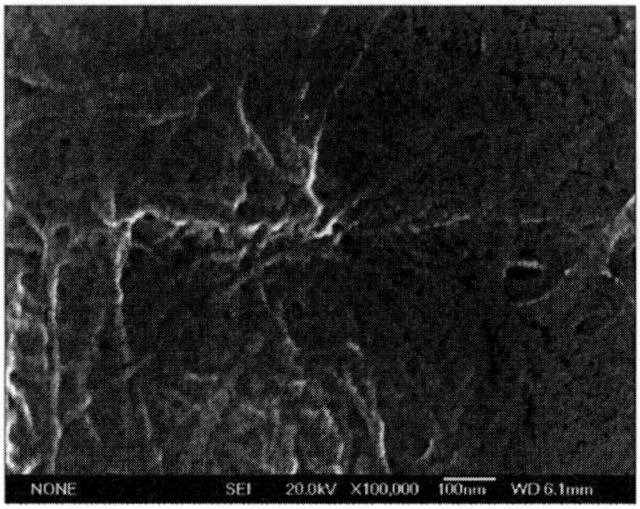

Figure 1: FESEM analysis of nanocellulose from sugarcane bagasse.

Optimization test was done after isolating nanocellulose from sugarcane bagasse. The influence of PVA, starch, solvent i.e. water and nanocellulose loading for the water absorption of the resulting biocomposite film was investigated by 2 level factorial design. An ANOVA analysis (Table 2) is performed to determine the effect of the factors and the interactions. The most significant range of amount for PVA, starch, water, and nanocellulose content were optimized to obtain the optimum mixture for the production of PVA/Starch/Nanocellulose film. Nineteen experiments were conducted as designed by 2 level factorial design while the water absorbing behavior were evaluated and acted as the response to the design. The relationship between the experimental results and independent variables was expressed as second order polynomial regression model which relates the water absorption (response) with all the variables (nanocellulose content, PVA content, starch content, and water content was calculated using design expert and represented as Eq. (2).

$$water\ absorption\ (Y) = \beta_0 + \beta 1A + \beta 2B + \beta 3C + \beta 4D + \beta 12AB + \beta 13AC + \beta 14AD + \beta 23BC + \beta 24BD + \beta 1234ABCD + \epsilon \quad (2)$$

Table 2: The experimental results of the parameters used in 2 factor factorial design.

Trials	Independent variables				Response
	Nanocellulose (g)	PVA (g)	Starch (g)	Water (mL)	Water absorption (%)
1	1	8	4	80	73.3
2	1	3	1	80	64.7
3	9	3	1	100	81.25
4	9	8	1	80	93.1
5	5	5.5	2.5	90	81.57
6	1	3	4	100	86.4
7	1	8	1	100	78.8

8	1	8	1	100	77.4
9	9	3	4	80	83.3
10	1	3	1	80	73.6
11	9	8	4	100	75
12	9	8	4	100	74.2
13	1	8	4	80	78.43
14	9	3	1	100	88.23
15	5	5.5	2.5	90	88.04
16	1	3	4	100	82
17	9	8	1	80	91.37
18	5	5.5	2.5	90	79.24
19	9	3	4	80	76

Table 2 tabulates the experimental values of responses in accordance to experimental design, out of 19 experiments, run 17 achieved the highest water absorption while run 2 had the lowest water absorption. There are many factors that can affect the final properties of the nanocomposite. One of the key parameters in achieving an improved performance of nanocomposites film is by having an evenly distributed dispersion of the materials. Processing PVA/Starch nanocomposites filled with nanocellulose is like mixing the materials with different scale dimension, which depends on many different interacting factors including the strong adhesion/bonding strength of the nanofiller with the matrix, type of material used, and geometry and orientation of the fibres [11].

Barrier properties of nanocomposites are sensitive to the microstructure and the interface between the matrix and filler. The interactions between the organic and inorganic phase at the interface determine the final behavior of the barrier performance of the composites [14]. The presence of void or free volumes in nanocomposite film increases the diffusion of water through these empty areas. Unlike oxygen molecules, water not only interacts with themselves, but also with the polymer matrices where they can form hydrogen bonds. The ability to form strong rigid structure is important in resisting water molecules to diffuse in the film.

By nature, PVA, starch, and nanocellulose are polar molecules that are hydrophilic, meaning water absorbing material. However, with the concept of nano scale dimension of fibres that has high surface area to volume ratio makes the nanocomposite film to have complex strong rigid network. This strong network may be attributed to the chemical interaction that involves PVA/starch/nanocellulose in that strong hydrogen bonding might occur among the nanocellulose structure of three hydroxyl groups on the surface with the structure of the starch and PVA in presence of water. The aspect ratio in terms of length to width ratio of the nanocellulose also plays a role in determining how strong is the forces between the chemical structures.

Table 3: Analysis of variance (ANOVA) for quadratic model of water absorption of PVA/S/SBNF.

Source	Sum of squares	DF	Mean squares	F-value	*p* value
Model	713.66	6	118.94	7.78	0.0014*
A	142.92	1	142.92	9.35	0.0099*
B	2.34	1	2.34	0.15	0.7024
C	24.55	1	24.55	1.61	0.2291
D	5.62	1	5.62	0.37	0.5557
AC	315.77	1	315.77	20.66	0.0007*
AD	222.46	1	222.46	14.55	0.0025*
Residual	183.45	12	15.29		
Lack of fit	25.63	2	12.81	0.81	0.4713
Pure error	157.82	10	15.78		
Cor total	897.11	18			

Standard deviation= 3.91, mean= 80.31, R-square= 0.7955, and adeq. precision= 9.727
**p* value < 0.05, indicates significant factor

Table 3 summarizes the statistical results obtained after analyzing by Design Expert software. Analysis of variance (ANOVA) was performed to find the effect and interactions of each variable. Both F-value and p-value were used to confirm the significance of each variable. The greater F- value indicates the estimated variation data is real and p values less than 0.05 suggested that the variable is significant. Based from Table 3 above, it can be seen that the model had a significant effect on the responses. The model F-value 7.78 and p-value 0.0014 implies the model is significant. The variables A, B, C and D are the independent variables nanocellulose, PVA, starch, and water content, respectively. AC and AD represents the interaction effects. As it is observed, the lowest p-values are obtained by nanocellulose loading interaction with amount of water (AD), followed by nanocellulose loading (A), and nanocellulose loading interaction with amount of starch (AC). Insignificant lack of fit, with p-value of 0.4713 is desirable and implies that the test is insignificant relative to pure error. High values of R^2 on the model of nanocomposite film (0.7955) clarifying the relation involving the response with the independent variables is highly correlated and that the model is highly significant.

Pareto chart in Figure 2 was used to distinguish between the factors that determine the importance of an effect and the greatest effects, considering the line on the chart at t-value of effect. There are two different t- value of effects which are Bonferroni limit line (t-value of effect= 3.294) and t-value limit line (t-value of effect= 2.200). Coefficients with t-value of effect above Bonferroni line are designated as certainly significant coefficients, and coefficients with t-value of the effect between Bonferroni line and t-limit are termed as likely to be significant, and coefficients below the t limit line is statistically insignificant. As depicted from the graph, factor A (nanocellulose loading) shows to have the highest effect on the water absorption, followed by D (amount of water), and B (amount of PVA). This shows that increasing number of nanocellulose will decrease the water absorption of the nanocomposites. The negative effects of the factors are the interaction AC (interaction of nanocellulose loading and starch), AD (interaction of nanocellulose loading and water), and C (starch).

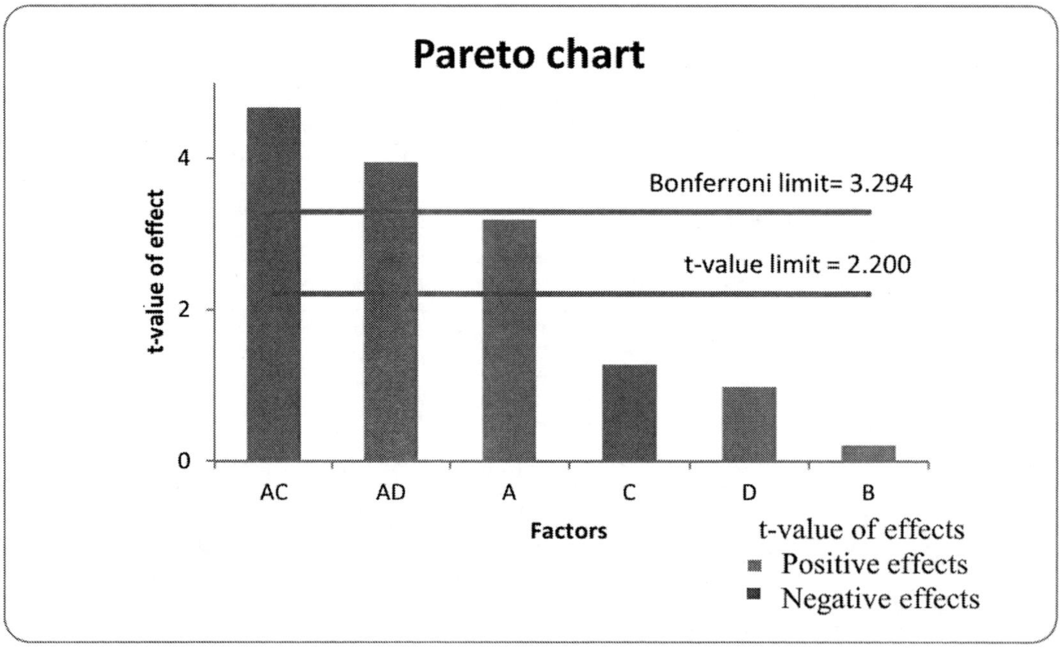

Figure 2: Pareto chart of four factors and their interaction effects on water absorption properties of PVA/starch/nanocellulose film; orange-coloured bar indicates positive effects of factors; the blue-coloured bar indicates negative effects of factors.

IOP Conf. Series: Materials Science and Engineering **368** (2018) 012005 doi:10.1088/1757-899X/368/1/012005

The reason for the negative effects of the interactions between nanocellulose with water and starch could be attributed by the chemical structure of each constituents and nanocellulose fibre morphology. Nanocellulose fibres morphology plays an important role in the mechanism of water absorption of film. The nanofibrils that form complex dense structure and smaller pores in nanometer range, increases the tortuosity within the film and thus decreases the permeability within the film [15]. Tortuosity is intended to describe the effort of a permeate e.g. oxygen or water in transmitting through the blocking material [16]. Nair et al., [17] reported that nanocellulose has a strong reducing effect on the water vapour diffusion due to its nanoscale dimension and rigid framework compared to cellulose fibres. Chemical structure of starch also plays a role in water absorption of the composite; this is supported by Zou et al., [18] reported with an increased content of starch, water absorption of PVA/starch composites increased. This may be attributed to the structure of starch with higher presence of OH group thereby increasing the site for water to bind to the OH groups.

Nanocellulose fibres bearing three hydroxyl groups present on the surface which makes it an interesting property. The relation between nanocellulose fibres and water absorption of the biocomposites has been reported by other researchers as well. The structures of the fibres play a role in diffusivity of the water to pass through. Kiani et al., [19] studied PVC reinforced with pulp fibres and concluded that water absorption increases with increasing fibre content due to the porous tubular structure, accelerating the diffusion of water into the polymer membrane. This shows porous structure of the fibres might accelerate the water permeability to the material. In contrast to a study by Abdul Khalil et al., [20] that incorporation of two fibre; woven jute fibre and EFB fibre in an epoxy composite film decreased the water absorption of the hybrid composites. This may be due to the packaging and arrangement of these fibres that filled up the voids during the formation of the composite film. As compared with only EFB fibres/epoxy composite the water absorption increases, due to presence of higher EFB fibres voids (8.6%) as compared to jute fibre (2.6%). Thus another factor that affects water absorption might be due to the void content in the polymer matrix. The greater voids presence, more water is absorbed into the composite film. Pothan [21] explained the water absorption of the fibres mechanism may be due to formation of micro-channels, providing a way for water to pass through the pores on the surface of the fibres. This shows those fibres with structure that gave higher amount of voids leads to higher amount of water absorption.

Physical structure of nanocellulose in terms of crystallinity also affects the water diffusion into the film. This statement is supported by Dufresne [22] where he reported that water transmission is affected by the degree of crystallinity, as crystalline phases block more transmission than amorphous phases. Sjöholm [23] also supported this statement, that the greater the degree of crystallinity the higher the barrier properties. The concept of having a rigid network that pack perfectly is like modelling the polymer chains forming a concise structure that impermeates the penetrating molecules [24]. Other factors such as lumen size, and adhesion of matrix fibre also effect water absorption behavior of nanocomposites [25]. Svagan et al. [26] studied the influence of nanocellulose on water sorption of starch/nanocellulose polymer. Nanocellulose was found to have strong effect on the diffusivity of water attributed by characteristics and geometry of cellulose, rigid fibre netwroks, and strong interactions between nanocellulose and starch matrix.

The goal of biocomposite film was to improve water resistance of the film, therefore the target values of response of water absorption percentage was the lowest values obtained from the experimental results. The acceptable values of desirability function were the values close to 100%. In this study, the water absorption property of the biocomposite with 1 % nanocellulose, 3 g PVA, 1 g starch, and 80 mL water had 83 % desirability. These levels of independent variables yield the lowest responses of water absorption with 69.41 %.

Table 4: Results for target values of the response that yields the lowest value of water absorption of nanocomposites film.

Constituents amount	Nanocellulose (%)	PVA (g)	Starch (g)	Water (mL)	Water absorption (%)	Desirability (%)
	1%	3	1	80	69.41	83

Figure 3: Photograph of optimized PVA/starch/nanocellulose nanocomposites placed on a background paper for demonstrating transparency.

Validation experiments were carried out to verify the accuracy of the model. Three set of experiments were carried out by using the optimal values of the factors. The experimental result showed water absorption of 69.9; while DOE software predicted the response to be at 69.41 %. It was found that experimental result differed with only 0.7 % error more than the predicted response. This reflects the accuracy 99.3% between the experimental and predicted results. This shows that 2 level factorial design model is accurate in predicting the water absorption of biocomposite film.

Table 5: Results of the validation experiments for water absorption of PVA/starch/nanocellulose film.

Validation	Nanocellulose (%)	PVA (g)	Starch (g)	Water (mL)	Water absorption		% Error
					Predicted	Actual	
	1	3	1	80	69.41	69.90	0.7

4. Conclusion

This study showed that statistical design can be used in evaluating the water absorption properties of the PVA/Starch/Nanocellulose thin film composites. Nanocellulose loading was identified by 2 level factorial designs as an important parameter for improving water absorption of nanocomposite film. Nanocellulose loading showed positive effect, whereas amount of starch, water, and PVA had a negative main effect. This suggests that the optimized content of PVA of 3g, content of starch 1g, content of water 80mL, and nanocellulose loading of 1% could be used for further characterization. This optimized condition shows strong adhesion and interfacial bonding between polar molecules of PVA starch blend and nanocellulose could be achieved resulting in minimum water absorption (64.1%). Optimization of these variables resulted in reduced cost, time, and material. Further studies could be focused on modeling the mechanism of water absorption studies of nanocomposites film.

Acknowledgement

The authors would like to thank Ministry of Higher Education (MOHE) and International Islamic University Malaysia (IIUM) for the research Grant FRGS16-044-0543.

5. References

[1] Othman S H 2014 Bio-nanocomposite materials for food packaging applications: types of biopolymer and nano-sized filler *Agric Agric Sci Proc* **2** 296-303

[2] Takasu A, Itou H, Takada M, Inai Y, and Hirabayashi T 2002 Accelerated biodegradation of poly (vinyl alcohol) by a glycosidation of the hydroxyl groups *Polymer 43* 227-231
[3] Lani N S, Ngadi N, Johari A, and Jusoh M 2014 Isolation, characterization, and application of nanocellulose from oil palm empty fruit bunch fiber as nanocomposites *J Nanomater* **2014** 13
[4] Guo R, Hu C, Pan F, Wu H, and Jiang Z 2006 PVA–GPTMS/TEOS hybrid pervaporation membrane for dehydration of ethylene glycol aqueous solution *J Membrane Sci* **281** 454-462
[5] Tang X and Alavi S 2011 Recent advances in starch, polyvinyl alcohol based polymer blends, nanocomposites and their biodegradability *Carbohydrate Polym 85* 7-16
[6] Fu S, Tian C, and Lucia L A 2017 Water Sorption and Barrier Properties of Cellulose Nanocomposites *Handbook of Nanocellulose and Cellulose Nanocomposites* 649-681
[7] Ng H M, Sin L T, Bee S T, Tee T T, and Rahmat A R 2017 Review of Nanocellulose Polymer Composite Characteristics and Challenges *Polym-Plast Technol* **56** 687-731
[8] Feldman D 2015 Cellulose nanocomposites *J Macromol Sci A* **52** 322-329.
[9] Zafeiropoulos NE (Ed.) 2011 *Interface engineering of natural fibre composites for maximum performance* Elsevier
[10] Yoon S D, Park M H, and Byun H S 2012 Mechanical and water barrier properties of starch/PVA composite films by adding nano-sized poly (methyl methacrylate-co-acrylamide) particles. *Carbohydrate Polym* **87** 676-686
[11] Kumar, R. (2014) Polymer-matrix composites : types, applications, and performance *Nova Science*
[12] Naguib H M, Kandil U F, Hashem A I, and Boghdadi Y M 2015 Effect of fiber loading on the mechanical and physical properties of "green" bagasse–polyester composite *J Radiat Res Appl Sci.* **8** 544-548
[13] Kumar A, Negi Y S, Choudhary V, and Bhardwaj N K 2014 Characterization of cellulose nanocrystals produced by acid-hydrolysis from sugarcane bagasse as agro-waste *Journal of Materials Physics and Chemistry* **2** 1-8
[14] Mittal V (Ed.) 2009 *Optimization of polymer nanocomposite properties* John Wiley & Sons
[15] Syverud K and Stenius P 2009 Strength and barrier properties of MFC films *Cellulose* **16** 75
[16] Matikainen L 2017 Nanocellulose as barrier coating deposited using a laboratory rod coater
[17] Nair S S, Zhu J Y, Deng Y, and Ragauskas A J 2014 High performance green barriers based on nanocellulose *Sustainable Chemical Processes* **2** 23
[18] Zou G X, Jin P Q, and Xin L Z 2008 Extruded starch/PVA composites: water resistance, thermal properties, and morphology *J. Elastomers Plast* **40** 303-316
[19] Kiani H, Ashori A, and Mozaffari S A 2011 Water resistance and thermal stability of hybrid lignocellulosic filler–PVC composites *Polym Bull* **66** 797-802
[20] Khalil H A, Jawaid M, and Bakar A A 2011 Woven hybrid composites: water absorption and thickness swelling behaviours *BioResources* **6** 1043-1052
[21] Pothan L A, Cherian B M, Anandakutty B, and Thomas S 2007 Effect of layering pattern on the water absorption behavior of banana glass hybrid composites *J Appl Polym Sci* **105** 2540-2548
[22] Dufresne A 2017 *Nanocellulose: from nature to high performance tailored materials* Walter de Gruyter GmbH & Co KG
[23] Sjöholm K 2010 Use of bio-based materials in fibre-based packaging. Master's thesis. Aalto University, School of Chemical Technology. Espoo 99
[24] Auvinen S and Lahtinen K 2008. Chapter 9: "Converted paper and paperboard as packaging materials", in Book 12: Paper and Paperboard Converting, 2nd Edition, J. Kuusipalo (ed.). Finnish Paper Engineers' Association 286- 332
[25] Mariatti M, Jannah M, Abu Bakar A, and Khalil H A 2008 Properties of banana and pandanus woven fabric reinforced unsaturated polyester composites *J Compos Mater* **42** 931-941.
[26] Svagan A J, Hedenqvist M S, and Berglund L 2009 Reduced water vapour sorption in cellulose nanocomposites with starch matrix *Compos Sci Technol* **69** 500-506

IOP Conf. Series: Materials Science and Engineering **368** (2018) 012006 doi:10.1088/1757-899X/368/1/012006

Sugar palm nanocrystalline cellulose reinforced sugar palm starch composite: Degradation and water-barrier properties

R A Ilyas[1], S M Sapuan[2, 3], M R Ishak[3] and E S Zainudin[2]

[1] Laboratory of Biocomposite Technology, Institute of Tropical Forestry and Forest Products (INTROP), Universiti Putra Malaysia, 43400 UPM Serdang, Selangor, Malaysia

[2]Department of Mechanical and Manufacturing Engineering, Faculty of Engineering, Universiti Putra Malaysia, 43400 UPM Serdang, Selangor, Malaysia

[3] Department of Aerospace Engineering, Faculty of Engineering, Universiti Putra Malaysia, 43400 UPM Serdang, Selangor, Malaysia

*Corresponding author: ahmadilyasrushdan@yahoo.com

Abstract. In this work, sugar palm nanocrystalline cellulose (SPNCCs) nanocomposites were prepared and used as a biodegradable reinforcement material to improve the water vapor barrier properties of the sugar palm starch (SPS)-based films. SPNCCs with different size based on hydrolysis time (30, 45 and 60 minutes denoted as SPS/SPNCCs-30, SPS/SPNCCs-45, and SPS/SPNCCs-60) were incorporated into SPS plasticizes with glycerol and sorbitol using solution casting method. Then the SPS and SPS/SPNCCs bionanocomposites were submitted to biodegradation by means of soil burial experiment and water vapor barrier test. The biodegradation test shows that SPS degrades very quickly than SPS/SPNCCs which lose 61.93% of its weight at the end of 7 days compared to the SPS/SPNCCs-60 bionanocomposite 52.61%. Adding 0.5 wt.% SPNCCs-60 loading significantly improve water vapor permeability (WVP) of the nanocomposite film by 19.94% compared with the neat film. This was ascribed to the high compatibility between SPNCCs and SPS matrices, which was supported by the field emission scanning electron microscopy (FESEM).

1. Introduction

Nowadays many researchers are developing fully biodegradable composites or biocomposite due to the continuing environmental catastrophe caused by non-biodegradable plastics. Generally, these biocomposites are composed of either natural matrices or natural fibres [1]. Thus in order to overcome this problems, natural biopolymers such a starch, polylactic acid (PLA) and polyhydroxyalkanoates (PHAs) are introduced as a potential alternatives to conventional plastics for packaging applications [2].

Starch is one of the potential candidates for the replacement of non-biodegradable plastic due to its ease availability, renewability, affordable and most importantly biodegradable [3]. Besides, these biopolymer also has been explored widely for the potential manufacture of products such as food and fruit packaging films, plastic shopping bags, flushable linter, medical delivery system and device [2,4]. Nevertheless, the utilization of starch polymer as a films have been reported to have low water barrier properties [5-6] . Thus, many studies have been conducted to improve the water sensitivity of starch-based materials without negotiating their biodegradability [7–9]. Consequently to overcome this limitations, the addition of nanosize natural fibres during the preparation of starch biopolymer film is an effective strategy for enhancement of the packaging films properties [10]. Nanosize natural fibres such as nanocrystalline celluloses (NCCs) and nanofibrillated celluloses (NFCs) can be extracted through the process of chemical and mechanical methods. Hydrolysis is one of the chemical methods that widely used to extract NCCs. Recently, NCCs has attracted a tremendous interest in material science study due its nanodimension, high crystallinity, high aspect ratio, high mechanical strength, low density as well as easy availability along with highly abundant [2, 11]. Moreover, the addition of NCCs has been reported to reduce the water vapor permeability of starch based film. This may due to the morphology of NCCs which increased the tortuosity in the starch based films leading to slower water vapor permeability processes, thus increasing their suitability for food packaging application [12].

Various types of nanosize cellulosic reinforcement have been determined and experienced in biopolymers. Interestingly, it was found that the compatibility between nanosize fibres and starch matrix is high, which is significant for improvement of the film water sensitivity [7]. Nanosize fibre or nanocrystalline cellulose (NCCs) can be isolated from biomass like jute [13], hemp [10], bamboo [14], corncob [15] and sugar palm [16]. Sugar palm is investigated to be a good source of cellulose fibres because of its high cellulose content, which then this cellulose was extracted to gain NCCs [16-17]. Sugar palm tree is versatile tree that grown in mostly tropical countries. Interestingly, the inner part of sugar palm stem contains starch which can be utilized as a biomatrix [18]. In this paper we developed a novel biodegradable bionanocomposites where the matrix (SPS) and nanofibre (SPNCCs) are derived from the similar natural source i.e sugar palm tree.

One of the potential application of SPS/ SPNCCs bionanocomposites is it can be used in packaging industries. The degradation is one of the indicator of packaging durability. Since there are no information on the biodegradability of SPS/SPNCCs bionanocomposites has been reported in the literature, therefore in this study is done to investigate how the addition of the various size of sugar palm nanocrystalline cellulose (SPNCCs) enhances the water vapor permeability and its effect on biodegradability of the sugar palm starch film and bionanocomposites. These basic data therefore are essential for the design and use of the resultant bionanocomposite.

2. Experimental procedure

2.1. Materials
Sugar palms fibre (SPF) and sugar palm starch (SPS) were extracted and collected from sugar palm trees located in Jempol, Negeri Sembilan (Malaysia). Sodium hydroxide, sodium chlorite (80 % purity), acetic acid, sulphuric acid (98 % purity), sorbitol and glycerol plasticizer were supplied by Evergreen Sdn Bhd (Semenyih, Malaysia).

2.2. Preparation of SPNCCs
Sugar palm fibres (SPF) were collected from different part of sugar palm trees (sugar palm frond, trunks, fibre and bunch) as it wrapped around the tree from top to bottom. The knife was used to remove the SPF from the tree. Then a Fritsch pulverisette mill was used for grinding and screening to gain an even size of SPF (2 mm). The cellulose preparation procedures were reported elsewhere [7, 16, 19]. Briefly, the SPF were extracted using two main processes known as delignification and mercerization. The initial and second processes of production of sugar palm cellulose (SPC) were conducted in accordance with ASTM D1104-56 (1978) and ASTM D1103-60 (1977) for the removal of lignin and hemicellulose, respectively, resulted in production of sugar palm celluloses (SPC). Colloidal suspensions of SPNCCs were prepared by acid hydrolysis treatment of SPC. The SPC were mixed with 60 wt% sulphuric acid aqueous solution at variable hydrolysis time of 30, 45, and 60 minutes with the ratio of the SPC to liquor was 5:100 (wt%). The excess sulphuric acid was then removed by centrifuge SPNCCs 5 times with 6,000 rpm, 10 min at 8 °C. Then SPNCCs were dialyzed with distilled water until a constant pH of 7 was gained. The resulting suspension were sonicated using ultrasonic treatment for 30 minutes [20]. Finally SPNCCs were freeze-drying to be used for nanocomposite film.

2.3. Preparation of the SPS/SPNCCs nanocomposite films

The fabrication of plasticizer sugar palm starch/ sugar palm nanocrystalline celluloses (SPS/SPNCCs) composite film was based on a solution casting methods. Starch, sorbitol, glycerol, SPNCCs and distilled water were mixed and sonicated together in order to obtain homogenous nanocomposite film [7]. Briefly, a solution of SPNCCs was prepared by mixing and sonicating it with 190 mL of distilled water with known concentration of SPNCCs (0.5 wt. % on the starch basis). The SPNCCs content was fixed at 0.5 wt. % based on the starch basis. Then, ten gram of SPS and plasticizer (30% on the starch basis) was mixed with the SPNCCs solution and stirred at 1000 rpm for 20 minutes at 85 °C in a disperser for the starch to be gelatinized. The ratio of plasticizer used is 1:1 with combination of sorbitol and glycerol ratio. Then the film-forming suspension was left cooling down and placed it under vacuum to remove air bubble inside the suspension before 45 g of the suspension was cast into each petri dish sized of 15-cm diameter. The petri dishes containing the film-forming solution were placed in an oven at 40°C overnight. SPS films prepared without SPNCCs served as the control (designed as SPS film). Whereas the nanocomposite film with different hydrolysis time of 30 min, 45 min and 60 min were denoted as SPS/SPNCCs-30, SPS/SPNCCs-45, and SPS/SPNCCs-60, respectively. The resulting films were kept in the desiccator at room temperature for one week to ensure the equilibrium of the water content in the films before doing any characterization.

2.4. Morphology analysis by FESEM

Surface characteristics of the SPS and SPS/SPNCCs films were determined by a FEI NOVA NanoSEM 230 machine (FEI, Brno-Černovice, Czech Republic) with an accelerating voltage of 3 kV in order to observe the degradation phenomenon after the biodegradability test. Before being tested by FESEM, samples were cut into 1cm × 1cm and then coated with gold to avoid charging using an argon plasma metallizer (sputter coater K575X) (Edwards Limited, Crawley, United Kingdom) [21].

2.5. Morphology analysis by TEM

The structure and diameter measurement of sugar palm nanocrystalline cellulose (SPNCCs) were determined by Hitachi H-7100 transmission electron microscopy (Hitachi, Tokyo, Japan). These image captured of the sample was obtained from high-resolution transmission electron microscopy (HRTEM). Before being tested by HRSEM, samples were placed on the surface of copper grid coated with a thin carbon film. After that the samples were stained with uranyl acetate for 1 min and allowed to dry at room temperature. The purpose of staining is to get better visualize and contrast image under a TEM.

2.6. Water absorption

Water absorption analysis was run in accordance to ASTM D 570-98. Firstly, the samples were dried for 24 h at 50 °C and cooled in desiccator to gain an even weight. Then the films were weighed and soaked in distilled water at room temperature. After a particular soaking period, the films were taken out of the water and wiped with a cloth and weighed again. The differences between the initial and final masses of the films were calculate using equation [22]:

$$Water\ absorption\ (\%) = \frac{M_{final} - M_{initial}}{M_{initial}} \times 100 \quad (1)$$

Where, $M_{initial}$ = mass before, M_{final} = mass final

2.7. Biodegradability test

Biodegradability was determined by measuring the weight loss of the bionanocomposite buries in compost soil under moisture controlled condition. The size of samples of 30 × 10 mm with triplicate of each samples were buried 100 mm below the surface of soil, which was regularly moistened with distilled water [1]. Each sample was dug out of the compost soil after being buried for 24, 48, 72 and 168 h, respectively, and then washed with water and dried to a constant weight at 60 °C in a vacuum oven. The weight loss was then determined using the equation below:

$$Weight\ loss\ (\%) = \frac{W_0 - W_t}{W_0} \times 100 \quad (2)$$

Where, W_0=weight before being buries, W_t= weight after being buried.

2.8. Water vapor permeability (WVP) test

The film samples were conditioned in a desiccator with a relative humidity of 50% at room temperature. The WVP test was investigated according to ASTM E96 (1995), with slight modifications according to Sanyang et al. [23]. The experiment was repeated thrice. Initially, the mouth of the cup which was 30 mm was filled with 20 g of silica gel. Circular film samples were cut and placed over the mouth of the cup, leaving about 3 mm vacuum to the top. Then it was sealed neatly. After that, the test cups were measured it mass before placing in a constant relative humidity chamber (25°C, relative humidity 75%). The mass gain values of the cups were measured periodically until the equilibrium state was reached. The mass increment of the test cups were recorded, and WVP was calculated as below,

$$WVP = \frac{(m \times d)}{(A \times t \times P)} \quad (3)$$

Where m (*g*) is the increment of test cup mass, *d* (mm) is the thickness of the films, *A* (m^2) is the area of film exposed, *t* (s) is the interval time of permeation, and *P* (Pa) is the water vapor partial pressure across the films. The results are in the unit of $10^{-9} \times g.s^{-1}.m^{-1}.Pa^{-1}$.

3. Experimental result and discussion

3.1. Physical Properties of Sugar Palm Nanocrystalline cellulose (SPNCCs)

The physical properties of the sugar palm nanocrystalline cellulose (SPNCCs) from different treatment was determined using Image J software. The images were collected from the TEM analysis.

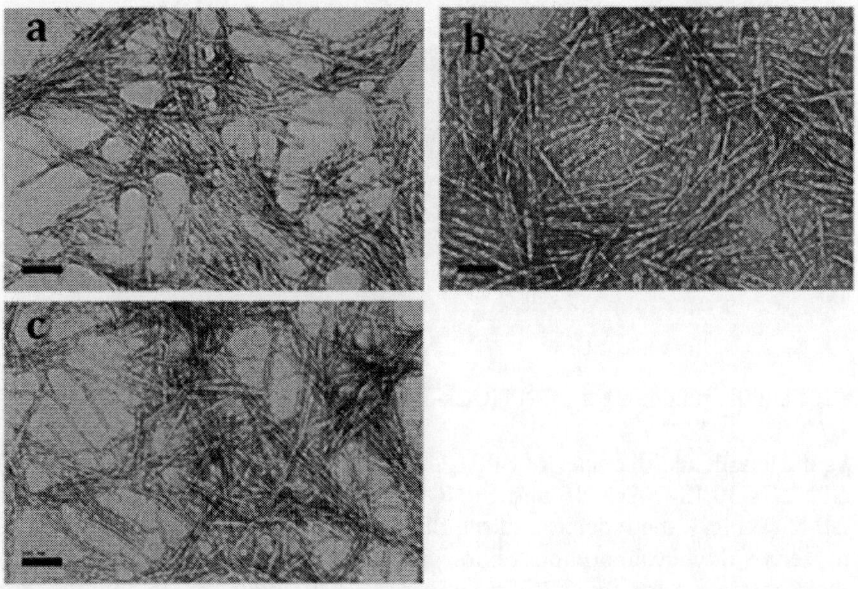

Figure 1. TEM micrographs of the (a) SPNCCs-30, (b) SPNCCs-45 and (c) SPNCCs-60.

Table 1. Physical Properties of SPNCCs-30, SPNCCs-45 and SPNCCs-60.

Fibres	Length (nm)	Diameter (nm)

Sugar Palm Cellulose (SPC)	-	11870
SPNCCs-30	175± 37.01	13±1.73
SPNCCs-45	130 ± 30.23	9± 1.96
SPNCCs-60	110± 33.69	7.5±1.35

Figure 1 and Table 1 show the TEM micrograph and physical properties of nanocrystalline cellulose, respectively. TEM micrograph of sugar palm nanocrystalline cellulose (SPNCCs) (Figure 1) revealed their nanometric dimension. As shown in Table 1, the average length of the SPNCCs-30, SPNCCs-45 and SPNCCs-60 were approximately 175± 37.01 nm, 130 ± 30.23 nm and 110± 33.69, respectively. Whereas the diameter of the SPNCCs-30, SPNCCs-45 and SPNCCs-60 were approximately 13±1.73 nm, 9± 1.96 nm and 7.5±1.35 nm, respectively. The diameter measured were similar to the nano-sized structures that were extracted from other agro-waste sources such as coconut husk (5.5 ± 1.5 nm) [24], rice straw (5.06 nm) [25], soy hulls (4.43 ± 1.20 nm) [26], sugarcane bagasse (4 ± 2 nm) [27], banana residues (5nm) [28] and smaller than microfibrils from wheat straw (10–80nm) [29], sisal fibres (30.9 ± 12.5 nm) [30] and sugarcane bagasse (30nm) [31]. These resultant images (Figure 1) revealed the aqueous suspensions contained sugar palm NCCs residing mostly of individual crystal and some aggregates.

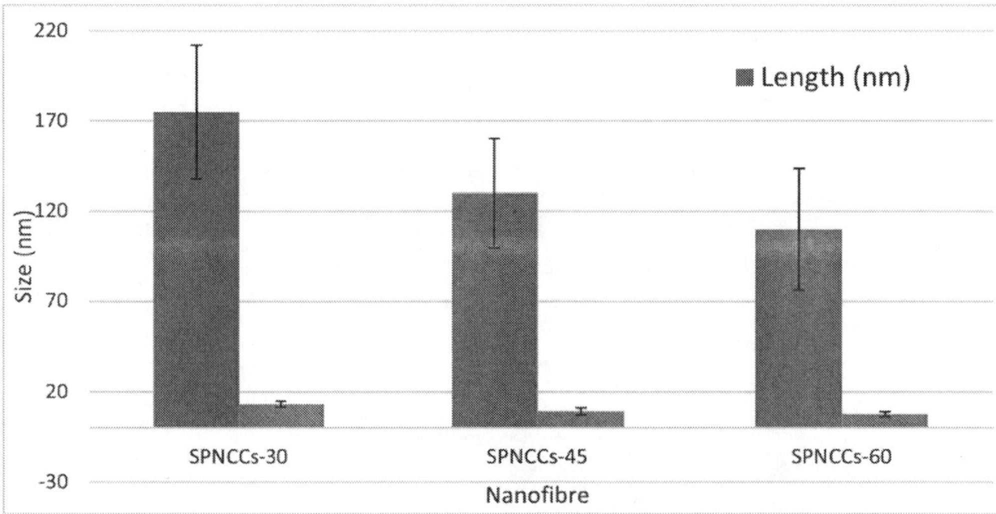

Figure 2. Length and diameter of SPNCCs-30, SPNCCs-45 and SPNCCs-60.

Figure 2 shows the length and diameter of SPNCCs-30, SPNCCs-45 and SPNCCs-60. The differences between the SPNCCs-30, SPNCCs-45 and SPNCCs-60 were in the reduction of size of length and diameter. Based on Table 1, the reduction of the SPNCCs-30 compared to SPC was 99.89%. This was attributed by the removal of the amorphous region via the hydrolysis treatment on SPC [16]. The process of the hydrolysis continued for the SPNCCs-45 and SPNCCs-60 for 45 minutes and 60 minutes, indicated the changes in the size of the SPNCCs which was reduce by 31% and 42% compared to SPNCCs-30, respectively. This was caused by the longer reaction time of H_2SO_4 via hydrolysis treatment on the fibres, which removed amorphous region from the nanofibres. Besides, longer hydrolysis time could irritate the structure (length and diameter) of SPNCCs. Thus, the longer the reaction times of hydrolysis, the smaller the length and diameter of the nanofibres [32].

3.2. Water absorption

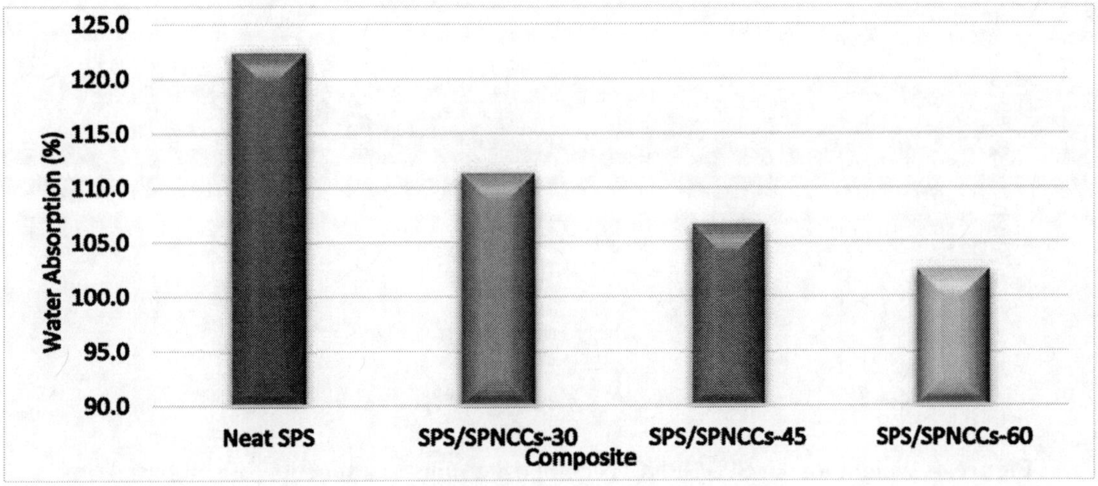

Figure 3. Water absorption of composite.

Water absorption is a significant study in SPS film due to water acts as a plasticizer. Figure 3 displays the percentage of water absorption of SPNCCs-30, SPNCCs-45 and SPNCCs-60 nanocomposite films. As been shown in Figure 3, it can be seen that the neat SPS gave the highest percentage of water absorption, followed by SPS/SPNCCs-30, SPS/SPNCCs-45, and SPS/SPNCCs-60, i.e. 111.3 %, 106.6 % and 102.5 %, respectively. This indicated that film reinforced with SPNCCs has lower hydrophilicity. Moreover, the water absorption of the film is high due to the presence of hydroxyl groups in film molecules. Film with higher concentration of starch is tend to absorb more water compared to film that have low concentration of starch [1]. Besides that, as the size of SPNCCs decreased, the percentage of the water absorption also decreased. Additionally, film with big particle size of SPNCCs exhibited greater tendency to absorb water which possibly due to a lower response to stress [33]. In this study, water absorption tests revealed that stabilizing the mixtures with SPNCCs decreased the water absorption capacity. This is due to the adhesion between the SPNCCs and SPS in which they tend to form strong hydrogen bonding via the physical interlocking with non-polar matrices and the OH- groups if hydrophilic polymer matrices [34-35]. The result of this experiment conducted are quite similar to those from previous work [1].

3.3. Biodegradation of bionanocomposites

Biodegradation is the decomposition of materials by the action fungi, bacteria and microorganism or by other biological means. Generally, the decomposition of polymer start when these microbial organism contact with the biodegradable polymer [36]. These microbial organisms transform the polymer through enzymatic or metabolic process that broken down the polymers into smaller compound which have lower average molecular weight. Therefore, this favors material's decomposition in the environment. When the biodegradation is complete, the process is called mineralization [37]. Figure 4 shows the weight loss of the SPS and SPS/ SPNCCs bio-nanocomposite after biodegradation testing. At the end of day 7, the weight of neat SPS had lost 61.94%, whereas the SPS/SPNCCs bionanocomposites had lost 56.88 %, 55.74 % and 52.61 % weight for SPS/SPNCCs-30, SPS/SPNCCs-45, and SPS/SPNCCs-60, respectively. The average degradation rate is 8.85 %/day, 8.13 %/day, 7.96 %/day and 7.52 %/day, respectively for the neat SPS, SPS/SPNCCs-30, SPS/SPNCCs-45, and SPS/SPNCCs-60, respectively. It is observed that the weight loss of SPS/SPNCCs composite is lower than that of the SPS at any given time points.

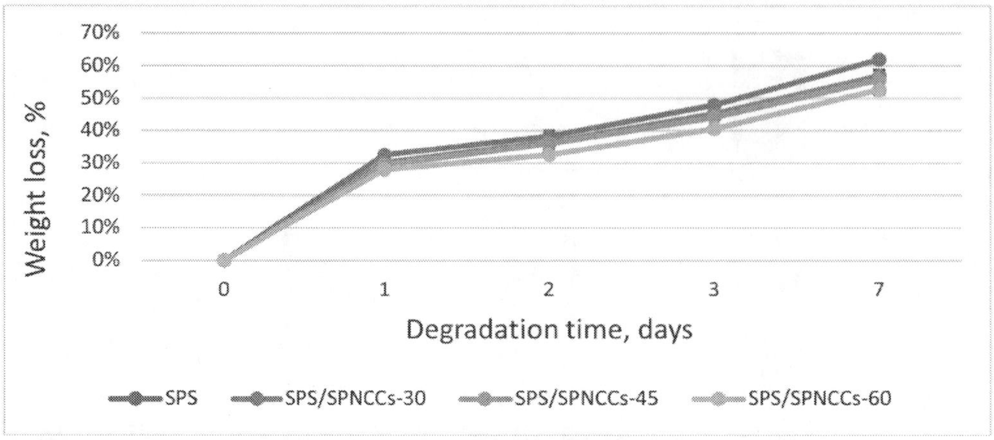

Figure 4. Weight loss of SPS/SPNCCs composite films as a function of soil burial time.

The control film, SPS film was totally degrade after 9 days while it took bio-nanocomposites films 12 days to degrade completely. The weight loss of for the neat matrix was higher compared to the bio nanocomposite films. There are two factors that may be attributed to this situation to occur which are water absorption by film and degree of crystallinity of SPNCCs in bio-nanocomposite films. SPNCCs facilitated the polymer matrix with a stabilization effect by creating a three-dimensional cellulosic network which restricted the chain mobility and reduced the number of hydroxyl groups, and thus reduced water absorption of SPS/SPNCCs nanocomposite films [38-39]. The weight loss for the control film, SPS was higher compared to the SPS/ SPNCCs bionanocomposite for the entire successive degradation test. This may be attributed to the physical properties of SPS which SPS absorb more water than SPS/ SPNCCs films, making it more exposed to microbial organisms attack [41]. These microbial organisms attack the SPS in the presence of water medium [42]. This can be related to the water absorption properties of starch films, where the water absorption for neat SPS was about 122.3 %, whereas for SPS/SPNCCs-30, SPS/SPNCCs-45, and SPS/SPNCCs-60 bionanocomposite films were 111.3 %, 106.6 % and 102.5 %, respectively. This properties could be attributed to the hydrophilic behavior of SPS [41]. Thus, the samples that have higher starch contents would have better biodegradability properties.

Generally cellulose is comprised of amorphous (less-oriented molecules) and crystalline region (highly oriented molecules). When it was treated with strong acid via hydrolysis process, the amorphous region broke up, thus producing NCCs with higher degree of crystallinity. The capability of cellulolytic microorganisms to degrade cellulose differs greatly with the physic-chemical characteristics of the substrate such as polymerization of cellulose and the degree of crystallinity [43], where the degree of crystallinity cellulose is major structural parameter [44]. It has been described in a previous work that crystalline region are more difficult to degrade [45]. Compared to control film SPS, SPS/SPNCCs contained higher crystallinity, which makes it a higher resistance to the microbial organism attacks than starch. This may attributed the difference in weight loss between SPS and the SPS/ SPNCCs bionanocomposite. Moreover, the variances in resistance to microbial organism attacks between SPS/ SPNCCs and starch SPS exhibited that in SPS/SPNCCs bionanocomposite, microbial organism attacks start with starch. In the case of SPNCCs/ starch, the amorphous region favors the microbial organism accessibility to the matric (mainly to the deterioration of starch) and SPNCCs have the minor role, as can be concluded from the difference in weight loss suffer by the SPS and the SPS/SPNCCs matrix [1].

When the microbial organisms consume the surrounding starch, the bionanocomposite film loses their structural integrity. Thus lead to the deterioration of the mechanical properties, hence allowing the attack of SPNCCs by microbial organism [46]. Indeed, the result gained herein shows that the SPS/SPNCCs bionanocomposite film will not cause any deterioration/ ecological impact, where in other words, the SPS/SPNCCs bionanocomposite film are fully biodegradable.

3.4. Water Vapor Permeability

Hydrophobicity are one of the most important criteria that should be emphasize in the process of material selection for food packaging application, since it may affect the performance of the end products whether the product is water-sensitive product or not. Therefore, films with low water vapor permeability are suitable for food packaging applications, including the reducing and preventing of moisture transfer between surrounding and the food [47]. Hence, minimizing the water vapor permeability (WVP) of SPS films is important for their tremendous potential application. Figure 5 shows the WVP of neat SPS and SPS/NCCs composite film as a function of soil burial time. It can be seen that the highest WVP is SPS film which is $9.58 \times 10^{-10} \times g.s^{-1}.m^{-1}.Pa^{-1}$. This is because of their hydrophilic nature, besides their sensitivity to moisture contain and surrounding humidity, a factor that is hard to control [48]. The reinforcement of starch with SPNCCs significantly improved the WVP of the neat SPS films. The addition of 0.5% SPNCCs into neat SPS films reduced their WVP value by 11.17 %. This decreased is due to the tortuous path caused by the presence of rigid crystalline structure and the dispersion of SPNCCs in the starch polymer, which prolongs or hinders the path of water molecules through the film matrix [7,49]. The WVP of the film showed an improved of 11.17 %, 14.72 % and 19.94 %, when altering the size of the SPNCCs reinforced bio matrix, SPS/SPNCCs-30, SPS/SPNCCs-45, and SPS/SPNCCs-60, respectively. This increments trending showed that the smaller the size of the nanofibres the harder the water to pass through the matrix SPS, thus improved the WVP of the neat SPS films.

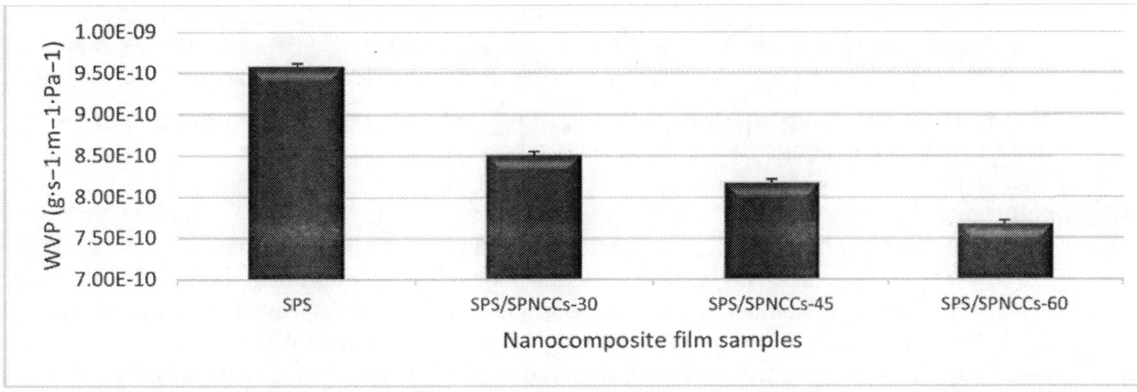

Figure 5. Water Permeability of SPS/SPNCCs composite films as a function of soil burial time.

3.5. Surface Morphology of Bionanocomposite Film

Figure 6 displayed the surface morphology of SPS-based films with and without the addition of SPNCCs after being degraded in soil test for 168 h. From the figure 6, it can be observed that most of the starch had degraded in both SPS and SPF/SPS bionanocomposite which was observed using FESEM. Nevertheless, this effect of degradation can be observed on the weight loss of the bionanocomposite, as explained earlier and observed in figure 4. Therefore it can be concluded that the addition of SPNCCs delays the degradation of the bionanocomposite. The micrograph of the neat SPS films showed a smooth and continuous surface with no trace of starch granular or cracks and agglomerations of SPNCCs. Similar observations were reported by Sanyang et al. (2016) and Dias et al. (2011) for neat sugar palm starch and rice flour films, respectively. Thus, the high dispersion of SPNCCs (Figure 6(b)) is a good sign of strong interfacial adhesion between the two components of the SPS-SPNCCs nanocomposite film. These good distributions translate into high WVP properties. Bilbao-Sainz et al. (2011) described that the functional properties of the composite improved when the fibres well-dispersed into the polymeric matrix. Besides, in Figure 6 (b) of neat SPS film after the process of degradation take place, the surface of the film become rough and wavy due to the microbial attack within the soil. Whereas in Figure 6 (b) of SPS/SPNCCs film observed the present of excessive numbers of SPNCCs distributed on the surface of SPS/SPNCCs film. This is due to the attacks of microbial on the surface of film which then revealed the image of SPNCCs that adhere within the film [1].

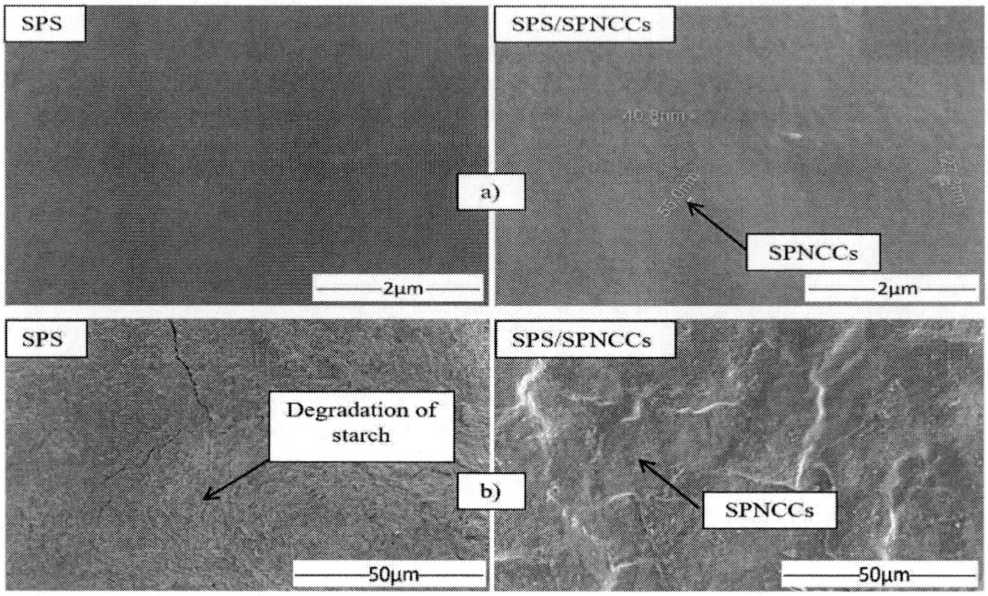

Figure 6. Surface morphology of SPS and SPF/SPNCCs bionanocomposite; a) before and b) after being buried.

4. Conclusion

An innovative bionanocomposite with combination of both matric and natural fibres are obtained from less utilize of the sugar palm tree was successfully developed. It was determined that the WVP effect on the SPS and SPS/SPNCCs bionanocomposite had brought a significant improvement when decrease the size of the SPNCCs reinforced bio matrix, at the end of 168 h of WVP test. Whereas, the biodegradation test revealed that the SPS degraded faster as compared to SPS/SPNCCs bionanocomposites. Thus, this bionanocomposite shows good WVP and biodegradable properties for packaging materials application.

Acknowledgments

The authors would like to thank the Universiti Putra Malaysia for the financial support through the Graduate Research Fellowship (GRF) scholarship. The authors are grateful to Dr. Muhammed Lamin Sanyang for guidance throughout the experiment. The authors also thank the Forest Research Institute Malaysia (FRIM) and Dr. Rushdan Ibrahim for their advice and fruitful discussions.

References

[1] Sahari J, Salit M S, Zainudin E S and Maleque M A 2014 Degradation Characteristics of SPF/SPS Biocomposites Fabrication of SPF/SPS biocomposites *Fibres* Text *East* Eur **22** 96–8

[2] Ilyas R A, Sapuan S M, Sanyang M L and Ishak M R 2016 Nanocrystalline cellulose reinforced starch-based nanocomposite: A review *5th Postgraduate Seminar on Natural Fiber Composites* (Serdang, Selangor: Universiti Putra Malaysiapp 82–7 (

[3] Sanyang M L, Sapuan S M, Jawaid M, Ishak M R and Sahari J 2016 Development and characterization of sugar palm starch and poly(lactic acid) bilayer films *Carbohydrate Polym* **146** 36–45

[4] Sanyang M L, Sapuan S M, Jawaid M, Ishak M R and Sahari J 2016 Recent developments in sugar palm (Arenga pinnata) based biocomposites and their potential industrial applications: A review *Renew Sust Energ Rev* **54** 533–49

[5] Sahari J, Sapuan S M, Zainudin E S and Maleque M A 2012 Effect of Water Absorption on Mechanical Properties of Sugar Palm Fibre Reinforced Sugar Palm Starch (spf / sps) Biocomposites *J Biobased Mater Bio* **6** 1–5

[6] Dias A B, Müller C M O, Larotonda F D S and Laurindo J B 2011 Mechanical and barrier properties of composite films based on rice flour and cellulose fibers *LWT - Food Sci Tech-Brazil* **44** 535–42

[7] Sanyang M L, Sapuan S M, Jawaid M, Ishak M R and Sahari J 2016 Effect of sugar palm-derived cellulose reinforcement on the mechanical and water barrier properties of sugar palm starch biocomposite films *BioResources* **11** 4134–45

[8] Sanyang M L, Sapuan S M, Jawaid M, Ishak M R and Sahari J 2015 Effect of Plasticizer Type and Concentration on Dynamic Mechanical Properties of Sugar Palm Starch Based Films *Int J Polym Anal Ch* **5341** 150622104055002

[9] Sahari J, Sapuan S M, Zainudin E S and Maleque M A 2013 Mechanical and thermal properties of environmentally friendly composites derived from sugar palm tree *Mater Design* **49** 285–9

[10] Cao X, Chen Y, Chang P R, Stumborg M and Huneault M A 2008 Green composites reinforced with hemp nanocrystals in plasticized starch *J Appl Polym Sci* **109** 3804–10

[11] Ilyas Rushdana A, Sapuan Salit M, Lamin Sanyang M and Ridzwan Ishak M 2017 Nanocrystalline Cellulose As Reinforcement For Polymeric Matrix Nanocomposites And Its Potential Applications: A Review *Curr Anal Chem* **13**

[12] Follain N, Belbekhouche S, Bras J, Siqueira G, Marais S and Dufresne A 2013 Water transport properties of bio-nanocomposites reinforced by Luffa cylindrica cellulose nanocrystals *J Membrane Sci* **427** 218–29

[13] Cao X, Ding B, Yu J and Al-Deyab S S 2012 Cellulose nanowhiskers extracted from TEMPO-oxidized jute fibers *Carbohyd Polym* **90** 1075–80

[14] Xie J, Hse C-Y, De Hoop C F, Hu T, Qi J and Shupe T F 2016 Isolation and characterization of cellulose nanofibers from bamboo using microwave liquefaction combined with chemical treatment and ultrasonication *Carbohyd Polym* **151** 725–34

[15] Silvério H A, Flauzino Neto W P, Dantas N O and Pasquini D 2013 Extraction and characterization of cellulose nanocrystals from corncob for application as reinforcing agent in nanocomposites *Ind Crop Prod* **44** 427–36

[16] Ilyas R A, Sapuan S M and Ishak M R 2018 Isolation and characterization of nanocrystalline cellulose from sugar palm fibres (Arenga Pinnata) *Carbohyd Polym* **181** 1038–51

[17] Ishak M R, Sapuan S M, Leman Z, Rahman M Z A and Anwar U M K 2012 Characterization of sugar palm (Arenga pinnata) fibres *J Therm Anal Calorim* **109** 981–9

[18] Ishak M R, Sapuan S M, Leman Z, Rahman M Z A, Anwar U M K and Siregar J P 2013 Sugar palm (Arenga pinnata): Its fibres, polymers and composites *Carbohyd Polym* **91** 699–710

[19] Ilyas R A, Sapuan S M, Ishak M R and Zainudin E S 2017 Effect of Delignification on the Physical, Thermal, Chemical, and Structural Properties of Sugar Palm Fibre *BioResources* **12** 8734–54

[20] Bondeson D, Mathew A and Oksman K 2006 Optimization of the isolation of nanocrystals from microcrystalline cellulose by acid hydrolysis *Cellulose* **13** 171–80

[21] Sheltami R M, Abdullah I, Ahmad I, Dufresne A and Kargarzadeh H 2012 Extraction of cellulose nanocrystals from mengkuang leaves (Pandanus tectorius) *Carbohyd Polym* **88** 772–9

[22] Edhirej A, Sapuan S M, Jawaid M and Zahari N I 2017 Cassava/sugar palm fiber reinforced cassava starch hybrid composites: Physical, thermal and structural properties *Int J Biol Macromol* **101** 75–83

[23] Sanyang M L, Sapuan S M, Jawaid M, Ishak M R and Sahari J 2016 Development and characterization of sugar palm starch and poly(lactic acid) bilayer films *Carbohyd Polym* **146** 36–45

[24] Rosa M F, Medeiros E S, Malmonge J A, Gregorski K S, Wood D F, Mattoso L H C, Glenn G, Orts W J and Imam S H 2010 Cellulose nanowhiskers from coconut husk fibers: Effect of

preparation conditions on their thermal and morphological behavior *Carbohyd Polym* **81** 83–92
[25] Thiripura Sundari M and Ramesh A 2012 Isolation and characterization of cellulose nanofibers from the aquatic weed water hyacinth—Eichhornia crassipes *Carbohyd Polym* **87** 1701–5
[26] Flauzino Neto W P, Silvério H A, Dantas N O and Pasquini D 2013 Extraction and characterization of cellulose nanocrystals from agro-industrial residue – Soy hulls *Ind Crop Prod* **42** 480–8
[27] de Morais Teixeira E, Bondancia TJ, Teodoro KBR, Corre^a AC, Marconcini JM and Mattoso LHC 2011 Sugarcane bagasse whiskers : Extraction and characterizations *Ind Crops Prod* **33** 66
[28] Zuluaga R, Putaux J-L, Restrepo A, Mondragon I and Gañán P 2007 Cellulose microfibrils from banana farming residues: isolation and characterization *Cellulose* **14** 585–92
[29] Alemdar A and Sain M 2008 Isolation and characterization of nanofibers from agricultural residues - Wheat straw and soy hulls *Bioresource Technol* **99** 1664–71
[30] Morán J I, Alvarez V A, Cyras V P and Vázquez A 2008 Extraction of cellulose and preparation of nanocellulose from sisal fibers *Cellulose* **15** 149–59
[31] Bhattacharya D, Germinario L T and Winter W T 2008 Isolation, preparation and characterization of cellulose microfibers obtained from bagasse *Carbohyd Polym* **73** 371–7
[32] Xie F, Pollet E, Halley P J and Avérous L 2013 Starch-based nano-biocomposites *Prog Polym Sci* **38** 1590–628
[33] Edhirej A, Sapuan S M, Jawaid M and Zahari N I 2017 Effect of various plasticizers and concentration on the physical, thermal, mechanical, and structural properties of cassava-starch-based films *Starch/Staerke* **69** 1–11
[34] Minelli M, Baschetti M G, Doghieri F, Ankerfors M, Lindström T, Siró I and Plackett D 2010 Investigation of mass transport properties of microfibrillated cellulose (MFC) films *J Membrane Sci* **358** 67–75
[35] Majeed K, Jawaid M, Hassan A, Abu Bakar A, Abdul Khalil H P S, Salema A A and Inuwa I 2013 Potential materials for food packaging from nanoclay/natural fibres filled hybrid composites *Mater Design* **46** 391–410
[36] Edhirej A, Sapuan S M, Jawaid M and Zahari N I 2017 Preparation and characterization of cassava bagasse reinforced thermoplastic cassava starch *Fiber Polym* **18** 162–71
[37] Tahri N, Bahafid W, Sayel H and El Ghachtouli N 2013 Biodegradation: Involved Microorganisms and Genetically Engineered Microorganisms *Biodegradation - Life of Science* (InTech)
[38] Khan A, Khan R A, Salmieri S, Le Tien C, Riedl B, Bouchard J, Chauve G, Tan V, Kamal M R and Lacroix M 2012 Mechanical and barrier properties of nanocrystalline cellulose reinforced chitosan based nanocomposite films *Carbohyd Polym* **90** 1601–8
[39] Noshirvani N, Ghanbarzadeh B, Fasihi H and Almasi H 2016 Starch-PVA Nanocomposite Film Incorporated with Cellulose Nanocrystals and MMT: A Comparative Study *Int J Food Eng* **12** 37–48
[40] Sahari J, Sapuan S M, Zainudin E S and Maleque M A 2013 Thermo-mechanical behaviors of thermoplastic starch derived from sugar palm tree (Arenga pinnata) *Carbohyd Polym* **92** 1711–6
[41] Sahari J, Sapuan S M, Zainudin E S and Maleque M A 2013 Effect of Water Absorption on Mechanical Properties of Sugar Palm Fibre Reinforced Sugar Palm Starch (SPF/SPS) Biocomposites *J Biobased Mater Bio* **7** 90–4
[42] Kiatkamjornwong S, Sonsuk M, Wittayapichet S, Prasassarakich P and Vejjanukroh P-C 1999 Degradation of styrene-g-cassava starch filled polystyrene plastics *Polym Degrad Stabil* **66** 323–35
[43] Wan Y Z, Luo H, He F, Liang H, Huang Y and Li X L 2009 Mechanical, moisture absorption, and biodegradation behaviours of bacterial cellulose fibre-reinforced starch biocomposites *Compos Sci Technol* **69** 1212–7
[44] Fan L T, Lee Y-H and Beardmore D H 1980 Mechanism of the enzymatic hydrolysis of cellulose: Effects of major structural features of cellulose on enzymatic hydrolysis *Biotechnol*

Bioeng **22** 177–99
[45] Alvarez V A, Ruseckaite R A and Vázquez A 2006 Degradation of sisal fibre/Mater Bi-Y biocomposites buried in soil *Polym Degrad Stabil* **91** 3156–62
[46] Żuchowska D, Steller R and Meissner W 1998 Structure and properties of degradable polyolefin-starch blends *Polym Degrad Stabil* **60** 471–80
[47] Sanyang M L, Ilyas R A, Sapuan S M and Jumaidin R 2018 Sugar Palm Starch-Based Composites for Packaging Applications *Bionanocomposites for Packaging Applications* 125–47
[48] Wilhelm H M, Sierakowski M R, Souza G P and Wypych F 2003 Starch films reinforced with mineral clay *Carbohyd Polym* **52** 101–10
[49] Slavutsky A M and Bertuzzi M A 2014 Water barrier properties of starch films reinforced with cellulose nanocrystals obtained from sugarcane bagasse *Carbohyd Polym* **110** 53–61

A high strength green nanocomposites based on wood veneer nanocrystalline cellulose and epoxy resin polymer

J Zaihan[1*], R Rafeadah[1], M A Mohamad Nasir[1], W S Hashim[1] and A Ishak[2]

[1]Forest Products Division, Forest Research Institute Malaysia (FRIM), 52109 Kepong, Selangor

[2]Faculty of Science and Technology, School of Chemical Sciences and Food Technology, Polymer Research Center (PORCE), Universiti Kebangsaan Malaysia (UKM), 43600 Bangi, Selangor, Malaysia

*Corresponding author: zaihan@frim.gov.my

Abstract. This study was conducted to utilize natural thin wood veneers, nanocrystalline cellulose (NCC) and epoxy resin to produce new bionanocomposites. The thin wood veneers were used as the basic structure element in the nanocomposites. The NCC was obtained from the chemically hydrolyzed sawdust with diameter of ~10-20 nm. In this study, NCC was mixed with 3-glycidyloxypropyltrimethoxysilane (GPS) before being dispersed homogeneously into the epoxy resin and thin veneer. The GPS which act as the coupling agent enhanced the interfacial adhesion between NCC and epoxy resin polymer. The Fourier Transform Infrared Spectroscopy (FTIR) results showed that NCC was successfully treated with GPS. The finding showed that the optimum value of GPS solution required to functionalize NCC was 0.5 wt/wt %. The results on tensile strength tests on the green nanocomposites also showed that the strength improved upon the incorporation of treated NCC into epoxy resin and thin veneer. By incorporating treated NCC (0.1%) into the thin wood veneer/epoxy resin polymer based nanocomposites, the maximum increment value of tensile strength was found to have a 23% improvement compared to the control sample without NCC. The enhancement in the strength properties was expected as NCC possessed high surface-to-volume and aspect ratio characteristics. Through this study new high strength green nanocomposites have been successfully developed.

1. Introduction

Nanotechnology offers potential opportunities for development of advanced wood- or lignocellulosic-based biocomposites. It is an alternative tool to manipulate and control processes and materials at nanoscale level. The technology is currently leading to new analytical technologies that provide a fundamental understanding of material behaviour at the nanoscale. It includes the incorporation of nanomaterials into biocomposites to achieve enhanced performance of the final products. It also leads to modifications of the wood and lignocellulosic raw material surfaces at the nanoscale. By exploiting these potential opportunities to identify, control, optimize material and process factors in real-time, products can be engineered to have highly specified performance characteristics with economic advantages. Nanocrystalline cellulose (NCC) is cellulose in crystalline form, which is extracted from biomass and processed into several forms such as solid flake, liquid and gel. The lateral dimension of nanocrystals is in the order of 10-100 nm and the length can be in the micrometer scale, both parameters depending on the preparation method [1]. Unlike other nanomaterials, NCC manufacturing is already a sustainable and viable bulk process. NCC can be generated in a vast amount each day, producing other viable by products such as glucose (for alternative fuel) and gypsum (for buildings) and their mechanical performance compares well with other materials [2-3]. Several main properties such as high strength, electro-magnetic response and a large surface area provide a basis for the manufacture of new and advanced materials using nanotechnology.

Over the last few years, there has been a great deal of interest in research and development works on nanocomposites in Malaysia. NCC is considered as one of the ideal nano-reinforcements for matrices (including water-soluble and water-insoluble polymer systems) and has already been incorporated into many polymer matrices to produce highly reinforced composites [4-5]. Epoxy resins denote to a class of polymeric materials whose applications are extensive and range from coatings, to adhesives to composite materials for electronics, sporting goods and others [6]. The inclusion of nanofillers into epoxy resins can lead to a combination of high stiffness and strength, and good fatigue resistance [7]. The use of NCC as reinforcing fillers in epoxy resins have been reported from previous studies. Rusli and Eichhorn (2008) investigated the stiffness of cellulose whisker/epoxy system using Raman spectroscopy and highlighted the importance of the interface between matrix and nanofiller [8]. Lu et al. (2008) recently reported that the modulus increased approximately 6 fold upon incorporation of 5% w/w microfibrillated cellulose into an epoxy matrix [9].

As worldwide demand for timber and lignocellulosic resources grows in order to compete with the needs of rapidly increasing populations, the management of this sustainable resource becomes a critical issue. There is a need to fully utilize the natural wood efficiently. Wood veneer refers to thin slices of wood, usually thinner than 3 mm and typically obtained by shaving, peeling or slicing thin layers of wood off a suitable log. These wood veneer products are commonly used as decorative surfacing for furniture, interior and exterior components of automobiles. Typically, fused decorative products for furniture and automotive applications are made by bonding thin sheets of high-quality wood to a core material such as plywood, metal or particleboard. The thin veneer (with average thickness – 0.3 mm) by its own, exhibits relatively poor strength. Hence, this study attempted to utilize natural wood veneer as a basic structure element in producing a new type of green nanocomposite. The aim of this study was to investigate the contribution of the NCC reinforced epoxy polymer properties and its bonding/network structure to the overall properties of the produced nanocomposites. By incorporating NCC, enhancement in strength property is expected as NCC possesses high surface-to-volume and aspect ratio characteristics to ensure effective incorporation with the veneer matrix. Silane was used as the coupling agent to promote interfacial adhesion via chemical bonding which is stronger compared to the possibly occurring hydrogen bonding between the hydrophilic functional groups, leading to the improvement of the composite properties.

2. Materials and Methods

2.1 Materials and chemicals

Sawdust was used as the raw material in producing NCC. Glycidyloxypropyltrimethoxysilane ≥98 % (GPS) and epoxy resin were produced by Sigma-Aldrich. Ultrathin wood veneers utilized in this study were supplied by a nearby supplier. All other chemicals used were purchased from Merck (M) Sdn. Bhd. The chemical structures of GPS ($C_9H_{20}O_5Si$) and epoxy resin are illustrated in Fig.1.

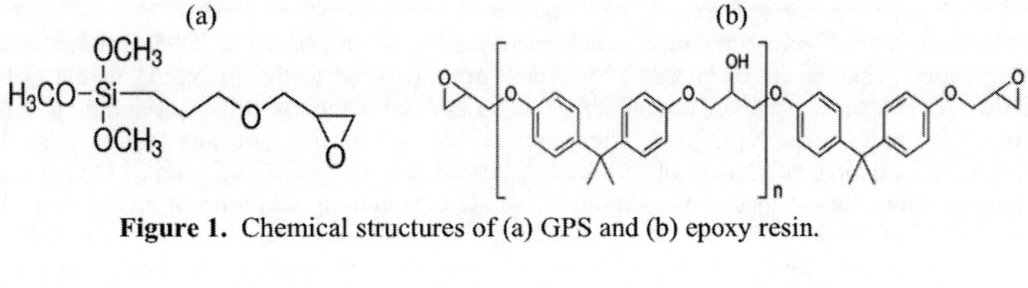

Figure 1. Chemical structures of (a) GPS and (b) epoxy resin.

2.2 Surface treatment of NCC

Nanocellulose was produced from sawdust using acid hydrolysis method reported elsewhere [10]. The solution was prepared by addition of 5 wt. % GPS, drop by drop to water- ethanol (80/20) (v/v %) under stirring. The pH of the solution was adjusted to 3~4.5 with diluted acetic acid to catalyze the hydrolysis process. Silane solution was stirred for 15 min and left for 1 hour to allow complete hydrolysis of the silane. After complete hydrolysis and silanol formation, NCC was dipped in solution and sonicated for 10 minutes to prevent the aggregation of NCC and to produce homogenous suspension. The suspension retained for two hours at room temperature. The suspension was then washed with distilled water using centrifuge to remove any excess silane from the NCC before lyophilization.

2.3 Production of NCC reinforced epoxy resin polymer and thin veneer
Treated NCC was dispersed in the curing agent of epoxy resin system (hardener). The mixture was sonicated to produce homogenously suspension. The admixture was then added into epoxy resin and stirred uniformly.

The treated NCC-epoxy polymer admixture was integrated into the thin veneer. The thin veneer was oven dried at 60°C. The treated NCC-epoxy admixture was cast on the veneer and distributed uniformly by hand roller. The treated veneer was pressed using cold press for 24 hours at 200 psi. Then, the veneer laid at ambient room temperature for curing process. A similar technique was used for the untreated NCC reinforced epoxy resin polymer/thin veneer composites. Nanocomposites with different untreated and treated NCC contents were prepared.

For comparison, the nanocomposite samples without the thin wood veneer were also prepared. The NCC-epoxy admixture was poured into a mold with dimensions of 170 mm x 170 mm x 3 mm and left cured at room temperature. These nanocomposites were produced using treated and untreated NCC, respectively.

2.4 Characterization

2.4.1 Fourier transforms infrared spectroscopy (FTIR). Each sample was mixed homogeneously with potassium bromide (KBr) and ground into powder using a mortar and a pestle. The mixture was transferred into a mold and was hydraulic pressed to form a 13 mm diameter KBr disc. The FTIR spectra were recorded in the range of 4000 to 450 cm^{-1} with a total accumulation of 16 scans employing a Perkin Elmer Spectrum 100 Series. All samples were analyzed in triplicate.

2.4.2 Tensile test. Tensile testing was carried out using Instron Universal Testing Machine according to ASTM D 638-91 with a cross head speed of 5 mm/min. The specimens were cut to dimensions of 160 mm x 13 mm. The tensile values were taken from an average of 10 specimens.

3. Results and Discussion
The nanocomposites investigated in this study were based on an epoxy resin formed by the reaction of an oligomeric difunctional diglycidyl ether of bisphenol A with an epoxide equivalent weight of 185-192 and a diethyl toluene diamine-based curing agent in the presence of wood veneer and NCC isolated from sawdust.

3.1 Fourier transforms infrared spectroscopy
FTIR analysis in Figure 2 attempted to characterize the spectra for NCC, GPS and treated NCC. The absorbance peaks in the range 3400-3300 cm^{-1} are attributed to the stretching of the OH groups of cellulose. The peaks observed around 2905 - 2899 cm^{-1} corresponds to C-H stretching in the spectra of both NCC and treated NCC [11]. The peaks at 1640–1650 cm^{-1} attribute to C=C stretching. The spectrum for treated NCC show absorbance at 1100–1200 cm^{-1} is Si –O-Si and Si-O-C stretching. GPS shows a strong absorbance at 845-820 cm^{-1} is C-O-C stretching vibration of epoxy ring. The peak at 1083 cm^{-1} is from the Si-O-CH_3 strong stretching vibration. The peaks observed around 1275- 1255 cm^{-1} corresponds to C-O-C stretching in the spectra of GPS. The strong absorbance of GPS shown in the treated NCC spectrum clearly indicates that NCC is successfully treated with GPS.

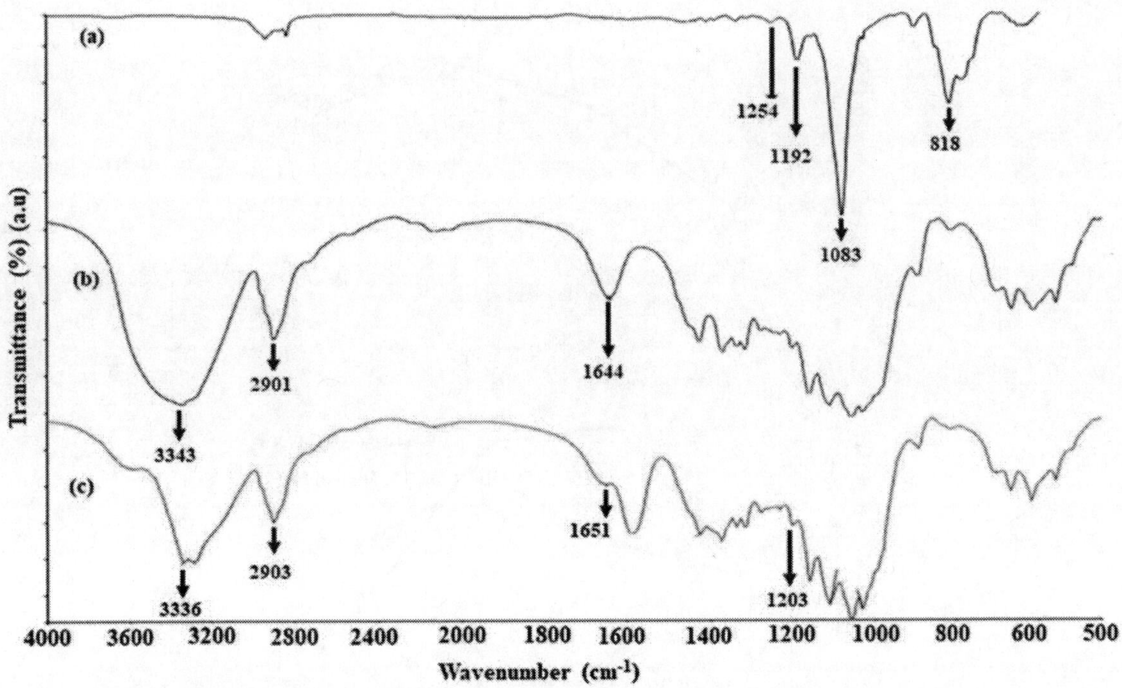

Figure 2. FTIR spectra of (a) GPS, (b) NCC and (c) treated NCC (NCC-GPS).

3.2 *Tensile properties*

The addition of GPS was to improve the compatibility and dispersion of nanocellulose into the polymer matrix. The tensile properties of the epoxy resin reinforced with GPS treated NCC composites are shown in Figure 3a. The finding shows that the tensile strength of the nanocomposites improved after modification with silane. The highest value is about 42 MPa at 0.5 wt/wt % of GPS solution utilized to treat NCC. The change in tensile strength is not significant for the nanocomposites with 1 wt/wt % of GPS solution compared to 2% wt/wt % of GPS solution after which it decreased. The advantages of silane coupling agents are: (i) they are commercially available in a large scale; (ii) they have alkoxysilane groups capable of reacting with OH-rich surface at one end, and (iii) at the second end, they comprise a large number of functional groups which can be tailored as a function of the polymer to be used. This is to ensure a good compatibility between the reinforcing element and the polymer or even develop covalent bonding between them [12].

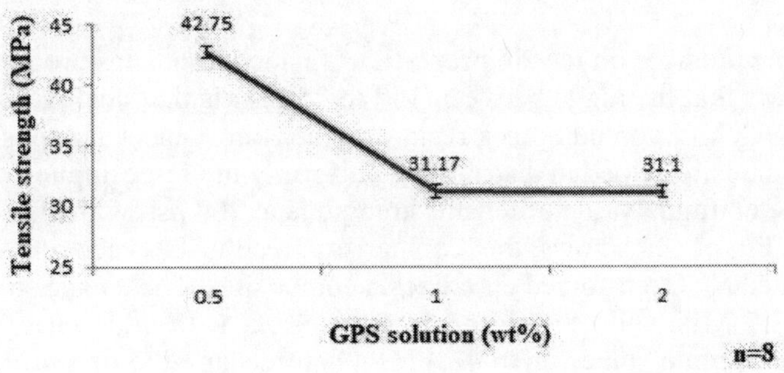

3(a)

IOP Conf. Series: Materials Science and Engineering **368** (2018) 012007 doi:10.1088/1757-899X/368/1/012007

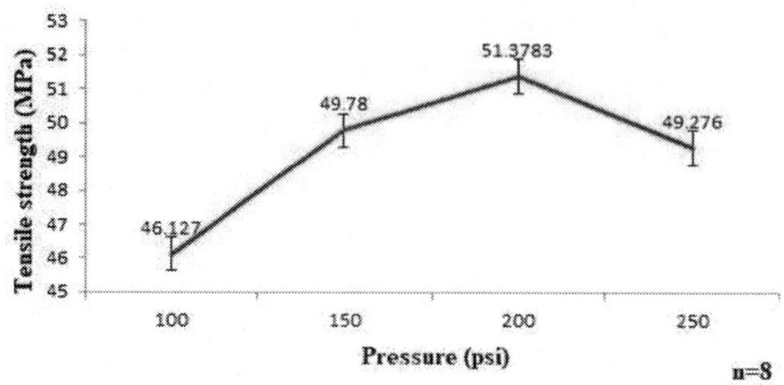

3(b)

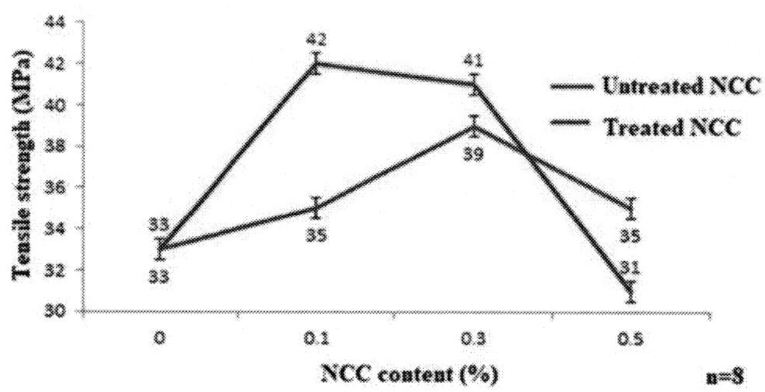

3(c)

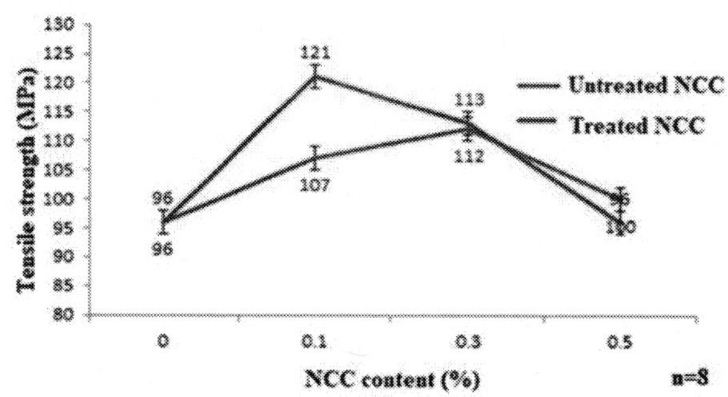

3(d)

Figure 3. (a): Tensile properties of nanocomposites at different percentage of GPS utilization. 3(b): Tensile properties of nanocomposites at different pressing pressure. 3(c): Tensile properties of untreated and treated NCC reinforced nanocomposites. 3(d): Tensile properties of untreated and treated NCC-epoxy resin polymer reinforced ultrathin wood veneer nanocomposites.

The effect of pressing pressure on tensile properties of nanocomposites produced is indicated in Figure 3b. The finding shows that the range between 150 to 250 psi is the ideal pressure range to develop the harmonized cured NCC reinforced epoxy resin polymer/thin veneer composites. The rigidity of this harmonized nanocomposite helped in distributing uniformly the force applied onto the nanocomposites. The result shows the optimum value of tensile strength is at 200 psi, with an 11.3 % improvement.

The finding in Figure 3c shows the comparison results between untreated NCC-epoxy resin composites and treated NCC reinforced epoxy resin composites. The tensile strength of neat epoxy resin was found to be 33.7 MPa and the reinforced composites with addition of 0.1 (w/w) treated NCC enhanced the tensile strength values up to 42.4 MPa before started to decrease to 41.2 and 30.9 MPa as the loading value increased, 0.3 and 0.5% respectively. The maximum increment value of tensile strength was found at 0.1 wt. % filler loading with a 26% improvement. On the other hand, the addition of untreated NCC into epoxy resin polymer has improved the tensile strength by a factor of 17% at 0.3

wt.% filler loading, Hence, the result reveals that the treatment of NCC using silane contributed to the improvement in interfacial bonding between the NCC and epoxy resin polymer [13-14]. This is due to the optimum reinforcing effect for the treated NCC in the nanocomposites was distinctly developed using lower amount of NCC content (0.1 wt.%) as compared to the untreated NCC content (0.3 wt.%) in the nanocomposites.

The effect on the tensile strength of untreated and treated NCC incorporated in ultrathin wood veneer-epoxy resin based composites was observed as a function of the NCC loading ranging between 0.1 and 0.5% (w/w) (Figure 3d). The thin wood veneer was used as the basic structure element in the nanocomposites. Similar finding was found for the optimum reinforcing effect for the untreated and treated NCC incorporated into thin wood veneer-epoxy resin based composites. The tensile properties of treated NCC incorporated into thin wood veneer-epoxy resin based composites increased by 23% at the optimum NCC loading of 0.1 wt. %. Whereas the addition of untreated NCC untreated into ultrathin wood veneer-epoxy resin based composites shows the increment of 22% of tensile strength at 0.3 wt.% NCC content. Beyond the optimum NCC loading, the tensile strength values plateau suggesting that the addition of more NCC above this threshold concentration does not help to improve the tensile properties. The reason for the value to plateau may be due to the potential aggregation of NCC particles after a certain concentration is reached, which results in no further improvement of mechanical properties.

4. Conclusion

Nanocrystalline cellulose (NCC) has a high aspect ratio characteristic that will improve the mechanical property of polymers. However, because of the hydrophilicity of NCC, it has to be modified chemically or physically to achieve the desired interfacial adhesion between the filler and the hydrophobic matrix. In this study NCC has been treated by 3- glycidoxylpropyltrimethoxysilane (GPS). The additions of treated and untreated NCC into epoxy resin based composites with and without thin wood veneer have improved the strength properties of the nanocomposites compared to the neat epoxy resin system. However, the addition of treated NCC rather than untreated NCC resulted in higher tensile. At low amount of NCC content (0.1%), the maximum increment values of tensile strength of thin wood veneer/epoxy resin polymer based nanocomposites was found to have a 23% improvement compared to the control sample without NCC. These green nanocomposites have the potential to be utilized to produce high strength laminated composite engineered products.

Acknowledgements

The author acknowledge the financial support from the Ministry of Science, Technology and Innovation (MOSTI) Malaysia under Science fund grant 03-03-10-SF0207

References

[1] Istomin A V, Demina T S, Subcheva E N, Akopova T A, and Zelenetskii A N 2016 Nanocrystalline Cellulose from Flax Stalks: Preparation, Structure, and Use *Fibre Chem+* **48** 199-201

[2] Hu S, Jiang F, and Hsieh Y L 2015 1D lignin-based solid acid catalysts for cellulose hydrolysis to glucose and nanocellulose *ACS Sustain Chem Eng* **3** 2566-2574

[3] Postek M T, Vladar A, Dagata J, Farkas N, Ming B, Sabo R, Wegner T H, and Beecher J 2008 September Cellulose nanocrystals the next big nano-thing?. In *Instrumentation, Metrology, and Standards for Nanomanufacturing II* (Vol. 7042, p. 70420D). International Society for Optics and Photonics.

[4] George J and Sabapathi S N 2015 Cellulose nanocrystals: synthesis, functional properties, and applications *Nanotechnology, science and applications* **8** 45.

[5] Tang L & Weder C 2010 Cellulose whisker/epoxy resin nanocomposites *ACS Appl Mater Inter* **2** 1073-1080

[6] May C (Ed.) 1987 *Epoxy resins: chemistry and technology*. CRC press. 485

[7] Shimokawa T and Hamaguchi Y 1981 Fatigue life distributions of notched graphite/epoxy composite specimens *Japan Society of Materials Science, Journal* **30** 373-379

[8] Rusli R and Eichhorn S J 2008 Determination of the stiffness of cellulose nanowhiskers and the fiber-matrix interface in a nanocomposite using Raman spectroscopy *Appl Phys Lett* **93** 033111

[9] Lu J, Askeland P and Drzal L T 2008 Surface modification of microfibrillated cellulose for epoxy composite applications *Polymer* **49** 1285-1296

[10] Rusli R and Eichhorn S J 2011 Interfacial energy dissipation in a cellulose nanowhisker composite *Nanotechnology* **22** 325706

[11] Sain M and Panthapulakkal S 2006 Bioprocess preparation of wheat straw fibers and their characterization *Ind Crop Prod* **23** 1-8

[12] Xie Y, Hill, C A, Xiao Z, Militz H, and Mai C 2010 Silane coupling agents used for natural fiber/polymer composites: A review *Compos Part A-Appl S* **41** 806-819

[13] Kargarzadeh H, Sheltami R M, Ahmad I, Abdullah I, and Dufresne A 2015 Cellulose nanocrystal: A promising toughening agent for unsaturated polyester nanocomposite *Polymer* **56** 346-357

[14] Sheltami R M, Kargarzadeh H, and Abdullah I 2015 Effects of silane surface treatment of cellulose nanocrystals on the tensile properties of cellulose-polyvinyl chloride nanocomposite *Sains Malaysiana* **44** 801-810

IOP Conf. Series: Materials Science and Engineering **368** (2018) 012008 doi:10.1088/1757-899X/368/1/012008

Start-up treatment of palm oil mill effluent (POME) final discharge using Napier Grass in wetland system

F A Ujang[1], N A Osman[1], J Idris[2], M I E Halmi[3], M A Hassan[1] and A M Roslan [1,4*]

[1]Department of Bioprocess Technology, Faculty of Biotechnology and Biomolecular Sciences, Universiti Putra Malaysia, 43400 Serdang, Selangor, Malaysia
[2]Faculty of Chemical Engineering, Universiti Teknologi MARA (UiTM), 94300, Kota Samarahan, Sarawak, Malaysia
[3]Department of Soil Management, Faculty of Agriculture, Universiti Putra Malaysia, 43400 Serdang, Selangor, Malaysia
[4]Laboratory of Biopolymer and Derivatives, Institute of Tropical Forestry and Forest Products (INTROP), Universiti Putra Malaysia, 43400 UPM Serdang, Selangor, Malaysia

*Corresponding author: ar_muhaimin@upm.edu.my

Abstract. Average of about 53 million m^3 POME is being produced yearly in Malaysia. The treated palm oil mill effluent (POME) will be discharged to the river. However, the POME final discharge still contains high oxygen demand (COD) and suspended solids (SS) that potentially to cause environmental issues. Therefore, the aim of this study is to treat the POME final discharge with Napier grass (*Pennisetum purpureum*) in the constructed wetland system. This study has been carried out to investigate the feasibility of the Constructed Wetland in polishing the final discharge according to river quality index. Besides, the propagation of Napier grass in the POME final discharge was evaluated. Four different physicochemical parameters such as chemical oxygen demand (COD), total suspended solid (TSS), ammonia and colour were analysed based on standard laboratory methods and procedures. Results revealed that the reduction of COD level by 71.57 %, TSS by 83.59 %, ammonia by 85.97 % and colour reduced by 87.62 % after 15 days treatment time. In term of propagation of plant, Napier grass showed an excellent propagation in POME final discharge by 13.94% increment. The overall results indicated that treatment of POME final discharge by using Napier grass constructed wetland was successful to reduce the concentration of the contaminants.

1. Introduction

Malaysia has abundant natural resources for commercial cultivation of crops such as rubber and palm oil. Malaysia is the second largest producer of palm oil in the world [1], the third largest for rubber and fourth largest for cocoa [2] Since palm oil industries are the largest industries in Malaysia, more than 13 million tonnes of crude palm oil is being produced yearly, covering 11 % of Malaysian land for plantation area.

The byproduct of the crude palm oil (CPO) production are, oil palm trunks (OPT), empty fruit (EFB), oil palm fronds (OPF), palm pressed fibres (PPF), palm kernel shells, palm kernel cake and liquid discharge palm oil mill effluent (POME) [2]. From the biomass mentioned above, POME is the most expensive and difficult waste to manage since it is generated in large volume in tonnes at a time [3]. POME is considered among the harmful waste that giving the detrimental impact to the environment if it is discharged untreatable [2,4]. POME is treated to final discharge and released to the nearby land or river since it is the easiest and cheapest way for disposal [3]. Although the treated POME released to the rivers meets the standard discharge limit set by the government [6], it still contains relatively high chemical oxygen demand (COD) and suspended solids (SS) content compared to the receiving water body [7]. The POME final discharge quality still does not meet the river water quality index as shown in Table 1. Thus, the discharge of POME final effluent wastewater may potentially cause the environmental problem if it is not being treated properly [4].

Table 1. Characteristic of raw POME, POME final discharge and river water [8].

Parameter	Raw POME (mg/L)	POME Final Discharge (mg/L)	River water (mg/L)
BOD_5	10,250-25,020	65	8
COD	15,000-59,700	520	30
SS	5000-18,000	217	40

A wetland system can be suggested to treat POME final discharge. A wetland system is more compatible as compared to another system to treat the POME final discharge. The advantage of wetland system is more cost efficient with fewer side effects compared to other the biological and chemical approaches [9] Wetland is an area that is saturated with water and an area between land and water that have characteristics of a distinct elements. Wetland has many benefits such as food or habitat for wildlife, flood protection, improve the water quality and can act as recreation site [10]. Wetland also can consider as natural environmental water treatment and it also environmental friendly [11].

Constructed wetlands are widely used in wastewater treatment throughout the world and vegetation from these systems can be used for raw material production [12]. Plant from constructed wetland should has ability to accelerate the development of microbial communities and promote ''living'' environments around roots (i.e., aerobic and anaerobic alternant environments). It is also assumed that plant play a central role in the sludge treatment wetlands (STWs) by preventing clogging, favouring dewatering and improving sludge mineralization [13]. The wetlands, either natural or constructed, is effective in treating biochemical oxygen demand (BOD), suspended solids, nitrogen, and phosphorus, and also for reducing concentrations of metals, organics, and pathogens [14]. Constructed wetlands have many advantages compared to conventional treatment techniques such as lower total lifetime costs and capital costs, environmentally friendly, tolerate fluctuating in flow and pollutant concentrations, capable of treating multiple or mixed contaminants, provide flood protection, have lower water and air emissions and secondary wastes, can be built to fit into the landscapes, provide habitat for plants and wildlife, provide educational and recreational opportunities [15].

Phytoremediation is a process that uses plants to degrade and remove contaminants from the environment [16]. Generally, phytoremediation of contaminants by a plant involves the following steps: uptake, translocation, transformation, compartmentalization, and sometimes mineralization [17]. Phytoremediation can degrade, remove, transform, or immobilize toxic compounds located in soils, sediments, and more recently in polluted ground water and wastewater in treatment wetlands [18]. The phytoremediation is recognized as a better green remediation and a possible method for the removal of wastewater pollutant [19] Phytoremediation technology has been used successfully for the remediation of wastewater contaminated and soils contaminated with various pollutants [20]. Phytoremediation is increasingly used as a technological complement for treatment of polluted water in different types of treatment wetlands [21]. Plants that are chosen to be used in the phytoremediation technique must possess several characteristics such as having high uptake of both organic and inorganic pollutants, and must be able to grow well and quickly in a range of different conditions of the polluted water conducted by (Darajeh *et al.*,2014) [22] such as *Pennisetum purpureum* (Napier grass). Napier grass (Figure 1) is perennial grass species found in the tropical and subtropical area around the world. This plant contains high morphological variation within the species and known as the fastest growing plant.

Figure 1. Napier Grass (common Napier).

Napier grass was introduced in Malaysia in 1921 from the East Africa and this grass currently the most popular fodder grass in the dairy and also in feedlot production systems [23]. A study has been conducted by Tayade *et al.*, (2005) [24] where Napier Grass (*Pennisetum clandestinum*) was planted together with broadleaf cattail (*Typha latifolia)* in a horizontal subsurface flow system for the primary-treated municipal wastewater treatment. The objective of this study was to reduce the residual pollutants from biologically treated POME final discharge using the constructed wetland according to river water quality index. The reductions in COD, BOD, TSS and ammonia values have been recorded to determine the effectiveness of constructed wetland.

2. Materials and method

2.1. *Start-up of wetland system using Napier grass*

The wetland system was set up as shown in Figure 2. Each tank was constructed from a plastic box approximately 57.5 cm in length, 40.5 cm in height and 38.5 cm in width. These tanks were filled with stones, coarse sand and fine sand with ratio 1:1:1 according to the finding in the parallel study. The ratio is referred to the volume of each sand. The wetland system effluent was collected for the analysis. Napier grasses have been chosen to be observed and to monitor the efficiency of the constructed wetland system in order to remediate and treat the POME final discharge. The planting stock was obtained from the UPM's farm and was planted at the Biomass Technology Laboratory, UPM. After 2 months, the Napier grass was transferred to the wetland systems. Unless stated otherwise, the study was done in triplicate.

2.2 Sampling

POME final discharge was collected from final process of POME treated ponds, Felda Pasoh Palm Oil Mill in Negeri Sembilan. The collected sample was stored in air tight plastic bottles at 4 $^{\circ}$C in the chiller to prevent any microbial biodegradation that can affect the experimental result. The POME FD Wetland tanks were topped up with 5L of POME final discharge every day. Whereas the control was set up with the rain water as the feed 5L every day. The outlet from the system was collected periodically from each tank on the sampling day 3, 6, 9, 12, 15 and 18. The measurement of Napier grass growth evaluation was carried out once per week.

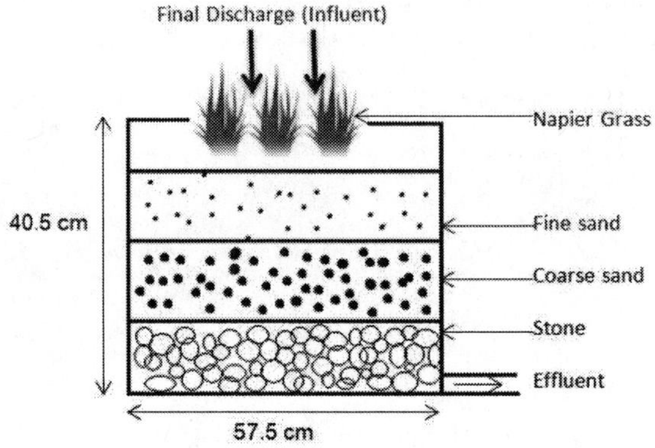

Figure 2. Constructed wetland system (this study).

2.3 *Analytical procedure*

All the water samples were analysed based on the APHA Standard Method for the Examination of Water and Wastewater [25]. COD was measured using COD Digestion Reagent vial. 2 ml of aliquot sample was added into the COD Digestion Reagent vial (Low range or High range were used). The vials containing sample were heated and refluxed for 2 hours. The vials were allowed to cool and measured using HACH DR2800 spectrophotometer (HACH, USA). TSS measurements were carried out by using a spectrophotometer DR 2800 (HACH, USA). 2 ml of sample was poured into sample cell and was measured at wavelengths of 810 nm. Colour measurements were carried out with a spectrophotometer DR 2800 (HACH, USA) at wavelength 455 nm. 2 ml of sample was added to the cell and the absorbance was taken. The amount of ammonia was determined by using Nessler method. 25 mL of the sample wasfilled to 25 mL of volumetric flask. 3 drop of each Mineral Stabilizer and Polyvinyl Alcohol Dispersing Agent were added to each flask. The sample and blank in the volumetric flask were inverted several times in every addition of reagent. Then 1.0 mL of Nessler Reagent was pipetted into the flask containing the sample. The flask was inverted again. 10 mL of the sample solution was poured into the sample cell and was inserted into the DR 2800. The wavelength used to measure the concentration of ammonia was 425 nm (HACH, USA). The Napier grass plants were evaluated on its growth efficiency by measuring their height.

3. Results and discussion

The system was run for 21 days, with sampling interval of 3 days. After 21 days, the system was stopped and the data were analysed. Figure 3 shows that the concentration of COD reduced from 453 mg/L to 128.8 mg/L from day 3 to day 15. The ability of Napier grass to reduce the concentration of COD has been proved by Klomjek *et al.,* [26] with the removal efficiency of 64 ± 6 %. In this study, the removal was 71.57 %. More interestingly with the addition of the same final discharge everyday, the concentration of COD keeps on reducing. This suggested that there was an increase in the performance of the wetland system in removing the COD.

Furthermore, it was reported that the use of Napier grass in the constructed wetland to treat sewage have reduced COD by 77.51 % [27]. Others also suggested that there was 42 % and 34 % of COD reduction in domestic wastewater for first and second of Napier constructed wetland over a three-month period [28]. The presence of microbes in the system could help to reduce the COD concentration. Previous works discovered that *Pseudomonas* sp is one of the microbes that can be found in the POME [29]. According to Abdirahman Elmi H. [29], the presence of *Pseudomonas aeruginosa* reduces to 58 % of COD value. Besides, the ability of *Pseudomonas* sp. to reduce COD concentration was also proved by Sonune *et al.*, [30] where 60 % of COD reduction from municipal wastewater was achieved.

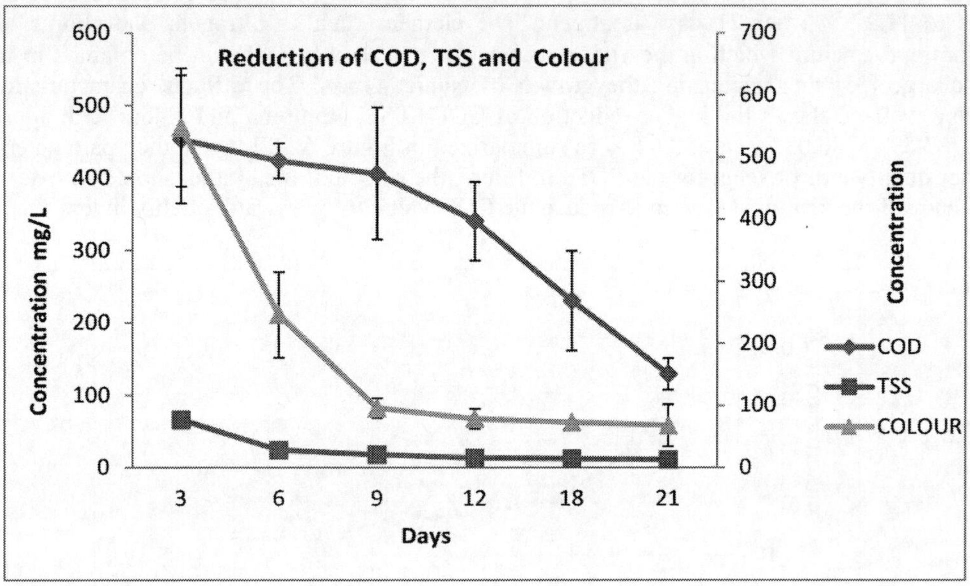

Figure 3. Reduction of COD, TSS and Colour after 21 days of treatment using constructed wetland system.

There is a reduction in the TSS values. It was shown that the concentration of TSS was consistently reduced from its initial values. The ability of Napier grass to reduce TSS in wastewater was proved by Dhulap *et al.*, [27], which reported that the use of Napier grass in the constructed wetland to treat sewage have reduced TSS by 55.17 %. Furthermore, there are 74 % and 82 % of TSS removal in domestic wastewater for first and second of Napier constructed wetland over a three-month period [28]. According to Vymazal *et al.*, [31], the removal mechanism for the suspended solids in the constructed wetland was by sedimentation and filtration. The particulates were filtered mechanically as water passes through the root masses of the Napier, sand and stones that have been constructed.

The concentration of colour also reduced from day 1 to day 21. The concentration of colour represent in Platinum-Cobalt (Pt-Co) At first, the concentration was recorded at 547.89±94.8 Pt-Co and it reduced to 67.83±33.706 Pt-Co that gave 87.62% removal of colour. Treatment of sewage through constructed wetland using Napier grass had successfully removed the colour and treated the water sample [27]. It was predicted that the colour removal is caused by a non-degradable organic material such as tannin that giving a tea colour to water sample [32] and by using wetland treatment, suspended solid was filtered out resulted in the concentration of colour. In addition, the presence of microbes in the system also helps to reduce colour level. This study was supported by previous study where 60 % of colour removal from POME was achieved by using *Pseudomonas aeruginosa* [29].

Figure 4 shows that the reduction of ammonia after treated in constructed wetland from 0.777 ± 0.102 mg/L to 0.109 ± 0.040 mg/L. This may be caused by nitrification process of the aerobic bacteria in the soil. The microbial metabolic pathways provide a wealth of processes that form or consume N_2O [33] The aerobic bacteria convert ammonia to nitrate. The process by the microorganism and contributes to the movement of nitrogen through the biogeochemical nitrogen cycle. Ammonia-oxidizing bacteria first oxidize ammonia to nitrate, followed by the oxidation of nitrate by the nitrite-oxidizing bacteria [34]. The nitrates produced by the oxidation of ammonia are a good fertilizer for the plants. This study proven that Napier Grass has shown better growth with the treated POME compares to the rainwater. Figure 5 shows that Napier grass with treated POME has increment in height by 13.94 % compares to rainwater that increases to 11.22 % after 21 days treatment. The element such as nitrogen, phosphorus, potassium, magnesium and calcium which is the vital elements for the plant growth can be obtained in the POME final discharge [35] thus increasing the growth of Napier Grass. The effluent characteristic after the constructed wetland shows the higher reduction of COD, TSS, ammonia and colour with approximately 71.57 %, 83.59 %, 86.03 % and 87.62 % as summarized in Figure 6. All the studied parameters meet the river water quality index except for the COD. In future, the period of the system should be extended to see the efficiency of the treatment system to reduce the COD value to river water quality index.

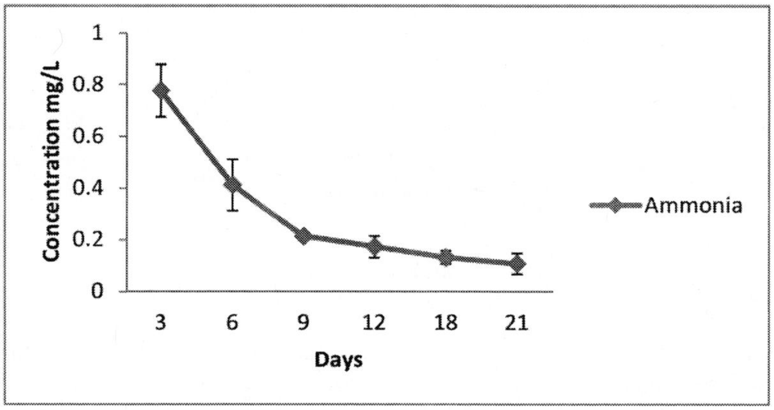

Figure 4. Ammonia Reduction after 21 days.

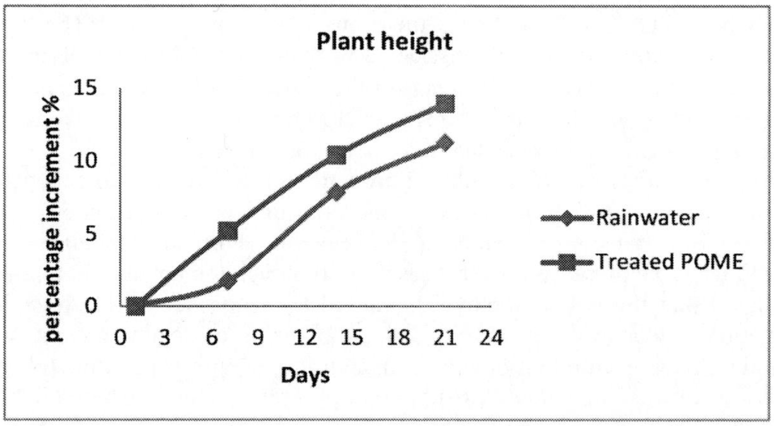

Figure 5. Percentage increment of plant height.

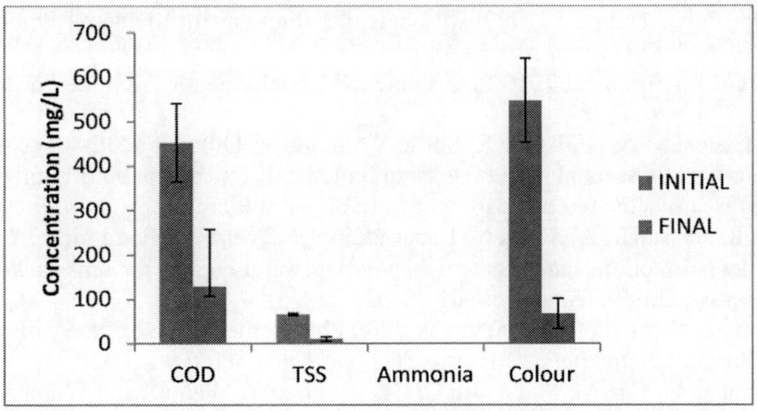

Figure 6. The initial and final concentration of COD, TSS, ammonia and colour. All value expressed in mg/L except for Colour that expresses in PtCo.

4. Conclusion

In this study, reduction of pollution concentration in POME final discharge through constructed wetland using Napier grass was performed. Four different physicochemical parameters were analyzed by using standard laboratory methods and procedure which are chemical oxygen demand (COD), colour, total suspended solid and ammonia. Results revealed that this system shows better water quality by reducing the level of COD by 71.57%, TSS by 83.59%, ammonia by 85.97% and colour reduce by 87.62%. The overall results indicated that treatment of treated POME by using Napier grass constructed wetland was successful to reduce the concentration of the contaminant. Besides, the propagation of Napier grass in the treated POME was evaluated. In term of propagation of plant, Napier grass showed an excellent propagation in treating POME and it could be vital elements for the plant growth. All the objective was achieved since all the pollutants were reduced and meet river water quality index except for COD. For recommendation, the system should be run in longer time to reduce the COD.

Acknowledgement
The authors would like to thank the Universiti Putra Malaysia for the technical and financial assistance in this study. The authors would also like to acknowledge Japan International Cooperation Agency through the SATREPS program for the training and financial help provided (Matching grant No. 6300156) to the first author in the course of the study.

References

[1] Gan P Y and Li Z D 2014 Econometric study on Malaysias palm oil position in the world market to 2035 *Renew. Sustain. Ene[1] Gan P Y Li Z D 2014 Econom. study Malaysias palm oil position world Mark. to 2035 Renew. Sustain. Energy Rev. 39 740–7rgy Rev.* **39** 740–7

[2] Singh R P, Ibrahim M H, Esa N and Iliyana M S 2010 Composting of waste from palm oil mill: A sustainable waste management practice *Rev. Environ. Sci. Biotechnol.* **9** 331–44

[3] Madaki Y S and Lau S 2013 Palm oil Effluent (POME) from Malaysia Palm Oil Mills: Waste or Resource *Int. J. Sci. Environ. Technol.* **2** 1138–55

[4] Rupani P F, Rajeev P S, Irahim M H and Esa N 2010 Review of Current Palm Oil Mill Effluent

(POME) Treatment Methods: Vermicomposting as a Sustainable Practice *World Appl. Sci. J.* **11** 70–81

[5] Igwe J C and Onyegbado C C 2007 A Review of Palm Oil Mill Effluent (Pome) Water Treatment *Glob. J. Environ. Res.* **1** 54–62

[6] Department of Environment 2010 Environmental Requirements: A Guide For Investors *Minist. Nat. Resour. Environ.* 1–78

[7] Ibrahim I, Hassan M A, Abd-Aziz S, Shirai Y, Andou Y, Othman M R, Ali A A M and Zakaria M R 2017 Reduction of residual pollutants from biologically treated palm oil mill effluent final discharge by steam activated bioadsorbent from oil palm biomass *J. Clean. Prod.* **141** 122–7

[8] Othman M R, Hassan M A, Shirai Y, Baharuddin A S, Ali A A M and Idris J 2014 Treatment of effluents from palm oil mill process to achieve river water quality for reuse as recycled water in a zero emission system *J. Clean. Prod.* **67** 58–61

[9] Lone M I, He Z, Stoffella P J and Yang X 2008 Phytoremediation of heavy metal polluted soils and water: Progresses and perspectives *J. Zhejiang Univ. Sci. B* **9** 210–20

[10] Zheng D, Chang Q, Gao M, She Z, Jin C, Guo L, Zhao Y, Wang S and Wang X 2016 Performance evaluation and microbial community of a sequencing batch biofilm reactor (SBBR) treating mariculture wastewater at different chlortetracycline concentrations *J. Environ. Manage.* **182** 496–504

[11] Sawal M 1986 General Standards for Discharge of Environmental Pollutants *Environ. Rules.* **2** 545–60

[12] Maddison M, Soosaar K, Mauring T and Mander Ü¨2009 The biomass and nutrient and heavy metal content of cattails and reeds in wastewater treatment wetlands for the production of construction material in Estonia *Desalination* **246** 120–8

[13] Vymazal J 2013 Plants in constructed, restored and created wetlands *Ecol. Eng.* **61** 501–4

[14] Vymazal J 2010 Constructed Wetlands for Wastewater Treatment *Water.* **2** 530–49

[15] Kamal S N B M 2009 Industrial Wastewater Treatment: Removal of Ammonia-Nitrogen and Orthophosphate in Palm Oil Mill Effluent Using Constructed Wetland

[16] Sanusi S N A, Abdullah S R S and Idris M 2012 Preliminary test of phytoremediation of hydrocarbon contaminated soil using paspalum vaginatum sw *Aust. J. Basic Appl. Sci.* **6** 39–42

[17] Schnoor J L, Light L A, McCutcheon S C, Wolfe N L and Carreia L H 1995 Phytoremediation of organic and nutrient contaminants *Environ. Sci. Technol.* **29**318A–323A

[18] Huang Q, Liu Y, Zheng X and Chen G 2012 Phytoplankton Community and the Purification Effect of Mangrove in the Mangrove Plantation-Aquaculture Coupling Systems in the Pearl River Estuary *Procedia Environ. Sci.* **15** 12–21

[19] Rezania S, Ponraj M, Talaiekhozani A, Mohamad S E, Md Din M F, Taib S M, Sabbagh F and Sairan F M 2015 Perspectives of phytoremediation using water hyacinth for removal of heavy metals, organic and inorganic pollutants in wastewater *J. Environ. Manage.* **163** 125–33

[20] Truu J, Truu M, Espenberg M, Nõlvak H and Juhanson J 2015 Phytoremediation And Plant-Assisted Bioremediation In Soil And Treatment Wetlands: A Review *Open Biotechnol. J.* **9** 85–92

[21] Zhang B Y, Zheng J S and Sharp R G 2010 Phytoremediation in engineered wetlands: Mechanisms and applications *Procedia Environ. Sci.* **2** 1315–25

[22] Darajeh N, Idris A, Truong P, Abdul Aziz A, Abu Bakar R and Che Man H 2014 Phytoremediation potential of Vetiver system technology for improving the quality of palm oil mill effluent *Adv. Mater. Sci. Eng.* **2014**

[23] Halim R A, Shampazuraini and Idris A B 2013 Yield and nutritive quality of nine napier grass varietas in Malaysia *Malaysian Soc. Anim. Prod.* **16** 37–44

[24] Tayade S T, Ojha A R, Kumar R and Singh R N 2005 Feasibility study of constructed wetland for treatment of municipal wastewater *Green Pages, Dir. Environ. Technol.*

[25] APHA/AWWA/WEF 2012 Standard Methods for the Examination of Water and Wastewater *Stand. Methods* 541

[26] Klomjek P 2016 Swine wastewater treatment using vertical subsurface flow constructed wetland planted with Napier grass *Sustain. Environ. Res.* **26** 217–23

[27] Dhulap V P and Patil S S 2014 Removal of Pollutants from Sewage through Constructed Wetland using Pennisetum purpureium *European Academic Research* **2** 543-558

[28] Tilak A S, Wani S P, Patil M D and Datta A 2016 Evaluating wastewater treatment efficiency of two field scale subsurface flow constructed wetlands *Curr. Sci.* **110** 1764–72

[29] Abdirahman Elmi H S 2015 Colour and COD Removal from Palm Oil Mill Effluent (POME) Using Pseudomonas Aeruginosa Strain NCIM 5223 in Microbial Fuel Cell *Int. J. Waste Resour.* **5** 3–5

[30] Sonune N A and Garode A M 2015 Bioremediation Potential of Bacterial Isolates for Municipal Wastewater Treatment *Curr. World Environ.* **10** 619–25

[31] Vymazal J, Brix H, Cooper P F, Haberl R, Perfler R and Laber J 1998 Removal mechanisms and types of constructed wetlands *Constr. Wetl. wastewater Treat. Eur.* 17–66

[32] Mohammed R R and Chong M F 2014 Treatment and decolorization of biologically treated Palm Oil Mill Effluent (POME) using banana peel as novel biosorbent *J. Environ. Manage.* **132** 237–49

[33] Butterbach-Bahl K, Baggs E M, Dannenmann M, Kiese R and Zechmeister-Boltenstern S 2013 Nitrous oxide emissions from soils: how well do we understand the processes and their controls? *Philos. Trans. R. Soc. B Biol. Sci.* **368** 20130122–20130122

[34] Zhang Xiaojing A4 - Liang, Yuhai A4 - Ma, Yongpeng A4 - Du, Jingjing A4 - Pang, Long A4 - Zhang, Hongzhong X A-Z 2016 Ammonia removal and microbial characteristics of partial nitrification in biofilm and activated sludge treating low strength sewage at low temperature *Ecol. Eng.* **93** 104-111–2016 v.93

[35] Muhrizal S, Shamshuddin J, Fauziah I and Husni M A H 2006 Changes in iron-poor acid sulfate soil upon submergence *Geoderma* **131** 110–22

A review on date palm (*phoenix dactylifera*) fibers and its polymer composites

W Ghori[1], N Saba[2*], M Jawaid[2] and M Asim[2]

[1]Department of Mechanical Engineering, College of Engineering, King Khalid University, Abha, Saudi Arabia

[2]Laboratory of Biocomposite Technology, Institute of Tropical Forestry and Forest Products (INTROP), Universiti Putra Malaysia, 43400 UPM Serdang, Selangor, Malaysia

*Corresponding author: naheedchem@gmail.com

Abstract. Date palm (Phoenix dactylifera) is the most significant agricultural crop, found abundantly in the Arabian Gulf in Saudi Arabia, Northern Africa, Pakistan, India, and the United States (California). Huge deposition of more than 15000 tons of date palm leaves and trunk are generated as wastes alone in Saudi Arabia. The date palm fibers from trunk and leaf possess considerable mechanical strength and modulus to many of the natural fibers and higher with respect to E-glass fibers and many aluminium alloys. Wide varieties of both thermosets and thermoplastics polymers have been modified by the reinforcement of date palm fibers derived from leaf and trunk. Date palm fibers represent a promising alternative substitute to synthetic fibers in polymer composite industries for both advanced structural and semi-structural applications. Present review article designed to be a comprehensive source of recent literature and study on date palm fibers origin, cultivation and its surface modifications. This article also intended to covers the recent findings in date palm fibers reinforced polymer composites with some interest on physical structure and chemical compositions, including their different commercial applications.

1. Introduction

The continuous growing demands and needs for minimizing the usage of synthetic fibers in polymer composites derived from petroleum products drive the concern of academician's towards the introduction of natural fibers in modern engineering endeavours. Natural fibers found naturally in all animals and plants, and exhibits superior, integrated, graded structures that commonly exhibit multifunctional performance [1]. Mimicry of these integrated microstructures and micro-mechanisms offers considerable potential to engineers in the design and continual improvement of composite material performance [2]. Natural fibers offers promising alternatives to costly synthetic fibers having an impact on environment [3]. Natural fiber reinforced polymer composites are leading competitors as component materials to improve the efficiency of many mechanisms, due to lightweight, high strength and effective stiffness. The effective utilization of natural fibers derived from renewable resources provides environmental benefits with respect to utilization of raw material and ultimate disposability. Currently, natural fibers as bio-fillers are gaining great attraction as they offer considerable advantages such as low cost, low density per unit volume, minimal tool wear, comparable specific strength and cost effective besides renewable, non-carcinogenic and degradable characteristics compared to synthetic fibers based polymer composites [4]. It may lead to the production of high durable consumer products from natural fibers that can be easily recyclable [5]. However, their wider exploitation are constraint due to water absorption tendency, relatively poor compatibility between polymer matrix and fibers, meagre performance under impact loading that leads to a significant reduction in strength, stiffness and stability [5-6].

Currently, plant fibers such as kenaf, coir, abaca, sisal, pineapple, cotton, jute, bamboo, banana, palmyra leaf, date palm, hemp and flex have achieved commercial success in the development of polymer composites, hybrid composites and green composites with varied types of polymer matrices.

Date palm fiber (DPF) is one of the waste materials in agriculture sector, which is widely grown in Saudi Arabia as well as Asia. Date (*dactylifera*) found to be the most essential tropical fruits available commercially throughout the world after pineapple and citrus. Date fruits and leaves are considered as waste materials of date trees and are extensively used for producing natural fibers [7]. The major constituents of DPF are same to that of other lignocellulosic fibers including holocellulose (60– 75 %), lignin (20 %), and ash (1.18 %) [8].

1.1 Natural Fibers

To the present situation, it is believed that source of petroleum based products are limited and uncertain, thus diverting the utilization of an alternative cheap, sustainable and easily available raw material. The countries growing plant and fruit are not for only agricultural purpose but also to generate raw materials like natural fibers for industries. Most of the emerging countries make business of lignocellulosic fibers for improving economic condition of poor farmers as much as country support. Natural fibers are the major classification of fibers, and can be further classified into vegetable, animal, and mineral fibers, displayed in Fig. 1.

Depending on their origin, plant fibers can be generally classified as bast, leaf, or seed-hair fibers [9]. In plants, bast and leaf fibers provide mechanical support to the stem or leaf, respectively, as is the case in date palm, flax, hemp, jute, or ramie. Fruit fibers are taken from fruits like dates, coconut (coir) fiber. Stalk fibers are obtained from husk and straw of crops like wheat, date palm, rice, barley, and so forth. Tree wood can also be used as fiber. Natural fibers have been used for a long period in many developing countries as cement materials [10].

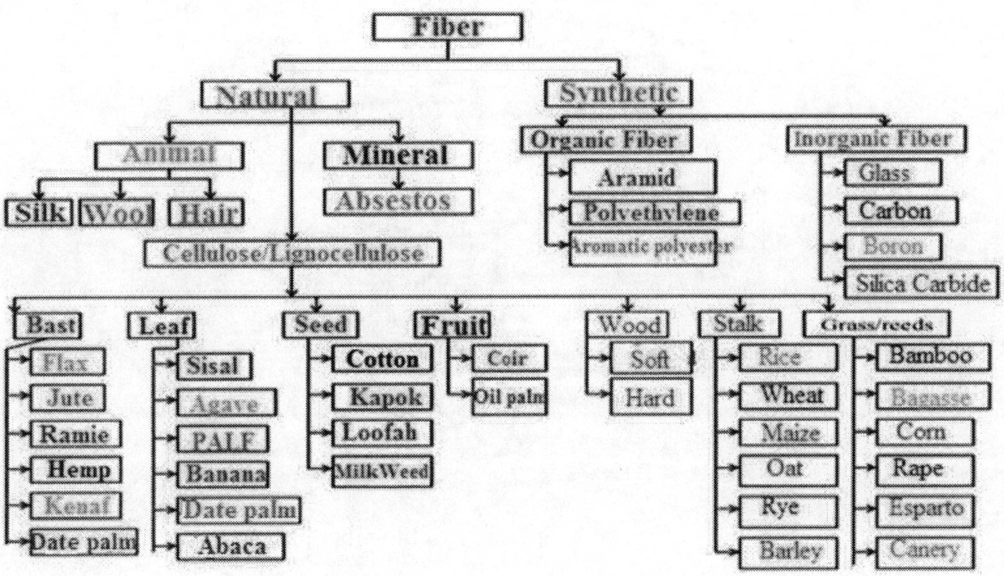

Figure 1. Classification of natural and synthetic fibers [11].

In general natural fibers have a hollow space (the lumen), as well as nodes at irregular distances that divide the fiber into individual cells. The surface of natural fibers is rough and uneven and provides good adhesion to the matrix in a composite structure. The low density, tensile strength and elongation at failure of natural fibers such as hemp, flax, ramie and synthetic fibers can compete with E-glass fibers properties and hence shows remarkable importance in composite technology [12]. The specific modulus and elasticity are better in natural fibers than the synthetic fibers [13]. A comparison of various parameters between natural and glass fibers is described in Table 1.

Table 1. Comparison between glass fiber and natural fibers [14].

Properties	Natural fiber	E-Glass fiber
Density	Low	Double
Cost	Low	High
Renewability	Yes	No
Recyclability	Yes	No
Energy Consumption	Low	High
Distribution	Wide	Wide
CO_2 Neutral	Yes	No
Abrasion To The Machine	No	Yes
Health Risk When Inhaled	No	Yes
Disposal	Biodegradable	Non Bio Degradable

Natural fibers consist of complicated and systematic internal cell wall structure, divided into three major parts[15]. The micro-fibril angle and arrangement inside the cell wall determine the fiber properties[16]. The mechanical strength of the fibers,(which determines the stiffness and tensile strength of composites) and dimensional stability of fibers depends on the lignin content[9]. The bio-fiber structure is shown in Fig.2.

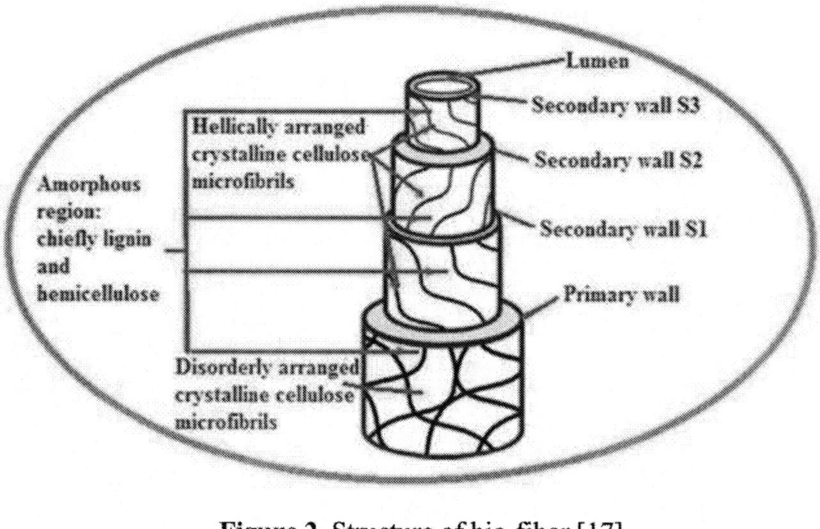

Figure 2. Structure of bio-fiber [17].

1.2 Polymer (Matrix)

Thermosets and thermoplastics are the major classification of polymer, where thermoset composite formulation is more multipart than thermoplastic. It depends on several parameters including, resin, curing agents, catalysts, and hardeners. The cross-linked structures are tough, creep resistant, and also tremendously resistant to solvents. Thermoplastics offer more advantage over thermoset polymers. Thermoplastics matrix composite possess low processing cost, flexible design, and easiness in moulding complexity. It is very simple method for processing of composites such as extrusion or inoculation moulding methods. A varieties of thermoplastic polymers like polyethylene, polypropylene, polystyrene and polyvinyl chloride (PVC) and their products are petroleum products are not entirely eco-friendly and non-biodegradable [18-19].

1.3 Natural fibers reinforced polymer composites

Polymers and their composites benefited modern industries and replaced conventional materials due to their high specific strength, light weight, cheaper, recyclability, biodegradability along long durability [20-21]. Natural fibers reinforced polymer composites is introduced for commercial reason, to provide the increasing of biodegradable and renewable materials. Natural fibers play a crucial role as they are used as engineered materials in composites, these advantages have attracted many scientists and researchers over the other fillers [7], [22], besides this they offered some restrictions and drawbacks [23-24]. In natural fiber composites, natural fiber provides tensile strength and stiffness and the matrix keeps the fibers together and transmits the applied shear forces. Reinforcements of fiber boost the property up to 80% because of the alignment property of fibers. Different components of plants and fruits obtained from agricultural crops and agricultural residues acts as the raw materials for fibers reinforced polymer composite industries. In literature review it has been reported that large varieties of natural fiber to reinforce different polymer matrices which includes wood [25-26], cotton[27], bagasse [28-29], rice straw [29-30], rice husk [31-32], wheat straw [33-34], flax [35], hemp [36-37], pineapple leaf [38] coir [39], oil palm [39-40], date palm [41], doum fruit [42], ramie [43], curaua [44-45], jowar [46], kenaf [47], bamboo [32, 48], rapeseed waste [49], sisal [50] and jute [51-52]. Moreover, synthetic fibers based polymer composites like glass fibers possess relatively high density with considerable high cost. Depending upon the chemical composition of fiber characteristics like recyclability, weather resistance, degradability, fungi attack are present in the natural fiber reinforced polymer composites [8, 53].

1.4 Date Palm Tree

The date palm, which belongs to the family Palmae (Arecaceae) is grown in tropical and subtropical regions since ancient period. Majorly dates are produced in the countries like Egypt (1,352,950 metric tons), Saudi Arabia (1,078,300 metric tons), Iran (1,023,130 metric tons), UAE (775,000 metric tons) and Algeria (710,000 metric tons) reports considerable improvement in industrialization of dates, worldwide production and its utilization. It plays a very significant role in the daily life of people of Arabian peninsula since last 7000 years [54].

Among date palm species, Phoenix dactyliferal is the tallest species with the height of more than 30 m and the fruit size ranging up to 100 mm × 40 mm size. A tropical date palm tree is characterized by trunk which is covered with greyish leaf bases and many offshoots which are present at its base. The fruits are juicy and nutritious [55]. Date palms have uniqueness that adapted them to various circumstances. Fig. 3 shows the young date palm tree with offshoots and date fruits. Date palm trees usually grown particularly in the sand, but it are not erinaceous since they have air spaces in their roots they can also grow well where soil water is close to the surface. Date palm rarely grows in saline conditions and their growth depends mainly on quality soil and water.

Brunch of date palm tree has elongated leaves which are pinnate divided and sharp needle fronds. These leaves are subtended by a cylindrical sheath of rough reticulate mass and fibrous material at its base which forms firm protective covering for terminal bud [56, 57]. Around 10-20 new leaves are produced annually which are adapted to both hot and dry conditions but it is not a xerophyte and copious water is needed [55-56]. Fig. 4 shows detailed morphological characteristics of date palm leaf highlighting length of leaf , thickness angle, leaflet length, thickness of rachis and number of leaflets [58].

Figure 3. Date palm tree (a) and Date fruit (b) [8].

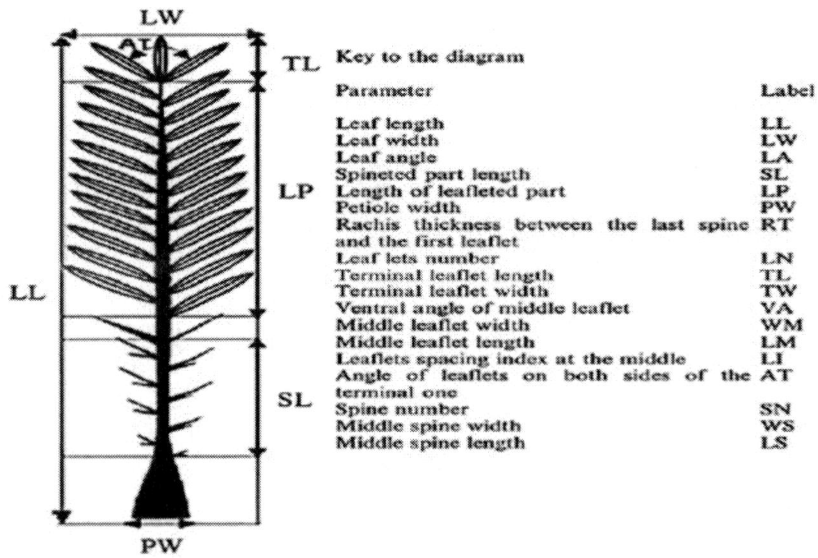

Figure 4. Detailed morphological traits of date palm tree leaf [58].

1.4.1 Extraction of Date Palm Fibers (DPF). There are different techniques opted for extraction of date palm fibers from leaves and trunk of date plant. The extraction of bast stem, leaf, and seeds from the plants can be done in a bundle form; therefore it is referred as fiber bundles. Extraction method of fibers is alike in both bast stem and leaf, while seed fibers have various methods like cotton lint extracted from ginning process. The extraction of strip fiber bundle is obtained from stem and leaf. Decorticator method is recommended for extraction process, followed by retting technique either dry retting and water retting, involving chemicals and biological treatment for removing fibers from stem [59]. Dew retting is most well-liked in Europe, but it's not accurate as water retting. Water retting technique is being used in Asian countries [60].

1.4.2 Date Palm Fiber. The worldwide production of date palm fiber annually is 42 % more compared to coir and 20-10 % more compared to hemp and sisal. Each stem of palm tree is surrounded by a mesh of single cross fibers which appears as a natural woven mat of fibers with different diameters. Normally this mat is separated and cleaned to make baskets and ropes. Figure 3A shows the shape of date palm fiber which is cylindrical in shape. The shape and structure of DPF resembles coir fiber, each DPF is an aggregation of 2-5μ sized multicellular fiber containing a central void [lumen] [61]. Due to this structure, surface modification is very much necessary to purify and clean the surface of the fibers from large amount of impurities and uncompleted growth, which may cause poor adhesion of fiber and matrix.

Extraction of fibers from date palm fiber is carried out by retting technique[8, 59] .Date palm fiber is more compatible natural fiber resource and constitutes a good chemical composition and possesses better mechanical strength than other natural fiber. DPF is multicellular lignocellulosic fiber containing polysaccharides cellulose (38-40 %), lignin in minor amount and a few minor components like fat, wax, pectin, inorganic substance, and so forth [8]. Many literatures have reported on cultivation of date palm fiber, their usage and purpose of utilizing date fruit and oil production. However, very little information available on the proper utilization of wastes of date palm trees particularly the leaves and the trunk. Separated date palm tree trunk and leaf rachis and fibers are illustrated in Fig 5.

Many researchers reported industrial applications of date palm fibers based on the property of these fibers in forming polymer composites for automotive components. These fibers are used as effective filler in both thermosetting and thermoplastics materials [8, 62-63]. The competitiveness of the date palm fibers in forming polymer composites suitable for automotive industrial components were reported by researchers [8].

Figure 5. Date palm tree trunk, leaf rachis (a) and (b-c) fibers [7].

1.4.3 Date Palm Tree Leaf Fibers. Traditionally, date palm leaves are utilized in fabricating baskets, mats, and ropes in many areas of the world. Competent utilization of this natural resource in synthesizing natural fiber-reinforced composite would have a high optimistic impact on the environment and improve the economic standard of working-class people [64]. Now, few works have been adopted by various researchers to develop polymer composite materials using fibers, obtained from different potions of date palm leaf. Mahdavi et al [7] have investigated the mechanical and morphological properties of wood plastic composites designed by reinforcing the fibers obtained from trunk, rachis, and petiole of date palm tree in high-density polyethylene (HDPE) matrix. The tensile strength of fibers obtained from leaf and stem of date palm tree has been investigated and compared with other natural fibers [65]. Researchers have studied the effect of different chemical modifications on date palm fiber surrounding the stems of date palm tree are shown in **Table 2** [66-67].

Table 2 . The average weight percentage of chemical composition of the date palm fibers from leaf (leaflet and rachis) [66-67].

Constituents	Cellulose	Hemicelluloses	Lignin	Ash	Extractive
Leaflet Leaf[a]	40.21 54.75a	12.8 20.00a	32.2 15.30a	10.54 1.75a	4.25 8.2a
Leaf	38.26	28.17	22.53	5.96	5.08

Note: a-value obtained from Sbiai et al. 2010 [67]

From their investigation, it has been clearly observed that 1 % of NaOH-treated date palm fiber shows the optimum mechanical properties, whereas HCl-treated date palm leaves show deterioration in mechanical properties. The average weight percentage of chemical composition of the date palm tree leaf and their fiber properties [66-67]

1.4.4 Chemical Constituents of Date Palm Fiber. Comparison of cellulose and lignin content of date palm fiber with that of other fibers demonstrates effectiveness and appropriateness of date palm fiber for being potential filler for natural fiber composites, shown in Fig. 6. Literature review illustrated that due to the presence of negligible amount of cellulose date palm fiber provides desired mechanical properties, lowered ability of absorbing water when compared to hemp and sisal. Additionally, it has greater content of cellulose than lignin, which has effective role in automotive industry [8].

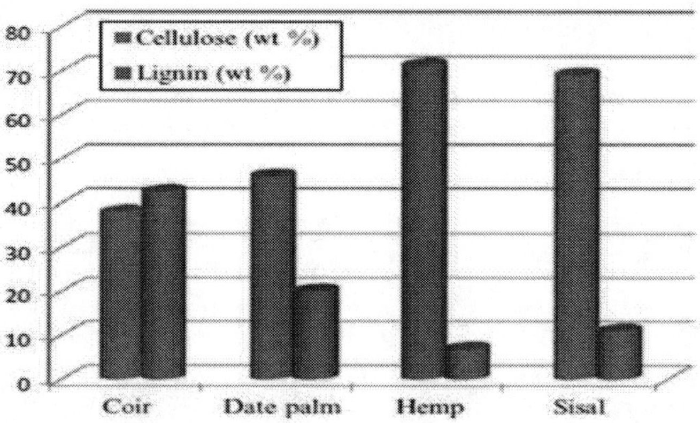

Figure 6. Comparison of cellulose and lignin contents between DPF and other natural fibers [8].

1.4.5 Properties of Date Palm Fiber. Properties such as density, fiber's length, diameter, thermal conductivity, aspect ratio (length/diameter), cost, and availability of the natural fibers are important to determine their compatibility for different natural fiber polymer composites industrial applications[8, 20, 68]. Date palm fiber is the most readily usable natural fibers having highest cellulosic content nearly 50 % as compared with other natural fibers. Fig.7 shows the density property comparison of date palm fibers and other natural types. Because of lower density of date palm fiber when compared to other natural fibers it is evident that it is accepted and reliable in the field of natural fiber composites.

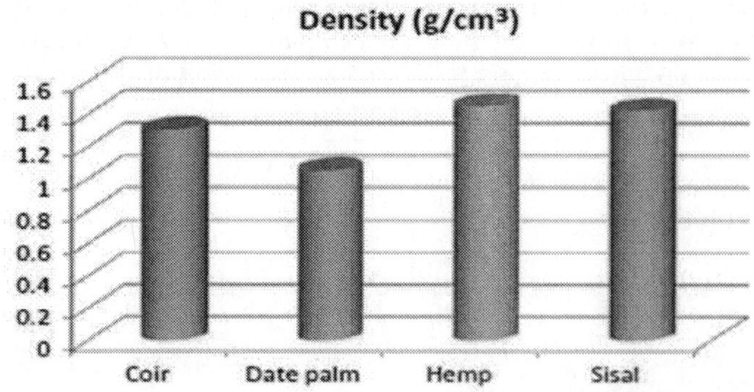

Figure 7. Density of date palm and other natural fibers [8].

Table 3 shows comparison of other physical properties of date palm fibers. Apparently, the physical properties of date palm fibers like diameter length, density and microfibril are quite efficient and display wide range of applications in the field of automotive and space industries [8, 21, 67]. Moreover, researchers reported that it has different industrial applications because it has optimum value of aspect ratio compared to other natural fibers [8].

Table 3. Comparison of physical properties of the date palm fibers with other natural fibers [8].

Properties	Fibers			
	Coir	**Date palm**	**Hemp**	**Sisal**
Density (g/cm^3)	1.15–1.46	0.9–1.2	1.4–1.5	1.33–1.5
Length (mm)	20–150	20–250	5–55	900
Diameter (μm)	10–460	100–1,000	25–500	8–200
Annual world Production	100	4,200	214	378
Cost Per Weight (USD/Kg)	0.3	0.02	1.2	1
Thermal Conductivity (W/mK)	0.047	0.083	0.115	0.07

1.4.6 Mechanical Properties of Date Palm Fiber. Research studies evaluates that factors such as chemical composition, structure, cell dimension, microfibrillar angle and defects influences the mechanical properties of natural fibers [53, 69-70]. In Table 4, mechanical properties of DPF and other natural fibers are listed. Young's modulus of DPF shows highest tensile strength on comparison with other natural fibers.

Table 4. Mechanical properties of various natural fibers [71]

Properties	**Diameter (μm)**	**Tensile Strength (MPa)**	**Young's Modulus (GPa)**	**Elongation at break (%)**
Jute	25–200	393–773	13–26.5	1.16–1.5
Flax	10–40	600–2,000	12–85	1–4
Sisal	50–200	468–640	9.4–22.0	3–7
Coir	100–450	131–175	4–6	15–40
Raw-date palm fiber	100–1,000	58–203	2–7.5	5–10

2. Challenges for Date Palm Fibers as Reinforcement

2.1 Date Palm Trunk Fibers

It has been reported in the literature, the use of stem fibers of date palm trees as reinforcement in polymeric materials. This research aims to study the potential of DPF by various treatments. The physical, mechanical and chemical properties obtained were compared to other common natural fibers. Evaluation of thermal stability, surface morphology, ultimate tensile strength and density upon chemical treatment of DPF were also done.

Treating DPF with different methods shows different properties of DPF. Researchers studied the effect of different treatments on DPF. They observed that the soda-treated fibers show highest increase in tensile strength and thermal degradation resistance whereas dioxin bleached fibers shows vice-versa. This difference will yield a variation of composites with different properties due to difference in interfacial adhesion between matrix and fiber [68].

2.2 Palm Fruit Stalk Fiber

Using ultrafine grinding technique micro fibrillated cellulose (MFC) was isolated from date palm fruit fiber stalks. These fibers were subjected to enzymatic pre-treatment with xylanase enzymes; it resulted in isolation of larger diameter of MFC molecules when compared with those isolated from untreated pulp. It was also observed that, the MFC molecules are obtained by enzymatic treatment has high density, tensile strength properties (dry and wet), lower water absorption and lower air permeability. MorFi analysis system are used to determine the percentage of fines at different grinding times which can be used as indicator to investigate the progress of fibers fibrillation into MFC [72]. It is an advanced rapid optical technique, which is found to be more reliable, more accurate and less expensive than the existing tools and could supply additional information on the fibers morphology, fine elements and the shives compared to Kajaani. MorFi brings great revolution in the area of paper and pulp industries with more statistical and reliable results on mechanical and chemical pulp fibers characterization, fiber biometry, fiber characteristics and fine elements.

2.3 Date Palm Leaf

Date palm leaf fiber show lower degree of compatibility with hydrophobic polymers due to its hydrophilic nature [8]. Existence of natural waxy substance on surface of fiber layer provides low surface tension, and hence does not allow a strong bond with polymer matrix. DPF can be effectively used as reinforcing component as it is lighter in weight and can be easily chemically modified to improve the efficacy of tensile and flexure strength in fiber reinforced biocomposites, to use as civil construction materials [41]. In future other strong fibers such as (coconut, sabai grass, elephant grass, sisal grass) can also be modified chemically and used as an alternative for heavy weight civil engineering materials as DPF.

2.4 Treatment of Date Fibers

To increase the bond ability and wettability between the cellulosic fiber and polymer matrix, different surface modifications such as chemical, physical and mechanical have been done. Date palm fibers like other natural fibers are treated in order to minimize the limitations and to amend the compatibility between date palm fibers and polymer in date palm reinforced polymer composites.

Many studies reveal that the effectiveness of the date palm fibers can be increased by chemical treatment [41, 62, 67-68, 73]. Alawar et al. [71] investigated the date palm fiber using different alkali treatments with a wide range from 0.5 to 5 % and acid treatment with 0.3 N,0.9 N and 1.6 N were used for 1h at 100 °C are used to modified the fiber surface . The results obtained showed and proved morphology improvement. Chemical treated fibers showed increase in tensile strength and considerable morphology advancement, whereas fibers treated with HCL were found to have negative impact on tensile strength and morphology. Fig. 8 shows effectiveness of the DPF upon chemical treatment where untreated fibers possess a weak outer layer that prevent strong bonding with the polymer matrix, while in treated DPF, the weak outer layer was removed during the treatment process leading to stronger bonding with the matrix present in the polymer composites.

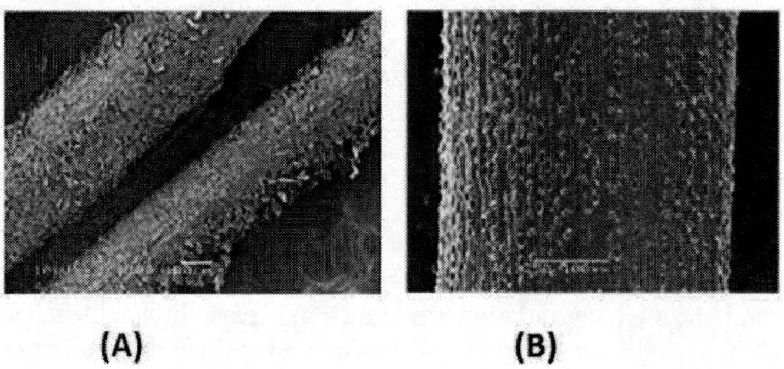

Figure 8. SEM of Untreated date palm fiber (A) and Treated date palm fiber with 1.5 % NaOH (B) [71].

Fig. 9 shows the effect of chemical treatment on the mechanical properties of date palm fibers. It depicts that proper chemical treatments enhances the tensile strength and Young's modulus of the

IOP Conf. Series: Materials Science and Engineering **368** (2018) 012009 doi:10.1088/1757-899X/368/1/012009

fiber. Fig.10 illustrates stress strain curves of DPF when treated with different concentrations of NaoH.

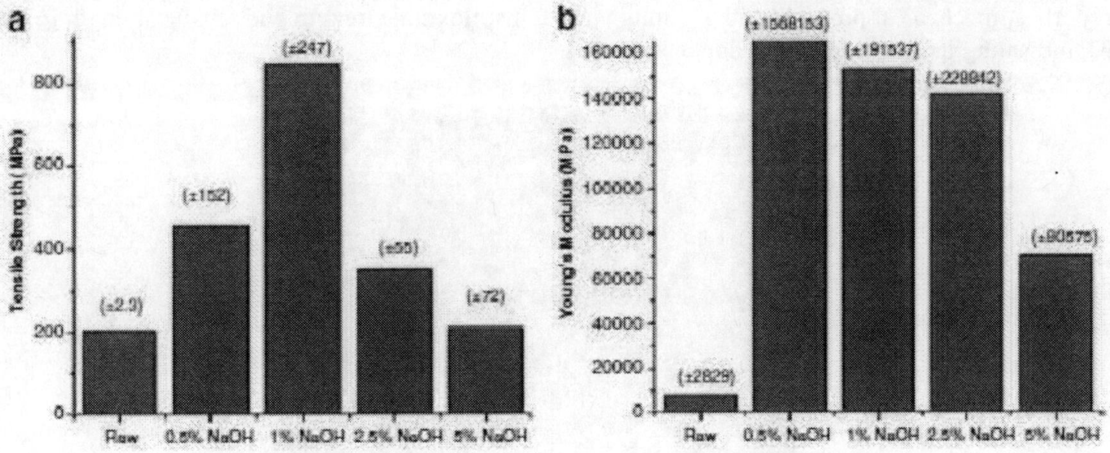

Figure 9. Effect of NaOH treatment on tensile strength (a) and Young's modulus of date palm fibers (b) [71].

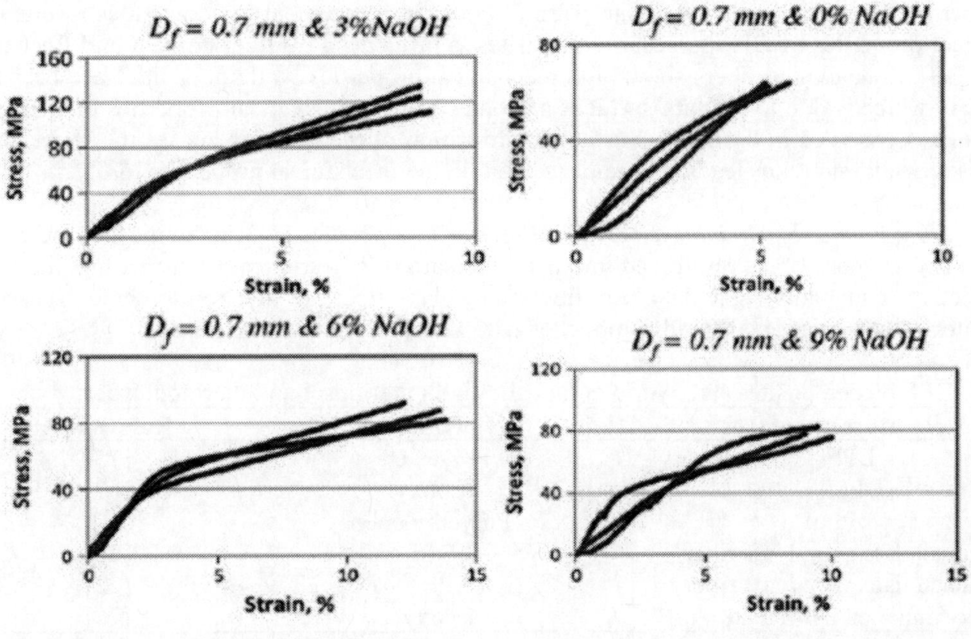

Figure 10. Stress/strain curves of date palm fiber with 0.7 mm diameter at different NaoH Concentration [73].

2.5 Matrices for Date Palm Fibers
In literature review it has been reported that matrices used for reinforcing date palm fibers for fabricating composites includes epoxy, polypropylene, polyester, high density and low density polyethylene (HDPE and LDPE) and ethylene terephthalate. Fig. 11 shows SEM micrograph of fracture surface of DPF/polyester composites. In other study unlike flax fibers and starch based

composites, DPF are combined with other types of fibers to make a complete, eco-friendly, biodegradable hybrid composite [74]. Most generally, it has been established that reinforcing of date palm fibers with different polymers exhibit considerable enhancements in, thermal and acoustical properties, mechanical properties like Young's modulus, flexural strength and tensile strength [62, 68, 73], indicating its wider industrial applications [8].

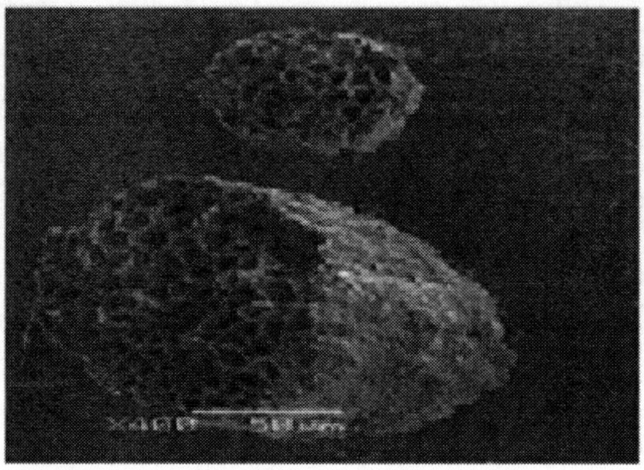

Figure 11. SEM micrograph of fracture surface of date palm fiber/polyester composite [75].

2.5.1 Date Palm Fiber Reinforced Polymer Composites. Table 5 display recent works on date palm fiber based composites. Using date palm fibers as reinforcement in composites resulted that the mechanical properties of DPF/polyester composites is influenced by fiber content and fiber treatment method [68]. Soda treated fibers show high mechanical properties while poor interface between fiber and matrix which is due to fracture surfaces attributes to low mechanical properties of DPF/polyester composites is observed in the study. Surface modification of the fibers increases the water absorption and improves adhesion thus leading to improvement in the mechanical properties [68].

2.5.2 Date Palm Fiber Reinforced Hybrid Polymer Composites. Hybrid composites are composites in which two or more fibers reinforced into a single matrix. Researchers develop a high-performance, cost-effective, and lightweight date leaf fibers and glass fibers as the reinforcement based hybrid composites. Composites and hybrid composites with DPLF are shown in Table 5.

Table 5. DPF biocomposites and hybrid composites with thermoset and thermoplastics.

Reinforcements	Polymers	References
DPF	Polyester	[75]
DPLF	Epoxy	[67]
DPF	Polyethylene	[7]
Bleached date palm leafs fiber	LDPE	[22]
Bleached date palm leafs fiber	PP	[22]
DPF/coconut cell particulate	Epoxy	[76]

Note: Low-density polyethylene-(LDPE); Polypropylene-(PP)

3. Applications

3.1 Date Palm Fiber Applications and Future Prospects

Date palms are multipurpose trees due to ancient cultivation, processing and utilization. Every part of the date palm can be used for one or more uses. Table 6 illustrates some uses of various parts of date palms. Date palm fibers presents several applications such as in textile, sports item, baggage, automobiles, cabinets, mats. Moreover, surface modified DPF are being introduced for making machine parts like transmission cloth, air-bag tying cords, conveyor belt cord and some cloths for industry uses. When the bundles of leaves are dried, they are known as "Barusti" which is used to make roofs, shades, enclosures and wall separators. Fishing traps are made up of ribs of these leaves [77]. DPF possess comparable mechanical and physical properties with respect to other natural fibers, making it suitable for its broader application as building and construction materials, textiles, furniture and automotive components besides as daily life materials.

Table 6. Diverse applications of various parts of date palms (http://hubpages.com/living/The-Various-Uses-of-Date-Palm-Tree).

Date Palm Part	Usages
trunk and wood	Timber, wood, or fuel
Trunk Fiber and Leaves	Baskets, cords, camel saddles, bags, crates, fans, food covers, twines, mats, trays, ropes, paper, and furniture
Dried leaves	Shades, roofs, separating walls, and enclosures
Leaf Ribs	Fishing boats and traps.
Leaf Base and Fruit Stalks	Fuel
Palm Pith	Date palm flour
Terminal buds (palm heart)	Salad and cooked vegetable
Date seeds	Livestock food and decoration beads
Date Seed Oil	Soap
Date fruit	Drugs and medical treatment for sore throat and cold, intestinal problems, fever, edema, cystitis, liver, and abdominal problems.
Groves	Environmental niches for local wildlife
Date Palms	Control of desertification and land reclamation and ornamental and landscape purposes

3.2 Applications of Date Palm Fibers Reinforced Polymer Composites and Hybrid Composites

Date palm fiber is one of the widely available natural fiber, due to its potentiality and competitiveness it has many industrial applications importantly in automotive ones. The peculiarity of this fiber is that it produces strong composites with different polymer matrices. To guarantee the marvellous properties of DPF and to ensure it to be used as ideal fiber in the automotive industry, various comparisons have been done between DPF and other commonly used fibers. In almost all comparisons, DPF was excellently chosen among different types as it has specific young's modulus to cost ratio criteria. It has been also found that the final mechanical properties of DPF and also its reinforced polymer composites can be increased by chemical treatment. The tremendous properties of DPF such as its performance, eco-friendly, cost effectiveness and its societal aspect has forced automotive industry to adopt DPF as it improve productivity and also it is highly sustainable. Moreover, adopting DPF has momentous environmental influence as waste management can be efficiently practiced. Natural fillers such as date palm fabric fiber and coconut shell particle filler can be used effectively as reinforcing materials for epoxy composites. It can be said that composites prepared from these fillers can replace synthetic fillers in some applications where high strength and stiffness is not the major concern.

4. Conclusions and Future Perspectives

Date palm fibers are the cheapest fibers among other natural fibers. Date palm fiber is very common in tropical regions and very easy to extract fibers from its leaves. Date palm fibers possess satisfactory definite Young's modulus to cost ratio criterion. Date palm fibers have very competitive properties toward enhancing the sustainability and productivity of automotive industry as well as solving and managing an environmentally waste problem issue. Date palm fibers is widely accepted in textile sector and already used in our daily life materials.

The mechanical properties of the composites can be increased by reinforcing the date palm leaf fibers into polymers. The use of date palm fiber in composite material is a new source of materials which can be economic, eco-friendly, and recyclable. Date palm fibers are successfully reinforced in epoxy, polypropylene and unsaturated polyester polymer to develop date palm fibers / polymer composites with improved properties. Hygroscopic nature of date palm fibers can be overcome by surface treatments with chemicals like NaOH, silane and others. Surface-treated date palm fibers reinforcement improves the mechanical and thermal properties of the polymer composites due to improved adherence between the date palm fiber and the polymer matrix.

In present article revealed that very limited study and research are reported in the literature improving the thermal properties including, electrical properties, thermal conductivity, dynamic mechanical analysis and modelling of date palm fibers reinforced polymer composites.

In future work, development of nano sized date palm fibers and its reinforcements in different varieties of thermosets and thermoplastics would be highlighted to fabricate nanocomposites and hybrid nanocomposites with enhanced properties. We also attribute that the development of biocomposites and hybrid composites by reinforcing date palm fibers in order to expand its applications also need major concerns.

Acknowledgement
We are thankful to Ministry of Higher Education for Providing HICOE Grant No: 6369108 to INTROP, UPM for doing this work.

References

[1] Curtis P T 1996 Multifunctional polymer composites. *advanced performance materials* **3** 279-293
[2] Kassner M E, Nemat-Nasser S, Suo Z, Bao G, Barbour J C, Brinson L C, Espinosa H, Gao H, Granick S, Gumbsch P, and Kim K S 2005 New directions in mechanics *Mech Mater* **37** 231-259
[3] Ma C C M, Tseng H T, and Wu H D 1998 Blocked diisocyanate polyester-toughened novolak-type phenolic resin: Synthesis, characterization, and properties of composites *J Appl Polym Sci* **69** 1119-1127
[4] Jawaid M H P S and Khalil H A 2011 Cellulosic/synthetic fibre reinforced polymer hybrid composites: A review *Carbohydr Polym* **86** 1-18
[5] Corbière-Nicollier T, Laban B G, Lundquist L, Leterrier Y, Månson J A, and Jolliet O 2001 Life cycle assessment of biofibres replacing glass fibres as reinforcement in plastics *Resour Conserv Recycl* **33** 267-287
[6] Abrate S 2005 *Impact on composite structures* Cambridge university press
[7] Mahdavi S, Kermanian H, and Varshoei A 2010 Comparison of mechanical properties of date palm fiber-polyethylene composite *BioResources 5* 2391-2403
[8] AL-Oqla F M, Alothman O Y, Jawaid M, Sapuan S M, and Es-Saheb M H 2014 Processing and properties of date palm fibers and its composites *Biomass bioenergy* 1-25
[9] Riedel U, Nickel J, and Herrmann A S 1999 High performance applications of plant fibres in aerospace and related industries. *German Aerospace Center (DLR)*
[10] Savastano Jr H, Warden P G, and Coutts R S P 2000 Brazilian waste fibres as reinforcement for cement-based composites *Cem Concr Compos* **22** 379-384
[11] Saba N, Tahir P M, and Jawaid M 2014 A review on potentiality of nano filler/natural fiber filled polymer hybrid composites *Polymers 6* 2247-2273
[12] Riedel U and Nickel J 2005 Applications of natural fiber composites for constructive parts in aerospace, automobiles, and other areas *Biopolymers Online*
[13] Kabir M M, Wang H, Lau K T, and Cardona F 2012 Chemical treatments on plant-based natural fibre reinforced polymer composites: An overview. *Composites Part B* **43** 2883-2892
[14] Asim M, Abdan K, Jawaid M, Nasir M, Dashtizadeh Z, Ishak MR, and Hoque ME 2015 A review on pineapple leaves fibre and its composites *International Journal of Polymer Science 2015*
[15] Madsen B 2004 *Properties of plant fiber yarn polymer composites: an experimental study*

[16] Bledzki A K and Gassan J 1999 Composites reinforced with cellulose based fibres *Prog Polym Sci* **24** 221-274

[17] Saba N, Jawaid M, Paridah MT, and Al-othman OY 2016 A review on flammability of epoxy polymer, cellulosic and non-cellulosic fiber reinforced epoxy composites *Polym Adv Technol* **27** 577-590

[18] Drzal L T, Mohanty A K, Burgueno R, and Misra M 2004 Biobased structural composite materials for housing and infrastructure applications: opportunities and challenges. In *Proceedings of the NSF housing research agenda workshop* 2 129-140

[19] Rout S K N J, Misra M, Mohanty A K, and Nayak S K 1998 Studies on Aqueous Co Polymerization of MMA onto Chemically Modified Coir Fibre and Its Effect on Mechanical Performance of Fibre. In *National Seminar on Polymer Research in Academy, Industry and R & D Organization, Calcutta, India*

[20] Alves C, Silva A J, Reis L G, Freitas M, Rodrigues L B, and Alves D E 2010 Ecodesign of automotive components making use of natural jute fiber composites *J Cleaner Prod* **18** 313-327

[21] Faruk O, Bledzki A K, Fink H P, and Sain M 2012 Biocomposites reinforced with natural fibers: 2000–2010 *Prog. Polym. Sci.* **37** 1552-1596

[22] Bendahou A, Kaddami H, Sautereau H, Raihane M, Erchiqui F and Dufresne A, 2008 Short palm tree fibers polyolefin composites: effect of filler content and coupling agent on physical properties *Macromol. Mater. Eng.* **293** 140-148

[23] Kim H S, Kim H J, Lee J W, and Choi I G 2006 Biodegradability of bio-flour filled biodegradable poly (butylene succinate) bio-composites in natural and compost soil *Polym Degrad Stab* **91** 1117-1127

[24] Tserki V, Matzinos P, and Panayiotou C 2006 Novel biodegradable composites based on treated lignocellulosic waste flour as filler. Part II. Development of biodegradable composites using treated and compatibilized waste flour *Composites Part A* **37** 1231-1238

[25] Ashori A and Nourbakhsh A 2010 Performance properties of microcrystalline cellulose as a reinforcing agent in wood plastic composites *Composites Part B* **41** 578-581

[26] Yang H S, Wolcott M P, Kim H S, and Kim H J 2005 Thermal properties of lignocellulosic filler-thermoplastic polymer bio-composites *J. Therm. Anal. Calorim.* **82** 157-160

[27] Raftoyiannis I G 2012 Experimental testing of composite panels reinforced with cotton fibers *Open Journal of Composite Materials* **2** 31

[28] Cao Y, Shibata S, and Fukumoto I 2006 Mechanical properties of biodegradable composites reinforced with bagasse fibre before and after alkali treatments. *Composites part A* **37** 423-429

[29] Habibi Y, El-Zawawy W K, Ibrahim M M, and Dufresne A 2008 Processing and characterization of reinforced polyethylene composites made with lignocellulosic fibers from Egyptian agro-industrial residues *Compos. Sci. Technol.* **68** 1877-1885

[30] Yao F, Wu Q, Liu H, Lei Y, and Zhou D 2011 Rice straw fiber reinforced high density polyethylene composite: Effect of coupled compatibilizating and toughening treatment *J Appl Polym Sci* **119** 2214-2222

[31] Chong E L, Ahmad I, Dahlan H M, and Abdullah I 2010 Reinforcement of natural rubber/high density polyethylene blends with electron beam irradiated liquid natural rubber-coated rice husk *Radiat Phys Chem* **79** 906-911

[32] Yang H S, Wolcott M P, Kim H S, Kim S, and Kim H J 2007 Effect of different compatibilizing agents on the mechanical properties of lignocellulosic material filled polyethylene bio-composites *Compos. Struct.* **79** 369-375

[33] Alemdar A and Sain M 2008 Biocomposites from wheat straw nanofibers: morphology, thermal and mechanical properties *Compos Sci Technol* **68** 557-565

[34] Le Digabel F, Boquillon N, Dole P, Monties B, and Averous L 2004 Properties of thermoplastic composites based on wheat-straw lignocellulosic fillers *J Appl Polym Sci* **93** 428-436

[35] Van de Weyenberg I, Ivens J, De Coster A, Kino B, Baetens E, and Verpoest I 2003 Influence of processing and chemical treatment of flax fibres on their composites *Compos Sci Technol* **63** 1241-1246

[36] Islam M S, Pickering K L, and Foreman N J 2010 Influence of alkali treatment on the interfacial and physico-mechanical properties of industrial hemp fibre reinforced polylactic acid composites *Composites Part A* **41** 596-603

[37] Kabir M M, Wang H, Lau K T, Cardona F, and Aravinthan T 2012 Mechanical properties of chemically-treated hemp fibre reinforced sandwich composites *Composites Part B* **43** 159-169

[38] Nadirah W W, Jawaid M, Al Masri A A, Khalil H A, Suhaily S S, and Mohamed A R 2012 Cell wall morphology, chemical and thermal analysis of cultivated pineapple leaf fibres for industrial applications *J Polym Environ* **20** 404-411

[39] Haque M M, Hasan M, Islam M S, and Ali M E 2009 Physico-mechanical properties of chemically treated palm and coir fiber reinforced polypropylene composites *Bioresour. Technol.* **100** 4903-4906

[40] Hassan A, Salema A A, Ani F N, and Bakar A A 2010 A review on oil palm empty fruit bunch fiber-reinforced polymer composite materials *Polym. Compos.* **31** 2079-2101

[41] Alsaeed T, Yousif B F, and Ku H 2013 The potential of using date palm fibres as reinforcement for polymeric composites *Mater. Design* **43** 177-184

[42] Arrakhiz F Z, El Achaby M, Malha M, Bensalah M O, Fassi-Fehri O, Bouhfid R, Benmoussa K, and Qaiss A 2013 Mechanical and thermal properties of natural fibers reinforced polymer composites: Doum/low density polyethylene *Mater. Design* **43** 200-205

[43] Feng Y, Hu Y, Zhao G, Yin J, and Jiang W 2011 Preparation and mechanical properties of high-performance short ramie fiber-reinforced polypropylene composites *J Appl Polym Sci* **122** 1564-1571

[44] Gomes A, Goda K, and Ohgi J 2004 Effects of alkali treatment to reinforcement on tensile properties of curaua fiber green composites. *JSME International Journal Series A Solid Mechanics and Material Engineering* **47** 541-546

[45] Silva R V and Aquino E M F 2008 Curaua fiber: a new alternative to polymeric composites *J Reinf Plast Compos* **27** 103-112

[46] Prasad A R and Rao K M 2011 Mechanical properties of natural fibre reinforced polyester composites: Jowar, sisal and bamboo *Mater. Design* **32** 4658-4663

[47] Tajeddin B, Rahman R A, and Abdulah L C 2010 The effect of polyethylene glycol on the characteristics of kenaf cellulose/low-density polyethylene biocomposites *Int J Biol Macromol* **47** 292-297

[48] Ismail H, Shuhelmy S, and Edyham MR 2002 The effects of a silane coupling agent on curing characteristics and mechanical properties of bamboo fibre filled natural rubber composites *Eur. Polym. J.* **38** 39-47

[49] Zabihzadeh SM, Omidvar A, Marandi MAB, Mirmehdi SM, and Dastoorian F 2011 Physical and mechanical properties of rapeseed waste-filled LLDPE composites *J Thermoplast Compos Mater* **24** 447-458

[50] Jacob M, Thomas S, and Varughese KT 2004 Natural rubber composites reinforced with sisal/oil palm hybrid fibers: Tensile and cure characteristics *J. Appl. Polym. Sci.* **93** 2305-2312

[51] Seki Y 2009 Innovative multifunctional siloxane treatment of jute fiber surface and its effect on the mechanical properties of jute/thermoset composites. *Mater. Sci. Eng., A* **508** 247-252

[52] Jawaid M, Khalil H A, and Bakar A A 2010 Mechanical performance of oil palm empty fruit bunches/jute fibres reinforced epoxy hybrid composites. *Mater. Sci. Eng., A* **527** 7944-7949

[53] Azwa Z N, Yousif B F, Manalo A C, and Karunasena W 2013 A review on the degradability of polymeric composites based on natural fibres *Mater. Design* **47** 424-442

[54] Chandrasekaran M and Bahkali A H 2013 Valorization of date palm (Phoenix dactylifera) fruit processing by-products and wastes using bioprocess technology–Review *Saudi journal of biological sciences 20* 105-120

[55] Jaradat A A and Zaid A 2004 Quality traits of date palm fruits in a center of origin and center of diversity *Journal of Food Agriculture and Environment* **2** 208-217

[56] Benkhalifa A 1999 Gestion de la diversité génétique du palmier dattier en Algérie. In *workshop constitution et organisation d'équipes de recherche scientifique dans les domaines de foresterie et des arbres fruitiers, Marrakech, Morocco* 13-15

[57] Dakheel A 2003 Date palm and biosaline agriculture in the United Arab Emirates. *The date palm from traditional resource to green wealth. Emirates Centre for Strategic Studies and Research, Abu Dhabi* 199-211

[58] Salem A O M, Rhouma S, Zehdi S, Marrakchi M, and Trifi M 2008 Morphological variability of Mauritanian date-palm (Phoenix dactylifera L.) cultivars as revealed by vegetative traits *Acta Botanica Croatica* **67** 81-90

[59] Foulk J A, Akin D E, Dodd R B, and McAlister III D D 2002 Flax fiber: potential for a new crop in the Southeast *Trends in new crops and new uses* 361-370

[60] Mwaikambo L 2006 Review of the history, properties and application of plant fibres *African Int J Sci Technol* **7** 121

[61] Rout J, Misra M, Tripathy S S, Nayak S K, and Mohanty A K 2002 Surface modification of coir fibers. II. Cu (II)-IO– 4 initiated graft copolymerization of acrylonitrile onto chemically modified coir fibers *J Appl Polym Sci* **84** 75-82

[62] Abdal-Hay A, Suardana N P G, Choi K S, and Lim J K 2012 Effect of diameters and alkali treatment on the tensile properties of date palm fiber reinforced epoxy composites *Int J Precis Eng Manuf* **13** 1199-1206

[63] Agoudjil B, Benchabane A, Boudenne A, Ibos L, and Fois M 2011 Renewable materials to reduce building heat loss: Characterization of date palm wood *Energy Build.* **43** 491-497

[64] Kaddami, H., Dufresne, A., Khelifi, B., Bendahou, A., Taourirte, M., Raihane, M., Issartel, N., Sautereau H, Gerard J F, and Sami N 2006 Short palm tree fibers–Thermoset matrices composites. *Composites Part A* **37** 1413-1422

[65] Rao K M M, Prasad A R, Babu M R, Rao K M, and Gupta A V S S K S 2007 Tensile properties of elephant grass fiber reinforced polyester composites *J. Mater. Sci.* **42** 3266-3272

[66] Mirmehdi S M, Zeinaly F, and Dabbagh F 2014 Date palm wood flour as filler of linear low-density polyethylene *Composites Part B* **56** 137-141

[67] Sbiai A, Maazouz A, Fleury E, Souterneau H, and Kaddami H 2010 Short date palm tree fibers/polyepoxy composites prepared using RTM process: effect of tempo mediated oxidation of the fibers *BioResources* **5** 672-689

[68] Al-Khanbashi A, Al-Kaabi K, and Hammami A 2005 Date palm fibers as polymeric matrix reinforcement: fiber characterization *Polym. Compos.* **26** 486-497

[69] John M J and Anandjiwala R D 2008 Recent developments in chemical modification and characterization of natural fiber-reinforced composites *Polym. Compos.* **29** 187-207

[70] Wong K J, Yousif B F, and Low K O 2010 The effects of alkali treatment on the interfacial adhesion of bamboo fibres. *Proceedings of the Institution of Mechanical Engineers, Part L: Journal of Materials: Design and Applications* **224** 139-148.

[71] Alawar A, Hamed AM, and Al-Kaabi K 2009 Characterization of treated date palm tree fiber as composite reinforcement *Composites Part B* **40** 601-606

[72] Hassan M L, Bras J, Hassan E A, Silard C, and Mauret E 2014 Enzyme-assisted isolation of microfibrillated cellulose from date palm fruit stalks *Ind. Crops Prod.* **55** 102-108

[73] Shalwan A and Yousif BF 2014 Investigation on interfacial adhesion of date palm/epoxy using fragmentation technique *Mater. Design* **53** 928-937

[74] Ibrahim H, Farag M, Megahed H, and Mehanny S 2014 Characteristics of starch-based biodegradable composites reinforced with date palm and flax fibers *Carbohydr Polym* **101** 11-19

[75] Al-Kaabi K, Al-Khanbashi A, and Hammami A 2005 Date palm fibers as polymeric matrix reinforcement: DPF/polyester composite properties *Polym. Compos.* **26** 604-613

[76] Rafeeq S N, Abdulmajeed I M, and Saeed A R 2013 Mechanical and Thermal Properties of Date Palm Fiber and Coconut Shell Particulate Filler Reinforced Epoxy Composite. *Indian Journal of Applied Research* **3** 89-92

[77] Chao CT and Krueger R R 2007 The date palm (Phoenix dactylifera L.): overview of biology, uses, and cultivation *HortScience* **42** 1077-1082

IOP Conf. Series: Materials Science and Engineering **368** (2018) 012010 doi:10.1088/1757-899X/368/1/012010

Effect of pretreatment concentration on pulp blending between oil palm empty fruit bunch and citronella leaf fibers in terms of pulp and paper properties

M Y Nur Aqeela [1], Z M A Ainun [1*] and M Jawaid [2*]

[1]Program of Pulp and Paper & Pollution Control, Laboratory of Biopolymer and Derivatives, Institute of Tropical Forestry and Forest Products, Universiti Putra Malaysia, 43400 UPM Serdang, Serdang, Selangor, Malaysia

[2]Laboratory of Biocomposite Technology, Institute of Tropical Forestry and Forest Products, Universiti Putra Malaysia, 43400 UPM Serdang, Selangor, Malaysia

*Corresponding authors: ainunzuriyati@upm.edu.my; jawaid@upm.edu.my

Abstract. Papers are one of the most used materials in food packaging industries. Additives need to be added in order to improve and prolong sensory properties of food. This is to maintain the food quality level before reaching to the end users. A preliminary study has been conducted to produce antimicrobial paper by blending the pulps of Citronella (*Cymbopogon nardus)* with oil palm empty fruit bunch (EFB). The papermaking was done according to TAPPI Standard Method. Morphological observation using Scanning Electron Microscope was carried out for all samples while physical, optical and mechanical paper properties were also determined. A simple antimicrobial test was also done by using 3 types of fungus namely as *Aspergillus brasiliensis* ATCC 16404, Gram positive *Escherichia coli* ATCC 25922 and *Salmonella choleraesuis* (subtype enterica). The morphological observation showed the difference between EFB paper and paper made of blended pulps. Citronella was found to affect the physical, optical and mechanical paper properties as expected. To be specific, the paper opacity was improved while the mechanical properties especially tensile and tear indices were dropped tremendously. However, the presence of Citronella in the paper sheet seems to have good potential as an antimicrobial agent as shown by the antimicrobial test

1. Introduction

Oil palm has tremendous characteristic that it is commonly recognized as "tree of life". This is due to all part such as fruits, trunks and leaves can be effectively utilized for living. Oil palm empty fruit bunches (EFB) are residues obtained from oil palm plantation area which increasing around 25 million tons by years [1]. The expansion of palm oil industry is expected to reach 80 million tons by 2020. Hence, the abundance of oil palm residues would be increasing especially EFB which is considered as renewable and readily available lignocellulosic materials [2]. EFB is characterized as lignocellulosic material which consists of long fibers in core and rich of silica [3]. It has been applied in many research that producing pulp and paper with high potential of being renewable natural resources in related industry. These fibers which can be transformed into commercial serviettes [4] and corrugated medium board [5] are proper alternative to reduce the dependency on the imported recycled fibers.

Cymbopogon nardus is a scientific name for citronella leaves which carries pleasant aroma and positive psychological effects on human beings [6]. The leaves are inedible which leads to huge abundance of crops. It is originated in tropical Asia countries such as Malaysia, Thailand, Indonesia, Singapore and India which are economically and physio-geographically rich with biodiversity and seasonal weather [7]. The extracted oil of *Cymbopogon nardus* can be used in some external applications such as, insect repellents [8], pesticide [9], fungicide and bactericide. Moreover, it has huge contribution in economic sector and medical industry which can be applied to synthesize a number of useful aromatic compounds and vitamin A [10]. Besides, *Cymbopogon nardus* is also utilized by soaps, detergent, cosmetics and food flavouring agent producers. In 2015, *Cymbopogon nardus* pulp was successfully prepared by using kraft and soda pulping methods [11]. Therefore, in this study *Cymbopogon nardus* was used by blending it with EFB pulps in order to produce potential antimicrobial paper for food packaging purpose.

Pulping is a crucial and important process in pulp and papermaking technology. It can be carried out by using 3 categories of pulping like mechanical, chemical, and thermal pulping or mixed of these 3 pulping. Acceptable pulps for papermaking are characterized by parameters such as bulky thickness, opacity, and light scattering coefficient [12]. Generally, pulps that are produced through mechanical pulping may obtain good paper opacity property [13]. On the other hand, paper which produced from chemical pulping normally has higher mechanical paper properties [14]. In most cases, pulp was produced mechanically due to lower cost and higher pulp yield factors [4]. Although the mechanical pulp yield was higher compared to chemical pulp, the amount of single fibers would be lesser because an amount of fibers still tend to be in bundle forms.

Mechanical pulp may form high surface roughness and blending paper will have low optical properties paper which can be improved by increasing the number of refining cycles and applying pre-treatment process [16-17]. Zhai et al. (2014) stated that pretreatment which was carried out prior to mechanical pulping process increased porosity, swelling, fibrillation, carbohydrate recovery and reduced lignin content in the papers [18]. In 2016, Johansson et. al, indicated that applying mechanical pulping would be economical and inexpensive method in pulp for paper [19]. Mechanical pulping offers good opacity and high yield of pulps compared to chemical pulping which lead to high profit value obtained by producer. In addition, mechanical pulping requires less chemical consumption compared to chemical pulping. Therefore, this pulping process is known as more eco-friendly production that contributes in forming biodegradable paper and reducing global pollution scale.

Many experiments showed that the lemongrass (*Cymbopogon citratus*) essential oil inhibited the mycelial growth of some pathogens: *Collectotrichum gloeosporioides* [20], *Fusarium verticillioides, Fusarium proliferatum, Fusarium graminearum* [21] and *Fusarium moniliforme* [22]. In 2011, Istianto and Emilda found out that essential oil of citronella grass (*Cymbopogon nardus*) could completely inhibit the mycelial growth of *Fusarium oxysporum* [23], which supported by Samerjai et al. [24]. Based on previous research, *Cymbopogon* family such as *Cymbopogon nardus* has well anti-microorganisms property and can be used in papermaking system. Therefore, this study was carried out in order to utilize the EFB pulp with Citronella pulp to provide a special paper with antimicrobial properties which is great beneficial for packaging paper industries.

2. Materials and Methods

2.1. Raw materials
Oil palm empty fruit bunch fiber was received from a local mill at Pahang in bale form while citronella leaves was harvested from Faculty of Agriculture, Universiti Putra Malaysia. The citronella leaves arrived in fresh forms stacked together in bundles which later dried until moisture content reached 5 % prior to pulp and paper making. Sodium hydroxide was purchased from R&M Chemicals.

2.2. Preparation, Testing and Blending of Pulps
Oil palm empty fruit bunch fibers were cut into 3-4 cm length prior to pre-treatment process. The pre-treatment was carried out by immersing the fibers in sodium hydroxide solution at 1 %, 2 %, 3 %, 4 % and 5 % of concentration. The process was held for 24 hours. The soaked fibers were then washed using tap water until the soapy condition getting lesser for about 30 mins. These washed fibers were then

refined using Refiner Mechanical Pulper (RMP) machine. The refined pulps were screened by using Somerville Screener.

On the other hand, Citronella leaves were cut into 2 to 3 cm prior to refining process using RMP machine. These refined pulps were beaten at a range of beating levels: 0, 1000, 3000 and 3500 revolutions using PFI Beater Mill. These beaten pulps were then tested their pulp freeness using Canadian Standard Freeness apparatus.

The pulp blending was carried out by applying a ratio of 20:80, Citronella on EFB pulps. The papermaking was carried out by using Handsheet Machine based on TAPPI Test Methods,1992 [25].

2.3. Characterization of paper

The characterization of paper was divided into three types namely as physical, optical and mechanical paper testing. The physical tests selected in this study were grammage (TAPPI Standard T410 om-08) [26] and bulk thickness (TAPPI Standard T411 om-97) [27] while optical paper properties consisted of paper opacity (TAPPI Standard T425 om-01) [28] and brightness (TAPPI Standard T 452 om-02) [29]. The mechanical properties were tensile (TAPPI Standard T404 cm-02) [30] and tear strength test (TAPPI Standard T414 om-98) [31].

2.4. Morphological observation

Morphological observation was carried out under Scanning Electron Microscope (SEM) JSM-7900 F, Schottky Field Emission. The samples were coated with gold by using Sputter Coater devise in order to avoid the sample surfaces from creating charging effects during scanning. The observation was done to observe the surface views of each fibers and papers. The magnification levels which were utilized during scanning were 50 to 1000 magnifications.

2.5. Antimicrobial test

Pathogenic microorganisms were obtained from the Institut Biosains, Universiti Putra Malaysia. The determination of antimicrobial properties of blended paper was carried out using two Gram-negative strains, *Escherichia coli* ATCC 25922 and *Salmonella choleraesuis*, and one filamentous fungi strain *Aspergillus brasiliensis* ATCC 16404. According to the Manual of Antimicrobial Susceptibility testing, the culture used for the antimicrobial test Gram-negative bacteria was Mueller Hinton broth (MHB), and Mueller Hinton Agar (MHA) fungus strains. Meanwhile, an inhibition growth assay using the technique of agar diffusion was applied to the plates containing MHA medium which were inoculated with Gram-negative strains, *Escherichia coli* ATCC 25922, *Salmonella choleraesuis* and one filamentous fungi strain *Aspergillus brasiliensis* ATCC 16404. The agar was drilled to form holes of about 0.1 cm, and then 10 g of raw fibers of EFB and citronella leaves was added into each hole separately. The incubation time for both *Escherichia coli* ATCC 25922, and *Salmonella choleraesuis* was 5 days. On the other hand, *Aspergillus brasiliensis* ATCC 16404 took 7 days. The antimicrobial activity was tested by the formation of a halo showing the inhibition zone.

3. Results and Discussions

3.1. Effect of sodium hydroxide concentration on EFB pulp properties

The pulp yield and freeness results were shown as in Table 1. Pretreating the EFB fibers with 5 % sodium hydroxide could produce the highest percentage of pulp yield. An increment of 15.87 % pulp yield was obtained compared to non-treated EFB fibers. The results approved that sodium hydroxide would increase the lignocellulose biomass pulp yield [32]. Besides that, pulp freeness results a decline pattern. Higher concentration of sodium hydroxide during pre-treatment process had successfully dropped the reading of pulp freeness about 60.53 %. This huge decrement showed that sodium hydroxide provides great effect in helping the defibrillation of fibers during refining process [34]. The fibers absorbed acceptable amount of water by draining 300-700 ml of water that is economical for papermaking process.

IOP Conf. Series: Materials Science and Engineering **368** (2018) 012010 doi:10.1088/1757-899X/368/1/012010

Table 1. Pulp yield and freeness of untreated and treated EFB pulps.

Concentration of sodium hydroxide, %	**0**	**1**	**2**	**3**	**4**	**5**
Pulp Yield, %	63	67	66	65	68	73
Canadian Standard Freeness, ml	760	520	450	400	360	300

3.2. Effect of beating level on citronella pulp property

The pulp freeness of Citronella after a series of beating level was presented in Table 2. The results proved that the Citronella leaves were more firm compared to EFB fibers. The pulp freeness of the Citronella showed 510 ml CSF reading after undergone 3,500 revolutions of beating level. A decrement of 37.42 % in terms of pulp freeness was achieved compared to the earliest sample without any beating. In every 500 revolutions of beating level, an amount of 50 ml pulp freeness value decreased. Kamoga (2015) reported that Citronella mechanical fibers which have been beaten until 1000 PFI revolutions recorded pulp freeness as high as 740 ml [36]. This showed that the fibers have less tendency to absorb water which may be related to less fibrillation occurred on the fiber surfaces. In other words, the citronella fibers were quite rigid and less flexible. Beating the pulp fibers using a laboratory blender can improve the structural or physical properties of paper [37].

Table 2. Pulp freeness of Citronella without and with a series of beating level.

	Beating level (revolutions)				
	0	1000	2000	3000	3500
Canadian Standard Freeness, ml	815	740	630	520	510

3.3. Effect of blending on the physical, optical and mechanical paper properties

Table 3 exhibited the results for physical, optical and mechanical paper properties which were made from EFB and blended pulps. The results of physical paper properties were divided into grammage, apparent density and bulk thickness. There was a decrement of 9.95 % in grammage which can be related to the less incorporation of citronella which might having fewer bonding with EFB pulps. As stated in earlier explanation, the citronella pulps were quite rigid and firm which difficult the bonding process. Hence in 1 meter square of paper, the weight of fibers in blended paper was lighter than in EFB paper due to pretreatment process that was applied onto EFB pulps [38] and improved its mechanical properties. In conjunction with grammage results, paper apparent density was found to have moderate difference between EFB and blended paper. There was slightly different of paper thickness between paper made from EFB and blended pulps. The thickness for both papers was 412.3 μm and 912.4 μm respectively. The citronella might be having coarser size of pulps which need wider areas to attach among the pulps due to free treatment onto pulps unlike the alkaline pretreated pulp [39].

The opacity of paper made from blended pulps was found out to be better than EFB paper which may be due to the presence of citronella pulps which filled the voids between fiber to fiber [40]. However, the brightness decreased 3.21% after adding the citronella. Different materials in terms of pulps will provide significant effect to the formation of paper included paper brightness.

The tensile strength of blended paper reduced tremendously while tear strength dropped 2.7 units. This decrement was expected due to the presence of citronella leaves which may have less quality bonding among each other. As stated in Table 2, the pulp freeness was considered as high that indicates the rigidity of fibers in absorbing water. Hence, this will affect the chances in bonding among the fibers and also with EFB fibers. Mechanical strength is affected by amount of fibers on one square meter paper [41]. As explained earlier, the grammage of blended paper is lesser than EFB paper. Therefore, it is proved that blended paper will obtain lower value of tear strength compared to EFB paper alone as shown in Table 3.

Table 3. Physical, optical and mechanical properties of EFB and blended paper.

Property	EFB paper	Blending paper
Physical properties		
Grammage (g/m^2)	36.38	32.76
Apparent density (g/cm^3)	0.243	0.223
Thickness (µm)	412.3	912.4
Optical properties		
Opacity (%)	44.60	50.24
Brightness (%)	36.09	32.88
Mechanical properties		
Tensile index (KNm/g)	28.58	-
Tear index (mN.m^2/g)	10.95	8.25

3.4. Effect of blending on paper morphological property

The cross cut views that blended paper was thicker than EFB paper as shown in Figure 1(a) and (b). As shown in Table 3, the thickness of blended paper was 2.21times than the EFB paper itself. Figure 1(c) and (d) presented that blended paper has less uniform focally on fiber shapes and sizes than EFB paper. This is because of the presence of citronella pulps in the paper systems.

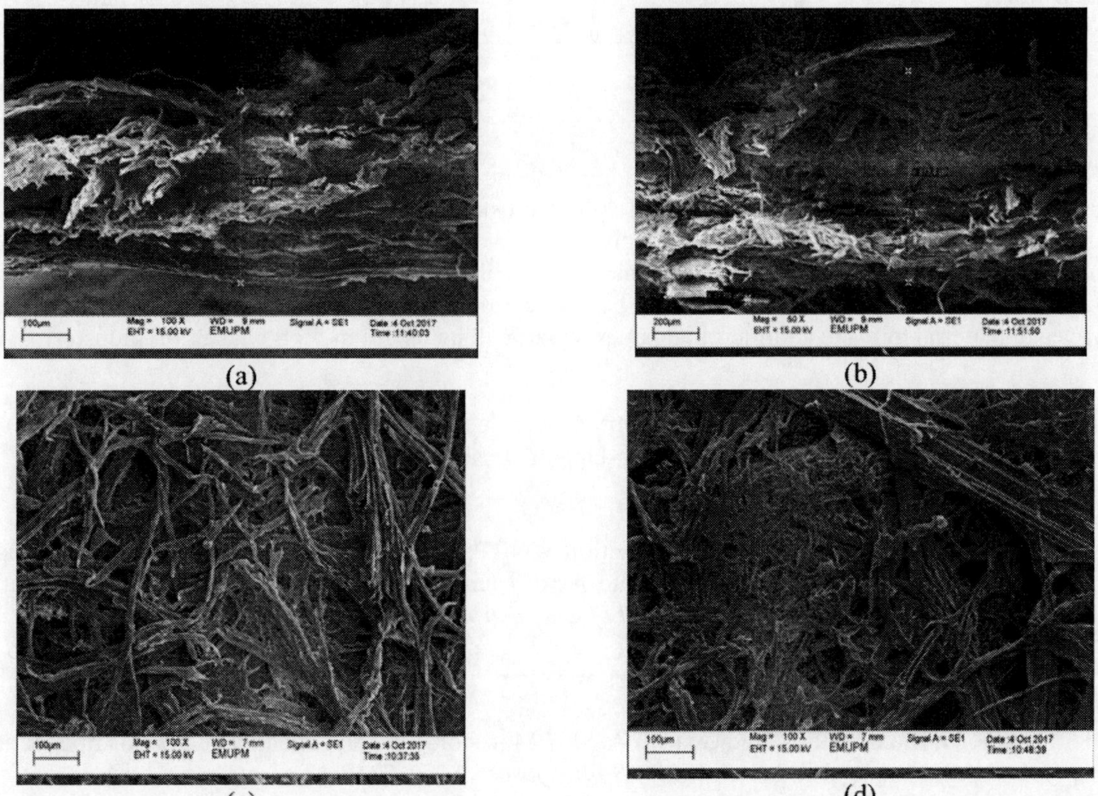

Figure 1. SEM observation of (a) cross cut of EFB paper at 100X magnification (b) cross cut of blended paper at 50X magnification (c) surface of EFB paper at 100X magnification, and (d) surface of blended paper at 100X magnification.

3.5. Effect of pulp blending on paper antimicrobial property

EFB fiber and EFB paper were investigated having low antimicrobial defense against *Aspergillus brasiliensis* ATCC 16404 but better resistant onto Escherichia *coli* ATCC 25922 and *Salmonella choleraesuis*. However, blended paper has low antimicrobial defense against Escherichia *coli* ATCC 25922 and *Salmonella choleraesuis* but vice versa onto *Aspergillus brasiliensis* ATCC 16404. Meanwhile, citronella fibers from raw materials have low stage of antimicrobial defences onto *Aspergillus brasiliensis* ATCC 16404, *Escherichia coli* ATCC 25922 and *Salmonella choleraesuis* but changed after blended with EFB pulp. The efficacy of the antimicrobial agent will depend on water activity of the sample [45]. These results are very important because they showed that blended paper had high potential acting as antimicrobial paper as long as improvement in ratio portion or particle size of citronella is revised The citronella fiber is showing antimicrobial agent property [42; 43]. In 2015, Nassar applied Escherichia *coli* ATCC 25922 on blended paper between lignocellulosic materials with recycled old newspaper having similar results [44]. Therefore, the blended paper needs to be improvised later on regarding on its ratio portion of blending and technique of applying the anti-microbial agent onto the paper.

Table 4. Anti-Microbial Test on Fungus and Gram Negative Strains

Type of sample	Fungus	Gram negative strains	
	Aspergillus brasiliensis ATCC 16404	*Escherichia coli* ATCC 25922	*Salmonella choleraesuis*
EFB paper	L	M	M
Blended paper	M	L	L
EFB fiber	L	M	M
Citronella fiber	L	L	L

L: low anti-microbial activity
M: medium of anti-microbial activity

4. Conclusion

Preliminary results showed that paper produced from citronella and EFB has high chances to be developed as antimicrobial paper provided some adjustments to the formulation can be revised and improvised. The pretreatment of EFB fiber gives impact onto pulp yield and permeability of water onto pulp produced as well as beating pulps by using PFI Mills. Blended paper of efb and citronella shows effectiveness against fungal attack and can be revised to improve the best formula for blending paper for waste material to make antimicrobial paper which will be useful in food packaging industry.

Acknowledgements
The authors are grateful to Malaysian Industry-Government Group for High Technology(MIGHT) for financial support of this work Under Newton-Ungku Omar Fund Grant No: 6300873

References

[1] Sudiyani Y and Hermiati E 2010 Utilization of Oil Palm Empty Fruit Bunch (EFB) For Bioethanol Production Through Alkali And Dilute Acid Pretreatment And Simultaneous Saccharification And Fermentation *Indonesian Journal of Chemistry* **10** 261-7

[2] Sharma A K, Anupam K, Swaroop V, Lal P S, and Bist V 2015 Pilot scale soda-anthraquinone pulping of palm oil empty fruit bunches and elemental chlorine free bleaching of resulting pulp *J Cleaner Prod* **106** 422-9

[3] Law K N, Daud W R W, and Ghazali A 2007 Morphological and chemical nature of fiber strands of oil palm empty-fruit-bunch (EFB) *BioResources* **2** 351-62

[4] Daud W R W, Wahid K A, and Law K N 2013 Cold soda pulping of oil palm empty fruit bunch (EFB) *BioResources* **8** 6151-60

[5] Khairul H A B, Rohaizu R, and Wan Rosli W D 2017 The Use of Oil Palm Empty Fruit Bunches (EFB) Fibers as Partial Replacement for Imported Recycled Fibers *Key Eng Mater* **737**

[6] Alka G and Reena 2016 Formation and application of citronella oil and lemon grass finish on Rambouillet wool and assessment of their physical properties and durability *International Journal of Basic and Applied Agricultural Research* **14** 88-93

[7] Bishoyi A K, Sharma A, Kavane A and Geetha K A 2016 Varietal Discrimination and Genetic Variability Analysis of Cymbopogon *Appl. Biochem. Biotechnol.* **179** 659-70

[8] Das K., Vasudeva C and Dang R 2016 Economical novel formulation and evaluation of herbal oils for mosquito and house fly repellent activities *Annals of Phytomedicine-An International Journal* **5** 91-6

[9] Ríos N, Stashenko E E, and Duque J E 2017 Evaluation of the insecticidal activity of essential oils and their mixtures against Aedes aegypti (Diptera: Culicidae) *Revista Brasileira de Entomologia*

[10] Weng D C J, Latip J, Hasbullah S A and Sastrohamidjoj H 2015 Optimal Extraction and Evaluation on the Oil Content of Citronella Oil Extracted From Cymbopogon Nardus *Malaysian Journal of Analytical Sciences* **19** 71-6

[11] Kamoga O L, Kirabira J B and Byaruhanga J K 2015 The potential of Cymbopogon nardus in the production of pulp for paper industry *In International Conference on Computing, Mechanical and Electronics Engineering* 9-10

[12] Dafaalla S 2015 Environmentally Friendly Pulping and Bleaching of Some Sudanese Acacias and Eucalyptuses (Doctoral dissertation, UOFK)

[13] Hosseini S E, and Abdul Wahid M 2014 The role of renewable and sustainable energy in the energy mix of Malaysia: a review *Int J Energy Res* **38** 1769-92

[14] Sixta H Ed 2006 *Handbook of pulp* (Vol. 1, pp. 343-5) Weinheim: Wiley-vch

[15] Daud W R W, Wahid K A, and Law K N 2013 Cold soda pulping of oil palm empty fruit bunch (EFB) *BioResources* **8**(4) 6151-60

[16] Harinath E, Biegler L T and Dumont G A 2013 Predictive optimal control for thermo-mechanical pulping processes with multi-stage low consistency refining *J. Process Control* **23** 1001-11

[17] Fatah I Y A, Khalil H P S, Hossain M S, Aziz A A, Davoudpour Y, Dungani R, and Bhat A 2014 Exploration of a chemo-mechanical technique for the isolation of nanofibrillated cellulosic fiber from oil palm empty fruit bunch as a reinforcing agent in composites materials *Polymers* **6** 2611-24

[18] Zhai R and Zhou X 2014 Enhanced effect of NaOH/thiourea/urea aqueous solution on paper strength of high yield pulp *BioResources* **9** 2154-66

[19] Johansson O, Fernando D, Fenitsius R, Daniel G and Ferritsius O 2016 Advancements in optical analysis yields new insight to mechanical pulping processes in an efficient and inexpensive way In *International Mechanical Pulping Conference 2016, IMPC 2016; Jacksonville United States 26 September 2016 through 28 September 2016* pp. 267-76 TAPPI Press

[20] Palhano F L, Vilches T T, Santos R B, Orlando M T, Ventura J A and Fernandes P M 2004 Inactivation of Colletotrichum gloeosporioides spores by high hydrostatic pressure combined with citral or lemongrass essential oil *Int J Food Microbiol* **95** 61-6

[21] Velluti A, Sanchis V, Ramos A J, Turon C and Marin S 2004 Impact of essential oils on growth rate, zearalenone and deoxynivalenol production by Fusarium graminearum under different temperature and water activity conditions in maize grain. *Journal of applied microbiology* **96** 716-24

[22] Nwachukwu E O and Umechuruba C I 2001 Antifungal activities of some leaf extracts on seed-borne fungi of African yam bean seeds, seed germination and seedling emergence *Journal of Applied Sciences and Environmental Management* **5**

[23] Istianto M and Emilda D 2011 Preliminary study of the activity of some essential oils against Fusarium oxysporum f. sp. cubense *Journal of Fruit and Ornamental Plant Research* **19** 111-21

[24] Samerjai C, Sanit S, Sukontason K, Morakote N, Wannasan A, Pereira R M and Sukontason K L 2016 Morphology of immature stages of flesh flies, Boettcherisca nathani and Lioproctia pattoni (Diptera: Sarcophagidae) *Acta Trop 163* 109-120

[25] TAPPI 1992 Technical Association of Pulp and Paper Industries Test Methods 1992-1993 TAPPI Press, Atlanta, Georgia

[26] TAPPI 2013 Grammage of paper and paperboard (weight per unit area) TAPPI T410 Om-08 1–14
[27] TAPPI 1997 Thickness (caliper) of paper, paperboard and combined board TAPPI T411 Om-97 1–4
[28] TAPPI 2001 Opacity of Paper TAPPI T425 Om-01 9–13
[29] TAPPI 2002 Standard Test Method for Brightness of Pulp, Paper, and Paperboard (Directional Reflectance at 457 nm) TAPPI T425 Om-02 15(Reapproved) 1–6
[30] TAPPI 1992 Tensile breaking strength and elongation of paper and paperboard (using pendulum-type tester) TAPPI T404 Cm-92 3–7
[31] TAPPI 1998 Internal tearing resistance of paper (Elmendorf-type method)
[32] Park J, Jones B, Koo B, Chen X, Tucker M, Yu J H and Park S 2016 Use of mechanical refining to improve the production of low-cost sugars from lignocellulosic biomass *Bioresource Technol* **199** 59-67
[33] Ghazali A, Wan rosli W D and Law K N K 2006 Alkaline Peroxide Mechanical Pulping (APMP) of Oil Palm Lignocellulosics: Part 2-Empty Fruit Bunch (EFB) Responses to Pretreatments *Appita Journal: Journal of the Technical Association of the Australian and New Zealand Pulp and Paper Industry* **59** 65
[34] Chen J 2017 *Mechanical Pretreatment of Lignocellulosic Biomass to Increase the Yield of Enzymatic Hydrolysis* (Doctoral dissertation, École Polytechnique de Montréal)
[35] Kamoga O L, Kirabira J B and Byaruhanga J K 2015 The potential of Cymbopogon nardus in the production of pulp for paper industry In International Conference on Computing, Mechanical and Electronics Engineering Singapore 9-10
[36] Noah A S, Alao O J, Adeniyi I M, Folahan A A and Ayodele E B 2014 Effect of beating on pulp fibre characteristics of empty fruit bunches (EFB) of Elaeis guineensis Jacq. (oil palm) in paper making Advances in Agriculture, Sciences and Engineering Research 4(6) 1651-58
[37] Razak A, Wahida N and Kalam A 2012 Effect of Fibre Size on the Tensile Properties of Oil Palm Empty Fruit Bunch Fibre Composites In *Advanced Materials Research* **576** 280-283
[38] Sharma N, Godiyal R D, Thapliyal B P and Anupam K 2017 Pulping and Bleaching of Hydro Distillation Waste of Citronella Grass (Cymbopogon winterianus Jowitt) for Papermaking *Waste Biomass Valorization* 1-11
[39] Rafidah D, Ainun Z M A, Hazwani H A, Rushdan I, Luqman C A, Sharmiza A and Jalaluddin H 2017 Characterisation of Pulp and Paper Manufactured from Oil Palm Empty Fruit Bunches and Kenaf Fibres *Pertanika J. Trop. Agric. Sci.* **40**
[40] Chauhan V S, Bhardwaj N K and Chakrabarti S K 2012 Effect of Particle Size of Magnesium Silicate Filler on Physical Properties of Paper *Canadian Journal of Chemical Engineering* **91** 855-61
[41] Sharip N S, Ariffin H, Hassan M A, Nishida H and Shirai Y 2016 Characterization and application of bioactive compounds in oil palm mesocarp fiber superheated steam condensate as an antifungal agent. *RSC Advances* **6** 84672-83
[42] Oussalah M, Caillet S, Saucier L and Lacroix M 2007 Inhibitory effects of selected plant essential oils on the growth of four pathogenic bacteria: E. coli O157: H7, Salmonella typhimurium, Staphylococcus aureus and Listeria monocytogenes *Food control* **18** 414-20
[43] Kumar A, Shukla R, Singh P and Dubey N K 2009 Biodeterioration of some herbal raw materials by storage fungi and aflatoxin and assessment of Cymbopogon flexuosus essential oil and its components as antifungal *Int Biodeterior Biodegradation* **63** 712-6
[44] Nassar M A, El Rahman, H M A and El Sahkawy M 2015 An optimum mixture of virgin rice straw pulp and recycled old newsprint (ONP) pulp and their antimicrobial activity *BioTechnology: An Indian Journal* ***11***

IOP Conf. Series: Materials Science and Engineering **368** (2018) 012011 doi:10.1088/1757-899X/368/1/012011

A study of mechanical and morphological properties of PLA based biocomposites prepared with EJO vegetable oil based plasticiser and kenaf fibres

S H Kamarudin[1,2*], **L C Abdullah**[1,3], **M M Aung**[4], **C T Ratnam**[5]

[1]Laboratory of Biocomposite Technology, Institute of Tropical Forestry and Forest Products, Universiti Putra Malaysia, 43400 UPM Serdang, Selangor
[2]School of Industrial Technology, Faculty of Applied Sciences, Universiti Teknologi MARA (UiTM) Shah Alam, Selangor
[3]Department of Chemical and Environmental Engineering, Faculty of Engineering, Universiti Putra Malaysia, 43400, Serdang, Selangor
[4]Radiation Processing Technology Division, Malaysian Nuclear Agency, 43000, Bangi, Selangor, Malaysia

*Corresponding author: sitihas_kam@yahoo.com

Abstract. A novel environmental friendly plasticized PLA kenaf biocomposites were fabricated by melt blending technique through a combination of epoxidized jatropha oil, one such non-edible vegetable oil material and renewable plasticizer. The mechanical and morphological properties of the composites were evaluated. The addition of EJO with 40 % kenaf fibre loading on PLA had improved the mechanical properties of the polymer composites. SEM micrographs clearly revealed the good physical interlocking for PLA and kenaf fibre at low loading levels (40 wt.%) while higher content (50 wt.%) of kenaf contributed to poor filler-matrix compatibility attributes that leads to filler-related failures which resulted voids. Alkalisation of kenaf fibre provides good interaction bonding between kenaf and PLA matrix which resulted on increased of value of mechanical properties of treated biocomposites compared with the untreated biocomposites.

1. Introduction

Recently, the growing environmental awareness together with the new rules and regulations have forced the industries to seek an alternative for environmentally and ecologically friendly materials for their products. The environmental impact caused by exploitation of plastics represent the largest proportion of domestic waste and contribute to the polymer pollution. Therefore, the focus on these day is developing environmentally friendly materials which are naturally degradable when disposed in the environment [1].

Biodegradable polymers have received much attention as they offer environmental benefits such as biodegradability, greenhouse gas emissions and renewability of the base material [2]. Poly(lactic) acid (PLA), a corn-based polymer is the most promising biodegradable polymer and have been introduced commercially in the market as an alternative to conventional synthetic polymers [3-4]. There are number of promising products in the market for this polymer, such as automotive parts from PLA since PLA based automotive parts emit less CO_2 compared to the other petroleum based-thermoplastics [5]. Other than that, PLA has high strength, thermal plasticity and biocompatibility [6]. This polymer provides good strength and easy processability in most equipment, however it is considered expensive and the physical properties such as brittleness limit the PLA polymer applications [7]. Hence, a way to improve the mechanical and thermal properties of PLA as well as to make it more cost effective material is by the addition of fibre or filler materials to the system [8].

Natural fibres such as flax [9-10], hemp [11], jute [12] or kenaf [13] have recently gained more attention in replacing synthetic fibres and been used for energy or biomaterials production for many years in order to replace or conserve fossil fuels. Moreover, the potential advantages of these natural fibres have been well documented such as low cost, good thermal properties, low density and non-abrasive processing [14]. Thus, it is crucial to utilize these natural fibres and incorporate them into biodegradable polymer such as PLA in order to improve the properties of PLA as well as effectively reduce the production cost by partially replacing the high price of PLA by low cost natural fibres without destroying the biodegradability of the polymer matrix.

Over the past few years, natural fibres reinforced polymer composites have attracted increasing attention and have been considered for various applications due to their lower cost, good mechanical properties and lightweight. The use of natural fibres offers the advantage of weight reduction (over 10-30%), good mechanical and manufacturing properties, high stability, no emission of toxic substances and overall reduction in cost. In the field of technical utilisation of natural fibres, kenaf reinforced composites are considered as one of the most important areas [2].

In the present study, kenaf fibre with 40 wt.% fibre content reinforced PLA composites were processed by melt blending and followed by hot and cold press method. However, the inferior fibre-matrix adhesion arises from the hydrophilic nature of the kenaf fibres have limited compatibility with the hydrophobic nature of PLA thermoplastic. The degree of bonding or adhesion between the fibre and the matrix at interface is extremely important as the modulus, strength and toughness of a composite relies on the chemical interactions or mechanical interlocking between the fibre and the matrix. Over the last decades, several approaches have been introduced towards enhancing interfacial adhesion including the use of MAPP and silane coupling agents [15], chemical modifications such as acid [16] and alkali [17]. Due to the nature of kenaf fibre, NaOH treatment was used to improve the interaction compatibility between PLA and kenaf fibre. The purpose of chemical modification of natural fibre is to improve interaction of fibre with polymer matrix. Several findings have revealed that the chemical treatments improve the properties of composites. By improving the fibre/matrix adhesion in this manner, the mechanical properties of the composite will be improved at the expense of a reduction in impact behavior [18].

Plasticiser produced from natural resources, are currently being employed in a biodegradable polymer, as a replacement for traditionally used phthalates in the plastic industry, which are characterised by their high toxicity [19]. Recent years, modified vegetable oils have gained a lot of interest as a renewable source of plasticiser for polymers (Robertson et al., 2010). Numerous research works have been reported on the use of epoxidized vegetable oil (EVO) as a plasticiser for the PLA composites [20-21]. There are few advantages on the use of epoxidized vegetable oil in polymer industry: biodegradable, environmentally friendly, renewable and possibly cheaper [22]. Furthermore, it has potential to be used as an additive for polymer industry that require properties, such as good lubricity, low vitality, high index viscosity and good solvency [23].

Research on composite materials based on polymers from vegetable oils has been developed and be used for industrial applications as well. The study and research in the literature dealing with the use of vegetable oil based resins in natural fibre reinforced composites are emerging from time to time [24-25]. Jatropha oil with higher content of monounsaturated oleic acid (C18:1) and polyunsaturated linoleic acid (C18:2) compared to palm oil and soybean oil which gives higher percentage of oxirane oxygen content (OOC) upon epoxidation process had been widely studied as plasticiser in PLA biocomposites. At least, to our knowledge, there is presently no published work on using EJO as plasticizer for PLA/natural fibre system. Hence, with the present work, it was aimed to introduce and propose the potential of EJO as a reactive plasticiser together with a combination of kenaf as natural fibre for a reinforcement to the PLA composite system.

2. Experimental

2.1 Material

Poly(lactic acid) (PLA) polymer resins of grade 2003 D was obtained from the Nature Works LLC Company from USA. Meanwhile, crude jatropha oil which is a non-food grade material was supplied by the Bionas Sdn Bhd, Kuala Lumpur, Malaysia and was used as received. Kenaf fibre in the size of 40 micrometer mesh was obtained from the National Kenaf Tobacco Board, Kelantan. Reagent grade hydrogen peroxide 30% and magnesium sulphate were supplied by Merck, Germany. Sodium hydroxide reagent grade and formic acid were supplied by Systerm.

2.2 Epoxidation of Jatropha Oil

The epoxidation of jatropha oil was carried out by "in-situ" epoxidation method as described by Goud et al., (2007) with a slight modification in the molar ratio of 1:0.6:1.7– organic acid: double bonds: hydrogen peroxide at temperature of 60 ^{0}C for 5 hours reaction time.

2.3 Fibre Treatment
Fibres were treated with alkali soaked in hot distilled water with continuous stirring at 60^{0}C for 4 hours to remove excess wax and other impurities Then, the treated fibre was washed with distilled water until reaching pH7 and dried at temperature 60 °C for 24 hours in the oven.

2.4 Preparation of PLA/Kenaf Treated/EJO (PLA/TK/EJO) Biocomposites
The PLA/KFT/EJO blends were prepared by the melt blending technique of PLA with 3 wt.% epoxidized jatropha oil and 40 wt.% ratio of treated kenaf fibre. The melt blending was carried out using a Brabender Internal Mixer (Germany) at 170 ^{0}C for 10 min and 50 rpm motor speed. After melt blending, the composites were removed and cooled at room temperature. The composite samples were then placed in the hot and cold press for a compression moulding with a pressure of 110 kg/cm at 170 ^{0}C for 11 min in order to produce biocomposite sheets of uniform 1mm and 3mm thickness. The biocomposites were then being analysed for their physical, chemical, mechanical and morphological properties characterizations. Prior to mixing process, PLA pellets and kenaf fibre were dried in oven at 60 °C for 24 hours to remove moisture content.

2.5 Characterisations
Fibres were treated with alkali soaked in hot distilled water with continuous stirring at 60^{0}C for 4 hours to remove excess wax and other impurities Then, the treated fibre was washed with distilled water until reaching pH7 and dried at temperature 60 °C for 24 hours in the oven.

2.5.1 *FTIR Analysis.* The structural changes of untreated and treated kenaf fibre were analysed using (Perkin-Elmer: Model 1000 Series) instrument equipped with a universal attenuated total reflectance (UATR) accessory. The spectra were recorded between 4000 cm^{-1} to 280 cm^{-1} wavenumber range.

2.5.2 Fibre Image Analysis. The untreated and treated fibres were mounted for longitudinal view onto a microscope slide. The diameter of the fibres were measured using an optical microscope model Olympus SZX12.

2.5.3 Tensile Properties. The tensile properties: tensile strength and tensile modulus were determined by Toyoseiki Universal Testing Machine (Model Strograph R1), with a grip attachment distance of 45 mm. Load of 1 kN was applied at constant cross head speed of 5 mm/min. The neat PLA and composites sheets were cut into dumbbell shapes prior to use in accordance to ASTM D638-10 standard. The width and thickness of the specimens were measured at three different points and the average was calculated. The resultant value for tensile properties of neat PLA and composites were taken from an average of seven samples. Error bar represents one standard deviation for all presented data.

2.5.4 *Surface Morphology Test.* The surface morphology of untreated and treated kenaf fibre, PLA/KF, PLA/KFT and PLA/KFT/EJO were examined using Scanning electron microscopy (SEM JEOL model JSM-6300F) with an accelerating voltage of 20 kV. Possible fibre pull out, fibre splitting, debonding, matrix cracking and fibre-matrix adhesion were analysed. The specimens were coated with gold prior to each test to avoid electrostatic charging during electron irradiation.

3. Results and Discussion

3.1 FTIR and morphology analysis of kenaf fibre and PLA/TK/EJO biocomposites
FTIR spectroscopy is a useful technique to study the functional groups of both untreated and treated kenaf fibre and to monitor the effect of fibre treatment and chemical changes of the lignocellulosic fibre before and after the treatment. Figure 1 shows the IR spectra for untreated and treated kenaf fibre. A broad peak can be observed at region around 3300 cm-1 indicating the presence of hydroxyl (OH) group in both but the intensity of the peak is lower for treated fibre due to the removal of hemicellulose and lignin. Alkali treatment reduces the hydrogen bonding due to the removal of the hydroxyl groups (-OH) by reacting with sodium hydroxide. The hydroxyl groups are also involved in hydrogen bonding with the carboxyl groups, perhaps also with the fatty acids that are available on the fibre surface of kenaf fibres. Thus, this was indicated by the reduction and broadening of the peaks around 3200 cm^{-1}. The absorbance peaks around 2900-2800 cm^{-1} were attributed

to stretching of C-H group. In addition, the vibration peak of C-H stretches at 2900 cm^{-1} in cellulose and hemicellulose presents in the untreated fibre becomes weak, indicating the part of the hemicellulose was removed [26].

A peak at 1729.83 cm^{-1} that present for untreated fibre disappeared for treated fibre. According to [27] and [28], the peak is corresponding to the C=O stretching of the acetyl group in hemicellulose. [29] assigned the peak as the ester linkage of carboxylic group in the ferulic and p-coumeric acid of lignin and/or hemicellulose. The disappearance of this peak for treated fibre confirmed the removal of lignin and hemicellulose during mercerization process. Mercerization removes the waxy layer, adhesive pectins and hemicelluloses that bind fibre bundles to each other and to the pectin and hemicellulose rich sheats of the core [30].

The intensity for peak at 1506.13 cm^{-1} for untreated fibre is decreasing after alkaline treatment process. The peak is associated to the C=C aromatic symmetrical stretching for lignin [31]. The absorbance between 1380-1320 cm^{-1} for both untreated and treated fibre were related to the bending vibration of C-H and C-O groups of the aromatic ring in polysaccharides. A peak at absorbance 1230.36 cm^{-1} for untreated fibre was found to be shifted to 1319.07 cm^{-1} after treatment process and the intensity of the peak also decreasing. According to [32], the peak is attributed to the aryl group in lignin, thus the change observed due to the removal of lignin after fibre treatment. Finally, the intense peaks around 1070-1010 cm^{-1} region were due to stretching of C-O and O-H groups [33].

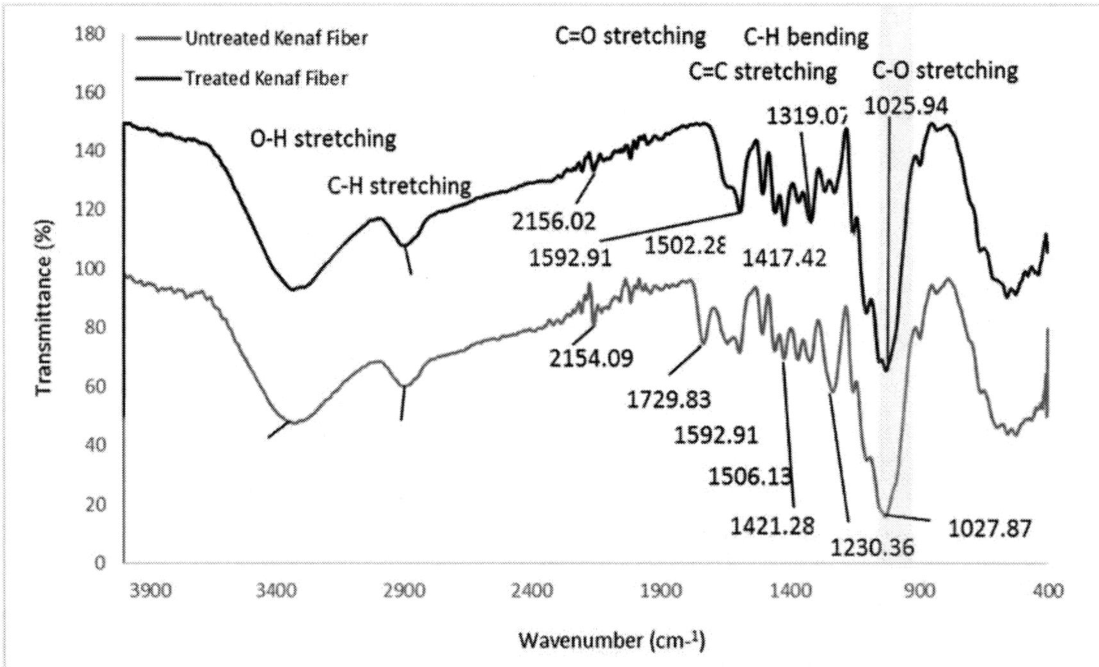

Figure 1. FTIR spectra of untreated and treated kenaf fibre.

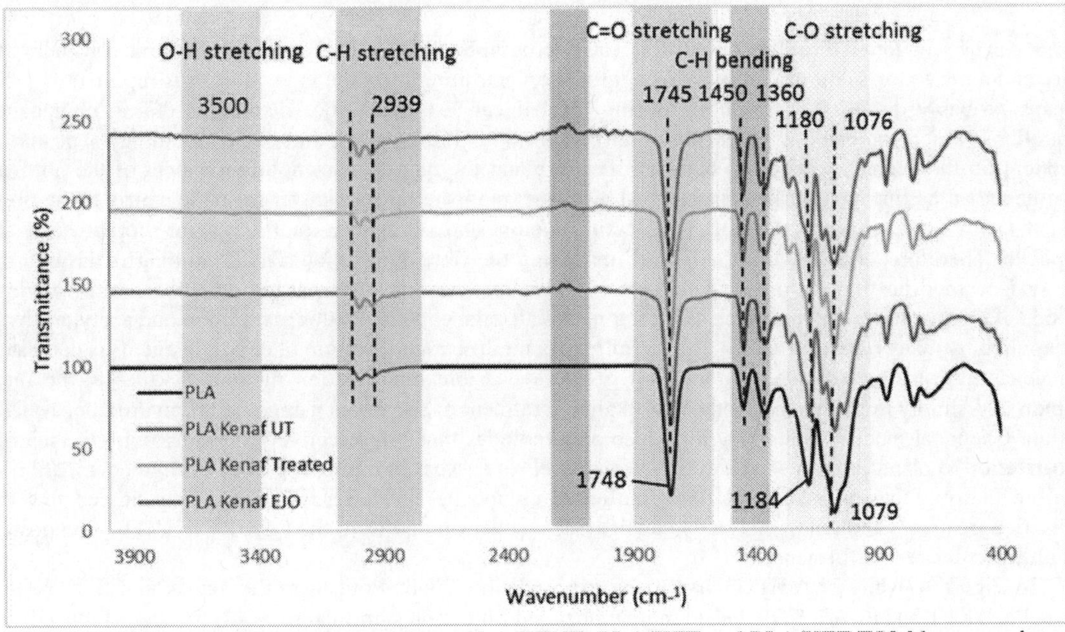

Figure 2. FTIR spectra of PLA, PLA/KFUT, PLA/KFT and PLA/KFT/EJO biocomposites.

Figure 2 displays the FTIR spectrum of PLA, PLA/Kenaf (Untreated), PLA/Kenaf (Treated) and PLA/Kenaf (Treated) /EJO samples. Biocomposite sample is assumed to have a combination of PLA and fibre absorbance peak. However as for all three samples besides PLA spectrum, they had approximately similar wavelength as compared to PLA's except the difference in the peak intensity in the C=O stretching and C-O stretching for esters. Those two peaks has shifted towards lower temperature, from 1748 cm-^{1} (neat PLA) to 1745 cm-^{1} in C=O stretching region and from 1184 cm-^{1} (neat PLA) to 1180 cm-^{1} and from 1079 cm-^{1} (neat PLA) to 1076 cm-^{1} in C-O stretching region indicated the miscibility and interaction between PLA, Kenaf and EJO took place. Furthermore, the shift may possibly due to an interaction between the hydroxyl group of PLA and epoxy group of EJO through hydrogen bonding (H-bonding) interaction, resulting in enhanced morphological properties which can be verified through SEM analysis in following Figure (Figure 4). A proposed possible site for interaction between PLA and EJO is shown in Figure 3.

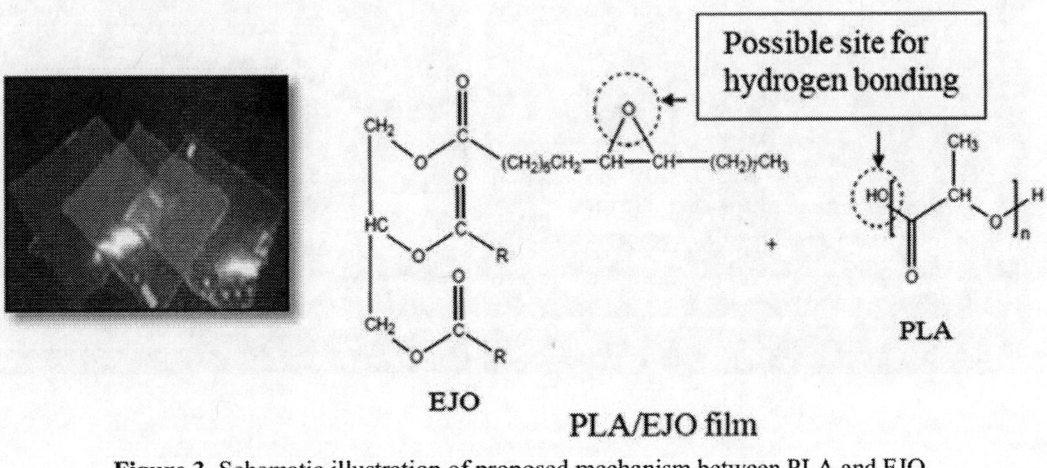

Figure 3. Schematic illustration of proposed mechanism between PLA and EJO.

In order to study the failure mechanism and to examine the possible fibre-matrix mechanism, the fracture surface of composites was studied using SEM [34]. Figure 4 shows the fractured surface micrographs of tensile specimen of neat PLA, PLA/KFUT, PLA/KFT and PLA/KFT/EJO samples respectively. It can be seen that the neat PLA presented an irregular and coarseness fracture surface, due to its highly brittle nature and failure occurs by micro-cracking where the cracks spread easily and freely. Based on the observation, a visible fibre pull-out can be seen for all three biocomposites. The micrograph of PLA/KFUT displays a notable gaps between a fibre and a matrix. In addition, presence of aggregation and fibre breakage is visible. It is reported by [35], the gaps probably due to the debonding during mechanical testing or poor dispersion during composites preparations which indicates a weak fibre-matrix adhesion. Fibre-matrix interfacial bonding is primarily dependent on the degree of adhesion between the fibre and the matrix. An important aspect of the ultimate performance of the final composite is influenced by the act of means stress transfer from the matrix to the fibre. Thus, it is important to understand these affecting factors and ultimately improving the properties of the composite. Therefore, one way to modify and improving the fibre–matrix interfacial bonding is through the fibre surface modification. In order to make the fibres become more compatible with the matrix, various methods of treatments on the fibre such as bleaching, alkali, silane, benzoylation, acrylation and acetylene have been applied by many researchers [36-38]. As for matrix modification, addition of coupling agents is necessary to enhance fibre-matrix adhesion. Among all of these possible modification methods available, the most common and simply implemented method is alkaline treatment of the fibres using sodium hydroxide (NaOH) solution. Chemical modifications may introduce new moieties that can effectively interlock with the matrix. Mercerization of plant fibres have achieved various levels of success in improving fibre strength, compatibility and fibre matrix adhesion in natural fibre reinforced composites [39]. Therefore, it is considered that the alkalisation treatment applied in this study can provide a better properties to the PLA/KFT/EJO biocomposites with ultimate desired performance.

In Figure 4, while PLA/KFUT had many exposed fibres pulled out from the matrix and filler related failure, PLA/KFT and PLA/KFT/EJO had more embedded fibres and significantly voids had been reduced. As it can be observed from the SEM micrograph of tensile fracture surface of PLA/KFT in Figure 4, the filler related failures were significantly reduced. It is possible to see the wax and impurities had been disappeared and removed after alkaline treatment. Surface treatment facilitates good adhesion between kenaf and PLA matrix through increased surface roughness. Furthermore, kenaf fibres are well trapped by the PLA matrix. Better stress transfer from the PLA matrix to kenaf fibre is expected through occurrence of the thin layer by the matrix linking the fibre surface to the matrix. Thus, improved adhesion leads to a significant increase in a tensile properties.

As for PLA/KFT/EJO, good interfacial adhesion and better compatibility between the interfaces was observed as there was better wetting out of the PLA/EJO onto kenaf fibres. Changes of surface topography and good fibre-matrix adhesion is required for effective stress transfer from matrix to the fibre [40]. It is worth noting that the enhancement in fibre-matrix adhesion explains the improvement in mechanical properties of PLA/KFT and PLA/KFT/EJO compared to the untreated one.

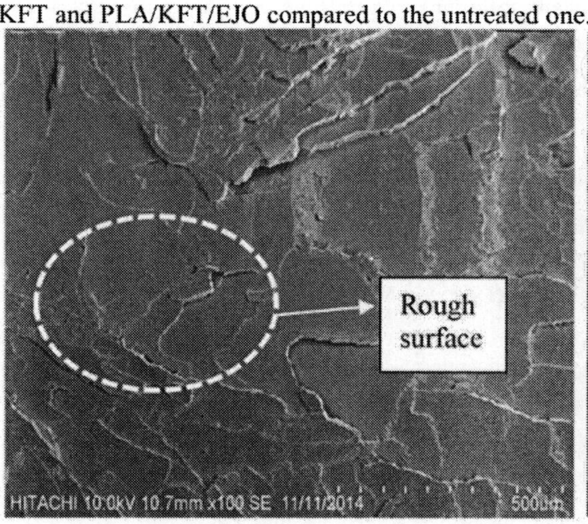

(a)

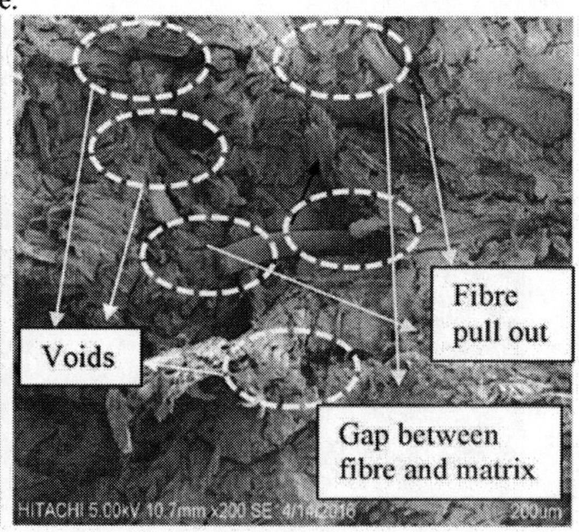

(b)

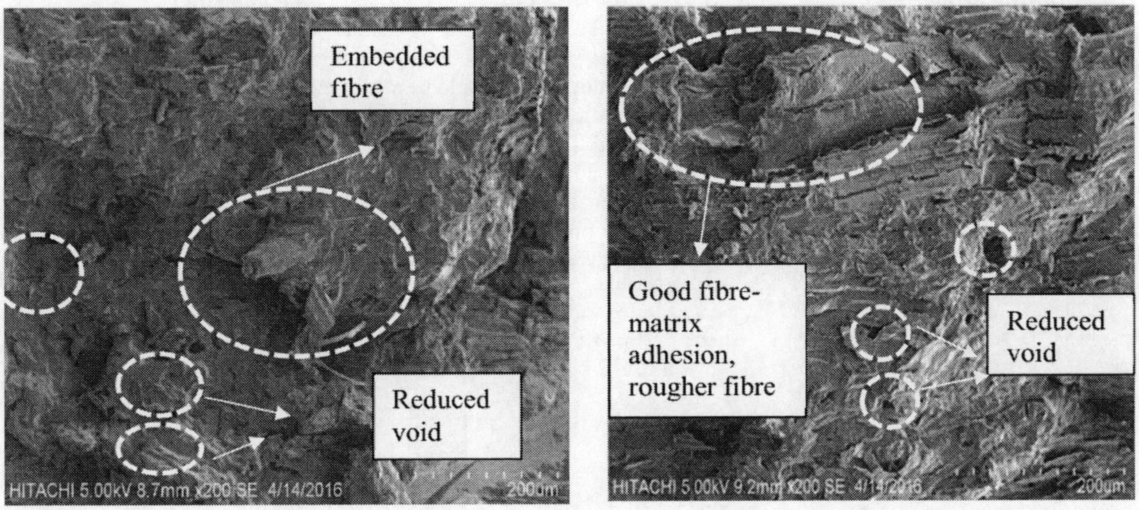

(c) (d)

Figure 4. Surface morphology of (a) PLA (b) PLA/KFUT (c) PLA/KFT and (d) PLA/KFT /EJO

3.2 Mean Diameter of Kenaf fibre

Figure 5 displays the image analysis diameter of untreated and treated kenaf fibres. As can be seen, the diameter of untreated kenaf fibres are larger than those of treated fibres. The measurement diameter of untreated kenaf fibres are between 0.12 to 0.22 mm while for the treated fibres, the diameter average values are between 0.081 to 0.14 mm. The results indicated that the diameter of kenaf fibres had been reduced after treated with alkali concentration.

This might be due to the phenomenon of fibre degradation and fibre delignification that occurred in kenaf fibre during mercerization treatment process. Kenaf fibre became rougher and finer due to the leaching of non-cellulosic materials like pectin, hemicellulose and lignin [41]. In addition, it could be explained by the process of alkalisation treatment which removed a great amount of lignin from the fibres. The diameter of the fibre had been reduced due to the removal of lignin. The alkali treatment removes certain hemicellulose and wax covering the external surface of the fibre and therefore it decreases the aspect ratio (length/diameter) values of fibres. The results are in agreement with another work done by [42] whom reported that 30 minutes alkali treatment with 1 % alkali solution at 90 °C contributed to fibre diameter reduction at 32 %. Similar findings have been observed by [43] whom emphasized the degradation of pectin and removal of lignin from the fibre cell wall that causing separation of the ultimate fibre or fibrillation which lead to the fibre diameter decreased by using 0 to 0.03 % alkali concentration.

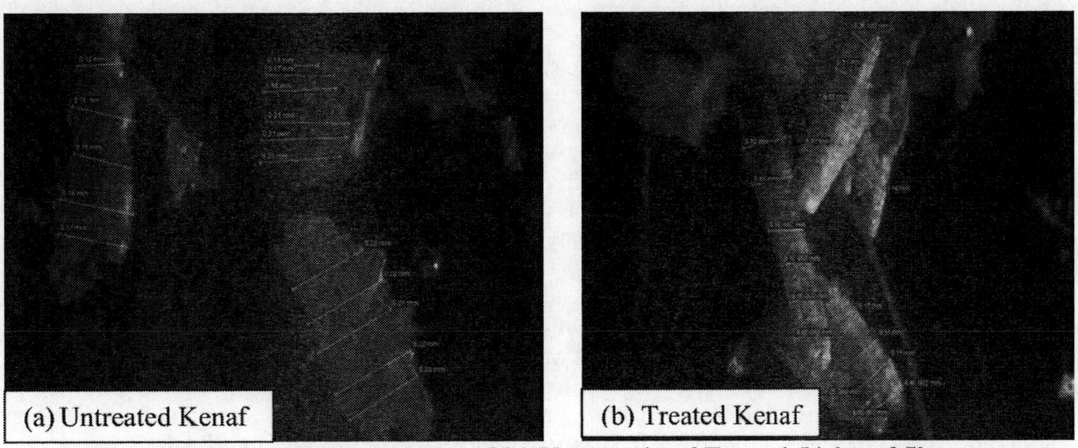

Figure 5. Diameter measurement of (a) Untreated and Treated (b) kenaf fibre.

3.3 Tensile properties of PLA, PLA/KFUT, PLA/KFT and PLA/KFT/EJO biocomposites

The results of tensile strength (TS) and modulus (TM) of neat PLA, PLA/KFUT, PLA/KFT and PLA/KFT/EJO are depicted in Figure 6. From the figure, it can be seen that the value of tensile strength for all three biocomposites are much lower as compared to the neat PLA. This decrement could be attributed to the weak interfacial interaction between PLA and kenaf fibre that leads to the inefficiency of stress transfer from PLA to kenaf fibre. Other than that, according to [44], a types of short fibre might have difficulties to support stress transfer from PLA and therefore lead to the strength reduction as fibre was added to PLA. Furthermore, it is also supported by the research statements from [45] and [46] whom claimed that this decrease is attributed due to the inability of the fibre, irregular shape to support stress transfer from the polymer matrix and poor interfacial bonding generate spaces between fibre and polymer matrix that leads to the weak structure in between them.

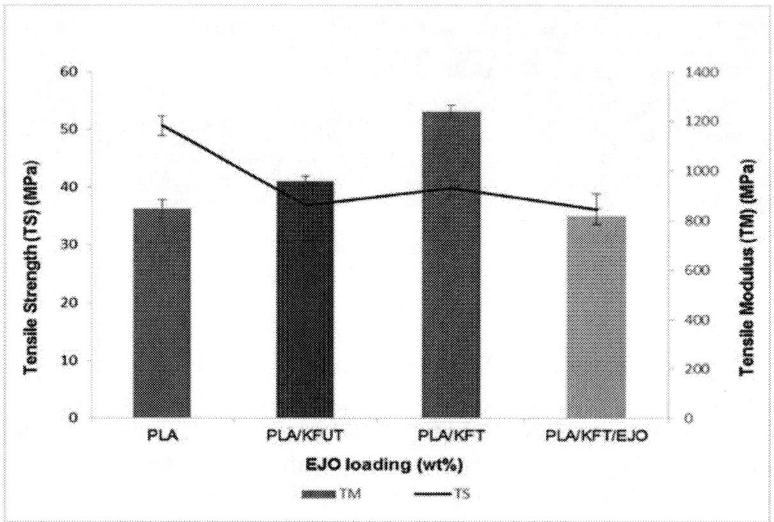

Figure 6. Tensile properties of PLA, PLA/KFUT, PLA/KFT and PLA/KFT /EJO biocomposites.

In this study, the tensile strength of composites reinforced with alkali treated fibres improves over that of composites reinforced with untreated fibres. It is clear that the alkalisation treatment of kenaf fibres results in a significant increase in strength over those of composites with untreated one. It is found that by adding treated kenaf fibre into PLA matrix, the tensile strength of biocomposites increased to 40 MPa for 40 wt.% kenaf fibre loading as compared with 36.83 MPa for the untreated fibres. This behaviour can be explained by considering the SEM micrographs presented in Figure 4. The image shows good fibre-matrix interfacial bonding by increasing the interfacial area of contact between the matrix and fibre. The PLA matrix can be seen adhering to the kenaf fibres subsequent to matrix failure. This evident indicate that alkali treatment produces strong bonds between the PLA matrix and kenaf fibre, with a free deposited surface fibres. This increase can be attributed to the removal of lignin and hemicellulose, which facilitates the rearrangement of fibrils along the direction of tensile deformation hence resulting in higher tensile strength [47-48].

EJO was used in this work in order to plasticise PLA and PLA/KF thus modifying the brittleness of neat PLA. The use of plasticiser had previously been reported for plasticising PLA which brings good result in its properties [49]. As the plasticiser introduced to the system, PLA became softer, more flexible, resulting in reduced strength and modulus. Thus, the value of tensile strength for PLA/KFT/EJO found to be lower than the value of PLA/KFT. However, the results for tensile strength did not differ significantly between PLA (36.83 MPa) and PLA/KFT/EJO (36.18 MPa). Introduction of EJO as a plasticiser into a PLA/KFT involves the addition of more free volume and so, increasing the flexibility of polymer chains, which in turns lowers the tensile strength and tensile modulus of the polymer [50].

On the other hand, as can be seen in Figure 6, addition of kenaf fibre to PLA regardless whether the fibre is treated or untreated resulted in a higher increase of composite modulus. The tensile modulus of biocomposite increased from 846.11 MPa for neat PLA to 958.21 MPa for PLA/KF/UT. The increment in tensile modulus with the addition of natural fibres was expected since stiffer behaviour of natural fibres is introduced into the polymer matrix [51-52]. A higher tensile modulus reflecting a rigid material, thus resists deformation for fibre composites.

As shown clearly, there is a significant increased in value of tensile modulus for PLA/KFT composite. Based on the result, PLA/KFT has higher tensile modulus compared to the untreated fibre composites due to the uniform fibre dispersion in PLA. Additionally, the improvement of modulus may indicate an improvement in effectiveness of oriented cellulose fibre due to the removal of the hemicellulose and lignin content which leads to better packing of cellulose chains. As a result, the fibres become relatively ductile after the removal of some hemicellulose and lignin content resulted in the improvement of stiffness of the fibre. According to [53], the alkalisation treatment have a lasting effect on the mechanical behaviour of natural fibres, especially on fibre stiffness. NaOH solution has ability to remove natural fats, waxes and impurities from the fibre surfaces [54]. Therefore, the alkalisation treatment resulting in opening of more hydroxyl groups and other reactive functional groups on the surface of the fibre [55]. The increase in tensile modulus for PLA/KFT compared to untreated fibre composites was likely attributed to the enhanced suface wettability between fibre and matrix causing from removal of impurities and waxy substances from the fibre surface and the creation of a rougher surface after the modification.

4. Conclusions

PLA matrix, kenaf natural fibres and non edible types of plasticiser namely epoxidized jatropha oil (EJO) were utilised to prepare biodegradable biocomposites and their physical, mechanical and morphological properties were studied and compared. From FTIR spectroscopy, it was found that the disappearance of C=O stretching of the acetyl group in hemicellulose and the decreasing in intensity of the peak after alkaline treatment process confriming the removal of hemicellulose and lignin for treated fibres. The shifted peak towards lower temperature in carbonyl stretching region of PLA/KFT/EJO indicated the miscibility and interaction between PLA, Kenaf and EJO had took place. By SEM observation, it is clear that impurities on untreated kenaf fibres had weakened the bonding strength between PLA and kenaf fibres by evidence of larger voids and notable gaps between fibre and matrix. The alkali treated fibres show a rougher surface topography resulted in better compatibility and good adhesion in both PLA/KFT and PLA/KFT/EJO biocomposites as compared with untreated fibre composites. For kenaf fibre diameter, it was observed that the surface treated fibre composites had lower diameter than the untreated fibre composites which concluded the fibre degradation and delignification had happened during alkalisation treatment. In terms of mechanical properties, kenaf core particles act as rigid filler and by introducing it into PLA matrix, the tensile modulus of PLA/Kenaf composites for both PLA/KFUT and PLA/KFT were significantly higher than those of the PLA matrix itself. The improvement in tensile modulus was much notable for treated kenaf fibre composites due to the improved interfacial interaction resulting in higher tensile stiffness. The promising properties of possible combination of PLA/KFT/EJO biocomposites interms of morphological and mechnical properties suggested that they could be used as a substitite and alternative to synthetic fibre composites. Futher works are recommended in order to improvise PLA/KFT/EJO biocomposites to be further developed as an ecologically friendly alternative to the petroleum-based-polymer that could be potentially functions in various research applications.

Acknowledgements

The authors wish to express their sincere thanks to Biocomposite Laboratory, Institute of Tropical and Forestry Products (INTROP), Universiti Putra Malaysia (UPM), Serdang and Loji Polimer, Radiation Processing and Technology Division, Malaysian Nuclear Agency, Bangi for excellent facilities. The authors are also grateful to the Biopolymer Research Group, Department of Chemical and Environmental Engineering, Faculty of Engineering, Universiti Putra Malaysia (UPM), Serdang for excellent support and cooperation. Thanks also towards Universiti Teknologi MARA (UiTM) as well as Ministry of Higher Education (MOHE) for financial support and assistance in order to complete this work.

References

[1] Yu L, Dean K, and Li L 2006 Polymer blends and composites from renewable resources *Prog Polym Sci* **31** 576-602

[2] Huda M S, Drzal L T, Mohanty A K, Misra M 2008 Effect of fiber surface-treatments on the properties of laminated biocomposites from poly (lactic acid)(PLA) and kenaf fibers *Compos Sci Technol* **68** 424-432

[3] Lim L T, Auras R, Rubino M 2008 Processing technologies for poly (lactic acid) *Prog Polym Sci* **33** 820-852

[4] Sinha Ray S, Yamada K, Okamoto M, Ueda K 2002 Polylactide-layered silicate nanocomposite: a novel biodegradable material *Nano Lett* **2** 1093-1096

[5] Pang X, Zhuang X, Tang Z, Chen X 2010 Polylactic acid (PLA): research, development and industrialization *Biotechnol J* **5** 1125-1136

[6] Yokohara T, Yamaguchi M 2008 Structure and properties for biomass-based polyester blends of PLA and PBS *Eur Polym J* **44** 677-685

[7] Cohn D, Salomon A H 2005 Designing biodegradable multiblock PCL/PLA thermoplastic elastomers *Biomaterials* **26** 2297-2305.

[8] Ochi S 2008 Mechanical properties of kenaf fibers and kenaf/PLA composites *Mech Mater* **40** 446-452

[9] Shanks R A, Hodzic A, Ridderhof D 2006 Composites of poly (lactic acid) with flax fibers modified by interstitial polymerization *J Appl Polym Sci* **99** 2305-2313

[10] Bodros E, Pillin I, Montrelay N, Baley C 2007 Could biopolymers reinforced by randomly scattered flax fibre be used in structural applications? *Compos Sci Technol* **67**462-470

[11] Masirek R, Kulinski Z, Chionna D, Piorkowska E, Pracella M 2007 Composites of poly (L-lactide) with hemp fibers: Morphology and thermal and mechanical properties *J Appl Polym Sci* **105** 255-268

[12] Plackett D, Andersen T L, Pedersen W B, Nielsen L 2003 Biodegradable composites based on L-polylactide and jute fibres *Compos Sci Technol* **63**1287-1296

[13] Lee B H, Kim H S, Lee S, Kim H J, Dorgan J R 2009 Bio-composites of kenaf fibers in polylactide: Role of improved interfacial adhesion in the carding process *Compos Sci Technol* **69** 2573-2579

[14] Mohanty S, Nayak S K 2006 Interfacial, dynamic mechanical, and thermal fiber reinforced behavior of MAPE treated sisal fiber reinforced HDPE composites *J Appl Polym Sci* **102** 3306-3315

[15] Yang H S, Wolcott M P, Kim H S, Kim S, Kim H J 2007 Effect of different compatibilizing agents on the mechanical properties of lignocellulosic material filled polyethylene bio-composites. *Compos. Struct.* **79** 369-375

[16] John M J, Anandjiwala R D 2008 Recent developments in chemical modification and characterization of natural fiber-reinforced composites *Polymer composites* **29** 187-207

[17] Han Y H, Han S O, Cho D, Kim H I 2007 Kenaf/polypropylene biocomposites: effects of electron beam irradiation and alkali treatment on kenaf natural fibers *Compos Interface* **14** 559-578

[18] Dittenber D B, GangaRao H V 2012 Critical review of recent publications on use of natural composites in infrastructure *Composites Part A* **43** 1419-1429

[19] Fenollar O, Sanchez-Nacher L, Garcia-Sanoguera D, López J, Balart R 2009 The effect of the curing time and temperature on final properties of flexible PVC with an epoxidized fatty acid ester as natural-based plasticizer *J Mater Sci* **44** 3702-3711

[20] Robertson M L, Chang K H, Gramlich W M, Hillmyer M A 2010 Toughening polylactide with the incorporation of polymerized soybean oil *Macromolecules* **43** 1807-181

[21] Mulla E A J, Suhail A H, Aowda S A 2011 New biopolymer nanocomposites based on epoxidized soybean oil plasticized poly(lactic acid)/ fatty nitrogen compounds modified clay: Preparation and characterization *Ind Crop Prod* **33** 23-29

[22] Alam J, Alam M, Raja M, Abduljaleel Z, Daas L A 2014 MWCNTs-reinforced epoxidized linseed oil plasticized polylactic acid nanocomposite and its electroactive shape memory behavior *Int J Mol Sci* **15** 19924-19937

[23] Lathi P S, Mattiasson B 2007 Green approach for the preparation of biodegradable lubricant base stock from epoxidized vegetable oil *Appl Catal B-Environ* **69** 207-212

[24] O'donnell A, Dweib M A, Wool R P 2004 Natural fiber composites with plant oil-based resin *Compos Sci Technol* **64** 1135-1145

[25] Dahlke B, Larbig H, Scherzer H D, Poltrock R 1998 Natural fiber reinforced foams based on renewable resources for automotive interior applications *J Cell Plast* **34** 361-379

[26] Liu W, Mohanty A K, Askeland P, Drzal L T, Misra M 2004 Influence of fiber surface treatment on properties of Indian grass fiber reinforced soy protein based biocomposites *Polymer* **45** 7589–7596

[27] Tserki V, Zafeiropoulos N E, Simon F, Panayiotou C 2005 A study of the effect of acetylation and propionylation surface treatments on natural fibres *Composites Part A* **36** 1110-1118

[28] Sgriccia N, Hawley M C, Misra M 2008 Characterization of natural fiber surfaces and natural fiber composites *Composites Part A* **39** 1632-1637

[29] Alemdar A, Sain M 2008 Isolation and characterization of nanofibers from agricultural residues–Wheat straw and soy hulls *Bioresource Technol* **99** 1664-1671

[30] Mwaikambo L Y, Ansell M P 2002 Chemical modification of hemp, sisal, jute, and kapok fibers by alkalization *J Appl Polym Sci* **84** 2222-2234

[31] Le Troedec M, Sedan D, Peyratout C, Bonnet J P, Smith A, Guinebretiere R, Gloaguen V, Krausz P 2008 Influence of various chemical treatments on the composition and structure of hemp fibres *Composites Part A* **39** 514-522

[32] Jonoobi M, Harun J, Mishra M, Oksman K 2009 Chemical composition, crystallinity and thermal degradation of bleached and unbleached kenaf bast (Hibiscus cannabinus) pulp and nanofiber *BioResources* **4** 626-639
[33] Nacos M K, Katapodis P, Pappas C, Daferera D, Tarantilis P A, Christakopoulos P, Polissiou M 2006 Kenaf xylan–A source of biologically active acidic oligosaccharides *Carbohyd Polym* 66 126-134
[34] Mathew A P, Oksman K, Sain M 2005 Mechanical properties of biodegradable composites from poly lactic acid (PLA) and microcrystalline cellulose (MCC) *J Appl Polym Sci* **97** 2014-2025
[35] Bax B, Mussig J 2008 Impact and tensile properties of PLA/cordenka and PLA/flax composites *Compos Sci Technol* **68**1601–1607
[36] Li X, Tabil L, Panigrahi S 2007 Chemical Treatments of Natural Fiber for Use in Natural Fiber-Reinforced Composites: A Review *J Polym Environ* **15** 25-33
[37] Sreekumar P A, Thomas S P, Marc Saiter J, Joseph K, Unnikrishnan G, Thomas S 2009 Effect of fiber surface modification on the mechanical and water absorption characteristics of sisal/polyester composites fabricated by resin transfer molding *Composites Part A* **40** 1777-1784
[38] De Rosa I M, Kenny J M, Maniruzzaman M, Moniruzzaman M, Monti M, Puglia D, Santulli C, Sarasini F 2011 Effect of chemical treatments on the mechanical and thermal behaviour of okra (Abelmoschus esculentus) fibres *Compos Sci Technol* **71** 246-254
[39] Kalia S, Kaith B S, Kaur I 2009 Pretreatments of natural fibers and their application as reinforcing material in polymer composites—A review *Polym Eng Sci* **49** 1253-1272
[40] Bachtiar D, Sapuan S M, Hamdan M M 2008 The effect of alkaline treatment on tensile properties of sugar palm fibre reinforced epoxy composites *Mater Design* **29**1285-1290
[41] Roy A, Chakraborty S, Kundu S P, Basak R K, Majumder S B, Adhikari B 2012 Improvement in mechanical properties of jute fibres through mild alkali treatment as demonstrated by utilisation of the Weibull distribution model *Bioresource Technol* **107** 222-228
[42] Saha P, Manna S, Chowdhury S R, Sen R, Roy D, Adhikari B 2010 Enhancement of tensile strength of lignocellulosic jute fibers by alkali-steam treatment *Bioresource Technol* **101** 3182-3187
[43] Mwaikambo L Y 2009 Tensile properties of alkalised jute fibres *Bioresources* **4** 566-588
[44] Ramli R, Yunus R M, Beg M D H, Prasad D M R 2012 Oil palm fiber reinforced polypropylene composites: Effects of fiber loading and coupling agents on mechanical, thermal, and interfacial properties *J Compos Mater* **46** 1275-1284
[45] Ismail H, Nizam J M, Khalil H P S A 2001The effect of a compatibilizer on the mechanical properties and mass swell of white rice husk ash filled natural rubber/linear low density polyethylene blends *Polym Test* **20** 125–133
[46] Yang H S, Kim H J, Son J, Park H J, Lee B J, Hwang T S 2004 Rice-husk flour filled polypropylene composites; mechanical and morphological study *Compos Struct* 63 305–312
[47] Gassan J, Bledzki A K 1999 Possibilities for improving the mechanical properties of jute/epoxy composites by alkali treatment of fibres *Compos Sci Technol* **59** 1303-1309
[48] Gassan J, Bledzki A K 1997 The influence of fiber-surface treatment on the mechanical properties of jute-polypropylene composites *Composites Part A* **28** 1001-1005
[49] Wadhi M M, Weliam R 2014 Effect of epoxidized sunflower oil on polylactic acid properties *Res Chem Intermediat* **40** 399-406
[50] Sears J K,. Darby J R 1982 The Technology of Plasticizers. John Wiley and Sons, New York
[51] Premalal H G, Ismail H, Baharin A 2003 Effect of Processing Time on the Tensile, Morphological, and Thermal Properties of Rice Husk Powder-Filled Polypropylene Composites *Polym-Plast Technol* **42** 827-851
[52] Zampaloni M, Pourboghrat F, Yankovich S A, Rodgers B N, Moore J, Drzal L T, Mohanty A K, Misra M 2007 Kenaf natural fiber reinforced polypropylene composites: a discussion on manufacturing problems and solutions *Composites Part A* **38** 1569-1580
[53] Agrawal R, Saxena N S, Sharma K B, Thomas S, Sreekala M S 2000 Activation energy and crystallization kinetics of untreated and treated oil palm fibre reinforced phenol formaldehyde composites *Mater Sci Eng, A* **277** 77–82
[54] Sreekala M S, Thomas S 2003 Effect of fibre surface modification on water-sorption characteristics of oil palm fibres *Compos Sci Technol* 63 861-869
[55] Islam M S, Pickering K L, Foreman N J 2011 Influence of alkali fiber treatment and fiber processing on

the mechanical properties of hemp/epoxy composites *J Appl Polym Sci* **119** 3696-3707

IOP Conf. Series: Materials Science and Engineering **368** (2018) 012012 doi:10.1088/1757-899X/368/1/012012

Natural fiber for green technology in automotive industry: A brief review

N Ramli[1], N Mazlan[1,2*], Y Ando[3], Z Leman[4], K Abdan[5], A A Aziz[2] and N A Sairy[2]

[1]Laboratory of Biocomposite Technology, Institute of Tropical Forestry and Forest Product (INTROP), Universiti Putra Malaysia, 43400 UPM Serdang, Selangor.
[2]Department of Aerospace Engineering, Faculty of Engineering, Universiti Putra Malaysia, 43400 UPM Serdang, Selangor.
[3]Graduate School of Life Science and Systems Engineering, Kyushu Institute of Technology, Japan.
[4]Department of Mechanical and Manufacturing Engineering Faculty of Engineering, Universiti Putra Malaysia 43400 UPM Serdang, Selangor.
[5]Department of Biological and Agricultural Engineering, Faculty of Engineering, Universiti Putra Malaysia, 43400 UPM Serdang, Selangor.

*Corresponding author : norkhairunnisa@upm.edu.my

Abstract. Fiber reinforced polymeric composites have been known and widely used because of their high specific strength and modulus compared to metals. In the last few years, biomaterials listed as a demand technology to be exploring by researchers especially in industrial purpose. This is push by environmental awareness and the over use of petrol resources lead to the development of new materials, called biocomposites, which will maintain a better future. This paper will be discussing about a brief review of natural fibers, use in automotive industry to achieve a green technology target in manufacturing of cars specifically. It's a fact that, related to weight reduction, the automotive industry can take advantages of using these materials, not only because of extinction of oil reserve, but because of high ability and importance of these materials itself in automobiles. Currently, most composites in the market are focused with long-term durability design while using nondegradable polymeric resins such as epoxies and high-strength fiber such as glass. All these materials prove to be a good characteristic of composite but still lack in environmental concern. This polymer and fiber are derived from petroleum, a nonreplenishable commodity. The momentum is to use biocomposites in common plastics to improve performance. Since the main purpose of this paper is to show a bio-composite which is suitable to replace the existing interior of automotive design, the work has focused on obtaining that bio-composite, taking account into the raw-materials cost reduction and the maintenance of the manufacturing process based on current scenario. The automotive industry is in their way to expand green technology in composites because the need is greatest. But producing the composites is energy intensive and polluting, while the durability of conventional composites, often seen as an advantage, is also their biggest challenge. Current fibers use in industry right now is difficult to dispose. They do not degrade naturally and could linger for generations.

1. Introduction

"The most environmentally friendly thing you can do for a car that burns gasoline is to make lighter bodies" (Henry Ford). Waste indiscriminate had been remind to us since school which is in early age. The concept of reducing, reusing and recycling activity is for decreasing rubbish in a landfill. Unfortunately this habit cannot be practicing because of attitude. Day by day it leads to a rise in health cases and environmental degradation problem. Regarding this matter, any solution crossed will be taken seriously to ensure action is louder than speaks. This will be including automotive industry because at the end life of automotive part will be landed in a landfill. Issues on recycling car components, is forcing the automotive industry to look at ways to lower the environmental impact of cars over their entire life cycle. Researchers do more studies aiming to help, by improving technologies to enhance the performance of plant fibre, a 'greener' alternative to glass fibre, for use in thermoplastic composite automotive parts. Natural fibres offer technical and environmental benefits when used to reinforce plastics [1].

At the end of 2010, total number of vehicles in Malaysia had reached amount of 21.25 million vehicles with an average of 12% increase of vehicle registration each year over the period of 5 years. If the estimation was to be continued, Malaysia will have as much as 31 million vehicles in the year 2020. As the automotive industry develops, its impact to the environment also increases. Thus, a proper solution of managing waste is needed to sustain the environment and reduce human impact towards nature [1]. Environmental concern as well as government legislation has motivated manufacturers in many countries to consider product life cycle issues and deal with product recovery at the end of the product life cycle. The automotive industry is one of the leading industries in this environmentally conscious manufacturing and product recovery [2].

2. Green technology to overcome waste disposal of car component

There is a dramatically surge in interest of eco-friendly materials and technology in industrial sectors and hybrid biocomposite from natural resources. All products have environmental impacts. It is either small or huge depending on the root causes and raw materials used in the production process. These will be including process or services which may involve energy consumption, natural resource depletion, solid waste generation, air pollution, land degradation and global warming issue. However, previous research stated that these impacts can be minimized by practicing sustainable design which helps to promote product quality and reduce cost while environmental degradation will be in a very minimum level to occur. Sustainable development is important, it plays huge role in industry sector as well as in our daily lives. This is to ensure both present and future needs fulfill with good quality of environment and preservation of natural resources can be well manage.

In addition, the biggest concern is on the end destination of automotive products where end-of-life of vehicle recycling is widely exploited over the world. The goal is to ensure the amount of waste generated from disposal process can be decreasing which lead to reduce environmental degradation. Therefore, end-of-life recycling is an important factor which must be included in product application for every product and material design. According to auto recycling statistics [3], most of car produced nowadays will be a good source of recyclable material. In fact, around 80 percent of a car can be recycled.

However, there are two major issues occurred for the current materials for internal car component. Both are high in cost and currently high of weight. Designing a combination of lightweight and strong biomaterials are strategies to develop advanced lightweight composite that can replace the traditional synthetic structures or metals which made as part of automotive components. Heavy composite panel structure will significantly influence the fuel consumption of a car and subsequently increase the carbon dioxide emission in air. In addition, composite made of glass fiber form sharp edges upon deformation and increase the concern related to safety issue.

The increasing demand for greener and biodegradable materials leading to the satisfaction of society requires a compelling towards the advancement of nano-materials science. The biodegradability of the natural fibers is considered as the most important and interesting aspects of their utilization in polymeric materials [4]. Designing a combination of lightweight and strong

IOP Conf. Series: Materials Science and Engineering **368** (2018) 012012 doi:10.1088/1757-899X/368/1/012012

biomaterials are strategies to develop advanced lightweight composite that can replace the traditional synthetic structures or metals which made as part of automotive components. Heavy composite panel structure will significantly influence the fuel consumption of a car and subsequently increase the carbon dioxide emission in air.

Natural fibers have been exploited for various purposes as reinforcing fillers for aircraft, automotive and construction materials. The most common natural fiber used in composites is bast fiber, such as hemp, jute, sisal and kenaf [5]. Natural fiber has high specific strength than glass fiber while it has a similar specific modulus. Another point of worthy consideration is the source available in our country and these natural fibers theoretically offer desirable specific strengths and modulus at a lower cost [6]. Kenaf is one of the natural fibers that can be used as reinforcement in making Polymer Matrix Fiber Composites (PMFC). This cellulosic source is known with both economic and ecological advantages as it can grows under a wide range of weather conditions. The strength and stiffness of kenaf fibers are provided by the cellulose components via hydrogen bonds and other linkages. It was reported that hybridized kenaf fiber potentially improved the damping property of the composite [6]. Natural fibers also offer remedy in neutralizing the discomfort generated from engines, exhaust systems, gears and wheels which is produced in form of noise, heat and electrostatic charges. The nonabrasive properties of natural fibres make it a good choice in term of safety as natural fibers do not form sharp edges upon deformation like glass fibers [7].

3. Natural fiber replacing polymer's function.

Recently, production of synthetic polymers based petrochemicals quite popular but it is not biodegradable and harmful. Polymers such as polyethylene and polypropylene can be prolonged in the environment for many years after their disposal [8]. Last two decades, significant advances in the development of biodegradable polymers have been study [9]. Basically, production of polymers which can biodegrade generally obtained via polymerization of agricultural-based raw materials. Variety forms of biodegradable polymer materials have been implemented and thus proven to have potential uses in industries. Many of these polymers are well suited for adhesive applications such as environmentally friendly packaging, recyclable envelope adhesives, carpet backing, and many other products that are eventually destined for the municipal waste disposal facility.

Flax and hemp which can be the source for biofibers have potential as a raw material which are of great importance for the production of various types of composites use in automotives, building materials, packaging, papers and furniture industries [10]. Biofibres as viable alternative to synthetic fibre as reinforcement in plastics because of low cost, lightweight, good mechanical performance and biodegradable properties. In fact, a better performance can be creating use biofibre with biodegradable resins. While in automotive industry, biocomposites found a vast number of applications [11]. Basically, applications for a car included doorpanels, seat backs, dashboards and package trays, head restraints and seatback linings. For example, Mercedes [12] used jute-based composites, applied for the door panels of the E-class, banana-fibre as composites for the A-class and other bio-composites for the S-class. The Araco Corporation in Japan present another way in 2003, created the Grasshopper as a fully electric vehicle. This vehicles totally made from plant-based composites mainly kenaf for its body [13].

In the last few decades, the use of natural fibers gained considerable attention like from both resources, renewable and non-renewable such as oil palm, sisal, flax, and jute to produce composite materials. The plants, which produce cellulose fibers can be classified into bast fibers (flax, hemp, jute, kenaf, and ramie), seed fibers (kapok, coir, and cotton), leaf fibers (abaca, pineapple, and sisal), grass and reed fibers (wheat, rice, and corn), and core fibers (hemp, jute, and kenaf) as well as all other kinds (roots and woot) [14]. As far as the concern, benefits of natural fibers are beyond compare to synthetic fibers where in term of its relatively renewable resources, biodegradability, low cost, less damage to processing equipment, low weight, improved surface finish of molded parts composite, being abundant flexibility during processing, good relative mechanical properties such as tensile modulus and flexural modulus, and minimal health hazards [15]. Thus, natural fibers widely use

IOP Conf. Series: Materials Science and Engineering **368** (2018) 012012 doi:10.1088/1757-899X/368/1/012012

because of those specific properties which can give more advatages to the users and even the producers.

Currently, automotive components produce from natural composites are already exist in manufacturing industry. In facts, those composites mainly based on polyester or Polypropylene and fibers like flax, hemp or sisal. The main reasons why natural composites were using are the highlight of weight reduction, and marketing rather than technical demands [16]. Germany stated as a leader in the use of natural fiber composites. The German auto-manufacturers, Mercedes, BMW, Audi and Volkswagen introduce natural fiber composites for interior and exterior applications. The first attraction is the inner door panel in the 1999 of the S-Class MercedesBenz, made in Germany, of 35% Baypreg F semi-rigid (PUR) elastomer from Bayer and 65% of a blend of flax, hemp and sisal. Eventhough this luxury automotive manufacturers are on board, that is not because they need to lower cost, in facts with the use of natural fiber composite, could be seen as evidence that industry need to do any initiative for environmental needs [17]. Back in 1996, Mercedes-Benz used an epoxy matrix with the addition of jute in the door panels in its E-class vehicles. Another paradigm of natural fiber composites' application appeared commercially in 2000, when Audi launched the A2 midrange car: the door trim panels were made of polyurethane reinforced with a mixed flax/sisal material. Toyota developed an eco-plastic made from sugar cane and will use it to line the interiors of the cars [18].

Volvo has started to use soya-based foam linings in its C70 and V70 models for their seats with natural fibers. To improve the quality of noise reduction they have also produced a cellulose-based cargo floor tray. In Western Europe, the yearly production of cars is up to 16 million vehicles that equate to an including usage of 80 000160 000 tons of natural fibers per year. German automobile companies like Daimler-Chrysler are continuing to lead the way, having a global natural fiber initiative program that benefits third nations by developing products made from natural fibers. One of the recent developments within the automotive industry has been the release of the Lotus Eco Elise. Another development was announced in 2008 at the EcoInnovAsia 2008 event held in Bangkok, Thailand, related to the Mazda 5. In this application, the manufacturer is using polylactic acid (PLA) in the interior consoles along with kenaf and PLA in the seat covers [19].

Sisal fibre is used traditionally to produce twine, ropes, carpets and bags. Natural fibres such as sisal are also increasingly used in composites in the automotive and construction sectors. About 5% of the composites manufactured in Europe incorporated natural fibres in 2012, the remainder mainly consisting of glass fibre composites [20]. The environmental benefits of biocomposites can be balance by the technologies while improving their thermal and dimensional performance provide an opportunity for their use in under the hood automotive applications [21]. In Japan, company of Hitachi Automotive Systems last year announced that it will produce a composite carbon/fiber drive shafts for the company Alfa Romeo, which will be built into the new Alfa Romeo Giulia car. These shafts are reduced by the noise of the central bearing by using carbon fibers which leads to greater comfort during the ride. In UK, door panels of the car Ford Mondeo are made of a mixture of kenaf fiber and polypropylene resins. The mass of the door is reduced from 5 to 10% by this way. The ability of fiber to absorb a large amount of moisture leads to greater comfort compared with synthetic materials. In Germany, higher performance applications are achieved with the inside panels of the Mercedes Benz E Class. Flax fibers were used in combination with epoxy resins. The biggest disadvantage of plant fibers are too large oscillations in quality, which depends on the growth rate, and absorption of moisture, which complicates the application [21].

4. Mechanical Properties

Mechanical properties of natural fibers will be depending on the factors like size, maturity as well as processing methods adopted for the extraction of fibers. Previous study stated that several properties such as density, tensile strength, modulus and other properties depend on the internal structure and chemical composition of fibers. It is known from the research that modulus of natural fiber decreases with increase of its diameter [23-24]. Regarding in the tabulated data from previous data collections, the strength of natural fiber is below the line, less value compare to glass fiber as can see in Table 5.1, while stiffness is on the same level. However, there is something special about natural fibers where it has low densities, which are up to 2 times lighter than glass, the resulting specific stiffness of natural fibers is substantially higher than the same parameter of glass fibers. Flax, hemp and jute fibers are particularly have better value of specific stiffness. Those factors are good explanation to be used as reinforcement in composites for automotive applications. Sisal, coir cotton and ramie are only useful if the price is cheaper and reasonable to buy in bulk condition. Wood fibers in the form of short fibers or flour are also used in composites processing, but mainly as filler. The properties of natural fibers presented in Table 1 are mostly for those fibers extracted by using dew-retting process, and could be regarded as conservative.

Table 1 Properties of some synthetic and natural fibers (Adapted from kavelin, 2005).

Fibers	**Tensile strength, MPa**	**Tensile modulus, GPa**	**Specific gravity**	**Specific strength**	**Specific stiffness**	**Failure strain, %**	**Price, Euro/kg**
E-glass	2500-3500	70-73	2.56	27	29	2.5-3.0	1.5-2.5
Carbon	2500-6000	220-700	1.75-1.9	116	400	1.4-2	30-50
Flax	500-900	50-70	1.4-1.5	33	50	1.3-3.3	0.5-1
Sisal	80-840	9-22	1.3-1.45	6	17	3-7	0.3
Jute	200-450	20-55	1.3-1.4	14	42	1.16-1.5	0.12-0.5
Hemp	310-750	30-60	1.48	20	41	2-4	0.5-1
Banana	530-750	7-20	1.4	5	14	1-4	0.5
Coir	130-175	4-6	1.15	3	5	15-40	0.25
Cotton	300-600	6-10	1.5	4	7	7.0-8.0	1.6-4.6
Silk	-	-	1.34	-	-	-	18.3-36.7
Wool	125-200	-	1.31	-	-	-	Up to 15.4

Natural fiber composites are often poorer in properties, mostly mechanical, compared to synthetic fiber composites. A possible solution to this issue is the use of natural fiber/synthetic fiber combination in polymer hybrid composites. Although the biodegradability of the composites is compromised by synthetic fibers, this is compensated by the improvement in their mechanical and physical properties. Hybrid composites use more than one kind of fibers in the same matrix and the idea is to get the synergistic effect of the properties of both fibers on the overall properties of composites. There has been a significant increase in research on natural fiber/synthetic fiber hybrid composites in recent years. Natural fibers are mostly hybridized with glass fibers because of their comparable properties and low cost. Some studies, however, have been done on hybridization of natural fibers with the more expensive carbon and aramid fibers. There is a considerable improvement in mechanical properties of these composites following hybridization, especially when synthetic fiber plies are used as skin and natural fiber plies are used as core. Various natural fiber surface treatments have been used to improve their interfacial adhesion with the matrices and, hence, their mechanical properties.

5. Future Prospective of Natural Fiber in Automotive Industry

Natural fiber give a lot of benefits to environmental advantages such as: reduced dependence on non-renewable energy/material sources, lower pollutant emissions, lower greenhouse gas emissions [25] increase energy recovery and end of life biodegradability of components [26]. Eventhough the advantages of natural fibers give specific benefit to environment, all these properties should be supported by quantitative analysis. In this regard, Joshia et al. mentioned three comparative Life Cycle assessment (LCA) studies to assess the overall environmental performances. This research lead to specific applications, analyzed in the study, natural fiber composites are environmentally suit to glass fiber composites on most midpoint indicators [27].

Drivers of superior environmental performance show from the comparative analyses are as follows

[25]:
- Less environmental impacts of natural fiber production compared to glass fiber production
- Substitution of matrix (synthetic polymer) by higher content of natural fiber in the composite
- Improvement of fuel efficiency and lower use phase emissions due to weight reduction
- Energy and carbon credits from end of life incineration of natural fibers [26].

6. Conclusion

The preservation of our environment requires that we stop developing materials that will, like many plastics, last indefinitely. Yet nature's way, to accept rapid degradation because there is continual renewal, is not an option. Industry, especially the automotive sector, which is an enormous user of bulk materials, would like a halfway house of reasonably long-lived materials that nevertheless degrade back into the environment when they are no longer needed. Reinforced plastics based on natural, mainly plantderived substances show promise of providing this and may turn out to be one of the material revolutions of this century.

The automotive industry is in the driving seat of 'green' composites because it is here that the need is greatest. Faced with pressures to produce fuel efficient, low-polluting vehicles, the industry has used fiber reinforced plastic composites to make its products lighter. But producing the composites is energy intensive and polluting, while the durability of conventional composites, often seen as an advantage, is also their Achilles' heel. Glass, carbon, and aramid fiber reinforced polyester, epoxy, and other similar resins are difficult to recycle and hard to dispose of. They do not degrade naturally and could linger for generations.

References

[1] Azmi M, Mat Saman M Z, Sharif S, Zakuan N and Mahmood S 2013 Proposed framework for end-of-life vehicle recycling system implementation in Malaysia *10.14279/depositonce-3753*
[2] Amelia L, Wahab D A, Haron C C, Muhamad N and Azhari CH, 2009 Initiating automotive component reuse in Malaysia *J Clean Prod* **17** 1572-1579
[3] LeBlanc R. (2016). Auto or Car Recycling Facts and Figures. Retrieved at https://www.thebalance.com/auto-recycling-facts-and-figures-2877933
[4] Cele H M, Ojijo V, Chen H, Kumar S, Land K, Joubert T, De Villiers M F R and Ray S S, 2014 Effect of nanoclay on optical properties of PLA/clay composite films *Polym Test* **36** 24-31
[5] Edeerozey A M, Akil H M, Azhar A B, and Ariffin M Z 2007 Chemical modification of kenaf fibers *Mater Lett* **61** 2023-2025.
[6] Akil H, Omar M F, Mazuki A A M, Safiee S Z A M, Ishak Z M, and Bakar A A 2011 Kenaf fiber reinforced composites: A review *Mater Design* **32** 4107-4121
[7] Akampumuza O, Wambua P, Ahmed A, Li W, and Qin X H 2017 Review of the applications of biocomposites in the automotive industry *Polym Composite* **38** 2553-2569
[8] Cheung H Y, Ho M P, Lau K T, Cardona F, and Hui D 2009 Natural fibre-reinforced composites for bioengineering and environmental engineering applications *Composites Part B* **40** 655-663
[9] Karlsson S and Albertsson, A C 1998 Biodegradable polymers and environmental interaction *Polym Eng Sci* **38** 1251-1253
[10] John M J and Thomas S 2008 Biofibres and biocomposites *Carbohyd Polym* **71** 343-364
[11] Njuguna J, Wambua P, Pielichowski K, and Kayvantash K 2011 Natural fibre-reinforced polymer composites and nanocomposites for automotive applications. In *Cellulose fibers: Bio- and nano-polymer composites* (661-700). Springer Berlin Heidelberg
[12] Carus M, Ortmann S, Gahle C H, and Pendarovski C 2006. Use of natural fibres in composites for the German automotive production from 1999 till 2005. *Nova-Institut, Hurth*
[13] Faruk O, Bledzki A K, Fink H P, and Sain M, 2012 Biocomposites reinforced with natural fibers: 2000–2010 *Prog Polym Sci* **37** 1552-1596
[14] Shalwan A and Yousif BF, 2013 In state of art: mechanical and tribological behaviour of polymeric composites based on natural fibres *Mater Design* **48** 14-24

[15] Saravana Bavan D and Mohan Kumar G C 2010 Potential use of natural fiber composite materials in India *Journal of Reinforced Plastics and Composites* **29** 3600-3613

[16] Puglia D, Biagiotti J, and Kenny JM 2005 A review on natural fibre-based composites—Part II: Application of natural reinforcements in composite materials for automotive industry *J Nat Fibers* **1** 23-65

[17] Koronis G, Silva A, and Fontul M 2013 Green composites: a review of adequate materials for automotive applications *Compos Part B-Eng* **44** 120-127

[18] Foisal A M, Ali M A, Byung S K and Jong I S 2009 Recent Developments in Natural Fiber Reinforced Composites *The Korean Society for Composite Materials* 22 4.

[19] Broeren M L, Dellaert S N, Cok B, Patel M K, Worrell E, and Shen L 2017 Life cycle assessment of sisal fibre–Exploring how local practices can influence environmental performance *J Clean Prod* **149** 818-827

[20] Pervaiz M, Panthapulakkal S, Birat K C, Sain M and Tjong J 2016 Emerging trends in automotive lightweighting through novel composite materials *Materials Sciences and Applications* **7** 26

[21] Jović D and Milićević J 2017 Influence of Application of New Material in Automotive Industry on Improving Quality of Life. *Center For Quality* 331-332

[22] Kavelin K G 2005 *Investigation of natural fiber composites heterogeneity with respect to automotive structures* (Doctoral dissertation, TU Delft, Delft University of Technology)

[23] Lilholt H and Lawther JM 2000 Natural organic fibers, In: Comprehensive composite materials, Vol.1 Fiber reinforcement and general theory of composites, Kelly, A.; Zweben, C.; Chou, T.-W. (eds.), *Elsevier*

[24] Van den Oever M J A, Bos H L and Van Kemenade M J J M 2000 Influence of the physical structure of flax fibres on the mechanical properties of flax fibre reinforced polypropylene composites *Appl Compos Mater* **7** 387-402

[25] Begum K and Islam M 2013 Natural fiber as a substitute to synthetic fiber in polymer composites: a review *Research J Eng Sci* **2278** 9472

[26] Sparnins E 2009 *Mechanical properties of flax fibers and their composites* (Doctoral dissertation, Luleå tekniska universitet)

[27] Joshia S V, Drzalb L T, Mohantyb A K and Arorac S 2004 Are natural fiber composites environmentally superior to glass fiber reinforced composites? *Composites Part A: Applied Science and Manufacturing 3 35*

IOP Conf. Series: Materials Science and Engineering **368** (2018) 012013 doi:10.1088/1757-899X/368/1/012013

Mechanical and barrier properties of kappa-carrageenan/cellulose nanocrystals bio-nanocomposite films

S Azizi[1] and R Mohamad[1,2*]

[1]Department of Bioprocess Technology, Bioprocessing and Biomanufacturing Research Centre, Faculty of Biotechnology and Biomolecular Sciences, Universiti Putra Malaysia, 43400 UPM Serdang, Selangor, Malaysia
[2] Laboratory of Biopolymer and Derivatives, Institute of Tropical Forestry and Forest Products, Universiti Putra Malaysia, 43400 UPM Serdang, Selangor, Malaysia

*Corresponding author: farizan@upm.edu.my

Abstract. The aim of this study is to report the effect of the addition of cellulose nanocrystals (CNCs) on the mechanical and barrier properties of Kappa-Carrageenan (KC) bio-nanocomposite films prepared through the solvent casting process. The characterizations of bio-nanocomposite films were carried out in terms of transmission electron microscopy (TEM), field emission scanning electron microscopy (FESEM), oxygen transmission rate (OTR), and tensile tests. TEM and FESEM results showed that at low loading levels, CNCs were dispersed homogenously in the Kappa-Carrageenan matrix. The tensile strength and modulus in films increased from 45.7 MPa to 108.4 MPa and from 295 MPa to 710 MPa respectively, when CNCs content went from 0 wt% to 3.0 wt%. The oxygen barrier property of Kappa-Carrageenan matrix was best enhanced at 3.0 wt% of CNCs loading. The enhanced properties attained by incorporating CNCs can be beneficial in various applications.

1. Introduction

Last decades, researchers have concentrated their investigations on the processing of nanocomposites to improve physical properties. In the vein of conventional composites, nanocomposites use a matrix where the nano-sized reinforcement materials are dispersed. The reinforcement is generally considered as a nanoparticle with at least one dimension lower than 100 nm. In the meantime, there is a simultaneous and increasing attention in developing application of bio-based products which can decrease the dependence on fossil fuel and shift to a sustainable materials basis [1-2]. This specific condition provides unique bionanocomposites possessing remarkable properties, which have been never found in traditional composites [3]. Hence, the next generation of materials in the future are biodegradable bio-nanocomposites [4].

The nano-sized reinforcements can obtain from various renewable resources depending on their biological origin, e.g., starch, cellulose, chitin [5-6] but main studies are devoted to cellulose fibers as this biopolymer is the most plentiful on earth. Cellulosic fibers are composed of assemblies of microfibrils [7], which form bundles and almost endless rods. Through acid and/or enzymatic hydrolysis, these microfibrils broke down into short crystalline rods or ''cellulose micro/nano-crystals." With a very high modulus of elasticity (138 GPa), and a high specific-surface-area, low density, low-cost, renewable nature, and recyclability, cellulose nanocrystals have merited substantial consideration for exploring new applications.

Cellulose as "filler" has been mostly used in bio-composites, which in some cases contributes to improve the stiffness but typically embrittles the polymer. The research of cellulosic nanocrystals as a reinforcing material in nanocomposite films was initially reported two decades ago [8]. Since that time significant literatures have been dedicated to nanocrystal cellulose which becomes a topical subject [9-12].

Kappa-carrageenan (KC) is one of the most promising biopolymers; it is composed of a linear chain of sulfated galactans and extracted from certain species of red alga. It exhibits good potential for development as a source of biodegradable films [13]. Though, KC films do not have noble mechanical and oxygen barrier properties, as they are essentially hydrophilic. Consequently, they might not be utilized for extensive applications, particularly in the food industry. Numerous strategies have been reported to overcome these drawbacks, for example blending with natural or synthetic polymers [14], blending with hydrophobic compounds [15] cross-linking [16] and formation of nanocomposite [17, 18]. In a study, the effect of using nanoclay on the properties of KC-based films was investigated [19]. They obtained improved the physical and mechanical properties for KC films. In a similar study, agar/kappa-carrageenan blend nanocomposite films were prepared using different contents of a natural MMT (Cloisite Na^+), and it was reported that the hydrophilic Cloisite Na^+ was the great compatible with an agar and kappa-carrageenan polymer matrix [20]. However important improvements in the properties of KC nanocomposite films have been reported in these researches, the films are still not comparable to synthetic plastic films. This specifies that more-efficient approaches or a combination of techniques are required to develop the films' properties to where they can substitute for non-biodegradable plastic films for packaging.

In this study cellulose nanocrystals obtained from acid hydrolysis of cotton cellulose with 50–80 nm in length and a diameter in the range of 5–10 nm were used to improve the mechanical and oxygen barrier properties of Kappa-carrageenan-based films. For these proposes the mechanical properties in terms of tensile strength, tensile modulus and elongation at break and barrier property in term of oxygen transmission rate (OTR) were assessed.

2. Materials and methods

2.1. Materials

All the chemicals were analytical grade and used as received without more purification. Cotton cellulose from filter paper (Q1, Whatman) was supplied by Fisher Scientific (Pittsburgh, PA, USA). Sulfuric acid (95%–98%, reagent grade) was purchased from Scharlau (Barcelona, Spain). Kappa-carrageenan with a 300 000 $gmol^{-1}$ average molecular weight was obtained from the Malaysian Nuclear Agency.

2.2. Extraction of CNCs

Extraction of the CNCs was carried out according to a previous work [21]. The cellulose powder harvested from one piece of filter paper (2 g) was hydrolyzed with a sulfuric acid solution (20 mL, 64 wt%) at 45 °C for 60 min. The resultant suspension was diluted 10-fold with cold water (4 °C) followed by centrifugation and dialysis until a neutral pH was reached. Finally, the sample was freeze-dried. The final product was cellulose nanocrystals with 50–80 nm in length and a diameter in the range of 5–10 nm.

2.3. Preparation of Nanocomposite

KC solution was prepared by dissolving 2 g of KC powder in 100 mL of distilled water at 82 °C under magnetic, for 30 min and subsequently cooled. The glycerol plasticizer (25 wt % on KC solid base) was then added to the KC solution while stirring for another 30 min. Nanocomposite samples were fabricated by dispersing different amounts of CNCs (0 wt%, 1 wt%, 3 wt% or 5 wt%) in 100 mL of distilled water for 1 h at room temperature. The obtained suspension was added to the KC solution, stirred for 1 h at room temperature and then sonicated for 30 min at 25 °C in a bath type ultrasound sonicator. The suspensions were then poured into glass plates and dried at 25 °C for three days, until the solvent was completely evaporated and a self-standing film was achieved.

2.4. Characterization
The morphologies of the sample films were studied by using a transmission electron microscope (HITACHI H-700, Japan) with acceleration voltage of 120 kV at room temperature. The ultrathin sections of bio-nanocomposite sample were prepared at −120 °C using a microtome equipped with a diamond knife, and then placed on a copper grid. The surface morphology of the sample films was further examined using a field emission scanning electron microscope (FESEM), (JSM-6360LA Philips), with an acceleration voltage of 5 kV. Before examination the specimens were mounted on a metal stub using carbon tape and then gold-coated under nitrogen atmosphere using a Bal- Tec SCD 005 sputter coater. The tensile properties of the sample films were measured as per ASTMD 638 test methods, using an Instron 4032 Universal Testing Machine. The samples, cut into dumb-bell shapes, employed a steel template and router machine. Seven specimens were tested and the average result of the values of five specimens was taken. Oxygen transmission rate of the sample films were evaluated by using of an oxygen transmission rate test (Labthink BTY-B1, China) instrument. The sample film with a diameter around 30 mm was placed in a cell and oxygen flowed on one side of the films, and the oxygen transmission rate was measured. This is the capacity of molecule gas that transfers through a surface per unit of time. Five specimens were tested and the reported results were average of the values of at least four OTR measurements.

3. Results and discussion

3.1. Morphology of the KC/CNCs bio-nanocomposite films
The polymer nanocomposites were prepared with different CNCs filler loading level (0 wt%, 1.0 wt%, 3.0 wt%, 5.0 wt%) in KC matrices. Figure 1 shows the rod-like cellulose nanocrystals were well dispersed into polymer matrix, when the low concentrations of CNCs were used. The best CNCs distribution was observed at 1.0 and 3.0 wt%, filler loadings. On the other hand, with the increase of CNCs contents (5.0 wt%) into KC hosts, their distribution became poor and most of the cellulose nanocrystals were agglomerated by hydrogen-bonded free hydroxyl groups.

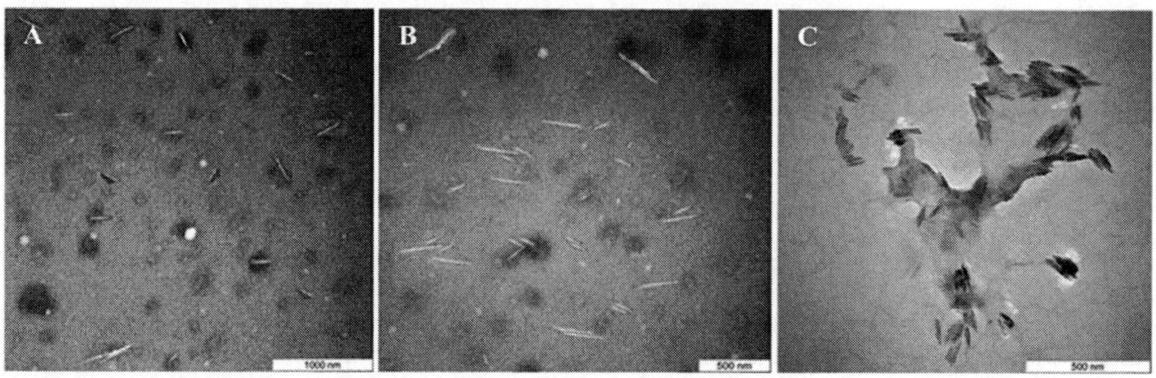

Figure 1. TEM images of KC/CNCs bio-nanocomposite films with 0.1 wt% (A), 3.0 wt% (B), and 5.0 wt% (C) CNCs contents.

3.2. Mechanical properties
Figure 2A and B display the effect of CNC concentration on the mechanical properties of the KC/CNC bio-nanocomposite films. The CNC with the KC matrix as the nano-filler exhibited an important reinforcing effect. As the CNC concentration increased, both the tensile strength and Young's

IOP Conf. Series: Materials Science and Engineering **368** (2018) 012013 doi:10.1088/1757-899X/368/1/012013

modulus improved considerably, but the elongation at break of the bio-nanocomposites declined. When the CNC content varied from 0 to 3.0 wt%, the tensile yield strength and Young's modulus enhanced from 295 to 710 MPa, representing a ~140 % enhancement, and the tensile strength increased from 45.7 to 108.4 MPa, indicating a ~137 % increase. This proposes that incorporating CNC into the KC matrix results in powerful interactions between the matrix and the filler, and thus confines the motion of the matrix and enhances rigidity [22]. Furthermore, increased crystallinity, which was donated by CNC, can also promote mechanical properties [23]. Further addition of CNC decreased the tensile modulus and strength, but this still stayed higher than that of the pure KC film. This decline is owing to interparticle interactions, which lead to the creation of weak points in the matrix. Instead, the elongation at break gradually decreased from 0 to 5.0 wt% of filler loading, with maximum decrease at 3.0 wt%. The elongation at breaks decreased from 8.3 to 2.5 wt%, representing a ~70 % reduction. This phenomenon can be described by the fact that the rigid filler network structure was created perfectly when the CNC concentration was 3.0 wt%.

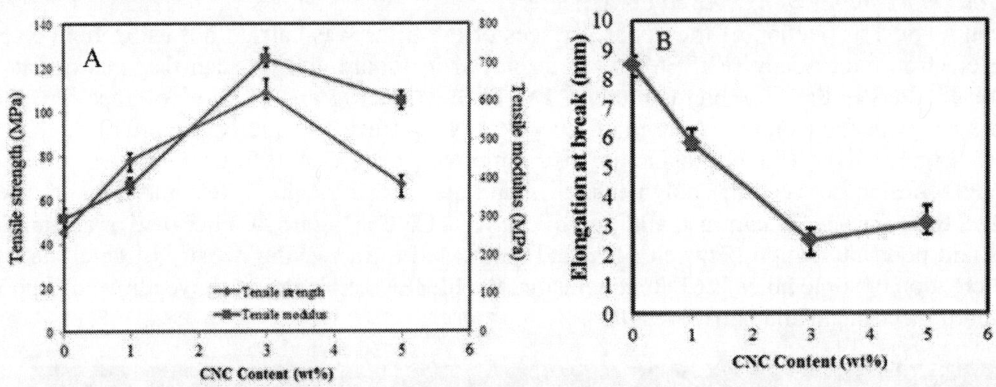

Figure 2. (A) Tensile strength and tensile modulus (B) elongation at break of KC/CNC bio-nanocomposites.

3.3. Oxygen transmission rate

Figure 3 showed the OTR versus percentage of cellulose nanocrystals. From figure 3, it is evidenced that the oxygen permeability is most extremely reduced, i.e. by ~80%, in the 3.0 wt% KC/CNCs compared to the control KC sample. The permeability of the bio-nanocomposite films is always lower than that of pure KC. These results propose that the combined CNCs make tortuous pathway for the diffusion of oxygen molecules.

When the low concentrations of CNCs were incorporated, the CNCs were distributed to a higher degree as revealed in the TEM micrograph, hence yielding a more effective barrier effect [24]. Thus an appropriate interaction between the KC and the surface of the CNCs can be happened. A weak interaction would lead to the being of hollows that permit the oxygen molecules to penetrate faster through the film. The penetrability is not declined more with a higher level of CNCs. This result can be owing to aggregation of the CNCs. The agglomeration observed beyond 3.0 wt% content may provide channels, or pores, in the film that allow further rapid diffusion [25].

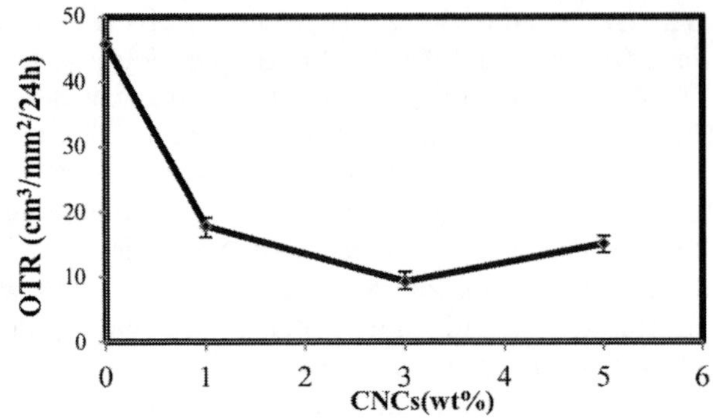

Figure 3. OTR of KC and its CNCs bio-nanocomposite films.

3.4. Surface morphology of KC-based films

The morphological inspection of the upper surfaces of the films was carried out using field emission scanning electron microscopy (FESEM) as is displayed in Figure 4. It is seen that incorporation of CNCs caused alters in the film microstructure. The FESEM micrograph of pure polymer film (Figure 4A) shows KC with some cracks. Adding of 3.0 wt% CNCs affects the microstructure of the KC films definitely (Figure 4B). The homogenous distribution of CNCs at 3.0 wt% content caused an appropriate adhesion between the polymer matrix and the surface of the CNCs and thus generated a smooth and tight surface. In contrast, addition of 5.0 wt% CNCs (Figure 4C) because of accumulation of CNCs and poor interaction between filler and matrix led to increasing density of crack deflection sites and creation of some holes [26]. In conclusion, the FESEM explanations have allowed supporting the measured mechanical and barrier results.

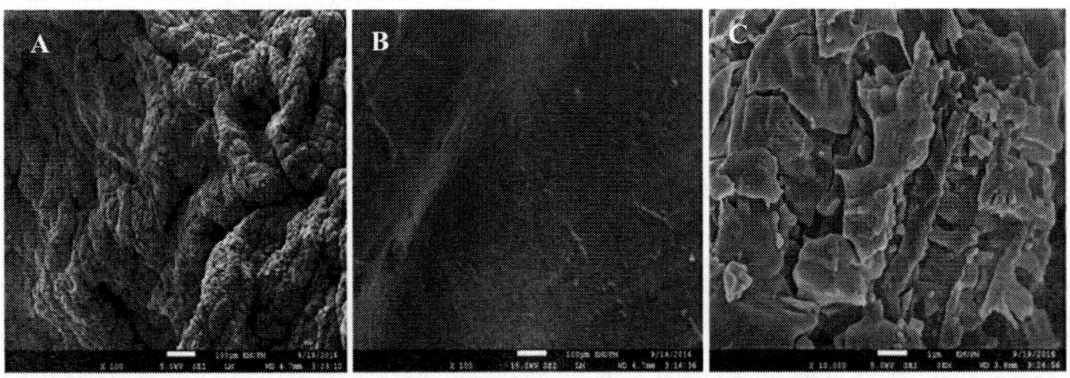

Figure 4. FESEM image of the upper surfaces of (A) KC, (B) KC/CNCs 3.0 wt% and (C) KC/CNCs 5.0 wt% films.

4. Conclusion

The reinforced bio-nanocomposites of KC matrix with cellulose nanocrystals were fabricated and potentiality of cellulose nano-sized filler was assessed on the structure and properties of resulted composites. The observation of TEM-based structures demonstrated that the CNCs were homogenously distributed at lower filler loading. While compared the control KC film and the KC/CNCs film, the bio-nanocomposite films displayed higher tensile and modulus strength; decreased elongation at break; and reduced oxygen transmission rate. This is owing to the parameters such as the nano-scale size effects of the CNC (high L/D), the high content of cellulose crystalline districts, the well dispersion of CNCs within KC matrix, and the strong interaction between CNC and KC matrix.

Acknowledgments
The authors are grateful to the Department of Bioprocess Technology, Faculty of Biotechnology and Biomolecular Sciences, for the laboratory facilities.

References
[1] Dufresne A 2008. Can. J. Chem 86 484–94
[2] Azizi S, Ahmad MB, Ibrahim NA, Hussein MZ and Namvar F 2014 Chin. J. Polym. Sci 32 1620-27
[3] Bras J, Hassan ML, Bruzesse C, Hassan EA, El-Wakil NA and Dufresne A 2010 Ind. Crops. Prod 32 627–33
[4] Habibi Y, Lucia LA and Rojas OJ 2010 Chem. Rev 110 3479–500
[5] Wang Y, Tian H and Zhang L 2010 Carbohydr. Polym 80 665–71
[6] Pandey JK, Chu WS, Kim WS, Lee CS and Ahn SH 2009 Composites: Part B 40 676–80
[7] OSullivan AC1997 Cellulose 4 173–07
[8] Favier V, Chanzy H and Cavaille JY 1995 Macromolecules 28 6365-7
[9] Dufresne A, Kellerhals MB and Witholt B 1999 Macromolecules 32 7396–01
[10] Habibi Y, Lucia LA and Rojas OJ 2010 Chem. Rev 110 3479–00
[11] Yang J, Han Ch-R, Zhang X-M, Xu F and Sun R-C 2014 Macromolecules 47 4077–86
[12] Gu M, Jiang Ch, Liu D, Prempeh N and Smalyukh II 2016 ACS Appl. Mater. Interfaces 8 32565–73
[13] Embuscado ME and Huber KC 2009 Films and Coatings for Food Applications Springer XIV, 416 p. 117 illus
[14] Martins JT, Bourbon AI, Pinheiro AC, Souza BW, Cerqueira MA and Vicente AA 2012 Food Bioproc Tech 1–12
[15] Shojaee-Aliabadi S, Hosseini H, Mohammadifar MA, Mohammadi A, Ghasemlou M, Ojagh SM, Hosseini SM and Khaksar R 2013 Int. J. Biol. Macromol 52 116–24
[16] Xu JB, Bartley JP and Johnson RA 2003 J. Membr. Sci 218 131–46
[17] Azeredo H 2009 Food. Res. Int 42 1240–53
[18] Quilaqueo Gutiérrez M, Echeverría I, Ihl M, Bifani Vand Mauri AN 2012 Carbohydr. Polym 87 1495–02
[19] Shojaee-Aliabadi S, Mohammadifar MA, Hosseini H, Mohammadi A, Ghasemlou M, Hosseini SM, Haghshenas M and Khaksar R 2014 Int. J. Biol. Macromol 69 282–89
[20] Rhim JW 2012 J. Food Sci 77 66–73
[21] Beck-Candanedo S, Roman M and Gray DG 2005 Biomacromolecules 6 1048
[22] Azizi S, Ahmad MB, Hussein MZ, Ibrahim NA, and Namvar F 2014 Int. J Nanomedicine 91909
[23] Yi XX, Ying QY, Jiu YX, Shi FY and Wang F 2013 Carbohydr Polym 98 1573–77
[24] Fortunatia E, Peltzer M, Armentano I, Torre L, Jiménez A and Kenny JM 2012 Carbohyd. Polym 90 948
[25] Paralikar SA, Simonsen J and Lombardi J 2008 J. Membr. Sci 320 248
[26] Azizi S, Ahmad MB, Ibrahim NA, Hussein MZ, and Namvar F 2014 Int. J. Mol. Sci 15 11040-53

Effect of fiber loading on mechanical and water absorption capacity of Polylactic acid/Polyhydroxybutyrate-co-hydroxyhexanoate/Kenaf composite

N I Ismail[1], Z A M Ishak[2]

[1]School of Material and Mineral Resources Engineering, Universiti Sains Malaysia, Engineering Campus,14300 Nibong Tebal, Pulau Pinang, Malaysia
[2]Cluster of Polymer Composite (CPC), Science and Engineering Research Centre (SERC),
Engineering Campus, Universiti Sains Malaysia 14300 Nibong Tebal, Penang, Malaysia

*Corresponding author: zarifin.ishak@googlemail.com

Abstract. Natural fibers have been found to be excellent reinforcing materials for preparing polymer matrix based composites. Environmental concerns also have resulted in replacing petrochemically derived polymer with the biodegradable renewable resource. In this study, mechanical properties and water absorption of polylactic acid (PLA)/ polyhydroxybutyrate co-hydroxyhexanoate (PHBHH) matrix with blending ratio 75:25 and its composites reinforced with fibers of kenaf have been studied. The composites, with different fiber loading (10, 20, 30 and 40 wt%) were prepared by melt blending techniques using an internal mixer at 50 rpm and 165°C for 8 minutes followed by a hot compression moulding technique. The results showed that tensile strength and modulus increased with increasing of fiber content. The water absorption behavior of biocomposites as a function of days was also investigated. It was found that the water absorption amount of biocomposites increased with increasing of fiber content and exposure time. It can be summarized that the recent developments in PLA/PHBHH/Kenaf show highly promising perspectives for the replacement of traditional petrochemical based polymers in applications such automotive to encourage materials degradation.

1. Introduction

Increasing concerns about the potentially damaging effects of nondegradable plastics on the environment, fast depletion of traditional sources of fossil fuels and the increasing cost of petrochemical feedstocks has raised new challenges in polymer research. It is now widely accepted that the use of long-lasting polymers of short lived applications is not sustainable. Among the vast number of biopolymers studied, two of the most promising are polyhydroxybutyrate-co-hydroxyhexanoate (PHBHH) and polylactic acid (PLA). Both of them are biodegradable, sustainable and biocompatible polyesters that can be produced from renewable resources. They are used in consumer products by several industrial sectors and have comparable thermal and mechanical properties to those of some conventional polymers. PLA is semi-crystalline polyester derived from lactic acid, obtained entirely from renewable resources such as corn, sugar beet and wheat. It is an aliphatic thermoplastic polyester that boasts a high modulus, high strength and good clarity. The glass

IOP Conf. Series: Materials Science and Engineering **368** (2018) 012014 doi:10.1088/1757-899X/368/1/012014

transition temperature of PLA is in the range of 55 °C – 60 °C while the melt temperature is in the range of 145 °C -160 °C. It has the great mechanical strength and easy to process. PLA is now produced on a large scale and used in various applications including packaging, agriculture and textiles. Even though PLA is compatible with many current processing techniques, the fact that it has a high glass transition temperature leads to brittleness in the final products [1]. In order to resist this problem, many researchers have examined a few ways in modifying PLA, such as copolymerization or composite formation [2]. The combination of PLA with other polymers may be the suitable methodology since this process allows the incorporation of new materials with enhanced properties. This technique also less expensive compared to chemical modifications or syntheses of tailor-made macromolecules.

PHB discovered by Lemoigne in 1925, is the most representative members of the PHA family. However, its high crystallinity and large spherulites turn the material to be stiff and rigid. PHB is only suitable for processing in a narrow temperature window due to unstable thermal during conventional melt processing. To overcome these limitations, some PHB copolymers have been biosynthesized by a variety of microorganisms. PHBHH was synthesized from oils and fats using Aeromonas sp. OL338 and Aeromonas sp. FA440. The shortcomings of conventional PHAs such as PHB and PHBV can be resisted by this new co-polyester [3]. PHBHH is more suitable for making films and other soft articles [4]. This is due to the unique characteristic of PHBHH which is substantially more flexible and ductile. The randomly distributed monomer units (hydroxyhexanoate) are excluded from (hydroxybutyrate) crystalline lattice, forming a short chain branch. This branch acts as a molecular defect hence disrupts the excessive regularity of the polymer chain. Then, the melt temperature (T_m) and crystallinity also will be reduced. The selection of individual components and their proportions must be carefully considered in a composite system since it will give effect on the physical, mechanical and thermal properties of the final materials.

Natural fiber, such as kenaf fiber can be incorporated into the polymer to achieve the desired properties and better texture in the resulting biocomposites. It is a basic philosophy to create new products that are eco-friendly and sustainable. Kenaf fiber has attracted much attention for this purpose because it offers both ecological and economic advantages. Kenaf fiber was used as reinforcement in the polymer composites because of disintegration and hydrophilic properties. Kenaf fiber also suitable as a filler in composite materials because it is not abrasive during processing, biodegradable, has low density and specific mechanical properties. Kenaf can grow under a wide range of climate conditions to a height of more than 5-6 m in 6-8 months [5].

There are three different mechanisms to figure about moisture diffusion in polymeric composites [6]. The first involves the diffusion of water molecules inside the micro gaps between polymer chains. The second one involves capillary transport into the gaps and flaws at the interfaces between the fiber and the matrix. It is caused by poor wetting and impregnation during the initial manufacturing stage. The third involves transport of microcracks in the matrix arising from the swelling of fibers, especially in the case of natural fiber composites. In this work, PLA/PHBHH/Kenaf composites were prepared via direct melt compounding. The mechanical and morphological properties of the composites were evaluated as a function of fiber loadings and water absorption rate. It can be hypothesized that a composite of the PLA, which is naturally hard and brittle, with PHBHH would effectively balance the shortcomings of the individual polymers. PHBHH is preferred rather than other PHAs conventional polymer such PHB or PHBV because this conventional polymers are thermally unstable during conventional melt processing and only suitable for narrow temperature processing. There is very little research has been reported regarding PLA/PHBHH and no systematic research published concerning the impact of natural fiber on these composites.

2. Experimental

2.1 Materials

A commercial linear and semicrystalline PLA (Ingeo™, 3052D) was provided in pellet form by NatureWorks LLC, USA. According to the supplier, PLA has a melt flow rate about 14 g/10 min and a density of 1.24 g/cm^3. The bacterial polyester, PHBHH, Aonilex was supplied by the Kaneka Corporation (Osaka, Japan). Mechanically retted Kenaf fiber (KF) in lengths 1.0 mm was supplied by Kenaf Fiber Industries Sdn. Bhd. (KFI), Malaysia.

2.2 Composite Processing

Before manufacturing the composite samples, the two biodegradable polymers were dried at 50°C overnight. Kenaf fibers were dried in a circulating air oven at 80°C for 12 h and were kept in desiccators prior to processing. PLA/PHB-HH/Kenaf composites were prepared by direct melt compounding using a twin roll counter rotating mixer (RTOI-55/20, POTOP, Guangzhou) at polymer blending ratio of 75/25 of PLA/PHBHH. The barrel temperature was set at 165°C and the screw speed was 50 rpm. The total mixing time was 8 minutes. It is started by adding PLA and PHBHH together and mixed for 3 minutes at 10 rpm then kenaf was gradually added and mixed for another 5 minutes at 50 rpm before being discharged. The mixtures produced were then compression molded in a hydraulic compression press model Kao Tieh Gotech Testing Machine Inc. (Taipei, Taiwan) for 5 minutes at 165°C with a pressure of 150 kg/cm^2. Prior to testing, the moulded specimens were kept in desiccators and conditioned at room temperature for at least 24 h.

Table 1. Designation and Composition of Materials

Designation	Materials	PLA (wt %)	PHBHH (wt %)	Kenaf (wt %)
PLA	PLA	100	0	0
PLA/KF10	PLA + Kenaf Fiber	90	0	10
PLA/KF20	PLA + Kenaf Fiber	80	0	20
PLA/KF30	PLA + Kenaf Fiber	70	0	30
PLA/KF40	PLA + Kenaf Fiber	60	0	40
PLA/PHBHH	PLA + PHBHH	75	25	0
PLA/PHBHH/KF10	PLA + PHBHH + Kenaf Fiber	67.5	22.5	10
PLA/PHBHH/KF20	PLA + PHBHH + Kenaf Fiber	60	20	20
PLA/PHBHH/KF30	PLA + PHBHH + Kenaf Fiber	52.5	17.5	30
PLA/PHBHH/KF40	PLA + PHBHH + Kenaf Fiber	45	15	40
PHBHH	PHBHH	0	100	0
PHBHH/KF10	PHBHH + Kenaf Fiber	0	90	10
PHBHH/KF20	PHBHH + Kenaf Fiber	0	80	20
PHBHH/KF30	PHBHH + Kenaf Fiber	0	70	30
PHBHH/KF40	PHBHH + Kenaf Fiber	0	60	40

2.3 Characterization techniques

2.3.1 Tensile and Flexural Tests

Tensile and flexural tests were carried out in accordance with ASTM D 638 and ASTM D 790 using a universal testing machine model Instron 3366 equipped with control system series IX. A crosshead speed of 5mm/min was used for both tests. 5 specimens were tested for every series of composites. All tests were done at room temperature.

2.3.2 Water Absorption Test

Water absorption tests were carried out according to ASTM D-570 specifications. Tensile specimens were cut from the compression-molded plates and used for the measurements of water absorption. After drying at 50°C for 24 hr to a constant weight to a precision of 0.001 g, the weight of the specimens before water immersion (W_d) were measured with a balance. The specimens were immersed in water at room temperature for 50 days. The weights of the specimens were measured at regular intervals and the amount of water absorption (M_t) was calculated according to Eq. (1):

$$M_t(\%) = \frac{W_w - W_d}{W_d} X\, 100 \qquad (1)$$

where W_d and W_w denote the weight of the dry material (the initial weight of materials before water immersion) and the weight of materials after water immersion, respectively. The surfaces of these specimens were thoroughly dried with tissue papers, and they were weighed immediately to determine the weight of the specimens (W_w). The specimens were immersed until they were saturated.

2.3.3 Scanning Electron Microscopy (SEM)

Scanning electron microscopy (SEM) was employed in order to observe the morphology of the sample's fracture surface from composites broken with a field-emission Scanning Electron microscope model Zeiss Supra 35VP. All samples were first coated with a thin layer of gold using a VG Microtech-Polaron Sputter Coater and analyzed.

3. Result and Discussion

Good strength properties of biocomposite materials can be produced when the fiber is uniformly dispersed and distributed in the matrix. PLA/PHBHH/Kenaf biocomposite tensile test sample were successfully prepared by melt blending techniques and then compression molded. The surfaces of all samples were smooth and showed brown in colour which indicating the uniform dispersion of kenaf within blends. Fiber loading and water absorption effect on mechanical properties were determined. The morphology of the fracture surface also was examined by scanning electron microscope.

3.1 Effect of Fiber Loading on Mechanical Properties

3.1.1 Tensile Properties. Figure 1 shows the effect of fiber loading on the tensile strength of PLA/PHBHH/kenaf fiber composites. In this study, PLA and PHBHH were used as the reference sample.

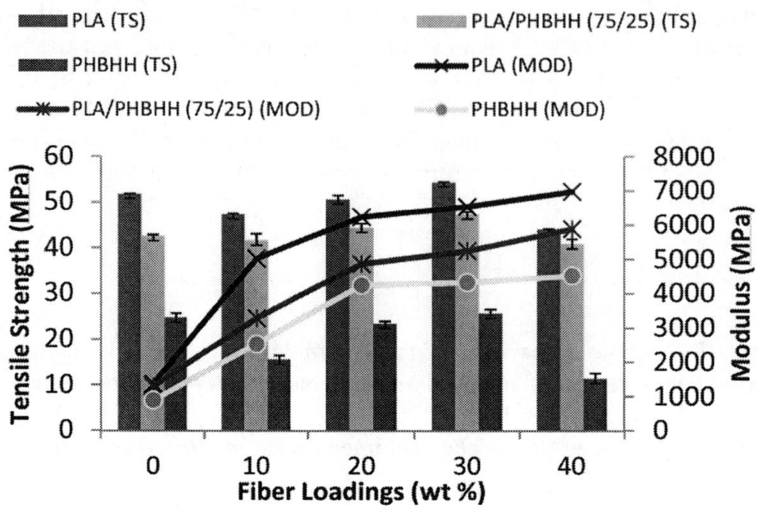

Figure 1. Tensile strength and tensile modulus of composites at different fiber loadings.

The neat PLA present the highest tensile strength results among all composites. Generally neat PLA show better tensile strength than composite PLA and these findings are in agreement with previous study [7]. In the figure, PLA/PHBHH/Kenaf composite present the tensile strength results with 47.28 MPa at 30 wt% of fiber loading. The tensile strength increased from 41.54 MPa, 44.35 MPa and 47.28 MPa with the increased of fiber loading and decreased thereafter. The strength increases until reaching the maximum filler loading at 30 wt% and decreased as more fiber was incorporated into the matrix. Suddenly dropped of strength can be seen at 30 wt% to 40 wt% of fiber loading from 47.28 MPa to 40.78 MPa.

According to the investigation, at 10 wt% of kenaf loading, the tensile strength of the materials were not improved indicating that there was poor dispersion and fiber distribution which lead to poor adhesion between the fiber and the matrix. This is due to the fact that the polymer resin unable to transmit and distribute the applied stress to the kenaf fibers resulting in lower strength. The fiber distribution and dispersion started to improve at 20 wt% to 30 wt% of fiber loading thus gives better strength compared to 10 wt% of fiber loading. Similar findings have been reported who incorporated 10 –30 wt % kenaf fibre into a PLA matrix [8]. They reported that the tensile strength of the composite increased with an increase in the amount of the fibers in the composite. However, at 40 wt% of fiber loading, the tensile strength of the materials decreased due to insufficient wetting of fiber by matrix. Other than that, this might also be caused by fiber-fiber interaction and agglomerates of fiber in the matrix since there are no polymer chains between the fiber layers. The presence of fiber agglomerates would promote weak points within the composite which would greatly affect its mechanical properties. The addition of kenaf will increase the modulus, but the higher fiber content will not improve the modulus in composites. Other factors that may contribute to the enhancement of the strength and stiffness of the biocomposite materials are porosity of the composites and chemistry of the natural fiber's surfaces [9]. The porosity arises due to inclusion of air during processing, limited wettability of fibers, lumens and other hollow features within fibers/fiber bundles (which may become closed during processing at high pressure) and due to the low ability of fiber to compact. Porosity in natural fiber composites has been shown to increase with fiber content, more rapidly once the geometrical compaction limit has been exceeded, depending on fiber type and orientation of fiber; flax/PP composites were found to have porosity increasing from 4 to 8 volume% as fiber content increased from 56 to 72 m% [10]. As mentioned earlier, its inclusion in models has been shown to give an improved prediction of strength and stiffness. Overall, the results obtained suggest an almost linear increase in the modulus with the addition of kenaf fiber. This may be due to the kenaf fibers imparting an overall increase in stiffness to the composite as the kenaf content is increased. As the fiber content in the composite increases, it restricts the mobility of the matrix, consequently the composite become stiffer and the tensile modulus increase. This kind of trend in the Young's modulus with increased fiber content in the composite was also observed in the case of polypropylene (PP) /wood fiber composites where the stiffness of the composite were found to be dependent on the fiber content and the homogeneity of the fiber dispersion [11].

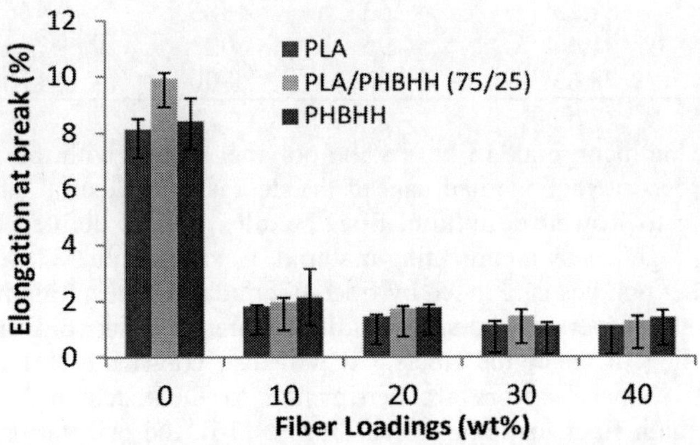

Figure 2. Elongation at break of composites at different fiber loadings.

The effect of various kenaf loadings on the elongation at break of the composite materials is shown in Figure 2. In general, the elongation at break properties of the composites decreased when 10 wt % of kenaf fiber was added and the value was gradually reduced with more addition of kenaf fiber and this trend has been expected. A reduction in strain at break with increasing fiber content in a polymer matrix is common in biocomposites [12]. PLA/PHBHH can elongate more, due to its elasticity. The blend is naturally brittle; adding more fibers in the composite should lower the strain at break and create stress-concentrated areas resulting in lower strength. It shows that the presence of kenaf fibers does not contribute to the elasticity or the final composite flexibility. The observed decrease in flexibility of the composites is likely to be related to the high stiffness of the composites. As we increased the fiber content, the composites became more rigid and the fiber restricted the stretching of composites. Therefore, the elongation at break decreased due to lower deformation behavior of the composites. There also reported that a rigid interface between the fibers and the matrix material reduces the deformability of the composites [13].

3.1.2 Flexural Properties. The effect of different kenaf loading on the flexural properties is shown in Table 2. Increasing the kenaf loading from 10 to 30 wt% resulted in an increase of the flexural strength of the composites. However, at 40 wt% kenaf loading, a reduction of about 16.32% in flexural strength was observed. This indicates that 30 wt% could be considered as the optimum fiber loading, and a further increase of kenaf loading would lead to inferior flexural strength. It is expected at 30% of fiber loading, good interfacial adhesion happened between kenaf and polymer matrix due to chemical interactions, such as hydrogen bonds. Hydrogen bonds are present in the composite between the hydroxyl groups in the kenaf fibers with the terminal hydroxyl groups [14] and the carbonyl groups of the ester groups of PLA [15]

Table 2. Flexural properties of composites at different fiber loadings

Composites	Flexural (MPa)					
	Strength			*Modulus*		
	PLA	*PLA/PHBHH*	*PHBHH*	*PLA*	*PLA/PHBHH*	*PHBHH*
0	115.80	85.45	38.32	3700	2700	1040
10	93.50	80.32	40.11	3967	2892	2243
20	96.17	81.23	41.63	4885	3445	2997
30	100.20	91.47	44.95	6021	4743	3493
40	88.46	78.63	25.80	6790	5110	3725

Insufficient wetting of the incorporated fiber by the polymer matrix with the presence of the fiber bundles and more fiber-fiber interactions had caused the decrease in flexural strength of 40 wt% of kenaf loading. According to Rowell definition, fiber bundles can be defined as a combination of individual fibers adhering together by a cementing material, usually lignin. The contact area between fibers and the matrix in fiber bundles is reduced by fiber-fiber adhesion resulting in poor stress transfer from the matrix phase to the dispersed fibers. In addition, inter-fibrillar voids also could be formed from the fiber agglomeration of entangled fibers. It will also contribute to the reduction of fiber-matrix interactions. All of these factors are detrimental to the mechanical performance of the composites, especially at high fiber loading. According to [16], the orientations of short fiber in a compression moulded specimen are usually in a combination of random-in-plane orientations of multi-layer composite laminates. Increasing the kenaf loading will increase the possibility of fibers in same lamina or different laminas orienting in a better direction to bear the bending load. However, there is an optimum fiber loading to be considered that will promote to fiber–fiber interactions or agglomeration. In this case, the fibers have the possibility of contacting each other either with fibers in the same laminate or fibers in neighbouring laminates leading to lower flexural properties.

The flexural modulus of the PLA/PHBHH/Kenaf composites shown to be linearly increased with increasing (0-40 wt%) kenaf content. High kenaf content combined with the stiffness of the fiber could explain the significantly enhanced flexural modulus of PLA/PHBHH/Kenaf composites. Kenaf might restrict the mobility of the polymer chain and thus hinder the ability of the polymer to deform. These observations are supported by [17], who concluded that good fiber dispersion in the composite system could be confirmed by observing the linear increase in the composite's modulus.

3.2 Effect of Water Absorption on Mechanical Properties

It was found that the water absorption amount of the composites increased with increasing the cellulose content. As it can be seen in the Figure 3, 10 days of soaking presented a rapid increase the level of water absorption then demonstrated a gradual increase in the water uptake, then the composites finally reached an equilibrium point, which is no more water uptake into biocomposites and the weight of biocomposites remains constant. As reported [18], the number of hydroxyl (OH) group in the composites increased when the cellulose content increased, resulting in increasing the water absorption due to these free OH groups come in contact with water and form hydrogen bonding, which results in weight gain in the biocomposites.

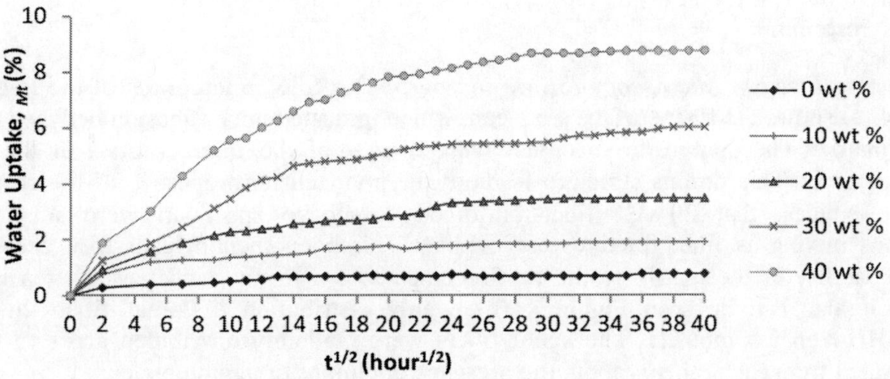

Figure 3. The percentage by weight of PLA/PHBHH/Kenaf composite after water absorption at different kenaf fiber loading

The higher the fiber loading is, the more hydrophilic sites exist in the composites and the higher water absorption of the composites. It is expected that the exposed fibers, which are located at or near the specimen's surface, absorb moisture faster than those in the interior parts (i.e., fibers that are entrapped or coated by the matrix). That is the reason behind the fast absorption of water in the composite with 40% KF since the composite has more fibers on the surface to absorb water faster. On the other hand, fibers in PLA/PHBHH/KF20 were mostly coated by the hydrophobic PLA and PHBHH matrix, which acts as a barrier for the diffusion of water, and, therefore, their absorption rates were slower.

3.2.1 Tensile Properties. The tensile properties of natural fiber composites decreased after water absorption because of the plasticizing effect of the water molecules, which weaken the interfacial bond strength between matrix and kenaf. When composites were exposed to moisture, the hydrophilic kenaf fiber starts swelling. As a result of fiber swelling, micro cracking of the brittle thermoplastic resin occurs. The cellulose content in kenaf fiber, then contributes to more water penetrating into the interface through microcracks induced by swelling of the fibers creating swelling stresses leading to composite failure. In other words, the higher hydrophilic character of fiber, which facilitates water absorption, promoting the degradation of the fiber–matrix interface, deleteriously affects the tensile strength of the composite. The lower tensile strength at higher fiber contents could be due to the lack of matrix that leads to inefficiency of stress transfer from the matrix to fibers and vice versa as reported [19].

Table 3. Tensile Properties of PLA/PHBHH/Kenaf after water absorption at different kenaf fiber loadings.

Composite	Tensile Strength (MPa)		Tensile Modulus (MPa)		Elongation at break (%)	
	Dry	*Wet*	*Dry*	*Wet*	*Dry*	*Wet*
0	42.45	41.32	1277	950	9.92	8.72
10	41.54	40.24	3277	1376	1.96	1.54
20	44.35	42.75	4852	2472	1.72	1.32
30	47.28	45.34	5237	2526	1.46	1.10
40	40.78	38.53	5876	2114	1.28	0.83

The decline in tensile modulus is expected because of the weakening of the fiber/matrix interface and the plasticization of the matrix and fiber. The normal trend for thermoplastic composites filled with solid fibers showed the tensile moduli of the composites increased as the fiber loading increased.

This is due to the restriction of the mobility and deformability of the matrix with the introduction of mechanical restraint.

3.2.2 Scanning Electron Microscopy (SEM) micrograph. A SEM micrograph of the fractured surface in Figure 4 (a) shows the fiber surface are clean, which indicated poor adhesion between the fibers and the PLA matrix. The composite specimen with 20 wt% of the filler consists of the considerable amount of pores. This porous structure leads to the poor tensile properties of the composites. It is rational to suggest that 20 wt% fiber-reinforced composite specimen were subjected to poor homogenous mixing as fiber fractures are evident. This phenomenon leads to a decrease in stress transfer capability of the matrix to the reinforcement. As a result, a decrease in tensile strength is observed. It also can be seen Figure 4 (b) evenly distribution of kenaf fibers in the 30 wt% PLA/PHBHH/Kenaf composite. The kenaf fibers were randomly distributed across a unit area and were separated from each other without the presence of clumps or agglomerates.

(a) (b)

Figure 4. The FESEM micrograph image of the kenaf distribution of the (a) 20 wt% and (b) 30 wt% PLA/PHBHH/Kenaf composite.

The fractured surface of PLA/PHBHH/ KF (40 wt %) is shown in Figure 5 in which a pullout fiber can be seen clearly. Thus, the fracture surface of the composite appeared to be dominated by pull-out damage rather than fiber breakage. In fact, there was some fiber debonding observed. The strong interface between fiber and matrix results in composite with high flexural strength. The failure indicates the separation between fiber and matrix. When the content of reinforced fiber reached above the certain value; these excessive fibers affect the even combination between matrix and fiber.

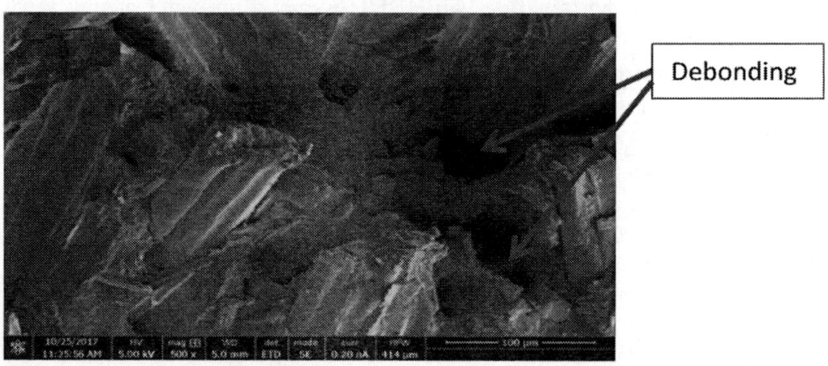

Figure 5. The FESEM micrograph image of the kenaf distribution on the fracture surface of 40 wt% PLA/PHBHH/Kenaf composite.

4. Conclusion

The effect of kenaf fiber loading and water absorption on the mechanical properties of PLA/PHBHH reinforced kenaf has been studied and concluded as follows. It shows that mechanical properties of composite increase when kenaf being incorporated into it. The reinforcement helps in stress transfer within the composite compared to the one without kenaf. However, the properties of the composite after water absorption are lower than dry composite due to weak interfacial bond between matrix and kenaf since plasticization effect occurs. Exposure to moisture results in significant drops in tensile properties due to the degradation of the fiber-matrix interface.

5. Acknowledgement

The authors would like to express special gratitude to Universiti Sains Malaysia and MyMASTER scheme for providing financial assistance.

6. References

[1] Saeidlou, S, Huneault, M A, Li, H and Park, C B (2012). Poly(lactic acid) crystallization, *Prog. Polym. Sci.* 1657-1677.

[2] Zhang J, Krishnamachari P, Lou J, Shahbazi A, (2010). Synthesis of poly(L(+) lactic acid) by polycondensation method in solution. Proceedings of the 2007 National Conference on Environmental Science and Technology' Springer, New York 3–8

[3] Lim, J S, Park, K, Chung, G S, and Kim, J H (2013). Effect of composition ratio on the thermal and physical properties of semicrystalline PLA/PHBHHx composites. *Materials Science and Engineering*, Elsevier

[4] Noda, I, Green, P R, Satkowski, M M, and Schechtman, L A. (2005). Preparation and Properties of a Novel Class of Polyhydroxyalkanoate Copolymers. *Biomacromolecules* 2005, 6, 580-586.

[5] Liu, Y, Labuschagne, M T.(2009) The influence of environment and season on stalk yield in kenaf. *Ind. Crop Prod.* 377–438

[6] Karmakar AC.(1997). Effect of water absorption on dimensional stability and impact energy of jute fibre reinforced polypropylene. *J Mater Sci Lett* 1997;16:462

[7] Oskman, K, Skirfvars, M, Selin, J F. (2003). Natural fibers as reinforcement in Polylactic acid (PLA) composites. *Compos. Sci. Technol.*,1317-1324.

[8] Ibrahim, N A, Ahmad, S N A, Yunus, W M Z W and Dahlan, K Z M (2009). Effect of Electron Beam Irradiation and Poly(vinyl pyrrolidone) Addition on Mechanical Properties of Polycaprolactone with Empty Fruit Bunch Fiber (OPEFB) Composite, *Journal of Express Polymer Letters*, 3(4): 226-234.

[9] Shibata, M, Ozawa, K, Teramoto, N, Yosomiya, R, and Takeishi, H. (2003). Biocomposites made from short abaca fiber and biodegradable polyesters. *Macromolecular Materials and Engineering* 288(1), 35-43.

[10] Pickering, K L, Aruan Efendy M G and Le T M (2016). A review of recent developments in natural fiber composites and their mechanical performance. *Compos Part A : Applied Sci and Manufacturing*, Vol **83** Pages 98-112

[11] Karmarkar, A, Chauhan, S S, Modak, J M, Chanda, M, (2007). Mechanical propertiesof wood-fiber reinforced polypropylene composites: effect of a novel compatibilizer with isocyanate functional group. *Compos. Part A* 38, 227–233.

[12] Mathew, A P, Oksman, K, and Sain, M. (2005). Mechanical properties of biodegradable composites from polylactic acid (PLA) and microcrystalline cellulose (MCC*), Journal of Applied Polymer Science* 97, 2014-2025

IOP Conf. Series: Materials Science and Engineering **368** (2018) 012014 doi:10.1088/1757-899X/368/1/012014

[13] Rozman, H D, Tan, K W, Kumar, R N, Abubakar, A, Mohd. Ishak, Z A, and Ismail, H. (2000). The Effect of lignin as a compatibilizer on the physical properties of coconut fiber-polypropylene composites, *European Polym.* J. 36, 1483-1494
[14] Bax, B, and Müssig, J. (2008). Impact and tensile properties of PLA/cordenka and PLA/flax composites, *Composites Science and Technology* 68, 1601-1607.
[15] Garlotta, D. (2001) A literature review of Poly(Lactic Acid). *J. Polym. Environ.* 63–64.
[16] Ahmad Thirmizir M Z, Mohd Ishak Z A, Taib R.M, Rahim S, Mohamad Jani S. (2011) Kenaf-Bast-Fiber-Filled Biodegradable Poly(butylene succinate) Composites: Effects of Fiber Loading, Fiber Length, and Maleated Poly(butylene succinate) on the Flexural and Impact Properties. *Journal of Applied Polymer Science*, Vol. **122**, 3055–3063
[17] Canigueral, N, Vilaseca, F, Méndez, J A, López, J P, Barberà, L, Puig, J, Pèlach, M A, and Mutjé, P. (2009). Behavior of biocomposite materials from flax strands and starch-based biopolymer, *Chemical Engineering Science* 64(11), 2651-2658.
[18] Tawakkal, I S M A, Talib, R A, Abdan, K, Chin, N L, (2012). Mechanical and physicalproperties of kenaf derived cellulose (KDC)-filled polylactic acid (PLA) composites. *BioResources* 7, 1643–1655.
[19] Anuar, H, Zuraida, A. and Fuad, F. (2010). Biodegradable PLA-Kenaf Fibre Biocomposite for Cleaner Environment.
[20] Afrifah, K A, Hickok, R A, Matuana, L M (2010) Polybutene as a matrix for wood plastic composites. *Compos. Sci. Technol.* 167–172
[21] Dhakal, H, Zhang, Z, Richardson, M. (2007) Effect of water absorption on the mechanical properties of hemp fiber reinforced unsaturated polyester composites. *Compos. Sci. Technol.* 1674–1683.
[22] Gupta B, Revagade N, Hilborn J, (2007). Poly(lactic acid) fiber: An overview. *Progress in Polymer Science*, 34, 455–482.
[23] Nishino, T, Hirao, K, Kotera, M, Nakamae, K, Inagaki, H,(2003). Kenaf reinforced biodegradable composite. *Composites Science and Technology* 63, 1281–1286.
[24] Yussuf, A A., Massoumi, I, Hassan, A, (2010). Comparison of polylactic acid/kenaf and polylactic acid/rise husk composites: the influence of the natural fibers on the mechanical, thermal and biodegradability properties. *J. Polym. Environ.* 18,422–429.

IOP Conf. Series: Materials Science and Engineering **368** (2018) 012015 doi:10.1088/1757-899X/368/1/012015

Nanofibrillated cellulose from Mahang gajah as thin film

L Jasmani[1*], S Adnan[1] and Z M A Ainun[2]

[1]Pulp and Paper Laboratory, Forest Products Division, Forest Research InstituteMalaysia (FRIM), 52109 Kepong, Selangor, Malaysia
[2]Laboratory of Biopolymer and Derivatives, Institute of Tropical Forestry and Forest Products, Universiti Putra Malaysia, 43400 UPM Serdang, Selangor, Malaysia

*Corresponding author: latifah@frim.gov.my

Abstract. Nanostructured cellulose has received tremendous attention due to its inherent unique properties. Being the most abundant organic polymer on earth, cellulose is the most suitable candidate for alternative replacement of petroleum-based products. Nanofibrillated cellulose (NFC) can be converted into thin films, aerogels and hydrogels. In particular, nanofibrillated cellulose suspension can form a neat and transparent thin film that has good film properties. Subsequently, this thin film can have numerous potential application areas such as in packaging, electronic etc. The objective of the study was to prepare nanofibrillated cellulose from local forest species namely Mahang gajah. After successful production of NFC, thin film was prepared from NFC and compared with commercial NFC. NFC was successfully produced from Mahang gajah possessing 19.7±17 nm width. Thin film produced from Mahang NFC showed comparable properties with commercial NFC such as thermal stability, porosity and wettability.

1. Introduction

The increasing demand for sustainable material is due to an increased drive for environmental-friendly product. Despite being the most abundant and renewable biopolymer on earth, it is only recently that cellulose has gained tremendous attention due to its conversion to nanostructured material or known as nanocellulose. Nanocellulose has inherent unique properties that include high strength, high surface area, non-toxic, biodegradable and ease of surface modification [1]. Owing to these characteristics, nanocellulose can be used as a strength enhancer in paper, fillers and additives to composites, super water absorbent material, flexible and transparent thin films, additives in cosmetics, oxygen barrier for food packaging and many more [2-3].

There are three types of nanocellulose namely nanofibrillated cellulose (NFC), nanocrystalline cellulose (NCC) and bacterial nanocellulose. The first two types of nanocellulose can be produced from any source of cellulose via chemical and/or mechanical treatments. Bacterial cellulose however is produced by bacteria of Glucanobacter genera such as *Acetobacter xylinum* [4]. NFC consists of alternating crystalline (highly ordered) and amorphous (disordered) regions. NFC is generally produced by delamination of wood pulp through mechanical pressure before or after chemical or enzymatic treatment. NFC can be converted into thin films, aerogels and hydrogels for use in various

IOP Conf. Series: Materials Science and Engineering **368** (2018) 012015 doi:10.1088/1757-899X/368/1/012015

industrial applications [5]. In particular, NFC suspension can form a neat and transparent thin film that has good film properties and subsequently can be used as substrate for organic electronic devices or solar cells [6]. Small micropore in the thin film produced from nanocellulose makes the light scatters less thus creating a transparent film-like paper. Many plastic materials used as a substrate have one important limitation i.e high thermal expansion coefficients which can be affected by high temperature during the preparation of flexible conductive film. On the other hand, NFC films exhibit good thermal properties as well as chemical stability and tunable optical characteristics [6]. Besides conversion to film, NFC can also be incorporated to other polymer matrices such as polyvinyl alcohol (PVA) [7], polylactic acid (PLA) [8] etc. functioning as strength additive.

Mahang gajah (*Macaranga gigantea*) is a common pioneering species from Rubiacea family that can be found easily on disturbed or open areas. Because of its abundance and easy availability, it finds application as a source of firewood or matchsticks [9]. In terms of its morphology and chemical properties, it has long fibre and high cellulose content thus can be a good source for cellulose [10]. Turning Mahang into nanocellulose can add its value and diversify its existing application. In this study, we described the preparation of NFC from Mahang gajah followed by fabrication of its thin film.

2. Materials and methods

2.1 Raw material
Mahang gajah (Macaranga gigantea) was obtained from FRIM campus in Kepong, Selangor. Sodium hydroxide and sodium sulfide were acquired from R&M Chemicals (UK), sodium chlorite from Merck (USA), sodium hypochlorite from R&M Chemicals (UK), glacial acetic acid from Ajax Chemicals (Australia), sodium bromide from Acros Organics (Belgium) and 2,2,6,6-methylpiperidine-1-oxyl (TEMPO) from Merck (Germany). Commercial NFC was acquired from The University of Maines, USA.

2.2 Primary processing, pulping and bleaching
Mahang gajah log was chipped using Taihei chipper followed by screening to obtain chips with suitable dimensions (2.5 x 2.0 x 0.3 cm) prior to pulping. Kraft pulping was chosen to prepare pulp for the production of NFC. In this study, the fibres were cooked in a rotary digester using 18% (sodium hydroxide and sodium sulfide) at 170°C for 31/2 hr. The ratio of fibre to liquor was at 1:6. After pulping, the pulp was bleached via elemental chlorine free bleaching process. Five step bleaching sequences namely DEDED (D denotes for chlorine dioxide and E is extraction by sodium hydroxide) was adopted. After a series of bleaching and washing, the bleached pulp was ready for use as a feedstock for NFC production.

2.3 Nanofibrillated cellulose preparation
Bleached pulp was fibrillated via PFI mill at 40,000 rpm. The pre-fibrillated pulp was then chemically pre-treated via TEMPO-oxidation process. TEMPO-oxidation method was followed and adapted according to method by Saito *et al.* [11]. To a suspension of NFC (2 wt%), NaBr (1.61 g), TEMPO (148.25 mg) and 1M NaOCl (12.75 ml) were added and continuously stirred. The pH of the suspension was maintained at pH 10 for 45 min. After that, a few drops of methanol were added to stop the reaction and the pH of the solution was brought down to pH 7 by adding 0.1 M HCl. The suspension was dialyzed against water overnight. After dialysis, the resulting suspension was sonicated for 30 min (20 % amplitude) and subsequently homogenized at 10,000 rpm for 2 h.

2.4 Thin film preparation
For NFC thin film preparation, a known volume of NFC suspension (0.5 wt%) was poured into a petri dish and left it to dry at room temperature. The thin film was then peeled off from the petri dish.

2.5 Characterization

The morphology of the sample was investigated using Stereo Microscope (Leica DM300 LED). Sample was stained with saffranin-o for better clarity. Scanning Electron Microscopy (SEM) (Quanta200) was also used for morphology and analysis was operated at accelerating voltage of 15 kV. Atomic Force Microscopy (AFM) imaging was carried out in tapping mode using Bruker Innova AFM. Thermogravimetric analysis (TGA) test was performed by TGA analyser (TA Instruments SDT Q600). Fourier Transform Infra-red (FTIR) spectra were recorded in the range of 4000-450 cm^{-1} from Perkin Elmer Spectrum 100 Series. Prior to FTIR analysis, sample was ground with potassium bromide to form a disc. Contact angle measurement was performed with contact angle analyser (First Ten Angstrom FTA1000). Porosity was recorded using Bendsten Porosity Tester. UV/Vis analysis was conducted via Jenway UV-Vis Spectrophotometer. Sample thickness was measured using L&W Micrometer.

3. Results and discussion

3.1 NFC production from Mahang gajah

After pulping, the yield for Mahang gajah was 39.9 % and Kappa number was 25.8. Morphology of Mahang gajah unbleached pulp was also determined. The fibres were cylindrical shape with tapered ends and their length varied from 1.22 to 2.50 mm with an average of 1.80 mm. The fibre length is considered long compared to other common Malaysian timbers as compiled and reported by Peh *et al.* [10]. The diameters of fibres ranged from 24 to 48 µm with an average of 35.6 µm. Figure 1 and 2 shows the photomicrographs of Mahang gajah pulp.

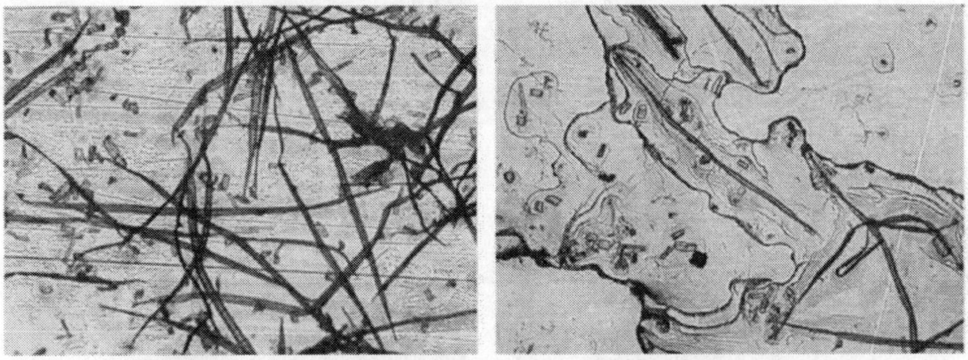

Figure 1. Micrograph of pulp at 4x magnifications.

Figure 2. Micrograph of pulp at 40x magnifications.

3.2 NFC morphology by SEM and AFM

NFC isolated from Mahang gajah pulp was subjected to morphology analysis by SEM and AFM. Figure 3 shows the images of NFC prepared from Mahang gajah. Even though SEM is not a suitable tool to study nanoscale material, nanofibrils can be identified as fine lines (thin fibrils) and the orientation was observed to be random-in-plane as shown in Figure 3 (a). Higher resolution microscope particularly AFM is essential to provide detailed information for assessment of cellulosic nanomaterial. Upon analysis using AFM, NFC was observed like aggregated fibrils with entangled network in agreement with observation from the literature [12]. NFC prepared from Mahang gajah displays diameter value of 19.7±17 nm and the length was assumed to be in several hundred micrometers due to the fibre network of which the length measurement is not possible. The nanoscale of the fibre diameter confirmed the reduction of size from the wood pulp. For commercial NFC, the width was reported as having nominal fibre width of 50 nm according to product specification sheet.

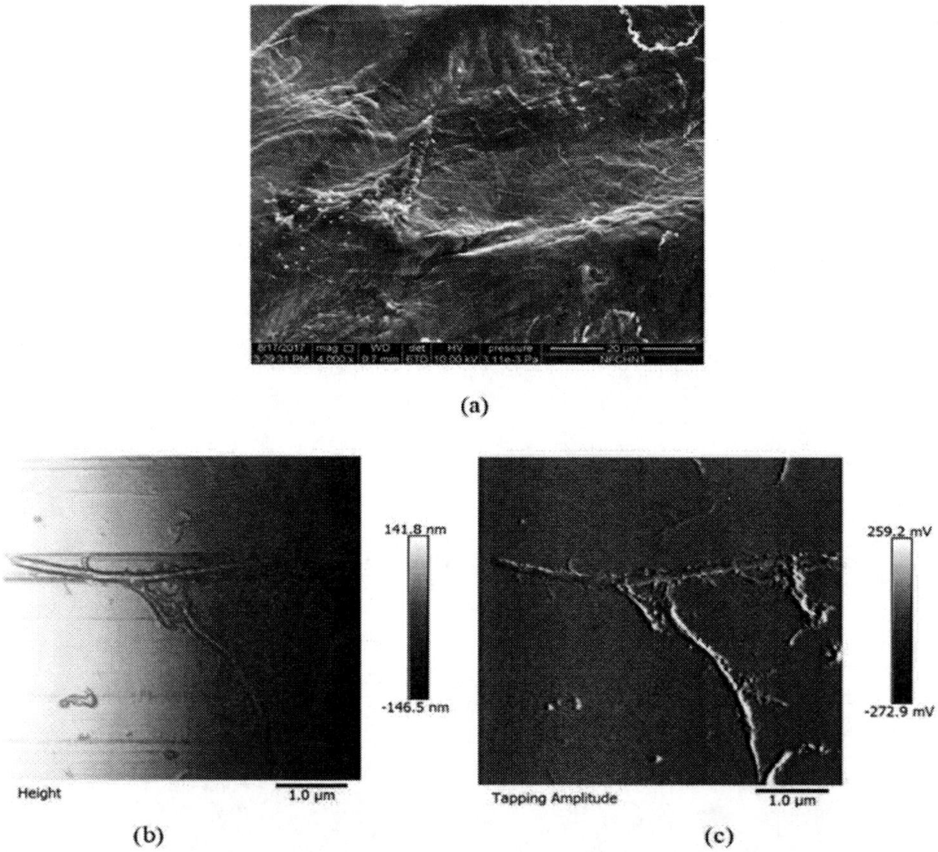

(a)

(b) (c)

Figure 3. Optical images of NFC by TEM (a) and AFM (b)(c).

FTIR was used to investigate the functionalities of the NFC. Looking at Mahang NFC, the typical cellulose spectra was observed. The broad peak between 3600-3000 cm^{-1} is due to the strong absorption of O-H stretching vibrations. Peak intensity between 3000-2800 cm^{-1} attributes to C-H and C-H_2 stretching vibrations. Stretching bands in the range of 1200-900 cm^{-1} are due to C-O and C-C. Primary alcohols in NFC exhibit strong absorption in 1075-1000 cm^{-1} whilst secondary alcohols absorb at 1150-1075 cm^{-1}. It is noted that the peak presence between 1635 to 1647 cm^{-1} was attributed to adsorbed water on the surface of the films.

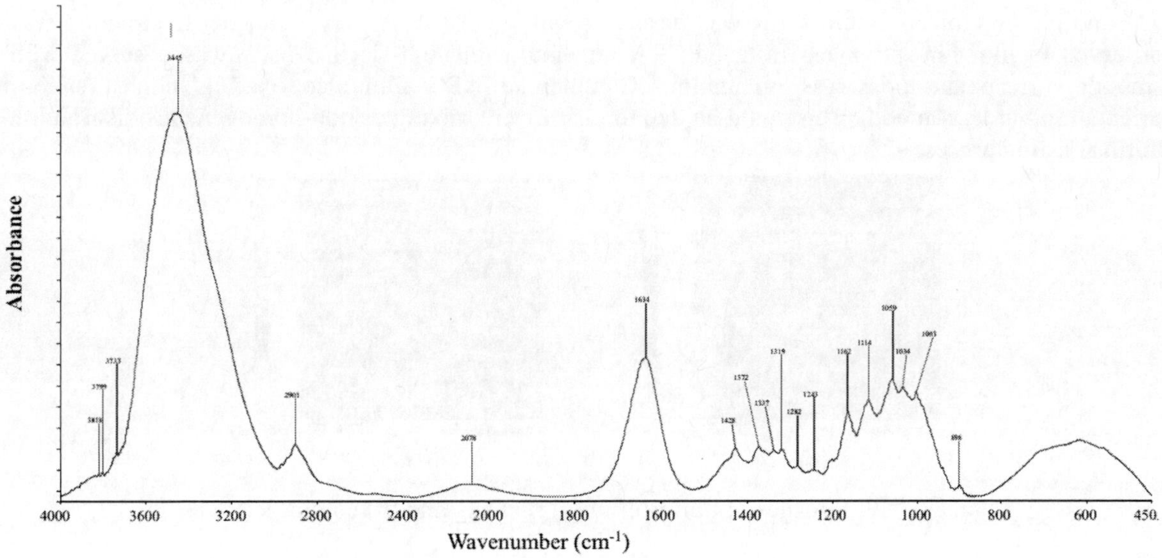

Figure 4. FTIR spectra for NFC from Mahang gajah.

3.3 Characterization of thin film

NFC was converted to thin film from Mahang gajah NFC. Thin film from commercial NFC was also fabricated for comparison purpose. The properties of film produced from both NFC were evaluated via several analytical tools to assess their performance. The film produced has thickness values in the range of 24-30 µm.

The thermal degradation characteristic of the Mahang gajah and commercial NFC was investigated using thermogravimetric analysis. The analysis was performed in a nitrogen environment at temperature between 25 °C to 500 °C. Thermal characteristics of fibrous materials depend on several factors such as chemical composition, molecular interaction, structure and degree of crystallinity [13]. For both samples, the first weight loss between temperatures of 50 to 200 °C was due to evaporation of adsorbed water on the sample surface. The second weight loss being the highest loss was observed in the temperature range of 270-370 °C. Mahang NFC however showed higher second weight loss than that of commercial sample. Above this temperature, the polymer experienced scission and molecular decomposition.

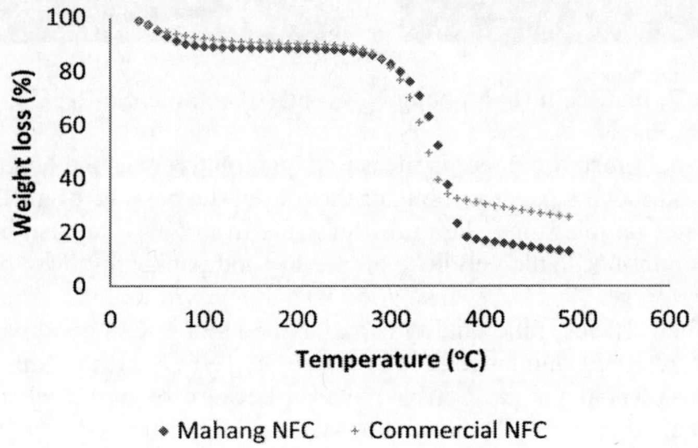

Figure 5. TGA thermogram for Mahang and commercial NFC thin films.

The morphology of NFC thin films was observed with an AFM. A very tight, dense structure was observed in thin film produced from NFC. A network with high fibril density was observed with smooth surface and crisscross orientation. Commercial NFC film also exhibits almost similar orientation but less smooth. This could be due to insufficient mixing or non-homogeneous distribution during casting process.

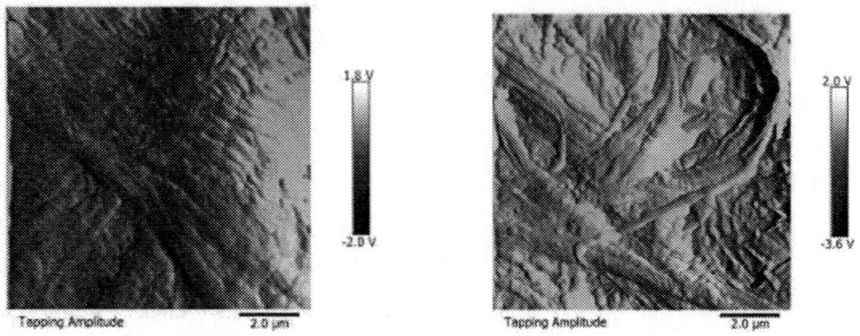

(a) (b)

Figure 6. AFM images of thin film of (a) Mahang and (b) commercial NFC.

Transparency of both films was assessed by first placing the film onto a paper with colored image and followed by measurement of light transmittance using UV-Vis machine. Film prepared from Mahang gajah and commercial sample were only transparent if the object was placed right at the bottom of the films as shown in Figure 7. If the film was moved 1 cm away from the object, the image became less clear. This is an indicative of low transparency of both films.

Light transmittance of the composite films was observed in the visible light wavelength range using UV-Vis machine. The Mahang NFC film recorded transmittance between 1 to 5 % at wavelength between 400 nm – 700 nm whereas commercial NFC had transmittance range between 1 to 15 %.

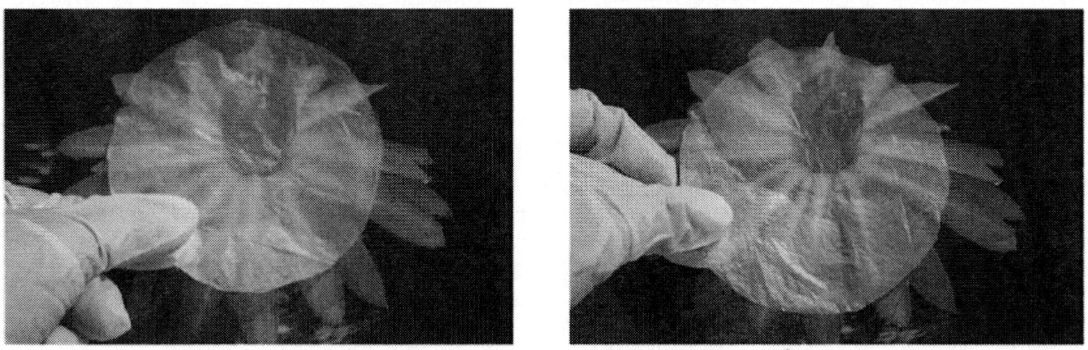

(a) (b)

Figure 7. Images of (a) Mahang NFC and (b) commercial NFC thin films.

In pulp and paper field, porosity of paper is measured quantitatively as the length of time it takes for a quantity of air to pass through a paper sample, or the rate of the passage of air through a sample using Bendsten porosity tester. In this study, the thin film made from NFC can also be treated as nanopaper as it was made from nanofibrillated cellulose alone. In food packaging, low porosity plays essential role to prolong food storage. This is because good packaging will not allow moisture to pass through and in contact with food. Hence, food quality is maintained along with good packaging. Low porosity also enhances water resistant and increase in wettability. NFC in thin film form or so called as nanopaper is a good potential for gas barrier material because of partial change of microporous to nanoporous in fibre network. Figure 8 shows that Mahang NFC film had slightly higher porosity than commercial NFC. An increased level of fiber makes the thin film denser, reducing the network of air passages and thus the porosity.

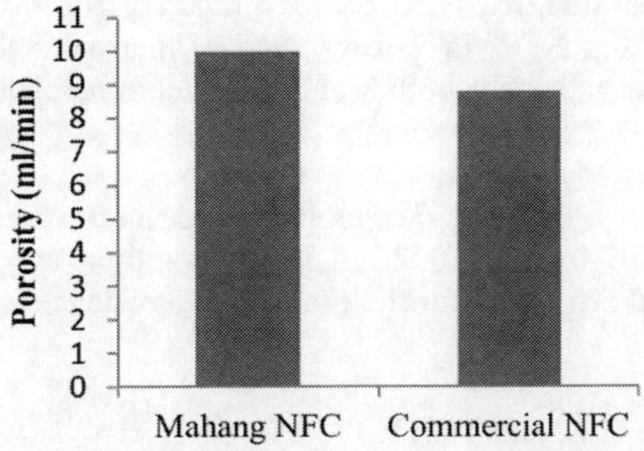

Figure 8. Porosity of Mahang and commercial NFC thin films.

Contact angle measurement was performed in order to assess wettability behaviour of NFC film. Water contact angle generally depends on structural properties of cellulosic based materials surface such as porosity, voids size, and interactions that occur between the surface and the water due to the free hydroxyl groups of cellulose and hemicellulose in fibers [14]. In this study, the smaller voids could be considered as explanation of the higher water contact angle values of the Mahang NFC film compared to commercial sample (Figure 9).

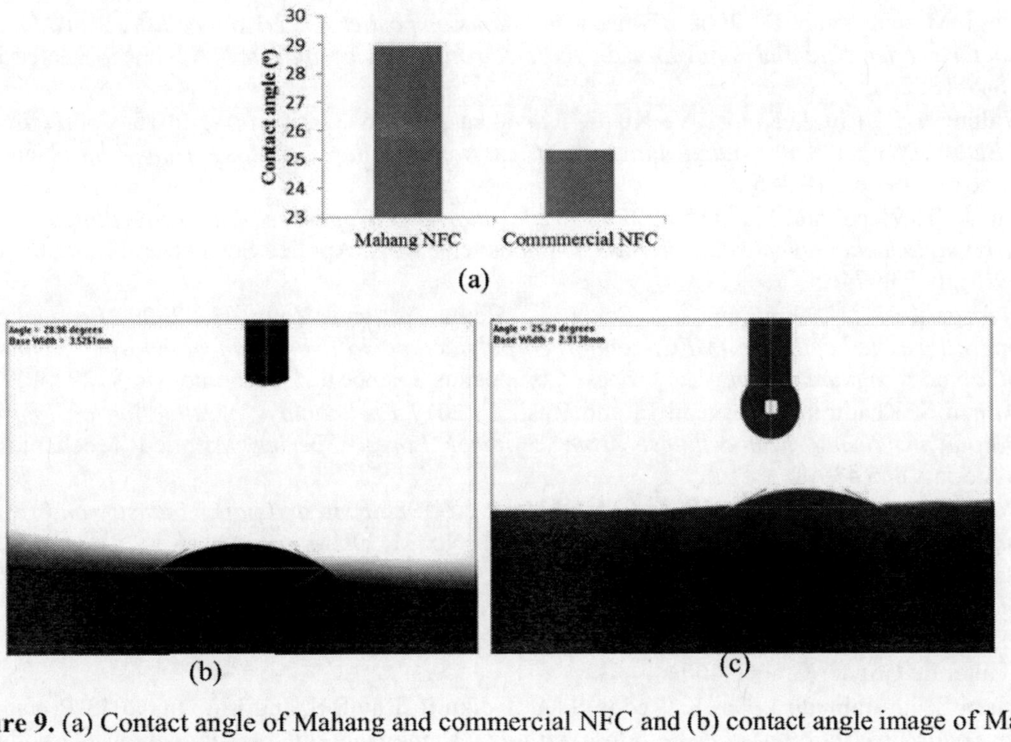

Figure 9. (a) Contact angle of Mahang and commercial NFC and (b) contact angle image of Mahang NFC and (c) commercial NFC.

4. Conclusion

In conclusion, it was demonstrated that NFC can be prepared from Mahang gajah and subsequently thin film can be fabricated from NFC. Properties such as thermal stability, porosity and wettability (contact angle) showed similar values for both Mahang and commercial NFC.

Acknowledgements
The authors thank Ministry of Science, Technology and Innovation (MOSTI), Malaysia for the funding under Science Fund (07-03-10-SF0226). LJ and co-authors are grateful to Pulp and Paper staff especially to Muhd Fazrul and Siti Hawa for their technical assistance.

References

[1] Peng B L, Dhar N, Liu H L, and Tam K C 2011 *Chemistry and applications of nanocrystalline cellulose and its derivatives: A nanotechnology perspective.* The Canadian Journal of Chemical Engineering **89** 1191-1206

[2] Moon R J, Martini A, Nairn J, Simonsen J, and Youngblood J 2011 *Cellulose nanomaterials review: Structure, properties and nanocomposites.* Chemical Society Reviews **40** 3941-3994

[3] Klemm D, Kramer F, Moritz S, Lindstrom T, Ankerfors M, Gray D and Dorris A 2011 *Nanocelluloses: A New Family of Nature-Based Materials.* Angewandte Chemie-International Edition **50** 5438-5466

[4] Lee K-Y, Aitomäki Y, Berglund L A, Oksman K, and Bismarck A, 2014 *On the use of nanocellulose as reinforcement in polymer matrix composites.* Composites Science and Technology **105** 15-27

[5] Nogi M and Yano H, 2008 *Transparent nanocomposites based on cellulose produced by bacteria offer potential innovation in the electronics device industry.* Advanced Materials **20** 1849-1852

[6] Valtakari D, Liu J, Kumar V, Xu C, Toivakka M, and Saarinen J J 2015 *Conductivity of PEDOT:PSS on Spin-Coated and Drop Cast Nanofibrillar Cellulose Thin Films* Nanoscale Research Letters **10** 386-396

[7] Lu, J., T. Wang, and L.T. Drzal, 2008 *Preparation and properties of microfibrillated cellulose polyvinyl alcohol composite materials.* Composites Part A: Applied Science and Manufacturing. **39**(5): p. 738-746.

[8] Nakagaito A N, Fujimura A, Sakai T, Hama Y, and Yano H, 2009 *Production of microfibrillated cellulose (MFC)-reinforced polylactic acid (PLA) nanocomposites from sheets obtained by a papermaking-like process* Composites Science and Technology **69** 1293-1297

[9] Adnan S, Khadiran T, Jasmani L, and Rusli R, 2017 *Preliminary Investigation on Improving Biopolymer using Nanocellulose from Tropical Forest Species.* Applied Mechanics and Materials **865** 43-48

[10] Peh T B, Khoo K C, Lee T W, and Mohd Nor M Y 1986 *Pulp and paper industry and research in Peninsular Malaysia.* Malayan Forest Records No. 31: FRIM

[11] Saito T, Kimura S, Nishiyama Y, and Isogai A 2007 *Cellulose nanofibers prepared by TEMPO-mediated oxidation of native cellulose.* Biomacromolecules **8** 2485-2491

[12] Dufresne A, 2012,*Nanocellulose, From Nature to High Performance Tailored Materials*: Walter de Gruyter GmbH. 460

[13] Maiti S, Jayaramudu J, Das K, Reddy S M, Sadiku R, Ray S S, and Liu D G 2013 *Preparation and characterization of nano-cellulose with new shape from different precursor.* Carbohydrate Polymers **98** 562-567

[14] Mashkour M, Afra E, Resalati H, & Mashkour M 2015 *Moderate Surface Acetylation of Nanofibrillated Cellulose for the Improvement of Paper Strength and Barrier Properties.* RSC Advances **5** 60179-60187

IOP Conf. Series: Materials Science and Engineering **368** (2018) 012016 doi:10.1088/1757-899X/368/1/012016

Recycling of waste rubber as fillers: A review

M Nuzaimah[1, 2*], S M Sapuan[1], R Nadlene[3] and M Jawaid[4]

[1]Department of Mechanical and Manufacturing Engineering, Faculty of Engineering, Universiti Putra Malaysia, 43400 UPM Serdang, Selangor, Malaysia
[2]Faculty of Engineering Technology, Universiti Teknikal Malaysia Melaka, 76100 Durian Tunggal, Melaka, Malaysia
[3]Faculty of Mechanical Engineering, Universiti Teknikal Malaysia Melaka, 76100 Durian Tunggal, Melaka, Malaysia
[4]Laboratory of Biocomposite Technology, Institute of Tropical Forestry and Forest Products (INTROP), Universiti Putra Malaysia, 43400 UPM Serdang, Selangor, Malaysia

*Corresponding author: nuzaimah@utem.edu.my

Abstract. Recently, in line with raising environmental concerns, researchers are now focusing on the uses of recycled product. Waste rubber is one of the waste material that has been recycled and being used in many area such in civil engineering, tyres production, polymer composite, energy source and many more. This paper highlighted the concern on the raising numbers of waste rubber products generated globally and how does this waste rubber can cause a serious threat to the environment. Re-use, recycle, recovery and pyrolysis is the common method in ensuring sustainable management of waste products and it keeps emerging. This paper will discussed briefly the development of recycling waste rubber into a composite material, the methods, characterisations and improvements that contributed by the waste rubber on the composites properties. Focus is on the development of composite filled with waste rubber tyres, waste rubber gloves, waste rubber condoms and few other waste rubber products. Incorporation of waste rubber fillers has improved the mechanical and physical properties of the composites. Thus, composites with waste rubber fillers has created a new sustainable material and has potential to be applied into various field.

1. Introduction

The constant increase of rubber waste has become major environmental problem for countries around the world. This is due to the demand of rubber products keep increasing, as for example the outgrow numbers of vehicle production and also high demand of rubber usage in industrial, medical, engineering, household products and many more. Due to that rising numbers of discarded rubber products, disposing these substances becoming a burden to the municipal council and lack of environmental awareness among citizens makes the problems of disposing rubber products worsen. Land fill dumping and open burning are among common, cheapest and easiest method of disposing waste rubber which leads to water, air and soil pollution. The poor disposal of rubber products globally becoming a potential threat to human health or even the living being and decrease the quality of environment [1-2]. Shu and Huang [3] reported that due to large volumes of rubber scraps, landfilling was no longer able to keep up with the increment as land cost are hiking up, thus disposing waste rubbers becoming a highly costly method.

Like other polymer materials that provides tremendous benefit to humankind, rubber is one of essential materials in many application due to its unique and impressive properties. Rubber exhibit outstanding behavior that is very high elasticity properties as it can be stretched to a few hundred percent and recover almost instantly to its original shape when force being released [4-5]. Extremely high elasticity behavior of rubber provides by remarkably low Young's modulus and high yield strain which resulted from the vulcanisation process [6]. Rubber having thermoset cross-linked structure which provides a very durable properties and high resistance to environmental agents. Due to these outstanding properties, rubber are widely used in application ranging from household, healthcare, industrial, military, automotive, civil and outer space application. However, these outstanding properties has a setback that causing rubber very difficult to degrade due to its complex structure and composition.

Rubber also held another valuable characteristics such as able to dampen or reduce vibration [4-5], better wet grip, relatively resistance to gas and water permeability and reduced swelling in hydrocarbon oils [7-8]. Another essential properties of rubber are high resistance to acids, alkali and chemical solution. It makes rubber an important materials in many critical environment including healthcare, food processing, chemical manufacturing, refining, automotive, electronics and semiconductor manufacturing, pharmaceutical and research laboratory [7-8].

The above mentioned properties of rubbers making rubber among the high demand materials as reported by the Malaysian Rubber Export Promotion Council (MREPC), the total world production of rubber in 2016 has increased to 26.9 million tonnes million tonnes if compare from 2010 [9]. International Rubber Study Group (IRSG) also forecasted that the world total rubber consumption will increase annually at an average 2.8% per annum from 2017 - 2025. In 2017, IRSG forecasted the world total rubber consumption is 28.05 million tonnes [10]. These numbers may raise concern among environmentalist as it indicates more rubber waste will be generated.

However, many literatures shows that a great progress has been made in sustainable management of waste rubber. An important effort taken by the researchers to ensure sustainable management of these rubber wastes are through reuse, recycling, recovery, and pyrolysis. Reuse the worn rubber (i.e: tyres) simply by retreading and reuse back. Recycling of waste rubber includes utilizing the discarded rubber into various applications such as in marine conservation program (as an artificial reef, erosion control, breakwaters and floatation devices). Other than that, recycling method includes grinding the waste rubber and then incorporate the rubber particles into matrices such as concrete or polymer and used for civil engineering, or being manufactured into brand new products such as athletic tracks and playground surface, as filler for rubberized composite and many more [5, 11]. Besides that, recovery is a method where the waste rubber is used as a fuel source for high temperature process such as cement kilns, steam production, electrical energy and etc. While, pyrolysis is to yield the basic chemicals in rubber waste by decomposing the rubber component such as steels, gas, oil and its filler such as carbon black [11–13]. Table 1 shows that some of the applications for waste rubber in our daily life.

1.1 Waste rubber types

Typically over 25 different types of rubber available such as natural rubber (NR), styrene-butadiene rubber (SBR), nitrile, ethylene-propylene-diene monomer (EPDM) rubber, fluorocarbon rubber, silicon and many more being used to produce numerous types of products [14]. These rubber generates various sources of waste such as scrap tires, inner tubes, discarded and rejected rubber glove, balloons, rubber bands, shoe soles, mattresses, hoses, seals, gaskets, diaphragms and so on. Table 2 shows the approximate amount of waste rubber being reused, recycled, recovery and landfill or stockpiled globally [14].

Table 1. Applications of waste rubber.

Application	Reference
as filler in flooring, shock absorbing mat, roofing materials as an alternative to fossil fuel	[2]
as lightweight fillers as an additives in Portland cement concrete, as modifiers in asphalt paving mixtures, as crash barriers, bumpers as an artificial reef	[3]
as energy source	[11–13]
as an energy source to produce electricity being process to recover the oil, steel, vinyl chloride and carbon use in civil application such as for playground and parking lot flooring as filler in road construction as modifier for asphalt	[12]
as a filler in concrete	[15, 16]
as tools for breakwater as floating purposes to produce mats, playground surfaces and trails for athletic used as fuel source	[11]
as filler in drainage structures	[17]
as filler-extender as an ion exchanger to clear the industrial water from heavy metals. as asphalt modifier as replacement for virgin rubber as replacement for carbon black	[18]
as source for mat as workability enhancer as carcass, sidewall and under-tread of tyres as inner liner for tubeless passenger tyre as inner tubes as semi-pneumatic tires as automotive floor mats as adhesive, sealing and tape compounds as filler in rubberized asphalt.	[19]

Table 2. Global estimation of waste rubber management.

	Estimated amount
Recycled	3 – 15%
Reused	5 – 23%
Recovery (energy)	25 – 60%
Landfill / stockpiled	20 – 30%

Tyres are among the major sources of waste rubber due to the large volume produced and problematic to handle due to their durability. Growing number of vehicles worldwide contributes to approximately 1.4 billion tyres being sold worldwide each year [2]. Each year, it is estimated that, in USA approximately 3.3 million tonnes tyres were generated, whereas in European Union (EU) is 2.5 million tonnes and UK alone is 0.5 million tonnes [14]. Meanwhile, Lo Presti [2] reported that, Europe produced 355 million tyres every year and estimated stockpiles were 5.5 million tonnes. Whereas in 2003, USA generated approximately 290 million scrap tyres. Aslani [20] on the other hand, mentioned that in USA, motorist discarded almost 303 million tyres each years and in 2009 alone approximately 594 thousand tons of scrap tyres were disposed in landfills. Illegal stockpiles of waste tyres were found at the residents' area which worrying nearby communities of potentially spreading of vulnerable diseases such as Leptospirosis, dengue and etc. Besides landfilling and stockpiles, waste tyres were also being burnt for disposal purpose and causing serious air pollution [8].

Meanwhile, waste rubber gloves also contributed to the amount of waste rubber generated worldwide. After usage, rubber gloves were discarded and dumped as garbage and become a threat to the environment. Rubber gloves was manufactured by fulfilled a strict specifications to ensure it is able to excellently function as a protective barrier to harmful substances. Thus, it has been specially formulated and manufactured with high strength properties and elasticity in order to minimize the possibility that gloves will rupture or tear under stress during used. The basic component in manufacturing glove is latex and additives such as vulcanised agent like sulphur, surfactant; to ensure homogenous and stable compound, accelerator agents to expedite curing process and pigments. All these ingredients has potentially being leached out while the gloves at the landfill and will pollute the soil and water source indirectly.

Furthermore, highly cross-linked thermoset structure of rubber material through vulcanisation process makes rubber glove possessed high resistance properties to acids, alkali, chemical solution and environment agents, which unfortunately it makes rubber glove very difficult to degrade [8]. The increment of waste rubber gloves worldwide relates to an increased demand of glove due to rising and greater concern on health awareness among world population. Societies are concern on safety and hygienic issue that leads to extra cautious on taking self-protection such as wearing glove in food preparation, humanitarian activities, gardening and many more. On top of discarded used glove; higher number of glove disposal also contributed by the high reject rate during the production process due to its strict production specifications and unstable nature of latex [21-22]. Beside tyres and gloves, other rubber waste that contributed to the rubber waste numbers globally is styrene butadiene rubber (SBR) for example from automotive scraps [23], waste rubber condoms [19-24] and industrial rubber waste such as latex foam based materials [25].

2. Development of waste rubber as a filler in a composites material

Many researches paying attention in recycling the waste material especially on polymer based material such as plastic and rubber which resulting a novel sustainable composites. For example, during the 1980s and 1990s significant development and application of utilizing waste tyres were blooming especially in civil engineering [3]. The technologies to recycle waste rubbers are continues to expand as more researchers started to pay attention to this area. Waste rubber were incorporated into various type of matrix such as concrete, latex, thermoplastic and thermoset materials [2, 3, 14, 15, 26–39]. Their investigations has proven that addition of rubber waste has enhanced the properties of the composites such as; improved tensile strength, modulus and elongation at break [19, 21, 23, 39, 40] soft phase of rubber provides flexibility to the composites [22], having better thermal insulation properties [24, 31] and improved impact resistance [41]. Based from this discoveries it shows that use of rubber waste into many potential applications have been proven successful, diversify and the studies are keep emerging. Importantly, the development of rubber based fillers composites benefited for a betterment of the environment. This section will briefly explain the reviewed and development of few researchers in utilizing the waste rubber products into various types of composites such as polymeric based composite and concrete based composite.

2.1 Waste rubber tyres as fillers for composites

IOP Conf. Series: Materials Science and Engineering **368** (2018) 012016 doi:10.1088/1757-899X/368/1/012016

The increment number of scrap tyres and stronger environment awareness worldwide makes all parties including federal regulations as well as researchers are giving attention to act for a sustainable approach towards waste tyres abundances. The waste tyres become a serious environmental issue due to their complex structure and component which makes it is difficult to degrade and it requires larger space in landfill and cause landfill limited area for other waste [42]. Many efforts has been conducted to minimizing the waste tyres stockpiling and dumping by treating the waste tyre as a source of sustainable materials.

A comprehensive review was done by Ramarad et al. [11] as they addressed relevant information regarding waste rubber tyres being utilized as a reinforcement material or filler in polymeric material to form a composites. From the review, it found that one of the key element to recycle the waste tyre includes; the rubber needs to be ground and shredded prior being incorporated into the polymer. The ground rubber was then go for surface modification in order to enhance adhesion between the rubber particles with the polymer matrix. Surface modification was conducted either physically or chemically and also by devulcanisation or reclamation process. In this review, they discovered that thermoplastic matrix is commonly being studied by many researchers while natural rubber matrix is also getting extensive attention. On the other hand, little research being conducted for other rubber types of rubber such as butadiene rubber and ethylene propylene monomer (EPDM) rubber. In their review, they highlighted that, waste rubber tyres in rubber matrix is easy to process and its composites acquires an acceptable properties. Review also highlights the lack of studies concentrating on dynamic mechanical, aging, thermal and swelling properties of waste tyre rubber polymeric blends.

On the other hands, Abu Jyadil et al. [31] conducted a development of polyester based composites filled with waste rubber tyre crumbs with an intention to be used as insulating material. They conducted up to 40% waste rubber crumb were added into the polyester and being characterized for its insulation properties such as thermal conductivity, water retention, density and thermal stability. They found that, an addition of rubber particles into the polyester matrix has potential to be used as a thermal insulator with the addition of waste rubber tyres has reduced the thermal conductivity and lower the water retention of the composites. The tensile strengths and compressive strength of the composite are higher than currently used insulating materials and it has similar strength to some of the construction materials [32].

Yehia et al. [18] has evaluated the capability of waste rubber to replace certain amount of virgin rubber in natural rubber mix. Their study shows that replacing 10-30% virgin rubber with waste rubber filler does not sacrifice the properties of the rubber mixture. They also found that incorporation of fine waste rubber powder with their surface being chemically treated into the asphalt concrete mixture has greatly improved the performance of the road pavement as the mixture is resistance to crack tip separation.

Reviewed conducted by Shu and Huang [3] reported that the used of waste rubber tyres into asphalt concrete is a successful effort as it improved properties and performance of the mixture. This successful outcomes is a result of good compatibility and interaction between waste rubber particles with the asphalt mixture binder. Lo Presti [2] mentioned in his review that, waste rubber tyres has been used in flooring for playground and sport stadium, in shock-absorbing mat and roofing materials. While, Yang [17] has conducted an investigation of using the scrap tyres as fillers in drainage structures such as for underground and horizontal drain and for ravine crossing. Moustafa and ElGawady [15] highlighted that scrap tyres being used as fillers in concrete to improve the properties such as ductility, damping and impact resistance.

2.2 Waste rubber glove as fillers for composites

The rubber glove industry has grown tremendously over the years to meet the world demands due to rising healthcare awareness globally. This is due to standard of life improved and healthcare reforms. Subsequently, demand for protective products of healthcare safety such as gloves, condoms, latex thread, etc. are also increased [19]. It can be relate with the 26% increment of Malaysia rubber glove export for the first quarter of 2017 that worth RM4 billion in sales if compared to the first quarter of 2016 which recorded RM3.2 billion [43].

High usage of rubber gloves globally has led to high disposal of used gloves. In addition to the higher number of glove disposal also contributed by the high reject rate during the production process due to strict production specification and unstable nature of latex [19, 40]. Due to their very complex structure and composition that contribute to excellent properties such tremendous elasticity, high resistance to acids, alkali, chemical solutions and environment agents, it makes rubber glove very difficult to degrade consequently creates a serious environmental problem. Glove's ingredients consist of latex (either it is natural or synthetic latex) and vulcanised agents (i.e: sulphur), surfactants, curing accelerator agents and pigments. Thus, dumping into landfill or any open area not only causing piling up but leaching off the ingredients of the glove components will pollute the water and soil.

Rubber glove that possessed many unique properties were beneficial in developing numerous new functional material instead of being treated as invaluable garbage and pollutants but rather a valuable source of sustainable materials. Most importantly, glove are relatively cheap and easy to get thus provide low cost of waste rubber gloves source that lead to low processing cost of the new material [39]. Elastic behavior of rubber glove can be also benefited into numerous field such as damping components, thermal and noise insulator and others. Few literatures has reported that composites with the waste rubber glove fillers shows better mechanical properties such as having better impact strength, better tensile properties and lower stiffness [8, 21, 44-45].

Besides that, there were few other findings that proved, fillers from waste rubber glove has improved the composites properties. For example, blending of filler from waste acrylonitrile butadiene rubber (NBR) glove with epoxidized natural rubber (ENR 50) and NBR rubber glove waste blended ENR 50 has shown improvement of tensile strength, modulus and elongation at break [39, 40]. A study mentioned that, they chose rubber glove specifically NBR waste gloves as a filler in their study because it has excellent puncture and tear resistance, with a reasonable impermeable to many types of chemicals. Thus, these properties will give additional value to the new develop material or composite. In addition using waste NBR glove helps to reduce cost of the material and also easy handling [39].

2.3 Waste rubber condoms as fillers for composites

A study of thermal performance on polymer composite with waste rubber condoms (also known as prophylactics) made from natural rubber as a filler in SBR rubber has been conducted by N.S. Saxena et. al [24]. The author suggested that, result from this study potentially can help in optimizing performance of cryosystem which properties of the composite in low temperature thermal properties can be used to further understand the thermal equilibrium process. The waste condoms were ground into polydispersed particles and later compounded with SBR using two roll mill. The composite was cured in hot press at 150°C and measured for thermal conductivity. The thermal conductivity of the composite was increased with increased of rubber content. A comprehensive review has been conducted by Rajan et al. [19] on recycling natural rubber based waste latex products that include waste condoms. The condoms was first powdered to a size 40 mesh and proceed for reclamation process prior added into raw rubber. This compound reported better in term of processing characteristic.

2.4 Other types of waste rubber as fillers for composites

Zanchet et al. [23] has developed a styrene-butadiene rubber (SBR) composite with utilizing SBR scraps from automotive extruded as its filler. The rubber scraps went for grinding and devulcanisation process through microwave exposure prior being incorporate into the virgin SBR matrix. Revulcanisation of the composite was later conducted by microwave exposure. The addition of SBR scraps giving better tensile strength and elongation if compared to control sample (SBR virgin without SBR scrap). The author also suggested that the increment of tensile strength and elongation due to the devulcanised SBR scraps promotes better interfacial adhesion with the SBR matrix.

Rajan et al. [19] reported that there was a work on using waste latex products as a filler in epoxidized natural rubber (ENR) which helps in reducing stickiness of ENR that give better processability during mixing process. Besides, addition of waste latex product helps reduce mixing time and improved product resistance towards aging. Mathew et. al [44] conducted evaluation on processing aspects, vulcanisation behaviour and dynamic mechanical analysis natural rubber prophylactic (NRP) waste blend with epoxidized natural rubber (ENR) with and without addition of marble powder. The Young's Modulus and tear strength of ENR increased with increased loading of NRP as well as with the presence of marble powder. But slight decreased at 40 phr of loading NRP. Damping behaviour affected only by the presence of marble powder. The glass transition temperature ENR increase with addition of 40 phr NRP. A study on recycling industrial rubber waste (latex foam waste) has been conducted by incorporating the waste into natural rubber compound for making rubber mat. Addition of the latex waste improved the mechanical properties of this rubberized compound which it has enhanced tensile strength and elongation [25].

3. Conclusion

Use of waste rubber as a fillers into composites, concrete mixtures and other matrices has become one of an attractive field for the researchers to investigate. It is because products that utilizing the waste rubber has high potential to benefit in countless field and at the same time it serve for a betterment of the environment. Recycling of rubber waste into many uses has been proven successful and the study are keep evolving. Researchers also interested in utilizing waste rubber as a reinforcing element in a polymer matrix as well as in rubber matrix. Waste rubber reinforced do provides acceptable mechanical and physical properties to the composites. Beside waste rubber tyres, waste rubber glove and waste rubber condoms, industrial rubber waste are also being recycled. As a conclusion, recycling of waste rubber into a new novel sustainable material has given many positive effect and studies and investigations in this field is rapidly progressing.

References

[1] Shah A A, Hasan F, Shah Z, Kanwal N and Zeb S, 2013 Biodegradation of natural and synthetic rubbers: A review *Int Biodeter Biodegr* **83** 145-157

[2] Presti D L 2013 Recycled tyre rubber modified bitumens for road asphalt mixtures: a literature review *Constr Build Mater* **49** 863-881

[3] Shu X and Huang B 2014 Recycling of waste tire rubber in asphalt and portland cement concrete: An overview *Constr Build Mater 67* 217-224

[4] bin Samsuri A 2010 Degradation of natural rubber and synthetic elastomers

[5] Imbernon L and Norvez S 2016 From landfilling to vitrimer chemistry in rubber life cycle **EUR** *Polym J* **82** 347-376

[6] McKeen L W 2014 Elastomers and Rubbers in *The Effect of Long Term Thermal Exposure on Plastics and Elastomers* 239–271

[7] Benazzouk A, Douzane O, Mezreb K, Laidoudi B, and Quéneudec M 2008 Thermal conductivity of cement composites containing rubber waste particles: Experimental study and modelling *Constr Build Mater* **22** 573-579

[8] Riyajan S A, Intharit I, and Tangboriboonrat P 2012 Physical properties of polymer composite: Natural rubber glove waste/polystyrene foam waste/cellulose *Ind Crop Prod* **36** 376-382

[9] Malaysian Rubber Export Promotion Council, "Industry Overview of World Rubber Production, Consumption and Trade in 2016," *MREPC*, 2018. [Online]. Available: www.mrepc.com/industry/industry.php

[10] "Latest World Rubber Industry Outlook," *International Rubber Study Group*, 2016. [Online]. Available: http://www.rubberstudy.com/news-article.aspx?id=5106&b=default.aspx

[11] Ramarad S, Khalid M, Ratnam C T, Chuah A L, and Rashmi W 2015 Waste tire rubber in polymer blends: A review on the evolution, properties and future *Prog Mater Sci* **72** 100-140

[12] Adhikari B, De D, and Maiti S 2000 Reclamation and recycling of waste rubber *Prog Mater*

Sci **25** 909-948

[13] Miranda M, Cabrita I, Pinto F, and Gulyurtlu I 2013 Mixtures of rubber tyre and plastic wastes pyrolysis: a kinetic study *Energy* **58** 270-282

[14] M. Forrest 2014 Overview of the World Rubber Recycling Marke in *Recycling and Re-Use of Waste Rubber* 17–18

[15] Moustafa A and ElGawady M A 2015 Mechanical properties of high strength concrete with scrap tire rubber *Constr Build Mater* **93** 249-256

[16] Sousa S P B, Ribeiro M C S, Cruz E M, Barrera G M, and Ferreira A J M 2017 Mechanical behaviour analysis of polyester polymer mortars reinforced with tire rubber fibres *Ciência & Tecnologia dos Materiais* **29** e162-e166

[17] Yang S 1999 Use of scrap tires in civil engineering applications

[18] Yehia A, Abdelbary E M, Mull M, Ismail M N, and Hefny Y 2012 October. New trends for utilization of rubber wastes. In *Macromolecular Symposia* (**320** 5-14). WILEY-VCH Verlag

[19] Rajan V V, Dierkes W K, Joseph R, and Noordermeer J W 2006 Science and technology of rubber reclamation with special attention to NR-based waste latex products *Prog Polym* **SCI 31** 811-834

[20] Aslani F 2015 Mechanical properties of waste tire rubber concrete *J Mater Civil Eng* **28** 04015152

[21] Ismail H, Galpaya D, and Ahmad Z 2009 The compatibilizing effect of epoxy resin (EP) on polypropylene (PP)/recycled acrylonitrile butadiene rubber (NBRr) blends *Polym Test* **28** 363-370

[22] Esmizadeh E, Naderi G, Bakhshandeh G R, Fasaie M R, and Ahmadi S 2017 Reactively compatibilized and dynamically vulcanized thermoplastic elastomers based on high-density polyethylene and reclaimed rubber *Polym Sci Ser B+* **59** 362-371

[23] Zanchet A, Carli L N, Giovanela M, Brandalise R N, and Crespo JS 2012 Use of styrene butadiene rubber industrial waste devulcanized by microwave in rubber composites for automotive application *Mater Design* **39** 437-443

[24] Saxena N S, Pradeep P, Mathew G, Thomas S, Gustafsson M, and Gustafsson S E 1999 Thermal conductivity of styrene butadiene rubber compounds with natural rubber prophylactics waste as filler *Eur Polym J* **35** 1687-1693

[25] Chuan S and Siang LG Cure and Physical Characterizations of Natural Rubber Blended with Recycled Latex-Foam-Waste

[26] Kakroodi A R, Kazemi Y, and Rodrigue D 2013 Mechanical, rheological, morphological and water absorption properties of maleated polyethylene/hemp composites: Effect of ground tire rubber addition *Compos Part B-Eng* **51** 337-344

[27] Ramarad S, Ratnam, C T, Khalid M, Chuah A L, and Hanson S 2017 Improved crystallinity and dynamic mechanical properties of reclaimed waste tire rubber/EVA blends under the influence of electron beam irradiation *Radiat Phys Chem* **130** 362-370

[28] Lopes D, Ferreira M J, Russo R, and Dias J M 2015 Natural and synthetic rubber/waste–Ethylene-Vinyl Acetate composites for sustainable application in the footwear industry *J Clean Prod* **92** 230-236

[29] Fukumori K and Matsushita M 2003 Material recycling technology of crosslinked rubber waste *R&D Rew Toyota, CRDL* **38** 39-47

[30] Thomas B S and Gupta R C 2016 A comprehensive review on the applications of waste tire rubber in cement concrete *Renewable and Sustainable Energy Reviews* **54** 1323-1333

[31] Abu-Jdayil B, Mourad A H, and Hussain A 2016 Thermal and physical characteristics of polyester–scrap tire composites *Constr Build Mater* **105** 472-479

[32] Abu-Jdayil B, Mourad A H I, and Hussain A 2016 Investigation on the mechanical behavior of polyester-scrap tire composites *Constr Build Mater* **127** 896-903

[33] Noimam N Z, Ismail H, Mohd Salleh M A A, Al Bakri Abdullah M M and Rosniza H, 2013 Thermal Properties of Different Recycled Acrylonitrile-Butadiene Rubber Glove (NBRr) Size

and its Blend Ratios on SBR/NBRr Blends In *Advanced Materials Research* **795** 377-382 Trans Tech Publications

[34] Guo S, Dai Q, Si R, Sun X, and Lu C 2017 Evaluation of properties and performance of rubber-modified concrete for recycling of waste scrap tire *J Clean Prod* **148** 681-689

[35] Chou L H, Lu C K, Chang J R and Lee M T 2007 Use of waste rubber as concrete additive *Waste Manage Res* **25** 68-76

[36] Aliabdo A A, Elmoaty A E M A, and AbdElbaset M M 2015 Utilization of waste rubber in non-structural applications *Constr Build Mater* **91** 195-207

[37] Fakhri M 2016 The effect of waste rubber particles and silica fume on the mechanical properties of Roller Compacted Concrete Pavement *J Clean Prod* **129** 521-530

[38] Ganjian E, Khorami M, and Maghsoudi A A 2009 Scrap-tyre-rubber replacement for aggregate and filler in concrete *Constr Build Mater* **23** 1828-1836

[39] Ahmad H S, Ismail H, and Rashid A A 2016 Tensile Properties and Morphology of Epoxidized Natural Rubber/Recycled Acrylonitrile-Butadiene Rubber (ENR 50/NBRr) Blends *Procedia Chem* **19** 359-365

[40] Salleh S Z, Ahmad M Z, and Ismail H 2016 Properties of natural rubber/recycled chloroprene rubber blend: Effects of blend ratio and matrix *Procedia Chem* **19** 346-350

[41] Riyajan S A, Intharit I, and Tangboriboonrat P 2012 Physical properties of polymer composite: Natural rubber glove waste/polystyrene foam waste/cellulose *Ind Crop Prod* **36** 376-382

[42] Sienkiewicz M, Janik H, Borzędowska-Labuda K, and Kucińska-Lipka J 2017 Environmentally friendly polymer-rubber composites obtained from waste tyres: A review *J Clean Prod* **147** 560-571

[43] Nation, "Rubber glove exports at RM4bil and set to grow," *The Star Online*, pp. 1–8, 22-May-2017

[44] Mathew G, Singh R P, Nair N R, and Thomas S 2001 Recycling of natural rubber latex waste and its interaction in epoxidised natural rubber *Polymer* **42** 2137-2165

[45] Mathew G, Nah C, Rhee J M, and Singh RP 2006 Multifiller-matrix adhesion effects in epoxidized natural rubber *Journal of Elastomers & Plastics* **38** 43-63

Effect of compatibilizer on morphological, thermal and mechanical properties of Starch-Grafted-Polypropylene/Kenaf fibers composites

N H Eszer [1], **Z A M Ishak**[1]*

[1]School of Materials and Mineral Resources Engineering, Universiti Sains Malaysia, Engineering Campus 14300 Nibong Tebal, Penang, Malaysia

*Corresponding author: zafirin.ishak@gmail.com

Abstract. Starch-grafted-polypropylene (starch-g-PP)/kenaf fibres composites were prepared by melt compounding and compression molding. Polypropylene-grafted-maleic anhydrides (MAgPP) were added at different weight contents (1, 3 and 5 wt. %) as compatibilizer. The morphological, thermal and mechanical properties of the composites were characterized using thermogravimetric analysis, scanning electron microscopy, tensile and flexural test. The results indicated that mechanical properties of the composites were improved by addition of compatibilizer. From TGA, the thermal stability of the composites with MAgPP was improved compared with the composites without MAgPP. Finally, scanning electron microscopy of fractured surface revealed better adhesion between the fibers and matrix with presence of compatibilizer.

1. Introduction

With the growing environmental awareness and shortage of natural resource, biodegradable and compostable materials are produced and used in increasing amounts in the areas of life and industry [1,2]. Therefore, the development of biocomposites based on biodegradable thermoplastic starch (TPS) and natural fibers is being widely explored, as an effort to reduce the plastic waste problem. However, the main disadvantages of TPS include unsatisfactory mechanical properties, particularly in wet or dry environments. TPS that were reinforced with natural fiber to improve the mechanical properties of the composites.

Lignocellulosic materials such as henequen, hemp, sisal, kenaf, coir, jute, palm and wood in their natural condition have been widely used as fillers for thermosetting and thermoplastic composites [3]. Lignocellulosic fibers also contribute to the composite with high specific stiffness and strength, a desirable fiber aspect ratio, biodegradability, renewable resources and a low cost per unit volume basis [3]. These fibers are mainly made of cellulose, hemicelluloses, lignin and pectins, with a small quantity of extractives [2]. Aside from such as low density, less machine wear during processing, no health hazards and a high degree of flexibility, kenaf fiber is also easy to be moulded into a product using conventional thermoplastic processing such as extrusion, compression moulding and injection moulding [4]. However, the incorporation of kenaf fiber into conventional polymer composites does not promise fully-biodegradable characteristics because the matrix phase is not capable of undergoing biodegradation. To produce a fully-biodegradable composite, a combination of natural fiber and biodegradable polymer matrix is needed.

In this study, a commercialized biphasic thermoplastic, Gaialene ® (supplied by Roquette, France) was used as a matrix to form an environmental friendly composite with kenaf fibers (KF). Gaialene ® is a starch-grafted-polypropylene (starch-g-PP) with the starch/polypropylene ratio of 52/48, which is designed for durable and semi-durable uses. According to the manufacturer, it is over 50% bio-based and the production yields a carbon footprint about 65% lower than that of the cleanest fossil polymers. Addition of natural fibers would also increase the bio-based portion of the material, making it more environmental friendly. In that case, a better interfacial adhesion is obtained between the matrix and filler because kenaf fiber is naturally hydrophilic. The matrix which is thermoplastic starch also is hydrophilic. Moreover, the additions of compatibilizer such as maleic anhydride (MA) between polymer and natural fibers are common solutions used to improve the interfacial adhesion of the composites [5,6]. In previous work, the improvement of tensile properties, hardness, creep stability and tensile impact stress after incorporation of KF into the starch-g-PP were reported by other researchers .In this study, starch-g-PP and kenaf fibers also used as the previous study but with addition of compatibilizer because the interaction between fiber and matrix influence the mechanical properties of the composites. Therefore, compatibilizer was used to improve the interfacial adhesion and mechanical properties of the composites.

2. Materials and methods

2.1 Materials

Starch-grafted-polypropylene with trade name (Gaialene ®) was obtained from Roquette, France. Density of the polymer is about 1.28 g/cm^3. The polymer was supplied in dry pellet and has melting temperature (T_m) of about 160 °C used as the matrix of the composites. Commercial maleated polypropylene, MAgPP (OVERAC CA 100) from Arkema with a melt flow index of 10g/10 min and density of 0.909 g/cm^3 was used as compatibilizer.

Kenaf fiber (KF) was supplied by Kenaf Natural Fiber Industries Sdn. Bhd. (KFI), Malaysia. In this work, kenaf fibers with 3 mm in length were used. The ash content in kenaf fiber is removed by sieving and washing the fiber. The average density of kenaf fiber is 1.4 g/cm^3.

2.2. Compounding conditions

Kenaf fiber was dried in a circulating air oven at 80 °C for 24 hours. Starch-g-PP also dried in circulating oven at 50°C for 12 hours to expel moisture before used for compounding. The kenaf fibers, starch-g-PP and MAgPP pellets were mixed by using an internal mixer with co-rotating double-winged rotors, operating at 175°C with rotor speed 60 rpm for 7 minutes and the weight percent of the kenaf fibers used was 30 wt.%, the composition were shown in Table 1.The composite compounds were compression moulded at 170 °C into specimens with dimensions of 120 × 12 × 3 mm^3 and 165 × 19 × 3 mm^3 for flexural and tensile specimens, respectively, by using a compression moulding machine (model Kao Tieh Gotech). The moulding cycle involved 5 minutes of preheating without pressure, 3 minutes of compression and 5 minutes of cooling.

IOP Conf. Series: Materials Science and Engineering **368** (2018) 012017 doi:10.1088/1757-899X/368/1/012017

Table 1. Formulations of the specimens

Materials designation	Starch-g-PP (wt.%)	Kenaf fibre (wt.%)	MAgPP (wt.%)
SGPP	100	-	-
SGPP/KF	70	30	-
SGPP/KF/1MAgPP	69	30	1
SGPP/KF/3MAgPP	67	30	3
SGPP/KF/5MAgPP	65	30	5

2.3. Mechanical testing
Tensile and flexural tests were performed on a Universal Testing Machine (Model Instron 3366). The specimens were kept in desiccators prior testing to avoid moisture absorption. Tensile and flexural tests were carried out in accordance with ASTM D638 and ASTM D570, respectively with a load cell of 10 kN and crosshead speed was 5 mm/min at room temperature. For tensile test, elastic modulus was evaluated deforming the specimens up to a strain of 1 % at a crosshead speed of 5 mm/min. The strain was measured with a gage length of 50 mm. Flexural test were performed in three-points bending mode with span length of 50 mm. At least five specimens were used for each test.

2.4. Thermal testing
Thermogravimetric analysis (TGA) analysis was performed on neat starch-g-PP, starch-g-PP/KF with and without compatibilizer in order to investigate the thermal stability of the composites. TGA analysis was carried out using TGA analyzer model Perkin Elmer Pyris 6. 5 – 10 milligrams sample was scanned at a heating rate of 10 °C/min from 30 to 620 °C in N_2 atmosphere.

2.5. Morphology
Field emission scanning electron microscope (FESEM) (model VPFESEM Supra 35VP) was used to examine morphology of fractured specimens. Prior to viewing the specimens using FESEM, the specimens were sputter coated with gold palladium to avoid electric charging during scanning.

3. Results and discussion

3.1 Mechanical properties

3.1.1. Tensile properties. Figure 1 shows tensile strength of neat starch-g-PP, starch-g-PP/kenaf fiber composites without and with different MAgPP content. The tensile strength of starch-g-PP increase from 10.68 MPa to 15.74 MPa with addition of the kenaf fiber loading. The increase indicated that the stress was transferred from the matrix to the fibers [7]. Addition of MAgPP to the starch-g-PP/KF composites resulted further increase in the tensile strength. It can be seen, in Figure 1, tensile strength of the composites increase with MAgPP content up to 3 wt.%. The improvement in the tensile strength is attributed to the increased adhesion between fiber and the matrix that facilitates more stress transfer through bonding to the fibers. The interface adhesion is due to the presence of MAgPP which reacted with the hydroxyl group present in the kenaf fiber surface. With further increasing MAgPP content, the tensile strength of the composites is slightly decreased, which is mainly attributed to the plasticizing effect of MAgPP molecules in the composites.

The modulus of the composites was affected by fibers, matrix and their interface. The tensile modulus of starch-g-PP was improved with addition of kenaf fiber, which is known that fibers act as rigid filler that increases stiffness. Tensile modulus is significantly increased with an increment of 232 % at 30 wt.% of fiber loading. It is well documented that, due to its relative high tensile modulus KF notably increases the tensile modulus of most polymer matrices [8-9]. From Figure 2, it is noted that the elongation at break decreased remarkably with addition of KF. It is understandable that the incorporation of KF with a relatively low elongation at break (1.6%) [8] to the starch-g-PP will reduce the overall elongation at break. However, slightly improvement of the tensile modulus and elongation at break of the composites with the MAgPP content is due to the increase of the interface adhesion.

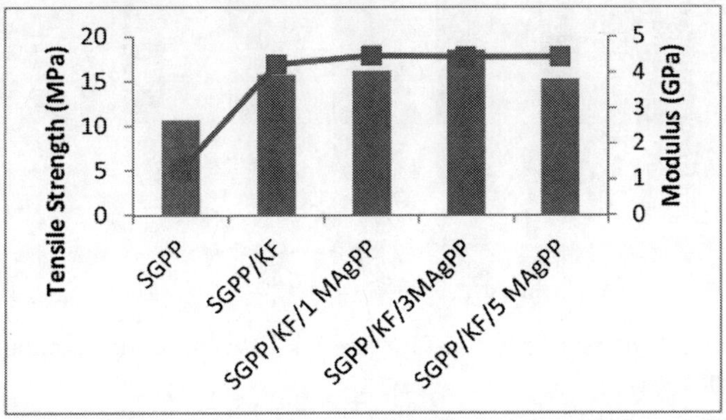

Figure 1. Tensile strength and modulus of neat starch-g-PP, starch-g-PP/kenaf fiber composites without and with different MAgPP content.

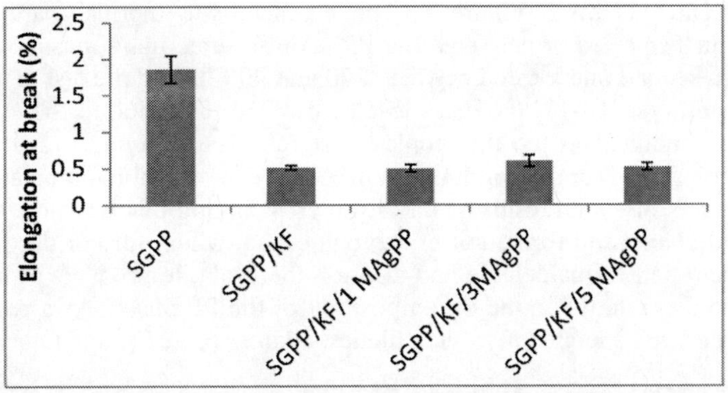

Figure 2. Elongation at break of neat starch-g-PP, starch-g-PP/kenaf fiber composites without and with different MAgPP content

3.1.2. Flexural properties. The flexural strength and flexural modulus for starch-g-PP/kenaf fiber composites without and with different MAgPP content was presented in Figure 3. The flexural strength of starch-g-PP increases from 19.52 MPa to 30.14 MPa with addition of kenaf fiber. The trends for flexural strength of the composites with MAgPP content are similar to the tensile strength of the composites. The flexural strength of starch-g-PP/kenaf fibre composites with 3% MAgPP content increased from 30.14 MPa to 37.08 MPa. The improvement is related to the better interface between the fiber and the matrix. The flexural modulus of starch-g-PP was improved with addition of kenaf fiber, which is known that fibers act as rigid filler that increases stiffness. Flexural modulus is significantly increased with an increment of 335 % at 30 wt.% of fiber loading. The slightly improvement of the flexural modulus of the composites with the MAgPP content is due to the increase of the interface adhesion.

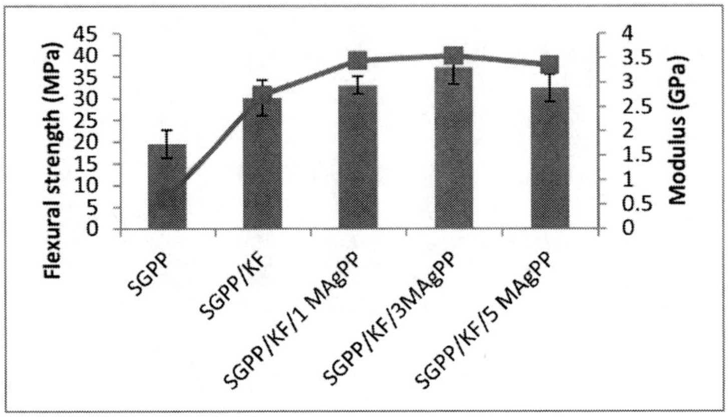

Figure 3. Flexural strength of neat starch-g-PP, starch-g-PP/kenaf fibre composites without and with different MAgPP content.

3.2 Thermal properties

The influence of MAgPP on the thermal stability of starch-g-PP/kenaf fibers composites were evaluated by TGA, and the results are shown in Figure 4. The characteristic thermal parameters of the thermal degradation are listed in Table 2. For neat starch-g-PP, exhibit three main mass loss steps, i.e. a first one below exhibit 250 °C; a second one located between 270 and 400 °C and the last one occur beyond 400 °C. As reported by some authors, [10-11] the mass loss below 250 °C is due to the evaporation of adsorbed water and plasticizers contained in the thermoplastic starch. The second mass loss is attributed to the degradation of starch compounds corresponding to a maximum degradation temperature of 350 °C [10]. The main degradation products of starch result mainly from crosslinking of glycosidic units after dehydration leading to unzipping of chains and formation of laevoglucosan, which further decomposes predominantly into furfural, formic acid and formaldehyde and also to other volatile products at low concentration [12]. The last degradation step is relative to the decomposition of the PP phase. As a result, aliphatic saturated and unsaturated products are generated involving dienes, alkanes (C_{12}, C_{13} and C_{14}) and alkenes.

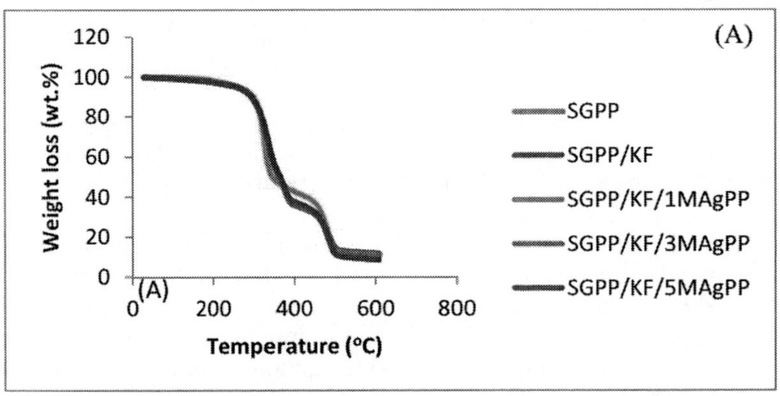

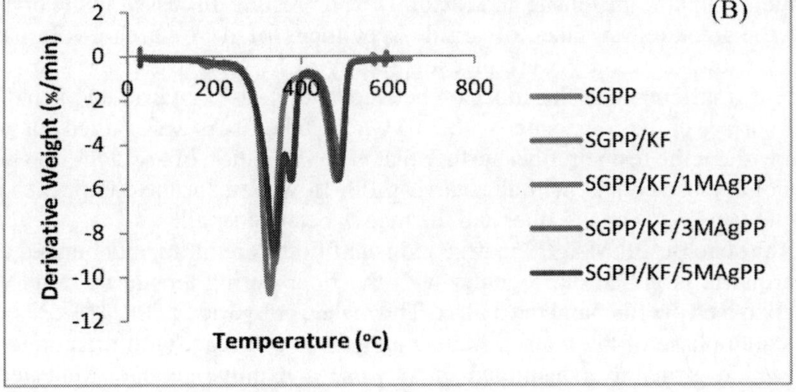

Figure 4. TGA curves of neat starch-g-PP, starch-g-PP/KF without and with different MAgPP content (A) TGA, (B) DTG

Table 2. TG characterization of the specimens

Specimen	T_{10} (°C)	T_{50} (°C)	T_{max} (°C)	Residue (%) (at 600 °C)
Starch-g-PP	301.03	347.47	327.51	10.71
Starch-g-PP/KF	299.95	363.23	336.98	11.36
Starch-g-PP/KF/1 MAgPP	299.34	363.56	376.67	11.29
Starch-g-PP/KF/3MAgPP	296.90	363.84	338.18	10.06
Starch-g-PP/KF/5MAgPP	297.36	366.19	484.13	8.00

Further, the thermal decomposition of starch-g-PP/KF composite is sharply lower than neat starch-g-PP indicating that introduction of the kenaf fibers reduces the thermal stability. The incorporation of KF in starch-g-PP matrix decreases the values of $T_{10\%}$ and $T_{50\%}$ corresponding to the degradation temperature at 10 and 50% mass loss, respectively. This can be explained by the lower stability of KF compared to the matrix, due to the presence of hemicelluloses that are very sensitive to thermal degradation at low temperature mass loss, respectively. The final decomposition temperature of the composite with MAgPP addition is increased compared with that of the composite without MAgPP. The decrease in percentage weight losses shows the slight improvement in thermal stability with the addition of MAgPP compatibilizer. This is due to the chemical bond between MAgPP with starch-g-PP matrix and kenaf fiber could enhance the interfacial adhesion.

3.3 Morphology

Two distinct phases are clearly noticeable in the biphasic neat starch-g-PP matrix, as shown in Figure 5 A and B. The continuous phase is polypropylene, while the dispersed phase is thermoplastic starch (TPS). It can be seen that TPS particles are homogenously dispersed in the PP matrix, and both phases are well bonded to each other [13,14]. The good compatibility between them could be attributed to the chemical affinity between TPS component in starch-g-PP and KF. In fact, both constituents contain hydroxyl groups and ether links in the glucose units and polysaccharides, which might enable the formation of hydrogen bonds [13].

It is well known that, without effective wetting on the fiber, strong interfacial adhesion cannot be achieved, and the lack of interfacial interactions results in poor mechanical properties for the composites.

IOP Conf. Series: Materials Science and Engineering **368** (2018) 012017 doi:10.1088/1757-899X/368/1/012017

Therefore, the SEM studies support the tensile and flexural testing results discussed in the previous section. The morphologies of the fractured surfaces of tensile specimens of neat starch-g-PP, starch-g-PP/KF composites without and with 3% MAgPP were investigated by SEM. The results are shown in Figure 5. From the images, there is a difference in the adhesion between starch-g-PP matrix and kenaf fiber. Figure 5C shows the fractured surface of the composites without MAgPP. The fibers were pulled out with relatively clean fiber surface and without the resin on fiber surface that is an indication of low adhesion between them. The kenaf fibers also not in close contact with the starch-g-PP. However, for the composites with MAgPP (shown in 7D and F), the gaps between the fiber and the matrix become small.

On the other hand, the starch-g-PP/MAgPP matrix and kenaf fibers are intimately bonded together. It is also observed that the matrix is pulled out together with the fiber during tensile fracture, which further substantiates adhesion between the fiber and the matrix. The maleic anhydride of the MAgPP compatibilizer can interact with the starch phase of the matrix. Maleic anhydride can react with the numerous hydroxyl groups of the starch which is mainly constituted of amylose and amylopectine. Maleated polyolefins compatibilizer was proven to effectively improve the interfacial adhesion between lignocellulosic fillers and polyolefins [15,16]. Morover, the interaction between the kenaf fibers and starch-g-PP/MAgPP matrix can be attributed to the formation of hydrogen bonds in the interfacial region, for instance, between the hydroxyl ({OH) groups of cellulose or its counterpart lignin in kenaf fibers with the anhydride groups in the MAgPP matrix. Figure 6 shows plasticizing effect of starch-g-PP/KF with different MAgPP loading. From Figure 6C, the plasticizing effect for starch-g-PP/KF/5MAgPP obvious compared to starch/KF/1MAgPP and starch-g-PP/KF/3MAgPP.

IOP Conf. Series: Materials Science and Engineering **368** (2018) 012017 doi:10.1088/1757-899X/368/1/012017

Figure 5. SEM micrographs of fracture surface of tensile specimen (A) 1.00 KX and (B) 5.00 KX neat matrix , (C,E) starch-g-PP/ KF (D,F) starch-g-PP/KF/3MAgPP

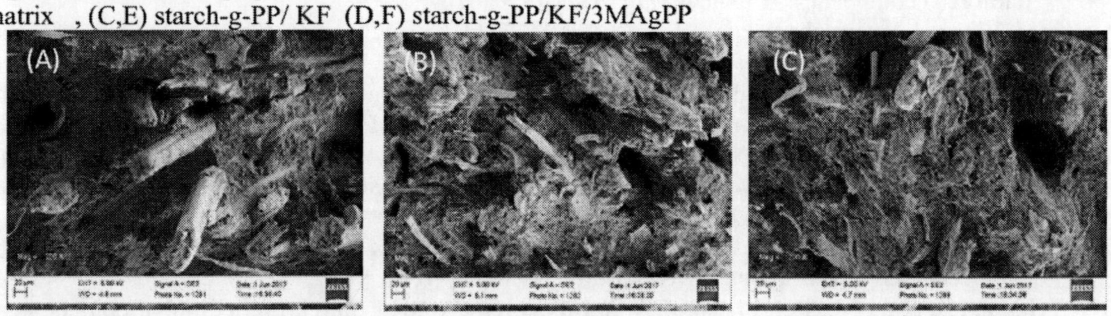

Figure 6. SEM micrographs of fracture surface of tensile specimen (A) starch-g-PP/KF/1MAgPP (B) starch-g-PP/KF/3MAgPP (C) starch-g-PP/KF/5MAgPP

4. Conclusions

MAgPP was used as a good compatibilizer for starch-g-PP/KF composites. The influence of different MAgPP content on the properties of the composites was studied. The mechanical properties of the composites were improved with the addition of MAgPP. Starch-g-PP/KF composite with 3 wt% MAgPP has tensile strength of 18.44 MPa and flexural strength of 37.08 MPa. The thermal stability of the composites increased with the addition of MAgPP. SEM examination on the tensile fractured surface of the compatibilized composites revealed the formation of the good fiber-matrix interfacial adhesion between KF and starch-g-PP matrix.

Acknowledgements
The author thanks Roquette Company for providing the starch-grafted polypropylene. This work is supprted by Ministry of Education by providing RUC Grant (Grant no.1001/PKT/8640012 and LRGS Grant (Grant no:203/PKT/6725002)

References
[1] Faludi G, Dora G, Renner K, Moczo J and Pukanszky B. 2013 Biocomposite from polylactic acid and lignocellulosic fibers structure-property correlations *Carbohyd Polym.* **92** 1767–75
[2] Yu T, Li Y and Ren J. 2009 Preparation and properties of short natural fiber reinforced poly(lactic acid) composites *T Nonferr Metal Soc* **19** 651–5

[3] Franco P J H and Gonzalez and A V 2005 A study of the mechanical properties of short natural-fiber reinforced composites *Compos Part B-Eng* **36** 597 – 608

[4] Zampaloni M, Pourboghrat F, Yankovich S A, Rodgers B N, Moore J, Drzal L T, Mohanty A K and Misra M 2007 Kenaf natural fiber reinforced polypropylene composites: A discussion on manufacturing problems and solutions *Compos Part A-Appl S* **38** 1569 – 80

[5] Arbelaiz A, Fernandez B, Cantero G, Llano-Ponte R, Valea A and Mondragon I. 2005 Mechanical properties of flax fibre/polypropylene composites Influence of fibre/matrix modification and glass fibre hybridization *Compos Part A – Appl S* **36**(12) 1637–44

[6] Bourmaud A and Pimbert S 2008 Investigations on mechanical properties of poly(propylene) and poly(lactic acid) reinforced by miscanthus fibers *Compos Part A – Appl S* **39**(9) 1444–54

[7] Huda MS, Drzal LT, Mohanty AK and Misra M 2006 Chopped glass and recycled newspaper as reinforcement fibers in injection molded poly(lactic acid) (PLA) composites: a comparative study *Compos Sci Technol* **66** 1813–24

[8] Akil H M, Omar M F, Mazuki A A M, Safiee S, Ishak Z A M and Abu Bakar A 2011 Kenaf fiber reinforced composites: A review *Mater Design* **32** 4107

[9] Ku H, Wang H, Pattarachaiyakoop N and Trada M, 2011 A review on the tensile properties of natural fiber reinforced polymer composites *Compos Part B-Eng* **42** 856

[10] Mano J F, Koniarova D and Reis R L 2003 Thermal properties of thermoplastic starch/synthetic polymer blends with potential biomedical applicability *J Mater Sci Mater Med* **14** 127–35

[11] Tomé L C, Fernandes S C, Sadocco P, Causio J, Silvestre A J, Neto C P and Freire C S, 2012 Antibacterial thermoplastic starch chitosan based materials prepared by melt mixing. *BioResources* **7** 3398–409

[12] Craig L and Beyler MMH. 1998 Thermal decomposition of polymers. SFPE handbook of fire protection engineering book. 1 ed. Section one: 110–31

[13] Hamma A, Kaci M, Mohd Ishak Z A and Pegoretti A 2014 Starch-grafted-polypropylene/kenaf fibres composites. Part 1: Mechanical performances and viscoelastic behaviour *Compos Part A – Appl S* **56**, 328

[14] Tessier R, Lafranche E and Krawczak P 2012 Development of novel melt-compounded starch-grafted polypropylene/polypropylene-grafted maleic anhydride/organoclay ternary hybrids. *Express Polym Lett* **6** 937–52.

[15] Keener T J, Stuart R K, and Brown T K 2004 Maleated coupling agents for natural fibre composites. *Compos Part A – Appl S* **35** 357 – 62

[16] Mohanty S, Nayak S K, Verma S K, Tripathy S S 2004 Effect of MAPP as a coupling agent on the performance of Jute–PP composites *Journal of reinforced plastics and composites* **23** 2047 – 2063

Physical and flammability properties of kenaf and pineapple leaf fibre hybrid composites

M Asim[1]*, M.T.Paridah[1]. M Jawaid[1], M Nasir[2] and N Saba[1]

[1]Laboratory of Biocomposite Technology, Institute of Tropical Forestry and Forest Products (INTROP), Universiti Putra Malaysia, 43400 UPM Serdang, Selangor, Malaysia
[2]Forest Product Division, Faculty of Forestry, Banda University, Banda, India

*Corresponding author: khanfatehvi@gmail.com

Abstract. Pineapple leaf fibre (PALF) was hybridized with kenaf fibre in order to achieve superior physical and flammability performance. The mixing effects of kenaf fibre and pineapple leaf fibre in phenolic composites are evaluated at various fibre ratios and investigated various physical properties such as density, void content water absorption, thickness swelling and flammability. Pure kenaf fibre composite showed lowest void content, water absorption and thickness swelling while pineapple leaf fibre revealed the opposite trend. After adding pineapple leaf fibre in kenaf composites, it absorbed more moisture and developed more void contents. Density of PALF composite was lower and kenaf fibre composite was highest but PALF/kenaf hybrid composite increased density with higher PALF. The flammability of hybrid composites were analysed by using UL-94 test method. Vertical and horizontal UL-94 test conducted to analyse fire resistance of composites from different angles. Phenolic resin is itself a fire resistant polymer and the percentage of polymer in all samples are fixed though ratios of both fibres pineapple leaf fibre and kenaf were change to analyse fire resistant of fibres and compatibility with polymers. The 70 % pineapple leaf fibre and 30 % kenaf fibre hybrid composites were very affective to improve the flammability of PALF/KF hybrid composites.

1. Introduction

As far as the environment is the concern, renewable resources of materials are the best alternative material of good economic potential and protection of nature, since, the renewable resources are characterized as biodegradable and recyclable, lignocellulosic fibres belong to this category [1]. Nowadays, natural fibres as reinforced material have been increased considerably due its low price compared to synthetic fibers [2]. Waste management of lignocellulosic and renewable materials and poor mechanical strength and thermal stability of polymers can be improved by developing natural fibres reinforced polymer composites [3-5]. Some plants fibres have potential to be applied in industries as raw materials such as pineapple, kenaf, coir, abaca, sisal, cotton, jute, bamboo, banana, Palmyra, talipot, hemp, and flex [6-7]. Pineapple leaf fibre (PALF) is one of the waste materials in agriculture sector, which is widely grown in Malaysia as well as Asia. After banana and citrus, pineapple (*Ananas comosus*) is one of the most essential tropical fruits in the world [8]. Natural fibres influenced mechanical strength due to chemical composition such as to cellulose, hemicellulose and lignin and its orientation and some other factors such as type of plant, environmental factors during growth as well as extraction methods used [9]. Usage of natural fibre can replace synthetic fibres to minimise health hazards, low density and high flexibility [10].

Recently, polymer hybrid composites have become the attraction into production of mechanical stable and renewable products [11]. Polymer hybrid composites are formed by incorporating two and more reinforcing materials into a polymeric matrix via-versa. The hybridization of two natural fibers or a synthetic fibre with a natural one can improve the physical, mechanical and thermal properties of composites [12]. The reinforcing materials provide a synergetic effect called a "hybrid effect" that determines new or enhanced properties [13]. The hybrid composites based on fiber fillers have become extremely important materials [14]. It is extensively used in aerospace, automotive, and construction industry etc. The main reason of using fibres as reinforced materials for fabricating the hybrid composites are their mechanical strength, lightweight, durability and competitiveness and eco-friendly [15].

Natural fibre reinforced polymer composites are sensitive to water absorption, humidity, dimension stability. Water absorption affects the mechanical strength and compatibility between fibre and matrix and leading towards poor stress transfer efficiencies from matrix to reinforcement [16]. Mechanism of water absorption in fibre composites is based on three ways; first method is diffusion method [17], water molecules flow into micro-cracks of polymer chains. Second method is capillary flow[18]; water flow along with the interface if interfacial bonding of fibre and matrix is weak. Last mechanism is moisture content; storage of water in micro-cracks present in both polymer and natural fibre [19]. Study of water absorption in natural fibre composite is necessary for different applications such as outdoor component, waste water treatment, packaging and building industry. In natural fibre reinforcement, fibre absorb water due to many reason such as temperature, amount of fibre, orientation of reinforcement, fibre characteristics, area of exposed surfaces, diffusivity, and surface protection [20].

Origin of most of the polymers are petroleum-based, due to this they are highly flammable/combustible [21]. Phenolic resins have been deeply studied for commercial purpose and still there is no substitute of phenolic in stability and heat-resistance quality [22]. Phenolic resin is very brittle in nature and due to this mechanical strength is very low and it is not biodegradable. Many natural fibres have been used as reinforced material to enhance the mechanical properties. Pineapple leaf fiber (PALF) was selected as reinforcement materials for phenolic resin due to the following reasons. First, PALF has good tensile strength (400–1600 MPa) and modulus (59 GPa) due to its high cellulose content 70–82% and high degree of crystallinity [8]. Second, since pineapple can be cultivated throughout the South-East Asia, it is possible for significant domestic production of PALF [23]. Third, PALF has already been used as a reinforcement to strengthen LDPE, rubber, thermosetting polyester, polyhydroxylbutyrate (PHB) and polyester amide [24]. Kenaf is comparatively very cheap and commercially available natural fibre to use as reinforcing material. Kenaf fibre (KF) showed very good compatibility with the different type of polymers such as epoxy and phenolic resin [25].

In this research, we have fabricated PALF and KF hybrid composites by using hand layup technique. It is assumed that kenaf will reduce water absorption due to low cellulosic content good compatibility with polymers. Phenolic resin will help to improve fire resistant properties of composites, where, PALF will improve compatibility due to its physical properties.

2. Materials and method

Novolac type Phenolic formaldehyde resin (Grade PH-4055) was supplied by Chemovate Girinagar, Banglore, India. Pineapple leaf fibres (PALF) (*Ananas comosus*) were harvested in Indonesia and kenaf were harvested from Malaysia and retted by immersion into water having 0.5% wt Urea. The chemical compositions of PALF and kenaf were collected from previous study [1] were examined in MARDI, Malaysia. The data of mechanical properties and fibre diameter were collected from the previous studies [23] shown in Table 1.

Table 1. Chemical composition, physical and mechanical properties of PALF and KF [1,23].

Properties	PALF	KF
Tensile strength (MPa)	290.61	282.60

Tensile modulus (GPa)	5.83	7.13
Cellulose content (%)	70.51	66.89
Hemicellulose content (%)	14.21	14.98
Lignin content (%)	2.93	6.85
Fibre diameter (μm)	78.80	83.50

2.1. *Fabrication of composites*
PALF and KF used as filler in the fabrication of composites. PALF and KF were ground into 0.8–1 mm by using grinding machine; fibres maintain 6–8 % moisture content. PALF and KF reinforced Phenolic composites were fabricated by using hand lay-up technique in 15 × 15 × 3 mm stainless metal plate. The 3 mm stainless steel plate was placed into hydraulic pressure hot press at 160 °C temperature. The stainless steel plate was removed from the press after 8 min and kept for cooling to room temperature, and then samples were cut for testing according to ASTM standard. The ratio of fibres and matrix of composites are tabulated in Table 2.

Table 2. Formulation of composites and hybrid composites

Type of composites and hybrid composites	PF (wt%)	PALF(wt%)	KF(wt%)
50% PALF	50	50	0
3P7K	50	15	35
1P1K	50	25	25
7P3K	50	35	15
50% KF	50	0	50

2.2. *Density*
Density was measured by using ASTM D 1895 standard. The density of the samples was calculated by using following equation

$$\text{Density (g/cm}^3) = m/v \quad (1)$$

Where, m represents mass of fibre/matrix and v represents volume of fibre/matrix.

2.3. *Void content*
For determination of voids in PALF and KF composites and its hybrid composites, ASTM-D-2734-70 [26] method was used. The void content was determined from the theoretical and experimental density of the composites through following equation:

$$\text{Void content} = 100/\,(R/D + r/d) \quad (2)$$

Where, R is the weight % of the resin in the composite;
D is the density of the resin matrix;
r is the weight % of the reinforcement in the composite;
d is the density of the reinforcement.

2.4. *Water Absorption Test*
Natural fibre composites sample were immersed into distilled water at room temperature. Water absorption was calculated by the equation 3 [27] for various period of time. All the samples were weighed before and after the soaking samples and recorded carefully. At the three intervals, water absorption was examined such as 3 day, 5 days and 7 days. After taking out samples from water, all the samples were wipe out to remove excess amount of water on the surfaces. The percentage equilibrium

water absorption was calculated as an average value of several measurements. The percentage of water absorption was calculated from Equation (3) using ASTM D570,

$$\text{Water absorption (\%)} = [(W_n - W_d)/W_d]X100 \quad (3)$$

where, W_n is the weight of composites samples after immersion and
W_d is the weight of the composite samples before immersion

2.5. *Thickness Swelling*

Ten samples of pure phenolic and each different PALF fibre loadings composites were prepared for the experiment of thickness welling. Thickness swelling was calculated according to ASTM D570 by using Equation (4)[27]. The thickness swelling experiment was conducted by measuring the swelling of the each sample. Measurement of the composite was taken before and after the soaking of composites. Samples were measured three time, first measurement was taken on third day of immersion second was fifth days of immersion and last was seven days of immersion.

$$\text{Thickness Swelling (\%)} = [(T_1 - T_0)/T_0]X100 \quad (4)$$

where, T_1 is the thickness after soaking and
T_0 is the thickness before soaking.

2.6. *Flammability test*

Flammability of the untreated and treated composites and hybrid composites were studied by a horizontal UL-94 test and a vertical UL-94 test according to ASTM D635 and ASTM D3801, respectively. For the horizontal UL-94 test, the specimen was fixed horizontally and a flame was applied to one end of the specimen. The time for the flame travel from 25 mm (from the end) to 100 mm was recorded. Then, burning rates of the composites were calculated as follows:

$$V = 60L/t \quad (5)$$

Where, V is the burning rates (mm/min),
L is burned length (mm), and
t is the time of burning (sec).

The UL-94 vertical test was carried by using a test specimen bar 125 mm long, 13 mm wide, and 3 mm thick. Specimen bars were suspended vertically over surgical cotton, and ignited with a Bunsen burner. The flammability types were classified from V-0 toV-2. The lower end of the specimen was exposed to the flame for 10 s and then removed. The time was recorded.

The specimen would be then assigned to class UL-94 V-0 if flaming time is less than 10 sec. The lowest flame retardancy rating is V-2 if flaming time is less than 50s and cotton indicator ignited by flaming particles or drops. V-1 is the intermediate class of V-0 and V-2, the flaming time should not exceed 30s and flaming drops could not ignite the surgical cotton below the specimen. The flammable materials which cannot be classified according to this method are marked with code NC (no classification, complete combustion of the specimen).

3. Results and discussion

3.1 *Density*

The theoretical and experimental densities of PALF and KF composites and their hybrid composites were calculated shown Table 1. The experimental density of PALF composites is lower (1.07 g/cm^3) than theoretical value (1.3 g/cm^3). However, the experimental density of KF composites is lower (1.25 g/cm^3) than theoretical value (1.27 g/cm^3) though this difference is very less in comparison of PALF composites. After addition of PALF at different fibre loading in KF composites, experimental densities reduced with inconsistency and at the same time the theoretical densities increased. Among the all hybrid composites, the 7P3K hybrid composites showed highest theoretical as well as experimental

IOP Conf. Series: Materials Science and Engineering **368** (2018) 012018 doi:10.1088/1757-899X/368/1/012018

densities. The difference in theoretical and experimental densities are due to air trap in composites during manufacturing process [28].

3.2 Void content
The density of the both, measured and theoretical value of PALF and KF composites and their hybrid composites are given in Table 3. From the table, it is clearly seen that the specific gravity of PALF and KF composites and their hybrid composites are vary in different fibre fractions composites, which indicated the contribution of fibres in density of composites. The void contents is the difference of theoretical and practical density of substance, the most common cause of voids is the incapability of the matrix to displace all the air which is trapped within the woven or chopped fibres as it passes through the matrix impregnation. The void content (%) of PALF and KF composites and their hybrid composite samples are presented in Table 3. The PALF composite showed 12.50 % void content, however, KF composites revealed very low void content (2.36 %). It is also observed that after adding PALF in to KF composites, the percentage of void content increased. PALF was added up to 70 % of total fibre loading and the void content also increased up to 11.11 %. The increment in void content is may be due to incompatibility between the phenolic resin and PALF and high percentage of fibre fraction were inadequate to mix with the polymer properly.

Table 3. Density and void content of PALF and KF-Phenolic composites and their hybrid composites

Weight Fraction of fibres (%)	**Theoretical density (g/cm^3)**	**Measured density (g/cm^3)**	**Void content (%)**
50% PALF	1.4	1.22	12.5
3P7K	1.3	1.13	13.07
1P1K	1.32	1.18	10.22
7P3K	1.35	1.2	11.11
50% KF	1.27	1.24	2.36

According to researchers [29-30], incomplete wetting out of the fibres by the matrix would lead to the formation of voids. During impregnation of fibres into the matrix or during manufacturing of fibre reinforced composites, the trapped air or other volatiles exist in the composites [31]. KF composites showed very less void content percentage among all composites and hybrid composites, it showed KF has very good interfacial bonding with Phenolic matrix and have very less hollow space. The reduction in void content is evidence of improvement in mechanical strength because the presence of void content in the composites significantly reduces the mechanical and physical properties of the composites [32].

3.3 Water absorption
Water absorption (WA) percentage of PALF and KF composites and PALF/KF hybrid composites were shown in Fig.1. Among the all types of composites, KF composite showed lowest WA percentage and the deference in WA percentage was very high. However, WA of rest of the composites and hybrid composites were very close to each other. PALF composites revealed lower WA than all hybrid composites though the difference was very less. Among the all hybrid composites, 3P7K and 1P1K hybrid composites showed the highest WA. However 7P3K hybrid composites revealed its WA in the middle of highest WA and PALF composites shown in fig.1. From the results, KF composites showed lowest WA due to lower content of cellulose and PALF composites should show highest WA due to higher content of cellulose.

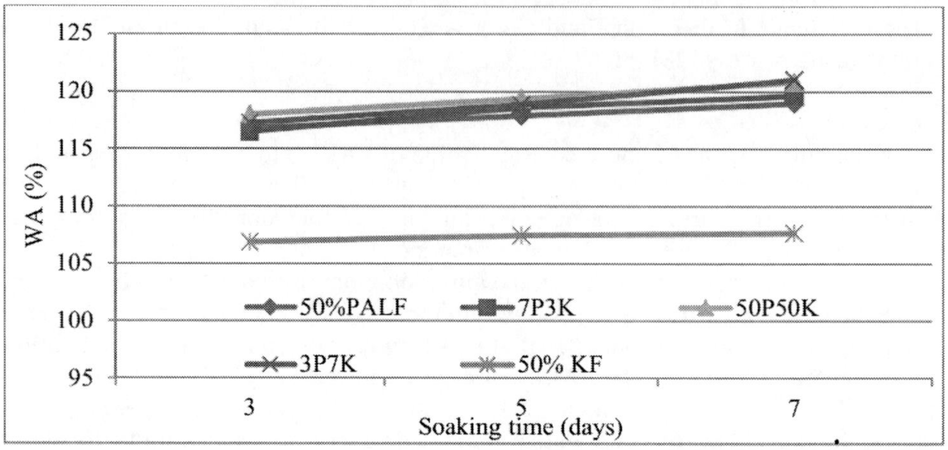

Figure 1. Water absorption of PALF and KF hybrid composites.

The expected results were to reduce the WA by adding KF in PALF composites; in the contrary the results showed that hybrid composites absorb more water content. By this research, it can be assume that in despite of having comparatively less quantity of cellulosic content, WA is increasing due to high void contents. The water molecules easily enter into the composites through such pores and these molecules are attracted by hydroxyl groups of natural fibres causing a rise of water absorption. This hydroxyl groups present in cellulose and hemicelluloses provide easy access to water in to composites. Besides that, there are many other factors such as porosity, void content, lumen size, and fibre-matrix adhesion also effect water absorption behaviour of the composites [33]. The inter-fibrillar spaces of cellulosic structure of natural fibres contain water molecules that crack and micro voids at the surface of composites [34]. The immersed composites into the water for long time, absorb water molecules through capillary action and fills the voids and cracks in the composites [27,35].

3.4 Thickness swelling

The natural fibre composites are highly hydrophilic in nature due to the presence of hydroxyl groups, which invite water molecules through hydrogen bonding [34]. These water molecules are the cause to build the moisture into cell walls and fibre/matrix interface. This is one of the main reason of dimensional changes of natural fibre reinforced polymer composites especially thickness and linear expansion due to reversible and irreversible swelling of the composites[36,37]. The Fig. 2 showed the TS of PALF and KF composites and their hybrid composites. KF composites showed very less impact of moisture on TS, with the immersing period the moisture swelling was increased.

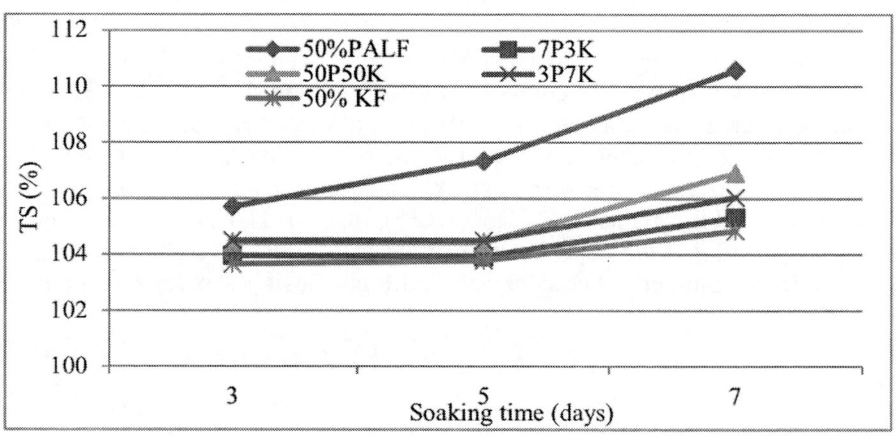

Figure 2. Thickness swelling of PALF and KF hybrid composites.

IOP Conf. Series: Materials Science and Engineering **368** (2018) 012018 doi:10.1088/1757-899X/368/1/012018

The hybrid composites 3P7K and 7P3K showed the higher TS values than KF composites but the pattern in dimensional changes are same affected by soaking periods. The hybrid composite 1P1K was showed more affected by last soaking period. The PALF composites showed highly prone to TS over the all soaking period of time. From these results, it is assumed that PALF has higher quantity of cellulosic materials and hydroxyl groups than KF, which absorb moisture content without hindrance. However, KF composite and KF hybrid composites showed less TS may due to have good interfacial bonding, less cracks and voids [28]. As can be seen from Fig. 2, the thickness swelling values of composites slightly increased with the increasing water exposure time. The swelling of the cell wall of fibres develop stress on the surrounding matrix and cause micro cracking, which reduce the mechanical properties of composite. As a consequence, the fiber–matrix adhesion is weak and the dimensional stability of composites particularly for outdoor applications will be greatly affected [34].

3.5 Under writer laboratory (UL-94)
The vertical and horizontal UL-94 tests were conducted to quantify and rank the flame retardancy of the composites. The results of the UL-94 tests for the PALF and KF composites and their hybrid composites are presented in Table 4. In Vertical UL-94, the flame retardancy of KF composite was very good and achieved V-0 grade because their combustion times were less than 10 seconds. After the addition of 30%, 50% and 70% PALF, the fire retardancy was reduced, and were considered in V-1 grade due to their combustion times were less than 30 seconds. On the other hand, PALF composites showed lower fire retardancy than KF composites and the combustion time was less than 13sec, considered in V-1 grade. No dripping was observed for any of the samples. Overall burning rate was low due to Phenolic matrix because of chemical structure of phenol resin is deformed and forms a layer of char on the surface of composites that protect the surface from heat [38].

Table 4. Horizontal and vertical burning rate of PALF and KF composites and PALF/KF hybrid composites

Sample	Vertical UL-94			Horizontal UL-94 (mm/min)
	Flaming drops	Cotton ignited	Classifications	
50% PALF	No	No	V-1	15.71
7P3K	No	No	V-1	18
1P1K	No	No	V-1	17
3P7K	No	No	V-1	15.56
50% KF	No	No	V-0	13.6

Horizontal UL-94 showed the combustion rate of KF composite lowest among all composites however after addition of PALF into KF composites, the combustion rate was increased shown in Table 4 Burning rate of 7P3K was highest it may be due to incompatibility of both fibres with matrix that exposed the uncovered cellulosic content of fibre to the fire.

4. Conclusion

- PALF composites showed lowest density and higher void content. After adding KF, density of hybrid composite increased and void content decreased.
- Water absorption of PALF composites was higher and after adding KF it did not show significant affect. Alone KF composites showed very low water absorption.
- Thickness swelling of PALF composites was highest and KF composite was lowest. After hybridisation of PALF and KF, thickness swelling was reduced.
- All hybrid composites showed good fire resistant properties. In vertical UL-94, KF composites showed better property. However, in horizontal UL-94, PALF composite showed slower burning rate.

5. References

[1] Asim M, Jawaid M, Abdan K, and Ishak M 2017 J. Polym. Environ. **1**
[2] Javni I, Zhang W, Karajkov V, Petrovic Z, and Divjakovic V 2002 J. Cell. Plast. **38** 229
[3] Hadjadj A, Jbara O, Tara A, Gilliot M, Malek F, Maafi EM, and Tighzert L 2016 Compos. Struct. **135** 217
[4] Azwa Z, Yousif B, Manalo A, and Karunasena W 2013 Mater. Des. **47** 424
[5] Otto GP, Moisés MP, Carvalho G, Rinaldi AW, Garcia JC, Radovanovic E, and Fávaro SL 2017 Composites Part. B**110** 459
[6] Yan L, Chouw N, and Jayaraman K 2014 Composites Part B **56** 296
[7] Bongarde U and Shinde V 2014 IJESIT **3** 431.
[8] Asim M, Abdan K, Jawaid M, Nasir M, Dashtizadeh Z, Ishak M, and Hoque ME 2015 Int J Polym Sci **1**
[9] George J, Sreekala M, and Thomas S 2001 Polym. Eng. Sci. **41**1471
[10] Saba N, Jawaid M, and Asim M. 2016 Nanoclay Reinforced Polymer Composites (Springer) pp. 1-28.
[11] Jawaid M and Abdul Khalil HPS. 2011 Carbohydr. Polym, **86** 1
[12] Asim M, Jawaid M, K. Abdan, M. R. Ishak and O. Y. Alothman 2017 j. renew. mater. **6** 38
[13] Beloshenko V, Voznyak Y, Voznyak A, and Savchenko B 2017 Composites Part B.**112** 22
[14] Sathishkumar T, Naveen J, and Satheeshkumar S 2014 J. Reinf. Plast. Compos **33** 454
[15] Gacitua W, Ballerini A, and Zhang J 2005 Cienc. Tecnol. **7** 159
[16] Chen H, Miao M, and Ding X 2009 Composites Part A **40** 2013.
[17] Shirangi M and Michel B 2010 Moisture Sensitivity of Plastic Packages of IC Devices (Springer). pp. 29-69.
[18] Espert A, Vilaplana F, and Karlsson S 2004 Composites Part A **35**1267
[19] Dhakal H, Zhang Z, and Richardson M 2007 Compos. Sci. Technol. **67** 1674
[20] Devi LU, Bhagawan S, Nair K, and Thomas S 2011 Polym. Compos, **32** 335
[21] Aji I, Sapuan S, Zainudin E, and Abdan K 2009 IJMME **4** 23
[22] Wang M, Wei L, and Zhao T 2005 Eur. Polym. J. **41** 903
[23] Asim M, Jawaid M, Abdan K, and Ishak MR 2016 J. Bionic. Eng.**13** 426
[24] Liu W, Misra M, Askeland P, Drzal LT, and Mohanty AK 2005 Polymer **46** 2710
[25] Asim M, Jawaid M, Abdan K, and Ishak M 2017 Fiber. Polym. **18** 940
[26] Rajulu AV, Chary KN, Reddy GR, and Meng Y 2004 J. Reinf. Plast. Compos. **23** 127
[27] Abdul Khalil HPS, Jawaid M, and Bakar AA 2011 BioResources **6** 1043
[28] Jawaid M, Abdul Khalil HPS, Khanam PN, and Bakar AA 2011 Compos. Sci. Technol. **19** 106.
[29] De Albuquerque A, Joseph K, de Carvalho LH, and d'Almeida JRM 2000 Compos. Sci. Technol. **60** 833
[30] Mariatti J, Nasir M, and Ismail H 1998 Preliminary studies of woven thermoplastic composites. MSc thesis, Universiti Sains Malaysia.
[31] Chua P, Dai S, and Piggott M 1992 Mater. Sci. Mater. **27** 919
[32] Jawaid M, Abdul Khalil HPS, Bakar AA, and Khanam PN 2011 Mater. Des. **32**1014
[33] Mariatti M, Jannah M, Bakar AA, and Khalil HA 2008 J. Compos. Mater. **42** 931
[34] Ashori A and Sheshmani S 2010 Bioresour. Technol. **101** 4717
[35] Ahmed KS and Vijayarangan S 2007 J. Appl. Polym. Sci. **104** 2650
[36] Abdul Khalil HPS, Issam A, Shakri MA, Suriani R, and Awang A 2007 Ind. Crops. Prod. **26** 315
[37] Nourbakhsh A and Ashori A 2010 *Waste Manag. Res.* **28** 51
[38] Chiu HT, Chiu SH, Jeng RE, and Chung JS. 2000 Polym. Degrad. Stab. **70** 505

Thermal properties of coir and pineapple leaf fibre reinforced polylactic acid hybrid composites

R Siakeng[1], M Jawaid [1], H Ariffin [2] and S M Sapuan [1]

[1]Laboratory of Biocomposite Technology, Institute of Tropical Forestry and Forest Products (INTROP), Universiti Putra Malaysia, 43400 UPM Serdang, Selangor, Malaysia

[2]Laboratory of Biopolymer and Derivatives, Institute of Tropical Forestry and Forest Products (INTROP), Universiti Putra Malaysia, 43400 UPM Serdang, Selangor, Malaysia

*Corresponding author: ramengmawii8@gmail.com

Abstract. This study examined the thermal behaviour of raw coir fibre (CF) and pineapple leaf fibre (PALF) reinforced Poly lactic acid (PLA) composites. Coir fibres were hybridized with PALF in order to achieved better thermal stability. The biocomposites contained 30% (vol.) fibres in the form of short fibres (1-2 mm), and the composite samples were made up by different fibre ratios. In this research, the composites were fabricated by hot press at 180 ^{0} C for 10 minutes pre-heat and 3 minutes press after melt compounding the fibres and polymer in Internal Mixer Brabender. The aim of this work is to investigate the effect of different fibre ratio on the thermal properties of CF/PALF/PLA hybrid composites by using thermogravimetric analysis (TGA). The effect of different fibre ratios in hybridization and thermal degradation of CF/PALF hybrid composites was analyzed and C1P1 hybrid composite showed highest thermal stability in all composites samples. Thermal properties of CF/PALF/PLA hybrid composites vary with fibre ratio probably due to difference in chemical composition of the two fibres. C1P1 hybrid composites (1CF: 1PALF) showed highest thermal stability all composites which could be due to optimum hybridisation of both fibres. The thermal degradation analysis showed that both CF and PALF slightly affected the thermal stability of PLA composites. Consequently, surface treatments of CF and PALF will be effective to further improve composite's thermal stability.

1. Introduction

Natural fibres (NFs) have been used for reinforcement for diverse polymer matrixes due to their various beneficial properties like low density and cost, biodegradability, high-specific strength and renewability[1, 2]. Environmental sustainability concern also drives researchers on utilizing NFs for reinforcement purpose in manufacturing composites to get comparable or even better composites with synthetic fibres such as glass, carbon, and aramid [3, 4]. Many NFs like jute, kenaf, hemp etc. are used in composite as these are cultivated with the main purpose of fibre production[5, 6]. In addition to these fibre plants there are other plants in which the fibre has secondary value such as banana, coconut and pineapple as these are cultivated for their fruits[6, 7]. However, these fibres are seldom used since they are discarded as agro-wastes [8].

Agro-waste fibres such as coconut coir and pineapple leaf fibres (PALF) draws great attention due to their low density and thermal conductivity to yield cost-effective and lightweight composite products[9]. Coir fibres have lower cellulose (36–43%) and hemicellulose (15.7 %) contents, higher lignin content (35.25 %) and microfibrillar angle (30–45) than other natural fibres which results in their relatively low mechanical strength[10]as well as highest elongation at break among popular NFs [11]. PALF is a smooth glossy fibre with high cellulosic content (70-82 %), hemicelluloses (18.8 %) and lignin (5-12.7 %) with high tensile strength and it has a softer surface than other NFs [12]. Selection of polymer depends on the requirements for specific applications and performances. A bio-based polymer that is used today, is poly lactic acid (PLA) [13]. It is made from polymerisation of lactic acid monomers which is derived from renewable sources such as corn starch; potato starch, sugarcane etc.[14].

Extensive research literature has covered the studies of natural fibre reinforced biodegradable composites. Coir fibres are one of the most used natural fibres as reinforcement in polymer composites due to their high structural and wear properties combined with availability [15].Jang et al. [16] studied the thermal stability and flammability of PLA/coir fibre composites and found that plasma treatment enhance the thermal properties of composites in addition to their decreased shrinkage with increasing the fibre content. Mir et al.[17] reported that coir fibres have high failure strain, providing better strain compatibility between the fibre and the matrix in short fibre reinforced polypropylene composites. Rosa et al. [18] investigated the mechanical, thermal and structural properties of biocomposites based on starch/ethylene vinyl alcohol/glycerol blends matrix and coir fibres in relation to the fibre content. The increase of fibre content resulted in decreased tensile strengths but considerably improved tensile moduli. Additionally, the fibre content slightly influenced the thermal stability of biocomposites without affecting its biodegradability.
Kaewpirom et al. [19] evaluated the tensile and thermal properties of PALF reinforced PLA green composites. Tensile modulus of the composites depends on PALF content in the composite and slightly affects the thermal stability. In addition, elongation at break was dramatically increased by incorporation of maleic anhydride coupling agent. Liu et al. [20] performed multiple properties evaluation on green composites from soy based plastic and PALF. The mechanical properties including tensile and, flexural properties and impact strength of the green composites increased with increasing fibre content and the presence of the compatibilizer (PEA-*g*-GMA). The addition of the compatibilizer also helps in fibre dispersion but decreased the moisture absorption. Many researchers have studied PLA composites reinforced with various NFs. The effect in PLA of various NFs such as jute [21], kenaf [22], coconut fibre [23-24], and PALF [19-20] has been extensively studied and is well understood.

Incorporating NFs into PLA composites enables the composites to overcome the high brittleness and improved thermal stability [25]. However, no report of PLA composites reinforced with coir fibre and PALF hybrid has been found. This work evaluated the influence of coir and PALF alone or together on the thermal properties of Coir fibre/PALF/PLA composites and hybrid composites. In this study, 30 vol.% untreated coir fibre and PALF were used as reinforcements in the PLA biocomposites.

2. Materials and method

2.1. *Materials*

Polylactic acid (PLA) pellets from Natureworks® with a molecular weight (M_w) of approximately 74,000 g/mol, density of 1.25 g/mol, glass transition temperature (T_g) of 55 to 60 °C and melting point (T_m) of 160 to 180 °C was purchased from TT Biotechnologies Sdn. Bhd., Malaysia.

Coir fibres (*Cocus nucifera*) were obtained from Innovative Pultrusion Sdn Bhd. Malaysia with an average density of 1.20 g/cm3. Pineapple leaf fibres (*Ananas comosus*) were procured from Tamil Nadu state of South India with an average density of 1.07 g/cm3. All fibres were used in as-received condition without any surface treatment.

2.2. *Composite fabrication*

Coir and PALF used as reinforcements in fabrication of composites were chopped into short fibres (1-2 mm) by using ring flakers machine; fibres were maintained at 6-8% moisture content. The composites were prepared by melt mixing in an internal mixer (Brabender-Plasticizer machine) All the five types of composites/hybrid composites were prepared through the following procedures. Composites were obtained by mixing 70% (35 g) of PLA at 180 °C for 3 min. Then, 30 % (15 g) of CF or PALF, previously dried at 80 °C for 24 h under vacuum, were added, and the mixture was blended at 180 °C for further 7 min, at a speed of 50rpm. Corresponding sample is coded as C30 and P30. Hybrid composites were obtained by mixing 70 % (35 g) of PLA at 180 °C for 3 min. Then, 30 % (15 g) of mixed coir and PALF in a specific proportion such as 1:1, 3:7 and 7:3 were added, and the mixture was blended at 180 °C for further 7 min, at a speed of 50 rpm. Composites were fabricated by using 15 x 15 x 3 mm stainless metal plate mould. The mould was placed into hydraulic pressure hot press at 180 °C temperature and was removed from the press after 15 minutes pre-heat and 5 minutes press kept for cooling in cold press for 3 minutes, and then samples were prepared for testing according to ASTM standard. The formulations of hybrid composites are described in Table 1.

Table 1. Formulation of Coir/ PALF/PLA hybrid composites

Designation	**PLA (Weight %)**	**Coir (Weight %)**	**PALF (Weight %)**
Pure PLA	100	_	_
C-30	70	30	_
P-30	70	_	30
C1P1	70	15	15
C7P3	70	21	9
C3P7	70	9	21

2.3. *Thermogravimetric Analysis (TGA)*

Thermal properties were characterized on the CF and PALF reinforced PLA composites by TGA machine Model (TGA Q 500 TA Instrument, USA). For each type of composites, 1 gram of powdered samples was used for all composites testing according to ASTM E- 03 (2003) [26]. The entire samples testing were carried out at the rate of 20 ^{0} C/ minutes in nitrogen gas flow under room temperature ranging from 30 ^{0}C to 600 ^{0} C. TGA helps to understand the thermal stability of the samples.

3. Results and discussion

Thermal stability of pure PLA as well as CF and PALF reinforced PLA composites and hybrid composites were investigated with thermogravimetric (TGA) analysis. Figure 1 and 2 shows the TGA and DTG curves for pure PLA, C 30 P 30, C1P1, C7P3 and C3P7 composites. Generally speaking, NFs degrade through two steps process and a pre- degradation stage which is the evaporation of absorbed moisture from the NFs at 50 to 100 ^{0}C. Since natural fibres are quite hydrophilic, they are subject to a dehydration process in which absorbed water is released. The first transition which occurs from 200 to 300 °C where the NFs undergoes hemicellulose degradation; finally at 400 to 500 °C, the NFs starts to decompose due to cellulose and lignin degradation [27].

Table 2.TGA characterization of neat PLA and Coir/PALF/ PLA composites

Designation	Maximum Degradation Temperature (^{0}C) T_m	Char Residue (%) at 600 ^{0}C
Pure PLA	277.81	1.05
C30	252.73	11.17
P30	287.23	5.53
C1P1	290.07	7.56
C7P3	288.64	1.14
C3P7	284.59	3.84

The temperature (T_m) corresponding to the maximum depomposition rate was used as the indicator of the PLA and composites thermal stability. Table 2 shows the T_m and final char residue of the composites against pure PLA. According to Fig. 1 and 2, the results of pure PLA and Coir/PALF/PLA composites do not show much difference, as the curves are relatively close to each other. It is seen from the TGA graph in fig 1 that for PLA and Coir/PALF/PLA composites and hybrid composites, the weight percentage drops significantly starting from 250 ^{0}C, which is mainly due to hemicellulose degradation.

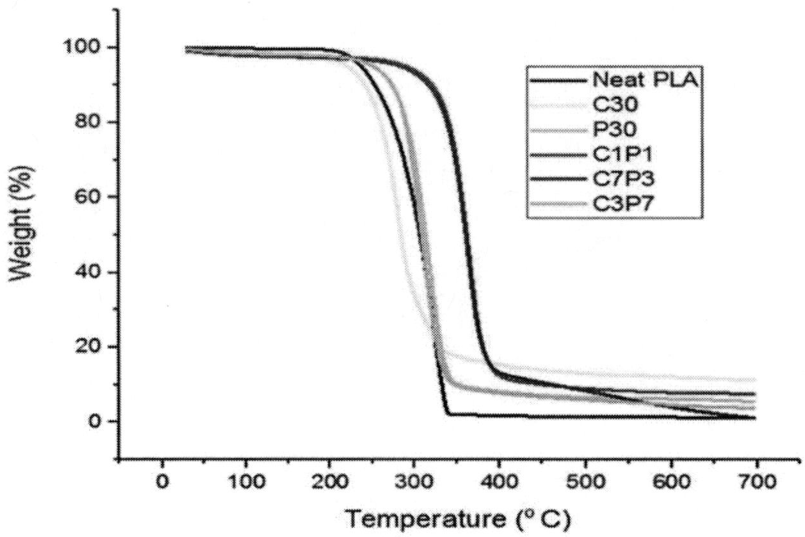

Figure 1. TGA curve of neat PLA and CF/PALF/PLA hybrid composites

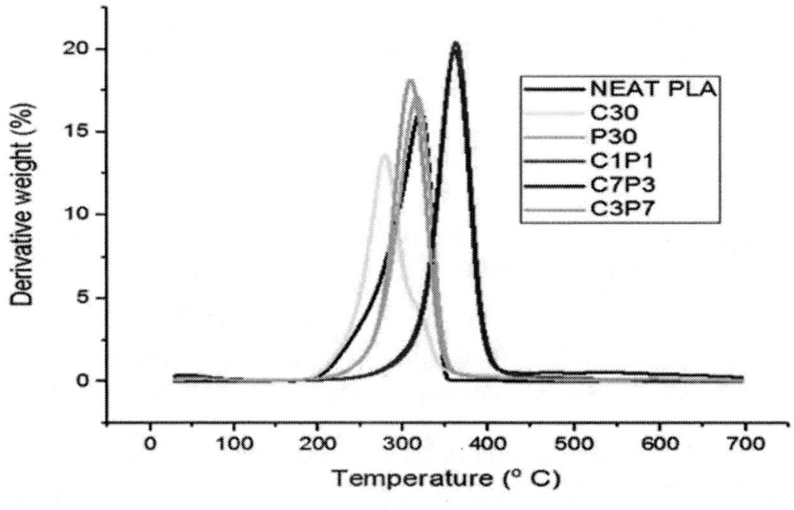

Figure 2. DTG curve of neat PLA and CF/PALF/PLA hybrid composites

Addition of CF to PLA decreases degradation temperature, for instance the maximum weight loss of C30 composite takes place at 252.73 ^{0}C compares to the neat PLA which shows maximum degradation at 277.81 ^{0}C which is in good agreement with the material characteristics of PLA composites reinforced with rice husk and kenaf fibres [28] as well as Phormium fibres [29]. On the other hand, incorporation of PALF into PLA matrix increases the thermal stability of PLA composites slightly as shown in figure 1 and 2, P30 shows T_m of 287.23^0 C . All the hybrid composites also reveals increased thermal stability compares to neat PLA such as 290.07^0 C, 288.64 ^{0}C and 284.59 ^{0}C in C1P1, C7P3 and C3P7 respectively. In all cases, there is more increase in the degradation temperatures when PALF is added. This could probably be explained by the difference in chemical composition of these two natural fibres. PALF has lower content of lignin [12] while coir fibre is rich in lignin [16] which justify highest percent of char residue in case of C30. Furthermore, the cellulose content in coir fibre is much lower than that of PALF. Hence, the thermal stability of PALF is slightly higher than that of coir fibre.

4. Conclusion

Two NFs namely; coir fibre and PALF were utilized to prepare PLA biocomposites and their thermal property were studied and compared. It was found that thermal degradation of pure PLA was faster with the incorporation of CF and slower by PALF reinforcement on the contrary. In particular C1P1 hybrid shows the best thermal stability where as C30 composites reveals fastest thermal degradation. In conclusion, PALF performed much better than CF in terms of improvement in thermal stability when compounded with PLA as a reinforcing fibre. The poor performances of coir fibre compare to PALF could be mainly ascribed the difference in chemical composition between coir fibre and PALF. The lower degradation temperature in CF filled composites as compared to PALF filled composites could be the leading character in the causes of composite's biodegradability. In order to improve the performance of coir and pineapple leaf fibres as reinforcements for PLA composite, chemical modification of coir and pineapple leaf fibres are recommended.

Acknowledgment
The authors are thankful to the Universiti Putra Malaysia for IPS Grant Vote No 9520200 and HICOE Grant Vote 6369108 for supporting this research work.

References

[1] Debnath S, Nguong C, Lee S 2013 A review on natural fibre reinforced polymer composites. *World Academy of Science, Engineering and Technology*. **11** 23-30.
[2] Summerscales J, Dissanayake NP, Virk AS, Hall W 2010 A review of bast fibres and their composites. Part 1–Fibres as reinforcements. *Composites Part A* **41**(10):1329-35.
[3] Sanjay M, Arpitha G, Yogesha B 2015 Study on mechanical properties of natural-glass fibre reinforced polymer hybrid composites: A review. *Materials today: proceedings* **2** (4-5):2959-67.
[4] Jawaid M, Khalil HA 2011 Cellulosic/synthetic fibre reinforced polymer hybrid composites: A review. *Carbohydr Polym* **86** (1):1-18.
[5] Mahapatra B, Mitra S, Kumar M, Ghorai A, Sarkar S, Kar C, et al 2012 An overview of research and development in jute and allied fibre crops in India. *Indian Journal of Agronomy* **57**:132-42.
[6] Faruk O, Bledzki AK, Fink H-P, Sain M 2012 Biocomposites reinforced with natural fibers: 2000–2010. *Prog Polym Sci* **37** (11):1552-96.
[7] Verma D, Gope P, Shandilya A, Gupta A, Maheshwari M 2013 Coir fibre reinforcement and application in polymer composites . *J. Mater. Env. Sci* **4** (2):263-76.
[8] Madurwar MV, Ralegaonkar RV, Mandavgane SA 2013 Application of agro-waste for sustainable construction materials: A review. *Constr Build Mater* **38**:872-8.
[9] Asasutjarit C, Charoenvai S, Hirunlabh J, Khedari J 2009 Materials and mechanical properties of pretreated coir-based green composites. *Composites Part B* **40** (7):633-7.
[10] Nam TH, Ogihara S, Kobayashi S 2012 Interfacial, mechanical and thermal properties of coir fiber-reinforced poly (lactic acid) biodegradable composites. *Adv Compost Mater* **21** (1):103-22.
[11] Nam TH, Ogihara S, Tung NH, Kobayashi S 2011 Effect of alkali treatment on interfacial and mechanical properties of coir fiber reinforced poly (butylene succinate) biodegradable composites. *Composites Part B* **42** (6):1648-56.
[12] Asim M, Abdan K, Jawaid M, Nasir M, Dashtizadeh Z, Ishak M, et al 2015 A review on pineapple leaves fibre and its composites. *Int. J. Polym Sci.* **2015** (1)
[13] Wool R, Sun XS 2011 *Bio-based polymers and composites*: Academic Press New York, NY.
[14] Tsuji H. Poly (lactic acid). Bio-based plastics: materials and applications. 2013:171-239.

[15] Arrakhiz F, Elachaby M, Bouhfid R, Vaudreuil S, Essassi M, Qaiss A 2012 Mechanical and thermal properties of polypropylene reinforced with Alfa fiber under different chemical treatment. *Mater. Des* **35**:318-22.

[16] Jang JY, Jeong TK, Oh HJ, Youn JR, Song YS 2012 Thermal stability and flammability of coconut fiber reinforced poly (lactic acid) composites. *Composites Part B* **43**(5):2434-8.

[17] Mir SS, Nafsin N, Hasan M, Hasan N, Hassan A 2013 Improvement of physico-mechanical properties of coir-polypropylene biocomposites by fiber chemical treatment. *Mater. Des* **52**:251-7.

[18] Rosa MF, Chiou Bs, Medeiros ES, Wood DF, Mattoso LH, Orts WJ, et al. 2009 Biodegradable composites based on starch/EVOH/glycerol blends and coconut fibers. *J. Appl. Polym. Sci.* **111** (2):612-8.

[19] Kaewpirom S, Worrarat C 2014 Preparation and properties of pineapple leaf fiber reinforced poly (lactic acid) green composites. *Fiber. Polym.* **15** (7):1469-77.

[20] Liu W, Misra M, Askeland P, Drzal LT, Mohanty AK 2005 'Green'composites from soy based plastic and pineapple leaf fiber: fabrication and properties evaluation. *Polymer* **46** (8):2710-21.

[21] Plackett D, Andersen TL, Pedersen WB, Nielsen L 2003 Biodegradable composites based on L-polylactide and jute fibres. *Compos. Sci. Technol.* **63**(9):1287-96.

[22] Huda MS, Drzal LT, Mohanty AK, Misra M 2008 Effect of fiber surface-treatments on the properties of laminated biocomposites from poly (lactic acid)(PLA) and kenaf fibers. *Compos. Sci. Technol.* **68** (2):424-32.

[23] Wu C-S 2009 Renewable resource-based composites of recycled natural fibers and maleated polylactide bioplastic: Characterization and biodegradability. *Polym. Degrad. Stab* **94** (7):1076-84.

[24] Dong Y, Ghataura A, Takagi H, Haroosh HJ, Nakagaito AN, Lau K-T 2014 Polylactic acid (PLA) biocomposites reinforced with coir fibres: Evaluation of mechanical performance and multifunctional properties. *Composites Part A* **63**:76-84.

[25] Nyambo C, Mohanty AK, Misra M 2010 Polylactide-based renewable green composites from agricultural residues and their hybrids. *Biomacromolecules* **11** (6):1654-60.

[26] ASTM E 1131 -03 2003 "Standard test method for compositional analysis by thermogravimetry", *ASTM International*, West Conshohocken, PA.

[27] Azwa Z, Yousif B, Manalo A, Karunasena W 2013 A review on the degradability of polymeric composites based on natural fibres. *Mater. Des*. **47**:424-42.

[28] Yussuf A, Massoumi I, Hassan A 2010 Comparison of polylactic acid/kenaf and polylactic acid/rise husk composites: the influence of the natural fibers on the mechanical, thermal and biodegradability properties. *J. Polym. Environ.* **18** (3):422-9.

[29] De Rosa I, Iannoni A, Kenny J, Puglia D, Santulli C, Sarasini F, et al. 2011 Poly (lactic acid)/Phormium tenax composites: Morphology and thermo-mechanical behavior. *Polym. Compos.* **32** (9):1362-8.

Preparation and characterization of modified and unmodified carrageenan based films

Ying Ying Tye*, Abdul Khalil HPS , Cheng Ying **Kok and Chaturbhuj K Saurabh**

School of Industrial Technology, Universiti Sains Malaysia, 11800 Minden, Penang, Malaysia.

*Corresponding author: tyying87@yahoo.com

Abstract. Seaweed is a green, abundantly available, and inexpensive source of polysaccharide that can be obtained from the sea. In this work, seaweed was introduced as a potential raw material for the production of a biopolymer-based film. A seaweed-derived biopolymer, carrageenan, with or without alkali modification was investigated for its physical, mechanical, and water vapor barrier as well as thermal properties. The effects of chemical modification on the properties of carrageenan were also further studied by Fourier Transform Infrared Spectroscopy (FTIR) analysis. Results showed that chemically modified carrageenan film exhibited better thermal stability, water vapor barrier property and less hydrophilic as compared to the native seaweed and unmodified carrageenan films. Nevertheless, its mechanical properties (except elongation at breaks) were among the lowest. FTIR analysis indicated that alkali modification of carrageenan capable of removing the hydroxyl and sulfate ester groups in carrageenan structure and meanwhile, forming 3,6-anhydrogalactose, which could improve the flexibility of the film. Moreover, the heating temperature also significantly impacted the properties of the film. As a result, by using carrageenan as a renewable feedstock with or without chemical modification was shown its feasibility on the production of biopolymer film possessed interesting properties.

1. Introduction

Seaweeds or known as marine macro-algae are multicellular plant-like organisms which generally found living attached to rocks in coastal areas. Since seaweed is abundantly available, able to grow in wide range of environments, cost effective and easy to cultivate in the natural environment as well as harvest year around, they are widely used as a human food source and herbal medicine. Instead of using seaweeds as a food, their excellent carbohydrate content also encouraged the industrial uses of seaweeds as a source of hydrocolloids (i.e. alginate, carrageenan, and agar) in the application for food, pharmaceutical and plastic industries [1].

Carrageenan is one of the seaweed-derived polymers obtained by extraction from red (Rhodophyta) seaweed. It composed of sulfated galactans, which have both gel forming abilities and viscosifying properties [2]. Therefore, it is commonly used as texturizing (or thickening) agents to control the functional properties of aqueous solution in food applications [3]. In general, carrageenan can be classified as Kappa (κ), Iota (ι) and Lambda (λ) according to their sulfate substitution pattern and 3,6-anhydrogalactose content, which affects the gel-forming properties. For kappa carrageenan, its structure consists of 25-30% of sulfate ester group and 28-35% of 3,6-anhydrogalactose while for Iota consists of 28-38% and 25-30% respectively. In Lambda structure have no 3,6-anhydrogalactose and contains 32-39% sulfate ester [4]. Since κ-carrageenan exhibits the smallest amount of sulfate esters than others commercial class of carrageenan, it has the least negatively charged, which allows less repulsion for intermolecular hydrogen bonding and thus, is likely to form a strong film with better gelling properties. Moreover, the presence of 3,6-anhydro-galactose in κ-carrageenan capable of reducing the hydrophilicity of the galactose residue and inverting the chair conformation from $^{1}C^{4}$ to $^{4}C^{1}$, which is also favorable to the gelation properties of carrageenan [5-6]. Therefore, κ-carrageenan possesses a good film-forming ability, which is in agreement with the gelling properties of κ-carrageenan.

Previous studies on carrageenan extracted from *Kappaphycus alvarezii* in hot water showed that they are mostly composed of strong κ-carrageenans gelling agents, significant amounts of low-molecular-weight galactans with κ-structure, and only small quantities of non-gelling carrageenans and agaroids [7]. However, carrageenan films have shown poor water vapor barrier properties due to the seaweed's hydrophilic nature [8]. Therefore, several chemical modifications have been proposed to modify the physicochemical properties of carrageenans. Alkali is one of the common and simple solutions used for the extraction of carrageenan as it induces chemical modification that leads to increasing gel strength in the final product [9-10]. Earlier studies on carrageenan are limited mostly to the characterization of the extracted carrageenan properties. There is no study reported on the impact of extraction temperature on the properties of carrageenan film. Hence, the purpose of this study was to develop the carrageenan based film from *Kappaphycus alvarezii*, in which carrageenan was extracted via aqueous extraction of carrageenan with or without alkali modification, and to determine the characteristics of the developed carrageenan based films. The effects of varied extraction temperature on the properties of unmodified and chemically modified carrageenan based films were also discussed through various characterizations.

2. Materials and methods

Dried raw red seaweeds (*Kappaphycus alvarezii*), which harvested from Sabah, were washed with tap water to remove any impurities found on their surface like salts and sand particles. After cleaned, they were completely soaked in distilled water and left for 5 hours at room temperature. The soaked materials were then cut into small pieces by using a knife. Small pieces of seaweed were distributed in a big plastic tray and then placed inside the oven at 40 ^{0}C for 3-4 days. The completely oven-dried samples further stored in air tight container until further use.

2.1. Preparation of seaweed samples

Before extraction, the dried seaweed was pretreated with ethanol and acetone to eliminate the organosoluble fraction and to ease the process of extraction and modification [3]. 25 g of oven-dried (OD) seaweed samples were weighed and then, soaked in 800 ml of distilled water with the mouth of beaker covered with aluminum foil and left overnight at room temperature. The next day, the soaked samples were blended for 15 minutes by using a blender, and later the volume was made up to one liter. Two liters of ethanol was added into the blended solution and left for half an hour. Then, the mixture was stirred simultaneously. After that, the seaweed slurry was filtered twice. The precipitate obtained by filtration was added with acetone until 1000ml and the mixture was stirred simultaneously for half an hour. Again, the seaweed slurry was filtered twice. Finally, the pretreated seaweed sample obtained was oven dried at 40 ^{0}C overnight before processing to extraction, modification, and precipitation of carrageenan.

2.2. Pretreatment of seaweed samples

Before extraction, the dried seaweed was pretreated with ethanol and acetone to eliminate the organosoluble fraction and to ease the process of extraction and modification [3]. 25 g of oven-dried (OD) seaweed samples were weighed and then, soaked in 800 ml of distilled water with the mouth of beaker covered with aluminum foil and left overnight at room temperature. The next day, the soaked samples were blended for 15 minutes by using a blender, and later the volume was made up to one liter. Two liters of ethanol was added into the blended solution and left for half an hour. Then, the mixture was stirred simultaneously. After that, the seaweed slurry was filtered twice. The precipitate obtained by filtration was added with acetone until 1000 ml and the mixture was stirred simultaneously for half an hour. Again, the seaweed slurry was filtered twice. Finally, the pretreated seaweed sample obtained was oven dried at 40 ^{0}C overnight before processing to extraction, modification, and precipitation of carrageenan.

2.3. Extraction of carrageenan

All the dried pretreated seaweed was placed in a 2000 ml beaker that containing distilled water (60 mL/g) at pH 7 and stirred until it dispersed in water by using a stirrer machine at 500 rpm for half an hour. The beaker was covered with aluminum foil and then placed in water bath at 60 °C and shaken at 55 rpm for 3 hours. After 3 hours, the beaker was taken out from the water bath. Double volume of ethanol was added to the solution and stirred simultaneously for half an hour. Then, the seaweed slurry was filtered by using a cloth filter and the precipitate was collected by filtration. Next, the obtained extracted carrageenan (precipitate) was placed in a plastic tray and oven dried at 40 °C for 2 days until it dried completely. Lastly, the extracted carrageenan (purified and unmodified carrageenan) was weighed and recorded before processing to the preparation of film. All the above steps for hot water extraction of carrageenan were repeated for a different condition of temperature at 80 °C water bath.

2.4. Extraction of carrageenan and alkali modification

Carrageenan from dried pretreated seaweed was extracted and chemically modified using 1M NaOH solution (60 mL/g). First, the mixture in the beaker was stirred at 500 rpm for half hour. Then, the beaker was covered with aluminum foil and then placed in the water bath at 60 °C and shaken at 55rpm for 3 hours. After 3 hours, the beaker was taken out from water bath and the solution was neutralized to approximately pH7 with HCl (2 M). Then, a double volume of ethanol was added into the solution and stirred with a spatula for half an hour. After that, the seaweed slurry was filtered through a cloth filter and the precipitate was collected by filtration. The obtained alkali extracted carrageenan (precipitate) was placed in a plastic tray and oven dried at 40 ^{0}C for 2 days until it dried completely. Finally, the alkali extracted carrageenan (chemically modified carrageenan) was weighed and recorded before processing to the preparation of film. All the above processes for alkaline treatment of carrageenan were repeated for a different condition of a temperature at 80 °C water bath.

2.5. Preparation of native seaweed, unmodified carrageenan and chemically modified carrageenan films

2% (w/v) of unmodified and chemically modified carrageenan solutions were prepared separately by soaking in distilled water overnight and blended for 15 minutes by using a blender (Panasonic MX-895 M). A 33.33% (w/w) concentration of glycerol based on the oven-dried weight of the sample was added to the blended slurry. Next, the mixture was stirred continuously at 500rpm for 4 hours by using a stirrer machine. After stirring for 4 hours, the mixture was transferred into a stainless steel tray with measurement of 32 x 9.5 x 1.5 cm and spread evenly and subsequently left to the oven dried at below 40 °C until dried completely for about 3-4 days. The films were stored in a desiccator with 50% relative humidity for at least 3 days before further characterization.

Moreover, a native seaweed film as a control was also prepared by mixing 2% (w/v) native seaweed solution with glycerol. Then, the mixture was heated to the different temperature (60 °C and 80 °C) in the water bath at 55 rpm for 3 hours and then, followed by drying as mentioned above.

2.6. Characterization of native seaweed, unmodified carrageenan and chemically modified carrageenan films

IOP Conf. Series: Materials Science and Engineering **368** (2018) 012020 doi:10.1088/1757-899X/368/1/012020

2.6.1. Film thickness. The thickness of biopolymer films was measured by using a precision electronic micrometer to the nearest 0.001 mm at 5 random locations on each film. The average thickness value for all the films was calculated and used in tensile strength and Young's Modulus calculations.

2.6.2. Mechanical testing. Mechanical strength and flexibility are basic requirements for a film to resist external stress and maintain its integrity during its application and subsequent shipping or storage. Typically, the mechanical resistance of biopolymer film is studied according to three essential parameters: tensile strength (TS), young's modulus (YM) and elongation at break (EAB). These parameters are determined based on stress–strain curves according to the norm of the American Society for Testing and Materials (ASTM). TS is defined as the maximum stress that a film can withstand when being stretched or pulled before failure. EAB is the maximum change in the length of a test specimen before breaking.

Tensile testing was conducted for each film specimen using the Miniature tensile Tester MT1175 (Dia-stron Instruments, United Kingdom) programmed with UvWin1000 software, in reference of ASTM D882-02. In tensile testing, at least five film specimens were cut in a rectangular shape with a dimension of 150mm x 2.5mm. The tensile strength, percentage of elongation at break and Young's Modulus of the film specimens were evaluated based on the generated stress-strain graphs. Each reported data was the mean value of three replicate.

$$\text{Tensile Strength, TS (MPa)} = \frac{\text{Maximum load (N)}}{\text{Cross−sectiona area of film (m2)}} \qquad (1)$$

$$\text{Elongation at Break, EAB (\%)} = \frac{\text{Final gauge length (mm)} - \text{Initial gauge length(mm)}}{\text{Initial gauge length (mm)}} \times 100 \qquad (2)$$

$$\text{Young Modulus (MPa)} = \frac{\text{Stress}}{\text{Strain}} = \frac{\text{Maximum load (N)} \div \text{Cross−section area of film (m2)}}{\text{Change in length (mm)} \div \text{initial gauge length (mm)}} \qquad (3)$$

2.6.3. Thermal analysis. Thermogravimetric analysis (TGA) was carried out to study the thermal stability or thermal degradation characteristic of the biopolymer films by using Perkin-Elmer TGA6 thermogravimetric analyzer. The film specimen was cut into tiny pieces and each sample had a mass between 8-10 mg. The tiny pieces of the sample were placed in the balance system and heated from 30 °C to 800 °C at a heating rate of 10 °C min^{-1} under a nitrogen atmosphere to avoid thermo-oxidative reactions.

2.6.4. Contact angle measurement. Static contact angle measurement on the surface of films was performed using a Contact Angle Measurement System, CAM 100/101/200 series Drop Shape Analysis contact angle meter whereby sessile drop technique has been applied. The film specimens were cut into square shape with a dimension of 20mm x 20 mm. A drop of test liquid was uniformly dropped on the film and images were recorded for the 30 s at a speed of 40 frames per second. The images were executed by using KSV Contact Angle Measurement system. Each type of film was repeated for three times and the mean value was calculated.

2.6.5. Water vapor transmission rate (WVTR) analysis. WVTR value of the film specimens was determined according to ASTM E96/E96M-05 method with slight modification. Films free of any defects such as pinholes, air bubbles or cracks were used for WVTR determination. Initially, the internal diameter of the beaker was measured for the calculation of WVTR. Then, the film specimen was sealed on the top of a 100 ml beaker which contained 60ml of distilled water by using a sealing tape and, it was recorded as initial weight. Next, it placed in the desiccator which had been maintained at room temperature and 0% RH. The amount of water transferred through the film and absorbed by the desiccant was determined as the weight loss. Each sample was weighed at an interval of 24 hours until a constant weight loss was obtained. Each type of film was repeated for three times and the mean value was calculated. The WVTR was expressed in g/m^2 per day and calculated by the following formula:

$$\text{WVTR} = \frac{\text{Constant weight loss (g)}}{\text{Area of circular mouth of beaker (m2)}} \quad (4)$$

2.6.6. Fourier Transform Infrared Spectroscopy (FTIR) analysis. FTIR spectra of the biopolymer films were analyzed by using an attenuated total refraction method (ATR) in an IR spectrometer. Film specimens to be analyzed were cut into square shape with a dimension of 10mm x 10mm and then oven dried at 60 °C for overnight. During analysis, the specimen was placed in the zinc selenide ATR cell. Prior to sample collection, a background spectrum was recorded and subtracted from the sample spectra. Infrared spectra of all the film specimens were obtained from Perkin-Elmer FT-IR 1600 spectrometer. A smoothing filter was applied to the spectra without losing peak resolution in order to reduce the apparent noise of the data correlated with Fourier self-convolution.

3. Results and discussion

Seaweeds or known as marine macro-algae are multicellular plant-like organisms which generally found living attached to rocks in coastal areas. Since seaweed is abundantly available, able to grow in wide range of environments, cost effective and easy to cultivate in the natural environment as well as harvest year around, they are widely used as a human food source and herbal medicine. Instead of using seaweeds as a food, their excellent carbohydrate content also encouraged the industrial uses of seaweeds as a source of hydrocolloids (i.e. alginate, carrageenan, and agar) in the application for food, pharmaceutical and plastic industries [1].

3.1. Mechanical testing

According to Table 1, the unmodified carrageenan based film obtained the highest TS and YM, followed by native seaweed and chemically modified carrageenan based films. It was believed that the unmodified carrageenan film possessed strong hydrogen bonding, which was capable of providing more affinity for the higher number of hydrogen bonds formed in the film. On the other hand, modified carrageenan based film attained the lowest TS and YM. This phenomenon was due to some of the hydroxyl and sulfate ester groups were released during alkali modification and thus, reduced the intermolecular hydrogen bonding of the film. Moreover, the removal of sulfate ester group capable of forming 3,6-anhydro-galactose, which was crucial in introduce gelling characteristics in the carrageenan [9]. Therefore, it was believed that the increase in the proportion of 3,6-anhydro-galactose increased the flexibility of the film. Hence, chemically modified carrageenan film obtained the highest value of EAB, as shown in Table 1.

Interestingly, the TS and YM of three different seaweed-based films were improved at low treatment (or extraction) temperature (60 °C), whereas EAB of these films was increased when treatment was conducted at elevated temperature (80 °C). The previous study reported that treatment or extraction temperature affected gel strength, which was higher when carrageenan was extracted at high temperature of about 80 °C. In contrast, the gel viscosity increased when the extraction temperature achieved 60 °C, and at higher temperatures this variable decreased due to the degradation of biopolymer [7]. Therefore, these findings verified that temperature significantly affected the mechanical properties of the films (Table 1).

Table 1. Mechanical properties, contact angle and WVTR of native seaweed, unmodified carrageenan and modified carrageenan based films.

Film specimens	Temperature (°C)	Tensile strength (MPa)	Young's Modulus (MPa)	Elongation at break (%)	Contact angle (°)	WVTR (g/m²/day)
Native seaweed	without	78.3	751.9	8.6	61.9	683.91

	heating					
	60	110.7	1002.6	18.5	67.2	646.63
	80	67.8	602.5	20.3	74.2	614.43
Unmodified carrageenan	60	146.4	1433.3	10.6	64.8	649.93
	80	118.3	1173.5	11.6	71.5	630.95
Chemically modified carrageenan	60	43.7	532.1	19.2	70.3	642.57
	80	15.6	115.4	33.2	82.5	583.09

3.2. Wettability analysis

The contact angle measurement was used to determine the surface hydrophobicity and wettability properties of biopolymer films. Generally, high contact angle indicated film surface with hydrophobic character. Or in other words, low contact angles corresponded to high wettability, whereas large contact angles corresponded to low wettability. Based on the results presented in Table 1, native seaweed film prepared without undergone heating process obtained the lowest value of contact angle (61.9°) due to the hydrophilic nature of seaweed. Nevertheless, the increase of contact angle was observed when the native seaweed underwent the heating process. Therefore, these findings verified that temperature significantly affected the wettability of the film, in which heating could reduce the hydrophilic nature of seaweed and form better gel strength, as discussed earlier.

According to Table 1, the contact angle of all films was increased with the increase of heating (or extraction) temperature. This phenomenon might be due to the removal of hydroxyl and sulfate ester groups at elevated temperature, which then resulted in increasing surface hydrophobicity of biopolymer film. This could be further verified by the analysis of FTIR. Therefore, 80 °C was a maximum temperature to increase the hydrophobic character of the film in the case of this study. Among the films, the alkali modified carrageenan based film obtained the highest value of contact angle. Previous study reported that κ-carrageenan capable of undergoing cyclization in the presence of 1 M sodium hydroxide at 80 °C, in which sulfate ester groups (very hydrophilic) were removed for the formation of 3,6-anhydro-bridge that could lead to a reduction of hydrophilicity of film [9]. Thus, the alkali modified carrageenan based film exhibited hydrophobic surface with low wettability, in comparison to other films.

3.3. Water vapor transmission rate (WVTR) analysis

Permeability is the contribution of diffusivity and solubility through the solid matrix, whereby in this research it was measured as water vapor transmission rate. WVTR is a method to measure the ease of moisture to penetrate and pass through a material [11]. The water vapor transmission of the carrageenan based film can be influenced by several factors such as the structure of polymer matrix, glycerol as plasticizer and surface wettability properties [12-13].

According to Table 1, it could be observed that the value of WVTR decreased with the increasing of the contact angle of all film. Since the native seaweed film prepared without undergone heating process obtained the lowest value of contact angle as mentioned earlier, it WVTR was the highest (683.91 g/m^2/day) among other films. This finding revealed that the water vapor was easy to diffuse through the film due to the hydrophilic character of native seaweed. In contrast, the decrease of WVTR was noted when film prepared by heating the native seaweed at 60 °C and 80 °C. Similar to others characterization studies, these finding also validated that temperature also significantly affected the WVTR of the film, in which heating would reduce the hydrophilic nature of seaweed and enhance hydrophobic characteristic of the film. Among the films, alkali modified carrageenan based films prepared at 80°C of extraction temperature attained the lowest value of WVTR. Since alkali could induce a chemical modification that improved the gel strength of the film due to the formation of anhydro linkage, the chemically modified carrageenan film became less hydrophilic. The water vapor diffusion across the film was reduced and thus, lowered the WVTR.

3.4. Thermal analysis – Thermogravimetric Analysis (TGA)

Thermogravimetric analysis (TGA) was performed to evaluate the thermal degradation of native seaweed, unmodified carrageenan and chemically modified carrageenan films at high temperature. TGA usually analyzed the materials that possessed either mass loss or gain due to the decomposition, oxidation or loss of volatiles like moisture. Figure 1 shows the weight decreasing pattern of all films caused by thermal decomposition. For native seaweed films, TGA thermograms illustrated that there were four significant stages of weight loss, whereas there were only three stages of weight loss for unmodified and modified carrageenan films.

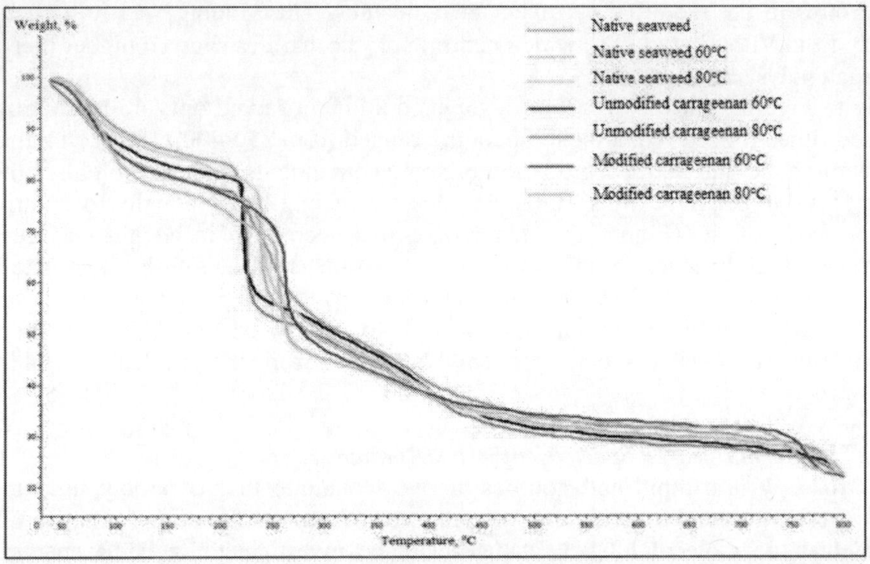

Figure 1. TG curves of native seaweed, unmodified carrageenan and chemically modified carrageenan based films.

Initially, the first stage of weight loss for all the films occurred in the temperature range of 60-100 °C (Figure 1). This phenomenon was due to evaporation of adsorbed water molecules trapped within the film matrix, which existed in H-bonded form with -OH groups of the galactose units along the polymer chain [14-15]. The weight losses at this stage for all native seaweed, unmodified carrageenan, and chemically modified carrageenan films were 20-22%, 19-20% and 16-17%, respectively. Native seaweed film attained the highest reduction in weight because of its hydrophilic nature which led to high water adsorption. In contrast, modified carrageenan film showed the lowest weight loss among the film. This phenomenon was probably due to the removal of –OH groups during alkali modification and thus, it could be further validated by FTIR and wettability tests.

In the second stage, all the films showed tremendously weigh losses along with different thermal degradation temperature, in which the native seaweed, unmodified carrageenan, and chemically modified carrageenan films achieved total weight losses of 21-25%, 46-50% and 58-61% at 130-180 °C, 200-240 °C, and 240-300 °C, respectively (Figure 1). This stage took place was due to the degradation of glycerol and initial decomposition of seaweed polymers [16, 17]. Furthermore, in the third stage, the weight losses of all films were further decreased, wherein native seaweed, unmodified carrageenan, and chemically modified carrageenan films achieved weight losses of 30-47%, 73-75% and 70-73% at 200-235 °C, 350-380 °C and 390-430 °C, respectively (Figure 1). This stage took place most probably related to the thermal degradation of seaweed or carrageenan polysaccharide chains [18, 19].

Based on the respective results from the second stage and third stage, the thermal degradation temperature was increased after extraction and modification of carrageenan. This revealed that the high temperature of thermal degradation resulted in the better thermal stability of the films prepared from unmodified and modified carrageenan. The unmodified and modified carrageenan films showed higher degradation temperature than native seaweed films as purification of carrageenan could lead to the removal of other thermally sensitive impurities like proteins, lipids, and etc. Furthermore, more hydrogen bonds formed in purified carrageenan films than native seaweed films which might also contribute towards the better thermal stability of the films [20, 21]. Nevertheless, modified carrageenan films were more thermally stable than unmodified carrageenan films. This was possibly caused by the presence of more charged groups (SO^{3-}) in unmodified carrageenan molecules than modified one did. After alkali modification, the formation of anhydro linkages indicated the reduction of charged groups in the modified carrageenan molecules. This finding was in agreement with the reported study from Villetti *et al* [22], which neutral polysaccharides showed higher thermal resistance than the charged polysaccharides.

According to Figure 1, there was a fourth stage could be observed only in the thermal analyses of native seaweed films. It occurred at the temperature ranged from 250-400°C with a total weight loss of 69-77%. This might be due to the degradation of some components such as proteins, lipid or minerals that could be found in native seaweed films. Besides, the weight loss gradually rose until reaching the temperature around 760 °C (Figure 1). This progressive decrement in weight was attributed to the further breakage of sub-products in the film matrix from the third or fourth stage. The char residues contributed by native seaweed films were ranged from 20% to 23% by weight, while for unmodified and modified carrageenan film were ranged from 17% to 19% by weight. The high amount of residue left in native seaweed films was most probably attributed from the residual of lipids, proteins or minerals.

3.5. Fourier Transform Infrared Spectroscopy (FTIR) analysis

Infrared spectroscopy is a rapid and non-destructive technique that is widely used to characterize different polysaccharides and to determine the presence of various functional groups. When chemical groups interact at the molecular level, changes can be investigated by FTIR spectra through the shifting of absorption bands. The FTIR spectra of native seaweed, unmodified carrageenan, and modified carrageenan based films are shown in Figure 2.

As it would be expected, after the extraction and modification process, the FTIR spectra were changed. According to the results of FTIR, all the spectra revealed a broad band ranging between 3100 and 3600 cm^{-1}, which was attributed to O-H stretching vibration from the hydroxyl group of polysaccharides which correspond to the hydrophilic character of seaweed. While, the broad band around 2800-3000 cm^{-1} was attributed to C-H stretching vibration [23]. Since native seaweed undergone heating process showed an increase in the percentage of transmittances, this information showed that heating process could remove some of the hydroxyl group from the seaweed (Figure 2). Nevertheless, no obvious changes in the IR spectra of hydroxyl group were observed for extraction temperature of 60 °C and 80 °C, wherein the native seaweed, unmodified carrageenan and modified carrageenan films prepared at 60 °C and 80 °C were 3329.14 cm^{-1}, 3332.99 cm^{-1}, and 3325.28 cm^{-1}.

FTIR spectra of film specimens showed the high intensity of band in the region of 750-1300cm^{-1} that corresponding to the carbohydrates region (Figure 2). A very strong absorption band in the region of 1210-1260 cm^{-1} was observed. This is due to the presence of S=O bond of sulfate ester groups. Moreover, the IR bands at around 840-850 cm^{-1} showed the presence of C-O-SO_3 of D-galactose-4-sulfate [3]. According to Figure 2, the peak of S=O bond was shifted from 1219.01 cm^{-1} (unmodified carrageenan film) to 1222.87 cm^{-1} (modified carrageenan film), whereas the peak of C-O-SO_3 was shifted from 844.82 cm^{-1} (unmodified carrageenan film) to 848.68 cm^{-1} (modified carrageenan film). This phenomenon was due to the removal of some sulfate ester groups and thus, formation of 3,6-anhydro linkage, which it was depicted by an IR band around 922 cm^{-1} that corresponding to the C-O bond of 3,6-anhydro-D-galactose [3]. Hence, this analysis revealed the changes in native seaweed, unmodified carrageenan and modified carrageenan based films mechanical, thermal and water barrier properties.

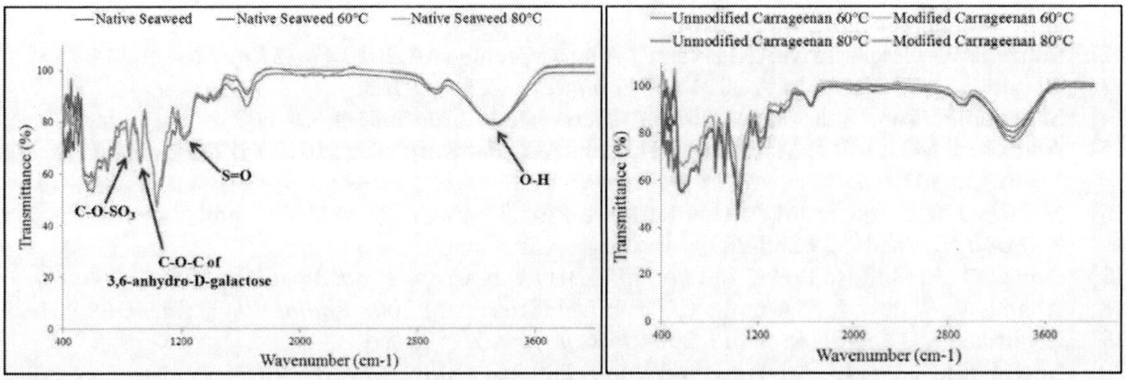

Figure 2. FTIR spectra of native seaweed, unmodified carrageenan and modified carrageenan based films.

4. Conclusion

Native seaweed, unmodified carrageenan, and alkali modified carrageenan based films showed differences in physicochemical, mechanical and water vapor barrier as well as thermal properties. Unmodified carrageenan was suitable to produce biopolymer films with the best mechanical properties which exhibited high tensile strength and Young's modulus; while modified carrageenan was suitable to produce biopolymer films with the best thermal stability and hydrophobic properties. Moreover, varying of extraction temperature also lead to significant changes in the properties of the film, wherein the increase of temperature reduced the tensile strength and Young's modulus, but improved the elongation at break and hydrophobicity agreed with the results of contact angle and WVTR. Therefore, it was believed that chemical modification of carrageenan could enhance all the properties of film provided that an appropriate extraction temperature was selected.

Acknowledgements

The authors gratefully acknowledge the Ministry of Higher Education for the Fundamental Research Grant Scheme – Malaysia's Rising Star Award 2015 (FRGS-203/PTEKIND/6711531).

References

[1] Abdul Khalil H P S, Tye Y Y, Saurabh C K, Leh C P, Lai T K, Chong E W N, Fazita M N, Hafiidz J M, Banerjee A and Syakir M I 2017a *Express Polym. Lett.* **11** 244-265
[2] Hernández-Carmona G 2013 *Functional ingredients from algae for foods and nutraceuticals food* ed Dominguez H (Mexico: Woodhead Publishing Limited) p 475–516
[3] Pereira L, Amado A M, Critchley A T, Van de Velde F and Ribeiro-Claro P J 2009 *Food Hydrocoll.* **23** 1903-1909
[4] Barbeyron T, L'Haridon S, Corre E, Kloareg B and Potin P 2001 *Int. J. Syst. Evol. Microbiol.* **51** 985–997
[5] Rhein-Knudsen N, Ale M T and Meyer A S 2015 *Mar. Drugs* **13** 3340-3359
[6] Balqis A I, Khaizura M N, RusslyA R and Hanani Z N 2017 *Int. J. Biol. Macromolec.* **103** 721-732
[7] Webber V, Carvalho S M D, Ogliari P J, Hayashi L and Barreto P L M 2012 *Food Sci. Technol.* **32** 812-818
[8] Nur Fatin Nazurah R, Nur Hanani, Z A 2017 *Carbohydr. Polym.* **157** 1479-1487
[9] Campo V L, Kawano D F, da Silva D B and Carvalho I 2009 *Carbohydr. Polym.* **77** 167-180
[10] Abdul Khalil H P S, Saurabh C K, Tye Y Y, Lai T K, Easa A M, Rosamah E, Fazita M R N, Syakir M I, Adnan A S, Fizree H M and Aprilia N A S 2017b *Renew. Sust. Energ. Rev.* **77** 353-362
[11] Pranoto Y, Rakshit S K and Salokhe V M 2005 *Lebensmittel-Wissenschaft und –Technologie -*

Food Sci. Technol. **38** 859-865
[12] Souza B W, Cerqueira M A, Teixeira J A and Vicente A A 2010 *Food Eng. Rev.* **2** 244-255
[13] Abdou E S and Sorour M A 2014 *Int. Food Res. J.* **21** 189-193
[14] Schlemmer D and Sales M J A 2010 *J. Therm. Anal. Calorim.* **99** 657-679
[15] Abdorreza M N, Robal M, Cheng L H, TajulA Y and Karim A A 2012 *LWT- Food Sci. Technol.* **46** 135-141
[16] Martins J T, Cerqueira M A, Bourbon A I, Pinheiro A C, Souza B W S and Vicente A A 2012 *Food Hydrocoll.* **29** 280-289
[17] Aouada F A, Mattoso L H C and Longo E 2011 *Ind. Crops Prod.* **34** 1502-1508
[18] Abad L V, Relleve L S, Aranilla C T and Dela Rosa A M 2003 *Radiat. Phys. Chem.* **68** 901-908
[19] Zohuriaan M J and Shokrolahi F 2004 *Polym. Test.* 23 575-579
[20] Li R, Fei J, Cai Y, Li Y, Feng J. and Yao J. 2009 *Carbohydr. Polym.* **76** 94-99
[21] Satyanarayana K G, Arizaga G G C and Wypych F 2009 *Prog. Polym. Sci.* **34** 982-1021
[22] Villetti M, Crespo J, Soldi M, Pires A, Borsali R and Soldi V 2002 *J. Therm. Anal. Calorim.* **67** 295-303
[23] Cerqueira M A, Souza B W S, Simões J, Teixeira J A, Domingues M R R M, Coimbra M A and Vicente A A 2011 *Carbohydr. Polym.* **83** 179-185

IOP Conf. Series: Materials Science and Engineering **368** (2018) 012021 doi:10.1088/1757-899X/368/1/012021

The glueline shear strength of *Peronema canescens* and Acacia hybrid

E Enggoh[1*], K C Liew [1], M H Sahri[1], M O M Khaidzir[2] and K Masseat[2]

[1] Forestry Complex, Faculty of Science and Natural Resources, Universiti Malaysia Sabah, 88400 Kota Kinabalu, Sabah, Malaysia

[2] Timber Engineering Laboratory, Forest Research Institute Malaysia, 52109 Kepong, Selangor, Malaysia

*Corresponding author: enggoh35@gmail.com

Abstract. The glueline shear strength is one of the common measures used to estimate potential performance of bonded wood joints. This study was conducted to determine the glueline shear strength of *Pernonema canescens* (Sungkai) and Acacia hybrid (A. hybrid). The wood were bonded with polyvinyl acetate (PVAc) and Emulsion Polymer Isocyanate (EPI) and these are for dry use in non-structural timber production. Sungkai and A. hybrid logs were obtained from Telipok and Ulu Kukut, Sabah respectively, were debarked, sawn and air-dried prior to being sawn to 20 mm thick, 60 mm wide and 150 mm long. After reaching the constant moisture content, the gluing process was carried out. 2 types of adhesives were used; EPI and PVAc. The glue was spreaded at a constant spreading rate of 200 g/m^2. 3 types of assembly combinations were made; Sungkai-Sungkai, A. hybrid-A. hybrid and Sungkai-A. hybrid. The assemblies were clamped for 24 hours, as recommended by the glue manufacturer. The test specimen preparation and the glueline shear strength test was conducted in accordance with the Manual for Testing Methods for Plantation Grown Tropical Timbers (Finger and Laminate Joints in Non-Structural Timber Production). The results showed that, the average glueline shear strength of Sungkai-Sungkai, A. hybrid-A. hybrid. and Sungkai-A. hybrid bonded with EPI are 7.48 N/mm^2, 6.58 N/mm^2 and 6.87 N/mm^2 respectively. While the average glueline shear strength of Sungkai-Sungkai, A. hybrid-A. hybrid and Sungkai-A.hybrid bonded with PVAC are 7.61 N/mm^2, 6.03 N/mm^2 and 7.60 N/mm^2 respectively. Based on the results obtained, the glueline shear strength for all assembly combinations exceed the minimum requirement of the shear strength for laminate joints, which is 5.32 N/mm^2. Thus, Sungkai and A. hybrid wood have the potential to serve as supplemental raw material for the non-structural timber production.

1. Introduction

To overcome the challenge related to shortage of raw material for the wood industries, through the NATIP 2009 – 2020, the government recommends to increase the utilisation of small diameter logs from plantation timber species and imported logs [1]. Thus, to increase the utilisation of small diameter logs, additional action such as wood bonding is required and vital. The bonding strength of bonded wood need to be assessed to avoid potential failure during its life service. One of the common measures used to estimates the bonding strength is the shear test [2]. Apart of strength test, percentage of wood failure is also occasionally used to estimate the bonding quality of bonded wood [3], with high percentage of wood failure indicates good adhesives penetration [4].

Both A. hybrid and Sungkai are fast growing tree species that have the prospective to serve as supplemental raw material for the wood industries [5, 6]. Thus, various and continuous research should be conducted including the study related to their gluing properties to ensure these species can be utilized efficiently and their usage can be expanded.

This study is conducted to determine the glueline shear strength and wood failure of A. hybrid and Sungkai bonded with polyvinyl acetate (PVAc) and Emulsion Polymer Isocyanate (EPI) for dry use in non-structural timber production.

2. Materials and experimental procedures

Two forest plantation tree species were used in this study namely Sungkai and A. hybrid. Sungkai and A. hybrid with diameter breast height ranging from 20 to 25 cm were obtained from Sabahpuri Nursery, Telipok and SAFODA Ulu Kukut, Kota Belud, Sabah respectively. The logs were sawn into a dimension of 50 mm x 100 mm x 120 m, and air-dried for about a month.

Figure 1. A. hybrid trees in Ulu Kukut, Sabah.

Figure 2. Sungkai trees in Telipok, Sabah.

Two types of adhesives were used, that is EPI and PVAc. Both adhesives were manufactured by GB Chemical (M) Sdn Bhd, Penang. Table 1 shows the properties of the adhesives used.

Table 1. Properties of EPI and PVAc.

Properties	EPI	PVAC
Viscosity (cps at 28°C)	8,000 – 12,000	12,500-15,000
Solid content (%)	36.5 – 40.5	36 - 40
pH	5.0 – 7.0	5.0 – 7.0

The air-dried rough sawn timber, were resawn into shorter strips of 20 mm x 60 mm x 150 mm. These shorter strips were air-dried until they reached constant weight and moisture content (MC) of 15±1%. Once these strips reached the constant weight and MC, the gluing process was carried out.

The EPI adhesive was mixed according to the manufacturer specification; 100 parts of resin was mixed with 15 parts of hardener (by weight). The type of hardener used was known as polymethylene polyphenyl polyisocyanate.

Single spread adhesive with constant spreading rate of 200 g/m^2 was applied onto each piece of the strip. Three different types of assembly combination were made for this test; Sungkai-Sungkai, A. hybrid – A. hybrid and Sungkai – A. hybrid. The adhesive spreading process was followed by a close assembly time (CAT) of 40 minutes as recommended by the manufacturer. CAT will allow the adhesive to penetrate into the wood before the pressing process. Later, the assemblies were clamped for 24 hours in a room temperature.

The fabrication of the test piece was in accordance with the Manual for Testing Methods for Plantation Grown Tropical Timbers (Finger and Laminate Joints in Non-Structural Timber Production) [7]. Six test specimen were obtained from each assembly.

The glueline shear test for dry use (dry condition) was conducted in accordance with the Manual for Testing Methods for Plantation Grown Tropical Timbers (Laminate Joints in Non-Structural Timber Production) at the Timber Engineering Laboratory, Forest Research Institute Malaysia, Kepong. The test was carried out by using the Shimadzu Universal Testing Machine with maximum loading capacity of 100 kN.

The basic density of wood samples for both species were also determined in accordance with the ASTM D 2395 – 14e1 [8]. While the percentage of wood failure were estimated accordance to ASTM D 5266 – 13 [9].

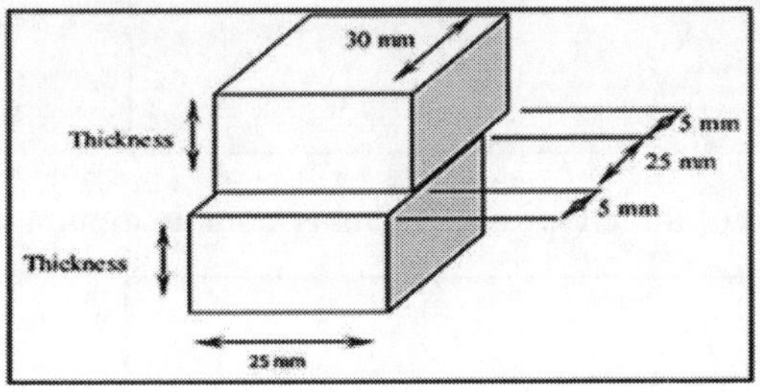

Figure 3. Test piece for shear test accordance to the Manual for Testing Methods for Plantation Grown Tropical Timbers [8].

Figure 4. Glueline shear test carried out using Shimadzu Universal Testing Machine.

3. Results and discussion

The average basic density of Sungkai and A. hybrid are 0.446 g/cm^3 and 0.539 g/cm^3, respectively. Based on the value obtained, A. hybrid has higher density compare with Sungkai. The average glue-line shear strength and percentage of wood failures for the three types of assembly combinations were shown in Table 2 and presented in Figure 4 and 5.

Table 2. The average glue-line shear strength and wood failure of three types of assembly combinations glued with EPI and PVAc.

Adhesive / Species combination	EPI		PVAC	
	Shear strength (N/mm^2)	Wood failure (%)	Shear strength (N/mm^2)	Wood failure (%)
Sungkai - Sungkai	7.48^a	79.60	7.61^a	90.20
A. hybrid – A. hybrid	6.58^b	8.80	6.03^b	18.80
Sungkai – A. hybrid	6.87^{ab}	28.80	7.60^a	57.30

*Min. requirement of shear strength (Dry condition) based on Manual for Testing Methods for Plantation Grown Tropical Timbers (Laminate Joints in Non-Structural Timber Production): 5.32 N/mm^2
*Min. requirement of wood failure (%) (Dry condition) based on Manual for Testing Methods for Plantation Grown Tropical Timbers (Laminate Joints in Non-Structural Timber Production): 30 %
* $P \leq 0.05$ (ANOVA), ^{a,b} the same letters indicate not significantly different (Tukey pos hoc test)

Based on the results obtained, the glue-line shear strength for all assembly combinations glued with EPI and PVAc exceeded the minimum requirement of 5.32 N/mm^2. However, for the percentage of wood failure, only Sungkai-Sungkai combination glued with EPI and PVAc and Sungkai – A. hybrid combination glued with PVAc exceeded the minimum requirement of 30 % of wood failure.

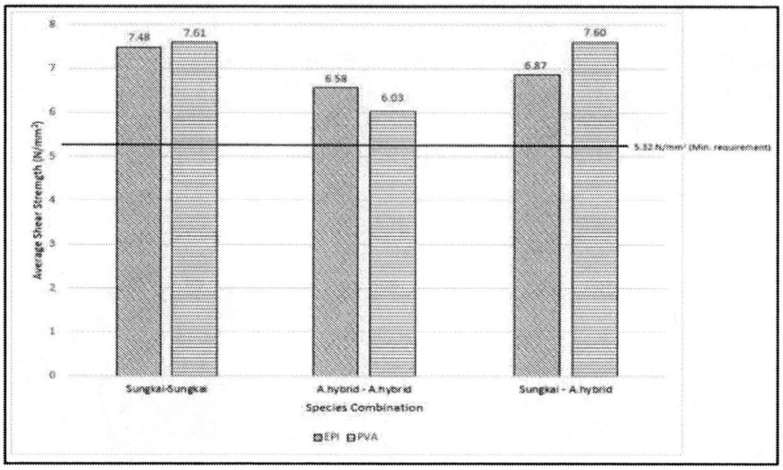

Figure 5. The average shear strength of three types of assembly combination.

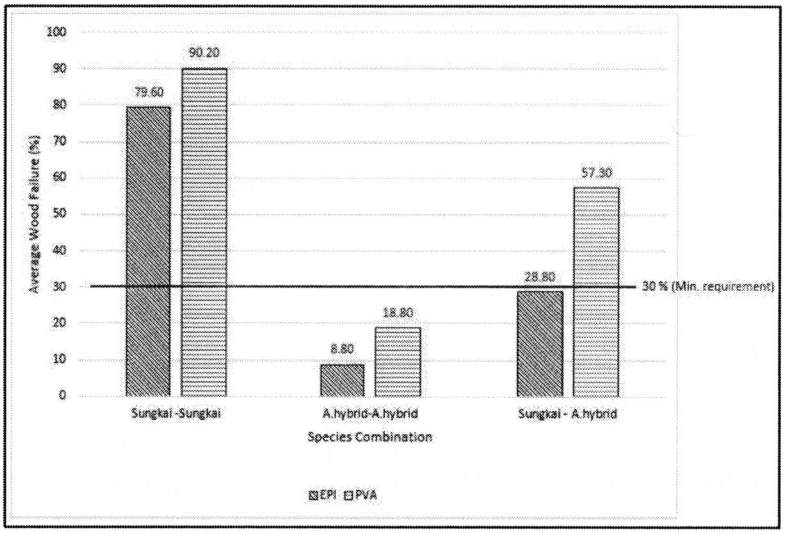

Figure 6. The average wood failure of three types of assembly combination.

In this study, for wood samples glued with EPI showed that, the highest glueline shear strength is from the Sungkai – Sungkai combination (7.48 N/mm^2), while the A. hybrid – A. hybrid combination has the lowest glue-line shear strength (6.58 N/mm^2). There was statistically significant difference between the groups as determined by one-way ANOVA ($p < .05$). A pos hoc Tukey test showed that the Sungkai-Sungkai group and A. hybrid-A. hybrid group differed significantly at $p < .05$, while the Sungkai-A. hybrid group was not significantly different from the two groups.

As for wood samples that was bonded with PVAc, the highest shear strength is obtained from the Sungkai-Sungkai combination (7.61 N/mm^2) and Sungkai – A. hybrid (7.60 N/mm^2). The lowest glueline shear strength was exhibited in A. hybrid-A. hybrid combination, which is 6.03 N/mm^2. The one-way ANOVA revealed that there was significant difference between the groups ($p < .05$). Based on the pos hoc Tukey test, there was no significant difference between Sungkai – Sungkai and Sungkai-A. hybrid group, however, the A. hybrid-A. hybrid group showed significant difference at $p < .05$ from the two groups.

Based on the average of wood failure, it shows that Sungkai-Sungkai combination has the highest percentage of wood failure, which is 79.60 %, whereas the A. hybrid-A. hybrid has the lowest average of wood failure, which is 8.80 %.This result shows that Sungkai-Sungkai combination bonded with EPI had a very good bonding compare to the other two types of species combination.

Similar to the samples glued by using EPI, the Sungkai-Sungkai combination glued with PVAc has the highest percentage of wood failure, which is 90.20 %, followed by Sungkai-A.hybrid combination with 57.30 % and A.hybrid-A.hybrid combination with 18.80 %.

The wood failure that occurred on the Sungkai-Sungkai assembly combination were deep wood failure, while most of the type of wood failure that occurred on the Sungkai – A. hybrid and A. hybrid- A. hybrid assembly combinations were shallow wood failure. However, for A. hybrid-A.hybrid assembly combination, the failure often occurred on the glueline.

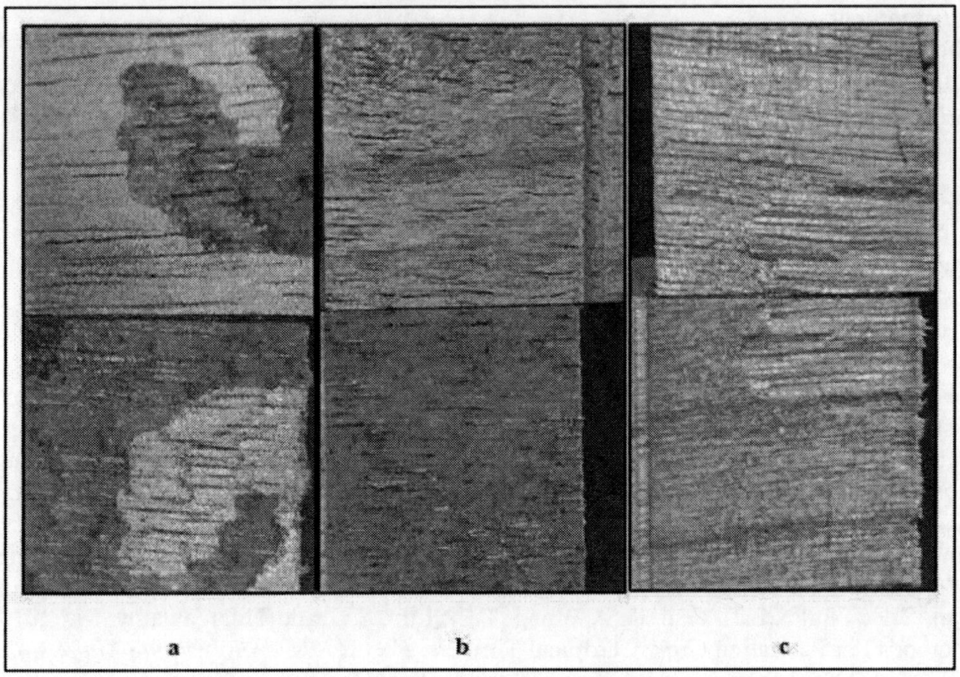

Figure 7 The type of wood failure occurred: **a)** shallow wood failure occurred in A. hyrbrid-A. hybrid **b)** glueline failure (0 % wood failure) occurred in A. hybrid-A. hybrid **c)** deep wood failure (100 % wood failure) occurred on Sungkai – Sungkai.

Although the glueline shear strength of A. hybrid-A.hybrid and Sungkai-A. hybrid exceeded the minimum requirement, however these assembly combinations failed to achieved the minimum requirement of the percentage of failure. Such results were also reported by Alamsyah et al, may be due to the presence of extractives on the surface of the A. hybrid wood [4, 10].

In general, the glueline shear strength and wood failure for sample bonded with EPI and PVAc exhibit similar result trend; combination of Sungkai-Sungkai had the highest glueline shear strength, followed by Sungkai-A. hybrid, while the A. hybrid-A. hybrid had the lowest glueline shear strength. The possible explanation for this occurrence could be due to the density of Sungkai and A. hybrid. Based on the basic density of both species, Sungkai had lower basic density compare with A. hybrid. As stated in various literature review and based on previous studies conducted [11-15] density is one of the factors that affected bonding performance of wood, with high-density wood is difficult to be bonded compare with low-density wood for numerous reason, such as; the thicker cell walls of high-density wood, smaller diameter of lumens that presence in high-density wood and the existence of higher extractives in high-density wood compare with low-density wood.

4. Conclusion

The glueline shear strength for all types of species combination bonded with EPI and PVAc met the minimum requirement of the referred manual, with the highest strength obtained from the Sungkai-Sungkai combination, while the lowest strength obtained from the A. hybrid-A. hybrid.

As for the percentage of wood failure, for species combination bonded with EPI, only Sungkai-Sungkai combination exceed the minimum value required. Whereas, for sample bonded with PVAc, Sungkai-Sungkai and Sungkai-A. hybrid combination met the requirement stated in the manual referred.

Hence, both Sungkai and Acacia hybrid showed the potential to serve as supplemental raw material for dry use in the non-structural timber production. However, further study is still required to fully understand the bonding performance of both species.

Acknowledgement
The author would like to thank the Forest Research Institute Malaysia, Kepong for the permission to conduct the shear test and allowing the author to use all facilities required for the test at the Timber Engineering Laboratory, Forest Research Institute Malaysia, Kepong, to SAFODA, Ulu Kukut and Sabahpuri Nursery Sdn Bhd for supplying the raw materials utilized in this research.

References

[1] Anon 2009 *National Timber Industry Policy 2009 - 2020* Putrajaya
[2] Vick C B 1999, *Wood Handbook: Wood As An Engineering Material*, "Chapter 9: Adhesive Bonding of Wood Material, pp 463, US Dept. Of Agriculture, Forest Service
[3] River B H and Vick C B Gillespie R H 1991 Wood as an Adherend, Marcel Dekker Inc., New York
[4] Alamsyah E M, Nan, L C, Yamada M, Taki K, and Yoshida H 2007 Bondability of tropical fast-growing tree species I: Indonesian wood species *J Wood Sci* **53** 40-46
[5] Suffian M, Ormondroyd G A, and Hale M D 2010 Comparisons of particleboard produced from Acacia hybrid and a UK commercial particleboard furnish from recycled wood *J Trop For Sci* 227-236
[6] Ahmad Zuhaidi 2015, National Seminar on Peronema canescens (29th – 30th April 2015), "*Peronema canescens* Jack: Potential Forest Plantation Species For Commercial Planting"
[7] Tan, Y.E., Lim, N.P.T., Gan, K.S., Wong, T.C., Lim, S.C. and Thilagawathy, M., 2010. Testing Methods For Plantation Grown Tropical Timbers. *Forest Research Institute Malaysia, Kepong.*
[8] ASTM D2395-14e1, Standard Test Methods For Density and Specific Gravity (Relative Density) of Wood and Wood-Based Materials, 2014, ASTM International, West Conshohocken, PA
[9] ASTM D5266-13, Standard Practice for Estimating the Percentage of Wood Failure in Adhesive Bonded Joints, 2013, ASTM International, West Conshohocken, PA
[10] Alamsyah E M, Yamada M, Taki K, Yoshida H, and Inai A 2006. Bondability of Tropical Fast-Growing Tree Species (II)-Malaysian Wood Species *Journal-Adhesion Society Of Japan 42* 499
[11] Frihart C.R, Hunt C.G 2010, Wood Handbook: Wood As An Engineering Msterial, "Chapter 10: Adhesives With Wood Materials Bond Formation & Performance"10-1 – 10-24
[12] Laiveniece L and Morozovs A 2014 Penetration depth of adhesive depending on applied pressure during gluing process. In *International Conference Materials, Methods and Technologies* 11-15
[13] Dunky M. 2003, Handbook of Adhesive Technology (2nd Edition), Editor: Pizzi.A & Mittal K.L, "Chapter 47: Adhesives In The Wood Industry" 883-951
[14] Scheikl, M., Walinder M. 2002, Cost Action E13: Wood Adhesion & Glued Products 1st Edition (Editor: Dunky, M., Pizzi, T. & Leemput, M.V.), "Chapter 4: Bonding Process-Theory of Bonding; Introduction" 89 – 90
[15] Sahri, M.H., Harun, J. and Jusoh, M.Z., 1986. The glue joint strength of four under-utilized Malaysian hardwood species [Lithocarpus; Pouteria malacensis; Parkia; Irvingia malayana].

Properties of paper incorporated with nanocellulose extracted using microbial hydrolysis assisted shear process

S Adnan[1*], A H Azhar[2], L Jasmani[1] and M F Samsudin[1]

[1]Pulp and Paper Laboratory, Biomass Technology Programme, Forest Products Division, Forest Research Institute Malaysia (FRIM), 52109 Kepong, Selangor, Malaysia.

[2]Faculty of Chemical Engineering, Universiti Teknologi MARA, 40450 Shah Alam, Selangor, Malaysia.

*Corresponding author: sharmiza@frim.gov.my

Abstract. In the past two decades, nocellulose is widely as a renewable source for various industrial applications. Nanocellulose in the form of nanocrystalline (NCC) or nanofibrillated (NFC) has many applications, mainly in enhancing the mechanical strength of composite materials, or as precursors for supercapacitors, aerogel, hydrogel and membrane fabrication processes. In this study, microbial hydrolysis combined with shear mechanical treatment was used as an alternative method to produce nanocellulose. Commercial cellulase enzyme from *Trichoderma reesei* (ATCC 26921) was used to hydrolyse bleached soda cellulose from *Macaranga*, a tropical pioneer forest species. The enzymatic hydrolysis was followed by enzyme deactivation, purification, homogenization and sonication. Resulting nanocellose was added at 1%, 2.5%, 5% and 10% loading into 60±3 gsm laboratory handsheets prepared according to TAPPI T205 method. The handsheets were tested for physical, mechanical and optical properties based on TAPPI T220 method. Results, among others, showed that the addition of nanocellulose up to 10% can reduce the air permeability of paper by 50%. This indicates the potential applications in food and pharmaceutical packaging in which degradation due to bacterial activities can be reduced and product shelf life can be prolonged.

1. Introduction

Cellulose is one of the most abundant renewable materials which mainly comes from plants such as wood, cotton, sugar bagasse, jute fibers and a variety of other natural products including bacteria, algae, fungi, invertebrates, amoeba and even in tunicates [1]. Recently, nanocrystalline cellulose (NCC) production from lignocellulosic materials attracts more and more attention due to their promising properties such as high mechanical strength (modulus 100-140 GPa), low density (1.6 - 3 gcm^{-3}), high aspect ratio, high surface area, chemical tenability, modifiable surface properties due to their reactive hydroxyl (-OH) side group, environmental sustainability and low cos [2-3]. Hence, NCC has been widely tested for diversed applications (polymer nanocomposites towards high-value added materials, enzyme immobilization, drug delivery and biomedical applications). During the 10th Malaysia Plan, Forest Research Institute Malaysia has studied several tropical forest species such as Mahang for the application of high efficiency processing technology towards juvenile wood. Mahang

(*Macaranga* spp.) belongs to the family Rubiaceae. There are about 27 species of *Macaranga* in Malaysia and 280 worldwide. Studies conducted in Peninsular Malaysia indicated they are common pioneering species and amongst the first to successfully colonise open or disturbed areas. The species is distributed throughout Malaysia and is locally abundant on disturbed sites of virtually any soil type [4, 5]. Its long fibres and light coloured wood makes it a commercially potential source of timber for pulp, paper and wood composite boards. In the past, it has been used as match sticks. Compared to other Malaysian timbers, *Macaranga* fibres have a high length to diameter ratio (L/D) of 52.2 and cellulose content of 46.5% [6]. These two properties indicate that *Macaranga* is a potential candidate for pure cellulose production with good properties. Thus, as an extension of the previous research findings, *Macaranga* cellulose was further exploited as a raw material for nanocellulose production.

Incorporation of nanocellulose in biocomposite products such as paper has been known to improve strength and surface properties [7]. Adding nanocellulose in other polymer films such as polyvinyl acid (PVA) and styrene-butadiene latex was reported to improve the modulus of elasticity but no significant improvement of tensile strength was observed [8]. However, other researchers have observed improvement of tensile strength (up to 30%) and thermal properties upon addition of nanocellulose in PVA composite [9, 10]. For textile such as polyester, incorporating nanocellulose has been reported to improve breaking load, increase onset temperature and reduce air and water [11]. Although the nanocellulose extraction processes can sometimes be tedious and costly, the improvement or benefits gained when incorporated into products can compensate the cost mainly due to the fact that most of the time, the dosage level of nanocellulose is below 10%.

In this work, the physical, optical and mechanical properties of paper added with nanocellulose are evaluated and the potential applications of the paper produced are recommended.

2. Materials and method

Logs from *Macaranga tanarius* trees were obtained from FRIM campus in Kepong, Selangor, Malaysia. Chemicals used were sourced from local Malaysian suppliers. Cellulase enzyme from *Trichoderma reesei* (ATCC 26921) was purchased from Sigma Aldrich Malaysia. *Macaranga* logs were processed into sawn timbers at FRIM sawmill before being chipped and screened into various sizes and the portion between 2 and 2.5 cm width and 3 to 8 mm thick was taken for chemical pulping. Soda pulping using 25% (wt/wt) sodium hydroxide (NaOH) was carried out in a 16L rotary digester at 170°C for a total of 3.5 hours (1.5 hrs heating time) with 1:6 wood-to-liquor ratio. The resulting pulp had a Kappa number of 14. This pulp was next bleached using a 5-stage elemental chlorine free (ECF) DEDED process with 3% Sodium chlorite ($NaClO_2$) for D stages and 2% NaOH for E stages. All bleaching stages were carried out at 70°C with reaction time of 120 minutes for stage 1, 60 minutes for stages 2 and 4 and 90 minutes for stages 3 and 5.

For the extraction of nanocellulose, bleached cellulose was mixed with 0.05M sodium acetate buffer at 5% consistency in a flask. 2.5% cellulase enzyme in liquid form was added and the mixture was placed in a shaking water bath at 25°C, 130 rpm and for 72 hours. The hydrolysis process was terminated by immersing the flask in a water bath at temperature above 80°C for 10 minutes. The residual liquid was filtered and the cellulose was thoroughly washed and later homogenised for 2 hours at 10,000 rpm. The homogenised suspension was filtered and the supernatant collected was sonicated, filtered and freeze-dried. Details of the nanocellulose extraction process are presented elsewhere [12-14]. The properties of nanocellulose were determined using Transmission Electron Microscope (TEM), Atomic Force Microscope (AFM) and X-Ray Diffraction (XRD).

Standard laboratory handsheets of 60±3 g/m^2 were prepared according to TAPPI T205 sp-02 method using unbeaten *Macaranga* bleached soda pulp obtained earlier. Bleached pulp was first disintegrated with NCC (dosage of 1%, 2.5%, 5% and 10%) before alum was added at 2% dosage as a retention aid in the stock chest. The pH of the stock was maintained between 5.5 and 6.5. Six most uniform pieces were selected for testing based on TAPPI T220 sp-01 method. Individual tests were carried out according to methods listed in Table 1. For air permeance, Lorentzen-Wettre Bendtsen type Smoothness and Porosity tester model 6 was used and samples were tested based on ISO 5636-3 isostatic method.

Table 1. Test method for preparation and testing of laboratory handsheets.

Properties	Method
Handsheet preparation	TAPPI T205 sp-02 [15]
Physical Testing	TAPPI T220 sp-01 [16]
Specific volume (bulk)	MS ISO 534: 2007 [17]
Air permeance	ISO 5636-3 1992 [18]
Opacity	MS ISO 2471: 2010 [19]
Brightness	MS ISO 2470-1: 2010 [20]
Tensile strength	MS ISO 1924-2: 2010 [21]
Tear strength	MS ISO 1974: 1999 [22]
Folding endurance	ISO 5626: 1993 [23]

3. Results and discussion

The results of pulping and nanocellulose extraction process are summarised on Table 2. The pulping yield is relatively low (typical hardwood timbers have pulping yield in the range of 40-50%) possibly due to the high sodium hydroxide level of 25% (wt/wt). The initial alpha-celllulose content of Mahang is about 46.5% [6]. The yield of only 35.4% and kappa number of 14 indicate that some of the cellulose have been degraded during pulping. Another 10% weight was lost during bleaching. Furthermore, it was observed that the nanocellulose yield or recovery from the bleached cellulose is very high compared to other works using enzyme reported in the literature between 63-84% [24, 25]. However, based on the XRD result (Figure 1) the crystallinity is marginally low compared to nanocellulose extracted using acid hydrolysis, typically in the range of 65-85%. Our previous work using acid hydrolysis showed crystallinity of 79% for NCC from *Acacia mangium* [10]. Other researchers have reported the crystallinity of nanocellulose obtained using enzyme pre-treatment in the range of (45-68%) [24-26]. The crystallinity of enzyme treated nanocellulose followed by mechanical treatment have shown reduction mainly due to the mechanical or hydrodynamic forces applied to the cellulose [27-28]. Figure 1 shows the diffractogram of *Macaranga* nanocellulose with the most intense peak for the 200 crystal plane ($2\theta = 22.5$). The peaks observed in the diffractogram correspond to typical crystalline character of cellulose Iβ [29]. The crystallinity measured for Macaranga nanocellulose NCC was 0.54 estimated using the Segal formula. The TEM and AFM images of the nanocellulose extracted are shown in Figure 2 (a) and (b) respectively. From the analysis of AFM image, the resulting nanocellulose has an average width of 8.1±1.4 nm and average length of 146±0.2 nm. The resulting width result agrees with the findings of other researchers who used enzyme pre-treatment in nanocellulose extraction [30-31]. The aspect ratio of NCC obtained in this work is smaller than NCC obtained in our previous work (18 vs. 26) [10]. In papermaking terminology, aspect ratio is also known as felting power and higher values have a positive effect of strength on paper strength [32]. This can also be applied to other products such as thin films.

Table 2. Overall *Macaranga* nanocellulose production results [33]

Properties	
Pulping yield	35.4%
Kappa number	14
Bleaching yield	90%
Nanocellulose yield	90.2%
Yield from wood	27.6%

Average nanocellulose length	146±0.2 nm
Average nanocellulose width	8.1±1.4 nm
Crystallinity	54%

The freeness of pulp stock at 3% consistency determined using TAPPI T 227 om-99 method is shown in Figure 3. The addition of 1 to 10% NCC showed less than 2.3% difference in the stock freeness. The specific volume or bulk of laboratory handsheets produced slightly decreased (by 13%) with higher addition of NCC (Figure 4). This could be explained by the fact that nanocellulose take less space and provide compact structure to the paper, thus reducing the volume while retaining the same mass. However, no significant changes were observed on the opacity and brightness of the paper (both change by less than 1%). These results are similar to works by Gonzalez [34]. Another researcher reported an increment of 2.8% and 3.2 % of opacity and brightness, respectively, with the addition of 10% to chemi mechanical pulp [35].

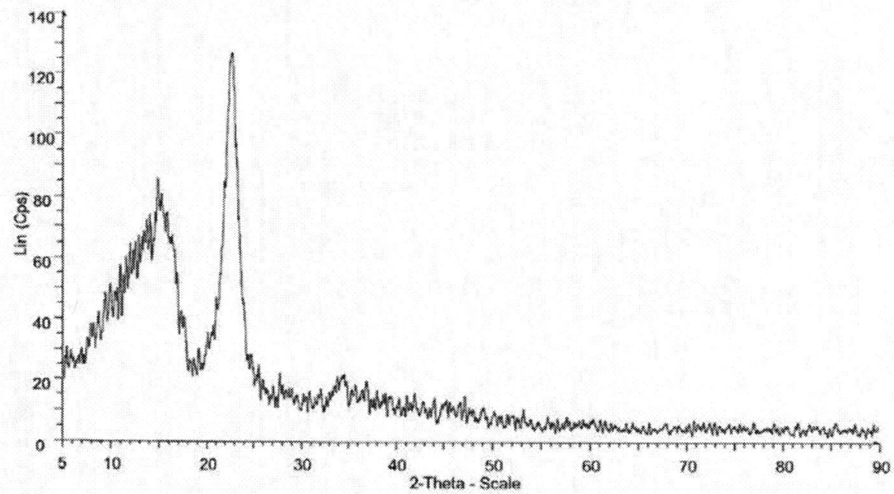

Figure 1. X-ray diffraction (XRD) diffractogram of nanocellulose.

All three strength tests conducted showed improvement of laboratory handsheets at higher NCC additions (Figure 5). The folding endurance improves by four-folds, tensile strength by 58% and tear strength by 35%. The enzymatic hydrolysis reaction cleaves the amorphous region of cellulose at the C-O-C bond on the cellulose chain, creating new chain ends with a new hydroxyl group on the surface of each as proposed by Schmidt [36]. Thus, each C-O-C bond broken will result in two additional hydroxyl groups. The presence of hydroxyl charge on the surface of nanocellulose is enhanced with higher dosage of nanocellulose addition. This factor, cumulatively increases the hydroxyl charge of the mixture which improves fibre bonding (intramolecular hydrogen bonds with neighboring glucose base unit) when the laboratory handsheets are dried.

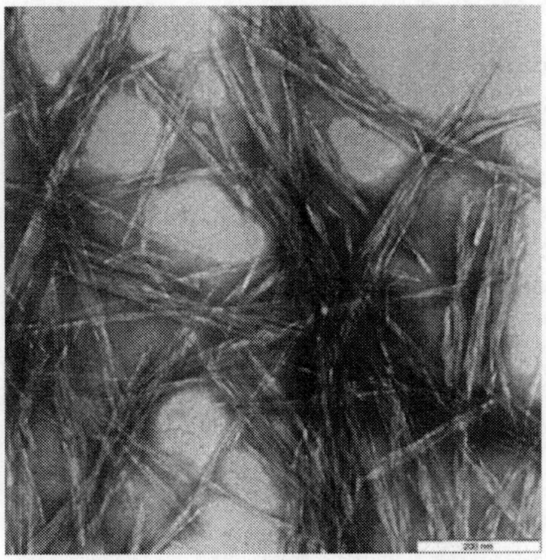

(a)

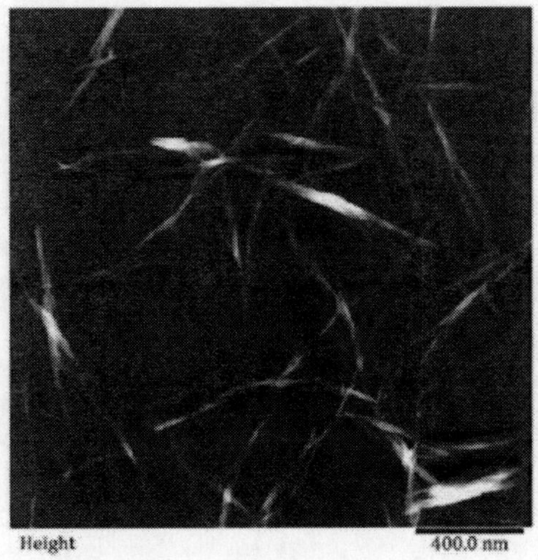

(b)

Figure 2. (a) TEM image of nanocellulose (b) AFM image of nanocellulose.

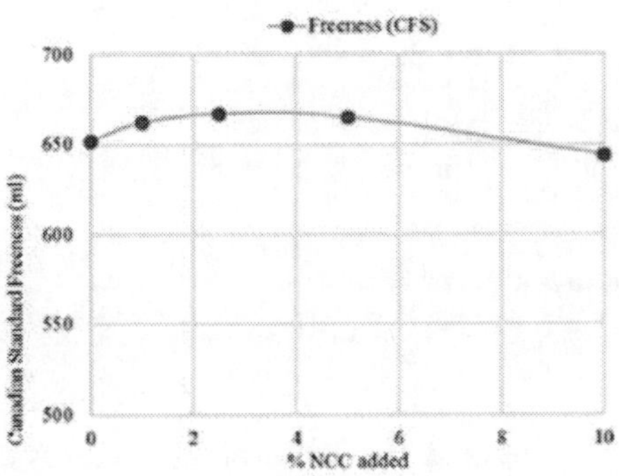

Figure 3. Canadian Standard Freeness of paper incorporated with nanocellulose.

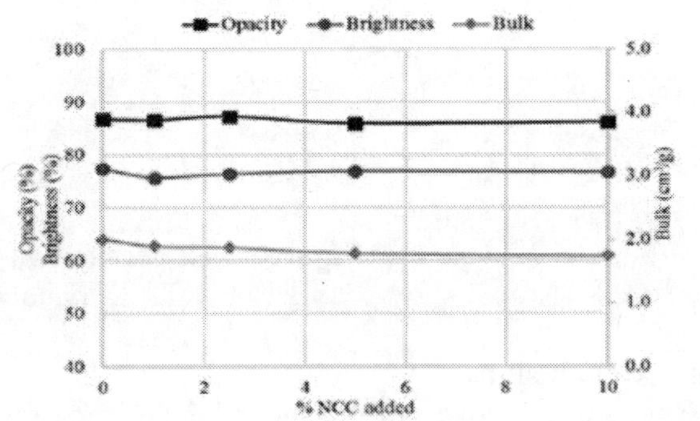

Figure 4. Bulk, opacity and brightness of paper incorporated with nanocellulose.

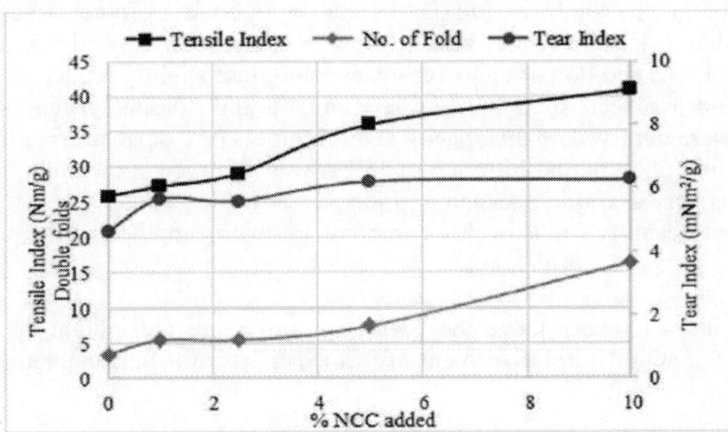

Figure 5. In this case simply justify the caption so that it is as the same width as the graphic.

The most interesting improvement of paper properties obtained in this study is the reduction of air permeance as shown in Figure 6. Increasing the dosage of nanocellulose to 10% reduce the air permeance by up to 53.7%. The conventional laboratory handsheet produced in this study has air permeability of 21.2 µm/Pa.s and this value was reduced to just 9.8 µm/Pa.s upon addition of 10% nanocellulose. The compact assembly of the cellulose/ nanocellulose in the paper provides longer routes for air molecules to diffuse through as explained by Nair et al. [37]. This result is very useful in packaging grade papers for food and pharmaceutical products as the reduction in air permeability will slow down microbial growth, thus lengthening the product shelf life.

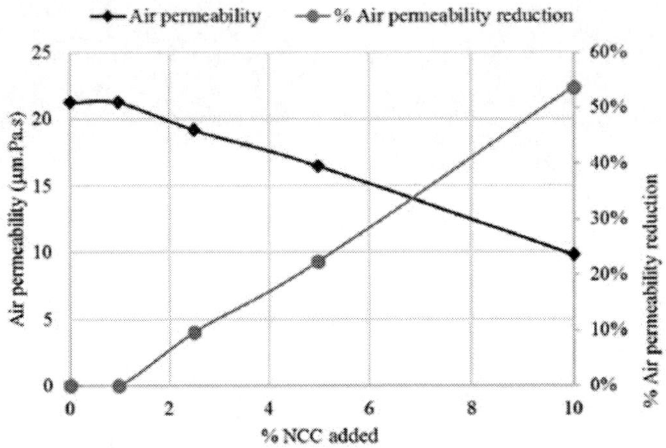

Figure 6. Air permeability of paper incorporated with nanocellulose.

4. Conclusions

This study shows that nanocellulose can successfully be extracted via microbial hydrolysis pretreatment followed by mechanical process. The nanocellulose recovery from the bleached cellulose obtained in this study is very high, exceeding 90%. The crystallinity is quite low compared to nanocellulose extracted using acid hydrolysis reported by others (54% vs. 80%). However, the observed crystallinity in this study is still in the range reported by other researchers (45-68%). Nanocellulose obtained in this study was added to paper and results showed that some of the paper properties are affected. The addition of up to 10% nanocellululose slightly reduces the specific volume and significantly improve tear strength, tensile strength and folding endurance. However, no significant changes were observed on the opacity and brightness of the paper. In terms of paper barrier properties, the reduction of air permeability up to 53.7% was observed upon addition of nanocellulose of 10%. This result is very useful in packaging grade papers for food and pharmaceutical products as the reduction in air permeability will slow down microbial growth, lengthening the product shelf life.

Acknowledgement

The authors would like to acknowledge the FRIM for providing the laboratory facilities and the Malaysia Economic Planning Unit, Department of Prime Minister for financial support under the 10th and 11th Malaysia Plans.

References

[1] Klemm D, Kramer F, Moritz S, Lindström T, Ankerfors M, Gray D, and Dorris A 2011 Nanocelluloses: a new family of nature-based materials *Angewandte Chemie International Edition* **50** 5438-5466
[2] Sun X, Wu Q, Lee S, Qing Y, and Wu Y 2016 Cellulose nanofibers as a modifier for rheology,

curing and mechanical performance of oil well cement *Sci Rep-Uk* **6** 31654
[3] Dufresne A 2013 Nanocellulose: a new ageless bionanomaterial *Mater Today* **16** 220-227
[4] Rahmad Z, Nik Fadzly N R, Mashhor M and Mohd Yunus Z 2008 The distribution of *Macaranga*, genus (family euphorbiaceae) in Penang Island, Peninsular Malaysia *J Bioscience* **19** 91–99
[5] Wan Mohd Shukri W A, Harun I, Nur Hajar Z S and Mohd Rizal S. 2007. Logged-over Forests Under the Selective Management System: A Case Study in Jengai Forest Reserve, Terengganu. In N.M Nik Zanariah, K.A.Sarifah & H. Nor Azman (eds). Highlights of FRIM'S IRPA projects 2006. Pp. 68-77.
[6] Peh T B, Khoo K C, Lee T W and Mohd Nor M Y. 1986. Pulp & Paper Industry & Research in Peninsular Malaysia. Malayan Forest Records No. 31. FRIM, Kepong.
[7] Johnson D A, Paradis M A, Bilodeau M, Crossley B, Foulger M, and Gelinas P 2016 Effects of cellulosic nanofibrils on papermaking properties of fine papers *TAPPI JOURNAL 15* 395-402
[8] Gagnon G, Rigdal R, Schual-Berke J, Bilodeau M and Bousfield DW. 2010. The effect of Nano-fibrillated cellulose on the mechanical properties of polymer films, *2010 TAPPI International Conf. on Nanotechnology for the Forest Product Industry,* Finland.
[9] Lee S Y, Mohan, D J, Kang I A, Doh G H, Lee S, and Han S O 2009 Nanocellulose reinforced PVA composite films: Effects of acid treatment and filler loading *Fiber Polym* **10** 77–82
[10] Jasmani L and Adnan S 2017 Preparation and characterization of nanocrystalline cellulose from Acacia mangium and its reinforcement potential *Carbohyd Polym* **161** 166–171
[11] Chattopadhyay D P and Patel B H 2016 Synthesis, Characterization and Application of Nano Cellulose for Enhanced Performance of Textiles *J Textile Sci Eng* **6** 2
[12] Adnan S, Jasmani L, Rusli R, Khadiran T, Ujang S and Saleh M. 2015. Nanocrystalline Cellulose from Tropical Forest Pioneer Species. Paper presented at *Nano Kebangsaan Conf. 2015*, Shah Alam, Malaysia
[13] Adnan S and Jasmani L. 2016. Environmentally-Friendly Method of Nanocellulose Extraction, in Mohd Isa MZ, Abd Shukor is (Eds.), *FRIM in Focus* June 2016, FRIM Publication, 2016, pp 8.
[14] Adnan S, Jasmani L, Rusli R, Khadiran T and Saleh M. 2016. Extraction of Nanocrystalline Cellulose from Local Forest Pioneer Species using Microbial Hydrolysis Assisted Shear Process, Paper presented at *Nano Kebangsaan Conf. 2016*, Technology Park Malaysia, 10-12 October 2016, Malaysia.
[15] TAPPI 205 sp-02. 2002. Forming Handsheets for Physical Tests of Pulp, TAPPI Press, Atlanta
[16] TAPPI T 220 sp-01. 2001. Physical Testing of Pulp Handsheets. TAPPI Press, Atlanta.
[17] Standards Malaysia. MS ISO 534: 2007. 2007 Paper and Board - Determination of Thickness, Density and Specific Volume
[18] ISO. ISO 5636-3 1992. 1992. Paper and board -- Determination of air permeance (medium range) - Part 3: Bendtsen method
[19] Standards Malaysia. MS ISO 2471: 2010. Paper and Board - Determination of Opacity (Paper Backing) - Diffuse Reflectance Method
[20] Standards Malaysia. MS ISO 2470-1: 2010. 2010. Paper, Board and Pulps - Measurement of Diffuse Blue Reflectance Factor - Part 1: Indoor Daylight Conditions
[21] Standards Malaysia. MS ISO 1924-2: 2010. 2010. Paper and Board - Determination of Tensile Properties - Part 2: Constant Rate of Elongation Method (20 mm/min)
[22] Standards Malaysia. MS ISO 1974: 1999. 1999. Paper – Determination of Tearing Resistance (Elmendorf Method)
[23] ISO. ISO 5626: 1993. 1993. Paper -- Determination of folding endurance.
[24] Martelli-Tosi M, Torricillas M D S, Martins M A, Assis O B G D, and Tapia-Blácido D R, 2016 Using commercial enzymes to produce cellulose nanofibers from soybean straw *J Nanomater*
[25] Mariño M, da Silva LL, Durán N and Tasic L 2015 Enhanced Materials from Nature:

Nanocellulose from Citrus Waste *Molecules* **20** 5908-23

[26] Qing Y, Sabo R, Zhu JY, Agarwal U, Cai Z and Wu Y 2013 A comparative study of cellulose nanofibrils disintegrated via multiple processing approaches *Carbohyd Polym* **97** 226– 34

[27] Tonoli GHD, Teixeira EM, Corrêa AC, Marconcini JM, Caixeta LA and LHC. 2012. Cellulose micro/nanofibres from *Eucalyptus* kraft pulp: Preparation and properties. *Carbohyd Polym* **89** 80-88

[28] Janardhnan S and Sain M M 2011 Targeted Disruption of Hydroxyl Chemistry and Crystallinity in Natural Fibers for the Isolation of Cellulose Nano-Fibers Via Enzymatic Treatment *BioResources* **6** 1242-1250

[29] Wada M, Nishiyama Y, Chanzy H, Forsyth T and Langan P 2008 The structure of celluloses. Advances in X-ray *Analysis* **51** 138

[30] Song Q, Winter WT, Bujanovic BM and Amidon TE. 2014. Nanofibrillated Cellulose (NFC): A High-Value Co-Product that Improves the Economics of Cellulosic Ethanol Production *Energies* **7** 607-618

[31] Mariño M, da Silva LL, Durán N and Tasic L. 2015. Enhanced Materials from Nature: Nanocellulose from Citrus Waste *Molecules* **20** 5908-5923

[32] Akgul M and Tozluoglu A. 2009. Some Chemical and Morphological Properties of Juvenile Woods from Beech (*Fagus orientalis* L.) and Pine (*Pinus nigra* A.) Plantations. *Trends in Applied Sciences Research* **4** *116-125*

[33] Adnan S, Khadiran T, Jasmani L, and Rusli R 2017 Preliminary Investigation on Improving Biopolymer Properties Using Nanocellulose from Tropical Forest Species. *Applied Mechanics and Materials* **865** 43-48

[34] González I, Boufi S, Pèlach M A, Alcalà M., Vilaseca F, and Mutjé P 2012 Nanofibrillated cellulose as paper additive in eucalyptus pulps. *BioResources* **7** 5167-5180

[35] Kasmani J E 2016 Effects of Ozone and Nanocellulose Treatments on the Strength and Optical Properties of Paper Made from Chemical Mechanical Pulp *BioResources* **11** 7710-7720

[36] Schmidt O. 2006. Wood and Tree Fungi Biology, Damage, Protection, and Use. Springer, Berlin, Germany

[37] Nair S S, Zhu JY, Deng Y, and Ragauskas A J 2014 High performance green barriers based on nanocellulose *Sustainable Chemical Processes* **2** 23

Determination of Agarwood volatile compounds from selected *Aquilaria* species plantation extracted by Headspace-Solid Phase Microextraction (HS-SPME) method

J Abd Majid[1,2*], AH Hazandy[2], MT Paridah[2], MA Nor Azah[1], J Mailina[1], S Saidatul Husni[1] and L Sahrim[1]

[1] Forest Research Institute Malaysia (FRIM), 52109 Kepong, Selangor, Malaysia
[2] Institute of Tropical Forestry and Forest Products, Universiti Putra Malaysia, 43400 UPM, Serdang, Selangor, Malaysia

*E-mail: majid@frim.gov.my

Abstract. A collection of inoculated heartwood of three *Aquilaria* species namely *A. crassna*, *A. sinensis* and *A. subintegra* were investigated for their volatile compounds. The volatile compounds were extracted using automated Headspace-Solid Phase Microextraction (HS-SPME) with fibre coating of 50/30 µm divinyl benzene/carboxen/poly-dimethylsiloxane (DVB/CAR/PDMS) and determined by gas chromatography/mass spectrometry (GCMS). The absence of proper scientific method in forming the agarwood has resulted in variations in agarwood quality and inconsistent grades in the market. Thus there is a need to quantify the amount of major volatile compounds found in the agarwood so that the quality can be scientifically verified. In this study, the commonly known volatile compounds present in agarwood sample were found but with varying amounts. At least three important compounds were obtained i.e. β-agarofuran, α-eudesmol and agarospirol. Based on the presence of these main volatile compounds, the study suggests that extraction by HS-SPME also useful instead of distillation process. The ability to identify key components in a short time can facilitate and shorten the analysis time. It further helps in developing the quality of agarwood industry standards.

1. Introduction

The *Aquilaria* tree content resinous heartwood commonly known as agarwood is a genus belonging taxonomically to the Thymelaeaceae. The agarwood tree produces resin as a defense mechanism against infection or injury. The resin is valued in many cultures for its distinctive fragrance, and is used as a principal component in incense and perfumes as well as in traditional medicine [1]. Agarwood chips produced significant aroma when it is heated or burned and they can be obtained from natural resources and nowadays it can be artificially induced [2]. However there are doubts about the quality of agarwood chips produced from the chemically induced tree. Usually, natural formed agarwood produce more resin and the price is higher than induced agarwood [2]. In addition, other uses of this product are constrained by limited supply and very expensive. One of the main reasons for the relative rarity and high cost of agarwood is the depletion of the wild resource [3]. This genus consists about 17 species, growing on particular sites of tropical Asia [3]. Among those species, five

IOP Conf. Series: Materials Science and Engineering **368** (2018) 012023 doi:10.1088/1757-899X/368/1/012023

are found in Malaysia, i.e. *Aquilaria malaccensis*, *A. sinensis*, *A. hirta*, *A. subintegra* and *A. crassna* [4].

Plantation for *Aquilaria* trees in combination with artificial agarwood-inducing methods serves as a way to supply agarwood and conserve of wild *Aquilaria* stock. Agarwood cultivation was increasing especially after several research studies showed that cultivated agarwood could provide feasible benefit [4]. Additionally, the existing agarwood-producing methods require a long time for agarwood formation, meanwhile, the resin produced in a short time in this inducing technique is deemed to be of inferior quality [5].

Studies on the chemistry of agarwood have reported presence of sesquiterpenes, chromone derivatives, sesquiterpene furanoids, tetradecanoic acid and pentadecanoic acid [6-10]. The sesquiterpenes and phenylethyl chromone derivatives are the main compounds in agarwood [11] and [12]. The sesquiterpene hydrocarbons contribute to the fragrant of wood aroma [13]. Based on the knowledge from the agarwood oil, studies had been carried out on the total volatile compounds from the agarwood chips using Headspace-SPME and by GC-MS [8, 14]. Chemical compounds were identified by matching their retention indices with literature values [9] and further verified by comparing with the Wiley and Adams library search data in GC/MS database. In this study, the chemical profiles of agarwood from different species of *Aquilaria* were examined using GCMS. The result is reported, compared and discussed in this paper.

2. Materials and methodology

Inoculated heartwood samples of three *Aquilaria* species namely *A. crassna*, *A. sinensis* and *A. subintegra* were investigated for its volatile compounds. *A. crassna* and *A. sinensis* samples were provided by RAL Plantation Sdn. Bhd, Kuala Kangsar, Perak meanwhile *A. subintegra* was obtained from Biobenua Teknologi Sdn. Bhd, Kuala Linggi, Melaka. The trees age were about 5 years old and all trees have been inoculated for 6 to 12 months.

2.1 Methods

2.1.1 Sample preparation. Samples are collected by cutting down the selected tree after being purchased from the planters. Infected wood known as agarwood removed from the white wood then the carving process was executed. In order to get agarwood, the dark wood will be separated from the white wood using sharp knives, a broken glass, or hooked knives and dried under the shade [15]. Under constant moisture content in room temperature, the samples were finely ground using basic analytical mill and ready to run the chemical analysis.

2.1.2 HS-SPME procedure. The agarwood volatile compounds were extracted using automated Headspace-Solid Phase Microextraction (HS-SPME) with fibre coating of 50/30 μm divinyl benzene/carboxen/poly-dimethylsiloxane (DVB/CAR/PDMS). Three replicates of 2 grams of each samples were prepared into 20 ml screw-top clear vial and tightly capped. The sample vials was introduced in an agitator for 10 min at incubation temperature of 80 °C. The extraction time was fixed at 10 min to trap volatile compounds. After sampling, the fibre was immediately injected into the GCMS injection port kept at 60 °C with a hold time of 10 min.

2.1.3 GCMS Analysis. The chemical constituents were determined by using gas chromatography/mass spectrometry (GCMS). Analysis was carry out using Agilent Technologies 7890A/5975C Series MSD with HP-5MS column (30 m x 0.25 mm ID x 0.25 μm film thickness). The programmed was set, initially at 60 °C for 10 min, increase to 180 °C for 1 min at 3 °C/min. Helium was used as carrier gas at a flow rate 1.0 ml/min and the ion source temperature was programmed at 280 °C. The chemical constituents were identified by mass spectral library (HPCH2205.L, Wiley7Nist05.L, and NIST05a.L). The results of the peak areas were expressed as peak counts.

IOP Conf. Series: Materials Science and Engineering **368** (2018) 012023 doi:10.1088/1757-899X/368/1/012023

3. Results and discussion

The volatile compounds identified in agarwood of three *Aquilaria* species namely *A. crassna*, *A. sinensis* and *A. subintegra* were shown in Table 1. Overall, a total of 20 compounds were identified in all of the woodchips of which 13 compounds from *A.crassna*, 17 compounds from *A. sinensis* and 17 compounds from *A. subintegra*. There are several common volatile compounds within these three species namely 4-phenyl-2-butanone, β-agarofuran, dihydroagarofuran, α-bulnesene, α-agarofuran, aromadendrene, 10-*epi*-γ-eudesmol, agarospirol, valencene, valerianol and α-eudesmol. Some of these compounds were major compounds in one species but not in other species. However, 9 similar compounds has been appeared in all study species which were 4-phenyl-2-butanone, α-guaiene, α-humulene, β-agarofuran, α-selinene, dihydroagarofuran, α-bulnesene, α-agarofuran and agarospirol exits either as major or minor compounds.

Table 1. Chemical compounds of *A.crassna*, *A.sinensis* and *A.subintegra* extracted by HS-SPME

No	Chemical compounds	Area (%)		
		A.crassna	***A.sinensis***	***A.subintegra***
1	4-phenyl-2-butanone	2.76	1.77	4.35
2	β-elemene	-	1.04	0.37
3	α-guaiene	0.97	3.43	0.59
4	α-humulene	0.59	3.23	1.51
5	β-agarofuran	8.55	2.13	1.9
6	α-selinene	0.88	5.44	1.88
7	ar-curcumene	-	6.07	5.39
8	dihydroagarofuran	8.41	5.44	3.96
9	α-bulnesene	6.17	5.36	1.26
10	α-agarofuran	4.06	4.91	1.7
11	Nerolidol	3.16	-	-
12	Aromadendrene	-	12.88	2.38
13	10-*epi*-γ-eudesmol	3.89	-	2.78
14	Agarospirol	14.85	2.48	2.92
15	Valencene	7.41	-	-
16	Valerianol	14.29	-	-
17	α-eudesmol	-	15.84	15.41
18	Santolol acetate	-	2.7	6.43
19	Cyclocolorinone	-	7.81	-
20	Aristolone	-	4.57	4.77

It is observed that temperature programmed at 80 °C, the fibre extracted most of the common major compounds in agarwood sample. Dihydroagarofuran was captured as one of major compounds in all three species. Agarospirol (14.85%) was the most abundance compounds for *A.crassna*, meanwhile α-eudesmol were the highest compounds in *A.sinensis* and *A.subintegra* with (15.84%) and (15.41%) respectively.

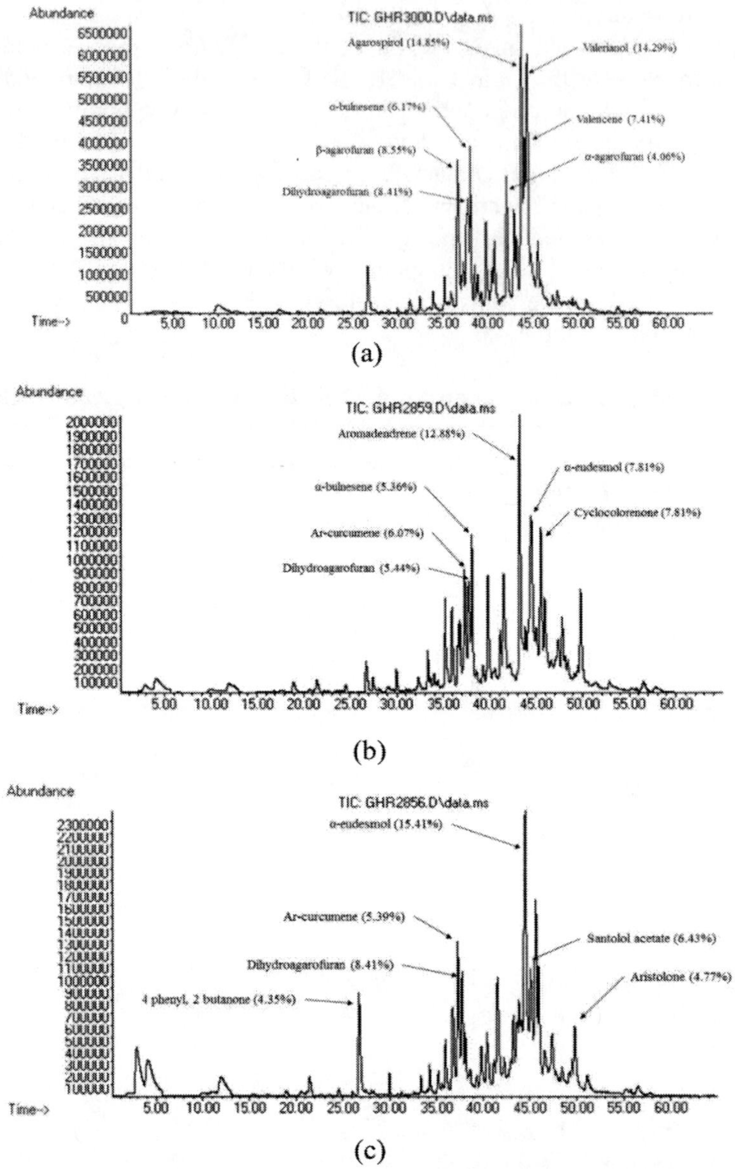

Figure 1. GCMS chromatogram profiles for each observed agarwood where (a) GCMS chromatogram of *A. crassna*, (b) GCMS chromatogram of *A. sinensis* and (c) GCMS chromatogram of *A. subintegra*.

All major volatiles compounds presence in this study as chromatogram profiles shows in (Figure 1). These presented volatile compounds from wood chips were similarly found in agarwood essential oils. (Table 2) shows the studied from other researchers indicate that there are some similarities and variations in the volatile compounds of several agarwood oils, such as 3-Phenyl-2-butanone, β-agarofuran, α-agarofuran, agarospirol and 10-*epi*-γ-eudesmol were some of the volatile compounds found occurring in all the oils studied [16].

Table 2. Some chemical compounds detected in some Malaysian agarwood oils.

Compounds	RI	Selangor (%)	Kelantan (%)	Pahang (%)	Terengganu (%)
3-phenyl-2-butanone	1249	1.50	5.77	7.80	0.79
α-guaiene	1448	-	0.67	-	-
β-agarofuran	1477	1.69	1.98	0.69	0.50
α-agarofuran	1553	4.83	2.96	1.48	1.57
Nor-ketoagarofuran	1557	2.09	-	-	-
10-*epi*-γ-eudesmol	1618	11.54	9.03	8.10	3.32
Agarospirol	1631	14.86	5.49	7.11	18.86
β-eudesmol	1649	-	-	-	5.74
Jinkoh-eremol	1650	10.62	7.70	6.31	-
kusunol	1659	18.94	-	-	-
Jinkohol II	1751	4.71	-	-	-

Source: [16]

4. Conclusion

The analysis of the agarwood by GCMS has been extracted using HS-SPME on three *Aquilaria* species namely *A. crassna, A. sinensis* and *A. subintegra* successfully obtained all the major volatile compounds in these samples. All those volatile compounds obtained from these *Aquilaria spp.* mostly found by others research studies on the chemistry of agarwood oils. They have reported presence group of aromatic terpene compounds including monoterpenes, sesquiterpenes and diterpenes [16-18]. Currently GC and GC/MS profiles are being used essential oils extracted from distillation process to assist the detection and authentication of agarwood. By using HS-SPME, the time of determination process of volatile compounds can be reduced in order to generate the GCMS profiles in agarwood without involved any distillation methods. Generating GC/MS profiles of agarwood samples of various species and from different sources will be helpful towards developing suitable chemical marker for grading agarwood. The presence of these major volatile compounds can be used as a guideline in assessing the consistent quality of agarwood obtained from plantations.

Acknowledgements
This work was done under RMK10 project (Classification of karas and ramin based on a combination of technology, chemometric analysis chemistry, electronic nose and image analysis characterization) in Natural Products Division, Forest Research Institute of Malaysia (FRIM). The authors would like to thank all Product Development Programme (PPH) research group for supporting this research. Special thanks to the Institute Tropical Forestry and Forest Products (INTROP) University Putra Malaysia for support the participating into WOBIC 2017 Conference.

References
[1] P Saikia, M L Khan 2012 *Aquilaria malaccensis* Lam. a red-listed and highly exploited tree species in the Assamese home garden *Current Science* **102** 546
[2] Xing Li Zhang 2012 Production of high-quality agarwood in *Aquilaria sinensis* trees via whole-tree agarwood-induction technology Chinese Academy of Medical Sciences & Peking Union Medical College *China Chemical Letters* **23** 727-730
[3] Pakamas C 2012 Overview of the Agarwood oil Industry. *IFEAT International Conference Proceeding. Singapore*: '*Essential Asia'* 131-138
[4] Suharti S, Pratiwi, Erdy Santosa and Maman Turjaman 2011 Feasibility Study of Business in Agarwood Inoculation at Different Stem Diameters and Inoculations Period. *Journal of Forestry Research* **8** 114 – 129

[5] Yangyang L, Huaiqiong C, Yun Y, Zheng Z 2013 Whole-tree Agarwood-Inducing Technique: An Efficient Novel Technique for Producing High-Quality Agarwood in Cultivated *Aquilaria sinensis* Trees. Chinese Academy of Medical Sciences & Peking Union Medical College, *China Molecules ISSN* 1420-3049 3087-88

[6] Ishihara M, Tsuneya T, Shiga M and Uneyama K 1991 Three sesquiterpenes from agarwood, *Phytochemistry* **30** 563-566

[7] Ng L T, Chang Y S and Azizol A K 1997 A review on agar (gaharu) producing *Aquilaria* species. *Journal of Tropical Forest Products* **2** 272-285

[8] Nor Azah M A, Chang Y S, Hassan N H, Ahmad A S. 2006. Investigation on *Aquilaria malaccensis* for the production of gaharu oils in Medicinal Plants- *Chemistry and Biological Activity*. Fasihuddin B.Ahmad, Zaini B. Asim, Laily B. Din and Ikram M. Said (eds) University Malaysia Sarawak 247-250

[9] Nakanishi T, Yamagata E, Yoneda K, Nagashima T, Kawasaki I, Yoshida, T, Mori H and Muira I 1984 Three fragrant sesquiterpenes of agarwood *Phytochemistry* **23** 2066-67

[10] Tamuli P, Boruah P, Nath S C and Leclercq P 2005 Essential oil of eaglewood tree: a product of pathogenesis. *Journal of Essential Oil Research* **17**:601-604

[11] Hashimoto K, S Nakahara, T Inoue, Y Sumida, M Takahashi, Y Masada 1985 *Chemical Pharmaceutical Bulletin* **33** 5088–91

[12] Ishihara M, Tsuneya T and Uneyama K 1993 Components of the volatile concentrate of agarwood *Journal of Essential Oil Research* **5** 283-289

[13] Arce A, Soto A, in N Benkeblia, P Tennant (Eds) 2009 "Tree and Forestry Science and Biotechnology," vol. **2** (Special Issue 1), *Global Sciences Books*, Ltd., UK 1–9

[14] Najib M S, N A Mohd Ali, M N Mat Arip, M Abd Jalil and M N Taib 2012 Classification of Agarwood using ANN *International Journal of Electrical Systems Research* **5** 20-34

[15] Zich F, Compton J 2001 Towards sustainable management of Papua New Guinea's agarwood resource. New South Wales: *A TRAFFIC Oceania report in conjunction with WWF South Pacific Programme*

[16] Nor Azah M A, Chang Y S, Mailina J, Abu Said A, Saidatatul Husni S, Nor Hasnida H & Nik Yasmin N Y 2008 Comparison of chemical profiles of selected gaharu oils from Peninsular Malaysia. *The Malaysian Journal of Analytical Sciences* **12** 338–340

[17] Chong S P, Osman M F, Bahari N, Nuri E A, Zakaria R and Abdul Rahim K 2015 Agarwood Inducement Technology: A Method for Producing Oil Grade Agarwood in Cultivated *Aquilaria malaccensis* Lamk *J Agrobiotech* Vol **6**

[18] Naef R 2011 The volatile and semi-volatile constituents of agarwood, the infected heartwood of *Aquilaria* species: A review *Flavour and Fragrance Journal* **26** 73-89

Oil Palm EFB/Kenaf Fibre Reinforced Epoxy Hybrid Composites: Dimension Stability Behaviours

F. Hanan, M. Jawaid*, M.T. Paridah

Laboratory of Bio Composites Technology, Institute of Tropical Forestry & Forest Product (INTROP), Universiti Putra Malaysia, Malaysia, 43400 UPM Serdang, Selangor, Malaysia.
*Corresponding: jawaid@upm.edu.my

Abstract. In recent years, natural fibres composites have gained increasing interest as a most promising material in different applications due to its attractive properties such as eco-friendly, cost-effectiveness and light weight. Though extensive research has been made on the performance evaluation of natural fibres composite materials, not much data is available on the dimension stability which restricts their use in exterior applications. In this study, hybrid composites were fabricated by hand lay-up technique by reinforcing oil palm empty fruits bunch (EFB) and kenaf fibre mats with epoxy matrix. Experiments are evaluated to study dimension stability that includes water absorption, thickness swelling, density and void content. Increasing the kenaf fibre into EFB composite totally improves the dimension stability of the hybrid composites thus indirectly leads to reduction in voids compared to oil palm EFB composites which is presented lowest percentage of dimension stability and higher void content.

1. Introduction

Natural fibres have been attracted for industries applications such as automotive compartment, sport equipment, construction materials and non-structural elements due to its renewable nature, low cost, easily availability, low density, biodegradable nature, easy fibre surface modification, wide availability and relative non-abrasiveness [1,2]. Previously, lot of works and developing of natural fibres composites have been done by combining using different types of fibres and various resins [3]. One of the natural fibres is oil palm empty fruit bunch (EFB) fibre that found a lot in Southern Asian countries and Malaysia is one of the biggest palm oil exporters in the world with having 50,000 km^2 of oil palm plantations and 423 palm oil mills operating in the country [4,5]. Oil palm fibre is hard and tough, which shows similarity to coir fibres. With average diameter 0.07 μm. of pores on fibre surface is useful for better mechanical interlocking with matrix resin in composite fabrication [6]. Despite the advantages, one of the important issues of natural fibre is the hydrophilic property of cellulose which impacts the weak interface bonding with hydrophobic polymer as a matrix [7]. Most of the drawbacks that have been identified can be overcome by effective hybridization of natural fibre with synthetic or natural fibres [8,9,10,11,12,]. Hybridization, especially where only variant natural fibre are combined, is fast receiving encouraging attention because it offers a range of properties that is quite difficult to obtain with single kind of reinforcement [13]. In the present study, oil palm EFB with kenaf fibre reinforced epoxy resin was prepared by using hand lay-up technique. Void content, density, water absorption and thickness swelling behaviors were studies. The main objective of this study is to investigate the effectiveness of hybridizing kenaf into oil palm EFB through dimension stability behaviors.

2. Experimental

2.1. Materials

Oil palm EFB mats were supplied by HK Kitaran Sdn. Bhd. (P.Pinang, Malaysia) while non-woven kenaf fibre mats were supplied by ZKK Sdn. Bhd. (Selangor, Malaysia) . The liquid epoxy resin DER-331 used in this study is based on epichlorohydrin and bisphenol A and the curing agent was epoxy hardener joint amine 905-3s. Silicon spray was used as a release agent. Resin, curing agent and silicon spray were procured from TAZDIQ Engineering Sdn. Bhd. (Selangor, Malaysia).

2.2. Preparation of the composites

Fabrication of oil palm EFB/kenaf fibre reinforced epoxy hybrid composites using stainless steel mould with dimension 300 mm x 300 mm x 5 mm. Hand-lay-up technique was chosen for making test sample. Silicon spray was coated with thin layer which act as a releasing agent. Epoxy resin and hardener uniformly mixed using ratio 2:1 for 15 minutes. Hybrid composites were prepared by hybridizing oil palm EFB and kenaf fibre with keeping the different weight ratio of oil palm EFB and kenaf fibre 4:1, 1:1, and 1:4 with total fibre loading at 50% by weight. Oil Palm EFB and kenaf fibres mat were stacking alternating in the mould, then were poured with resin matrix. The composite was pressed with a hand roller to eliminate any entrapped air bubbles. The mould was closed for curing process using hot press machine at pre pressed 5 minutes, compressed for 15 minute at 120oC. Hybrid composites were then cooled in cold press for 5 minutes. Single oil palm EFB and kenaf fibre composites also prepared as a control samples.

3. Characterization

3.1. Void Content

For determination of void contents in composites, ASTM D2734 method was used. The void content was determined from the theoretical and experimental density of the composite through Equation (1),

$$VoidContent(\%)=\frac{P_{theoretical}-P_{experimental}}{P_{theoretical}}$$

$$Where\ P_{theoretical}=\frac{1}{[(W_f/P_f)+(W_m/P_m)],}$$

And W_f is the fibre weight fraction, W_m is the matrix weight fraction, P_f is the fibre density and P_m is the matrix density.

3.2. Density

Density was measure by using the ASTM D1895 standard. The density of the samples was calculated by using following Equation (2),

$$Density(g/cm^3)=m/v$$

Where m is the mass of the composites, and v is the volume of composites.

3.3. Dimension Stability Test

The dimensional stability tests involved thickness swelling and water absorption according to ASTM D 570. Before testing, the weight and thickness of each sample were measured. Five samples of each type of composites were immersed in distilled water at room temperature. After 24 hours, the sample was taken out gently blotted with filter paper to remove excess water on the surface. The weight and thickness samples were recorded and were continue for several days until a constant value of the samples was obtained. The percentage of water absorption was calculated from equation (3) and for the thickness swelling was equation (4),

$$WaterAbsorption(\%) = \frac{W_n - W_d}{W_d}$$

Where W_n is the weight of composites samples after immersion and W_d is the weight of the composite samples before immersion.

$$Thickness\ Swelling\ (\%) = \frac{T_1 - T_0}{T_0}$$

Where T_1 is the thickness after soaking and T_0 is the thickness before soaking.

4. Results and Discussion

4.1. Void Content and density

Table 1 shows the void content and density of oil palm EFB/kenaf fibre reinforced epoxy hybrid composites. Oil palm EFB showed higher voids and low density respectively. It is clearly showed that oil palm EFB composites attributed to poor fibre-matrix dispersion and wetting of the fibres occur. According to Jawaid et al. [14] the most common cause of voids is the incapability of the matrix to displace all the air which is entrained within the fibre as it passes through the matrix impregnation. High voids in the composite leads to lower fatigue resistance, greater susceptibility to water absorption, and increased variation in mechanical properties [15]. In addition, oil palm fibre are porous and the mat is loosely packed hence a large amount of resin would be squeezed out of the mat during pressing molding. The percentage of void content reduced when additional of kenaf fibre into oil palm EFB fibre which is contributed by higher density of kenaf compare to oil palm EFB fibre. Kenaf fibre mats are tightly packed and also more compatible towards the epoxy resin. As the result, slightly fewer amounts of voids and density percentage are present in the hybrid composites between hybrid parameter respectively.

Table 1: Void content and density of Oil Palm EFB/Kenaf Fibre Reinforced Epoxy Hybrid Composites

Composites	Void Content (%)	Density (g/cm^3)
Pure EFB	8.40	1.02
EFB: Kenaf (1:4)	4.43	1.10
EFB: Kenaf (1:1)	6.15	1.07
EFB:Kenaf (4:1)	3.20	1.16
Pure Kenaf	2.50	1.21

4.2. Water Absorption

In general moisture diffusion in a composite depends on factors such as volume fraction of fibre, voids, viscosity of matrix, humidity and temperature [16]. Figure 1 shows the trend of water absorption of oil palm EFB/kenaf fibre reinforced epoxy hybrid composites. As can been seen, the water absorption increased with immersion time. Pure oil palm EFB composite indicates highest among the different type of composites. This exhibits high porosity or presence of voids on the surface of oil palm EFB composites. According to Chinomso et al. [17] study, amount of water absorbed at any given fibre loading of oil palm EFB composites and the hydrophilic nature of oil palm EFB fibre is believed to be responsible for the water absorbed by the composites. Khalil et al. [8] stated that oil palm EFB fibres contain a higher proportion of hemicelluloses, owing to the free hydroxyl group present which are higher water absorption and hydrophilic nature fibre. According to Azman et al. [18], hydroxyl group is the main chemical entity for the attraction of water molecules in the natural fibres which is absorb water through formation of hydrogen bonding. Hybridization of kenaf fibre into oil palm EFB fibre decreased water absorption of the oil palm EFB fibre. This is mainly attributed by packed arrangement of hybrid composites and the less hydrophilic nature of kenaf fibre as compared to oil palm EFB fibre.

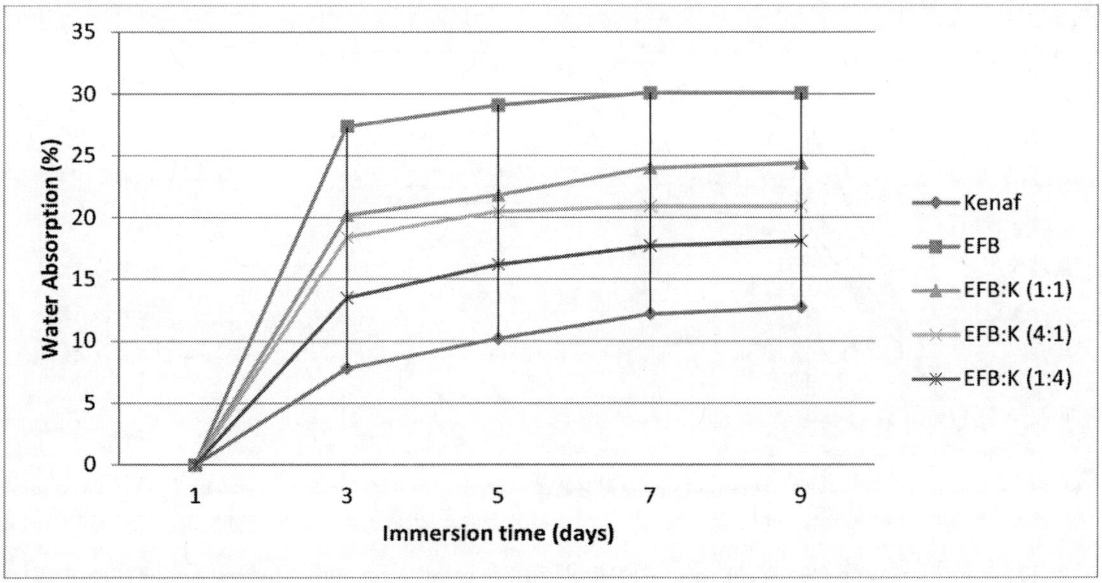

Figure 1: Water absorption (%) of oil palm EFB/kenaf fibre reinforced epoxy hybrid composites.

IOP Conf. Series: Materials Science and Engineering **368** (2018) 012024 doi:10.1088/1757-899X/368/1/012024

4.3. Thickness Swelling

Thickness swelling is closely related to the dimensional stability of the composite. When the composite is exposed to different humidity and temperature, stress is induce in between the layers of composite resulting in degradation of fibre matrix. The uptake of the water changed the thickness and affects the dimension of the composite [5]. Thickness swelling was carried out for several days until a constant weight was obtained. From the figure 2, it is observed that thickness swelling of oil palm EFB composite was highest among the different type of composites. It also showed the thickness swelling values of composites increased with an increased of water absorption time. By increasing the exposure time of composites to water, a significant amount of water absorbed, resulting in the swelling of fibre. Similar trend with water absorption, oil palm EFB indicates hydrophilic properties that responsible for the changes in the dimension of cellulose based composites particularly in the thickness and linear of the composite. In contrast, pure kenaf fibre composite shows moderate thickness swelling and slightly different with the hybrid composites. Hybridizations of oil palm EFB with kenaf fibre caused a remarkable reduction in thickness swelling rate. This can be expected to depend on the exposure of lignocellulosic fibre on the surface of composite [8].

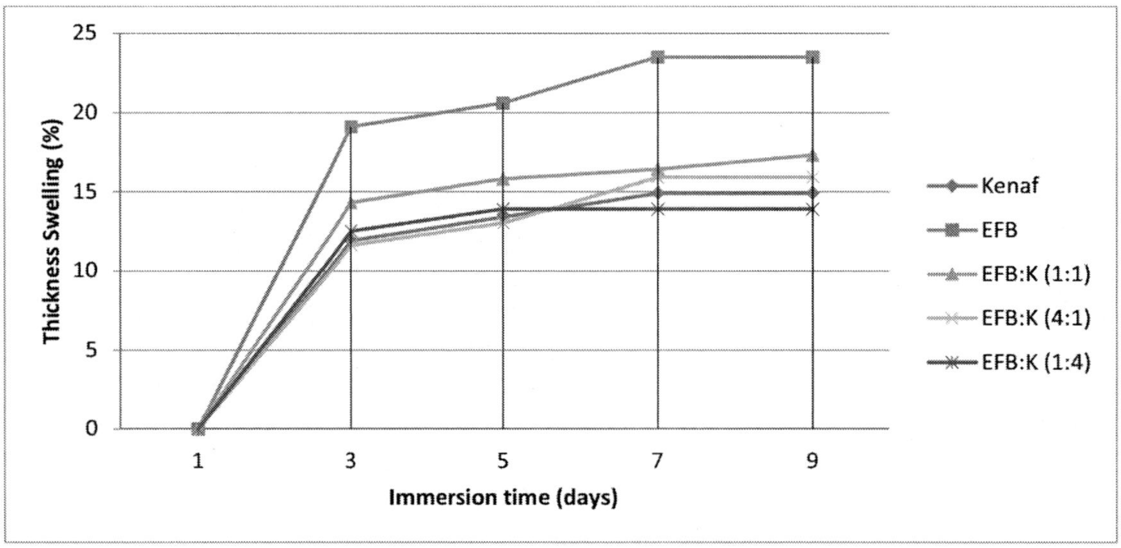

Figure 2: Thickness swelling (%) of oil palm EFB/kenaf fibre reinforced epoxy hybrid composites.

5. Conclusions

Oil palm EFB composites showed the highest percentage of all dimension stability behaviors. Hybridization of kenaf fibre into oil palm EFB composite significantly reduced void content and increased the density of hybrid composites. Increasing the kenaf fibre in the composite shows improvement the amount of thickness swelling and water absorption of the hybrid composites decreased due to packed arrangement of hybrid composite and less hydrophilic nature of kenaf fibre compare to oil palm EFB fibre.

Acknowledgments

We are thankful to Ministry of Higher Education for Providing HICOE Grant No: 6369108 to INTROP, UPM for doing this work.

References

[1] Shaji J., Sreekumar P.A, Jose M.K, Debora P., Sabu T., and Kuruvilla J. "Oil palm microcomposites: Processing and mechanical behavior," *Soc. Plast. Eng.*, **50**, pp. 1854-1863, 2010.

[2] Reza M., Jamaludin M.Y., Abdul R.M.S., and Sayed H.H. "Tensile properties of kenaf fibre due to various conditions of chemical fibre surface modifications," *Constr. Build Mater.*, **55**, pp. 103-113, 2014.

[3] Jawaid M. and Abdul Khalil H.P.S. "Hybrid composites made from oil palm empty fruit bunches/jute fibres: Water absorption, thickness swelling and density behaviours," *J. Polym. Environ.*, **19**, pp. 106-109, 2011.

[4] Tye. Y.Y., Lee. K.T., Wan. N.W.A. and Leh C.P." The effect of various pretreatment methods on oil palm empty fruit bunch (EFB) and kenaf core fibres for sugar production," *Procedia Environ. Sci.*, **20**, pp. 328-335, 2014.

[5] Teh W. F., Nurin S.S.N.A., Abdul Khalil H.P.S., Mohamad H.M.K., and Syakir M.I." Water Absorption and thickness swelling of oil palm empty fruit bunch (OPEFB) and seaweed composite for soil erosion mitigation." *J. Phys. Sci.*, **28**, pp. 1-17, 2017.

[6] Shinoj S., Visvanathan R., Panigrahi S., and Kochubabu M.,"Oil palm fibre (OPF) and its composites: A review," *Ind. Crops Prod.,* **33**, pp. 7-22, 2011.

[7] Reza M., Jamaludin M.Y., and Abdul R.M.S." A review of structural performance of oil palm empty fruit bunch fibre in polymer composites," *Adv. Mater. Sci. Eng.*," pp. 1-9, 2013.

[8] Abdul Khalil H.P.S., Jawaid M., Abu B. A." Woven hybrid composites: Water absorption and thickness swelling behaviours," *Bioresour.*, **6**, pp. 1043-1052, 2011.

[9] Abu B., Ahmad M. A., and Kuntjoro W. "The mechanical properties of treated and untreated kenaf fibre reinforced epoxy composite," *J. Biobased Mater. Bio.*, **4**, pp. 159-163, 2010.

[10] Alves, C., Ferrao P.M.C., Silva A.J., Reis L.G., Freitas M., Rodrigues L.B., and Alves D.E. "Ecodesign of automotive components making use of natural jute fibre composites," *J. Clean. Prod.*, **18**, pp. 313-327, 2010.

[11] Jawaid M., and Abdul Khalil H.P.S."Cellulosic/synthetic fibre reinforced polymer hybrid composites: A review," *Carbohydr. Polym.*, **86**, pp. 1-18, 2011.

[12] Sapuan S.M., Harun N., and Abbas K.A."Design and fabrication of a multipurpose table using a composite of epoxy and banana psedostem fibres," *J. Trop. Agric.*, **45**, pp. 66-68, 2007.

[13] Aji I.S., Zainudin E.S., Khalina A., Sapuan S.M.,"Studying the effect of fibre size and fibre loading on the mechanical properties of hybridized kenaf/PALF-reinforced HDPE composite," *Reinf. Plast. Compos*, **30**, pp. 546-553, 2011.

[14] Jawaid M., Abdul Khalil H.P.S., Abu B.A., and Noorunisa P.K."Chemical resistance, void content and tensile properties of oil palm/jute fibre reinforced polymer hybrid composites,' *Mater. Des.*, **32**, pp. 1014-1019, 2011.

[15] Jawaid M., Abdul Khalil H.P.S., Hassan A. and Elgorban A." Bi-layer hybrid biocomposite: chemical resistance and physical properties," *Bioresour.*, **7**, pp. 2344-2355, 2012.

[16] Dhakal H.N., Zhang Z.Y., and Richardson M.O.W.,"Effect of water absorption on the mechanical properties of hemp fibre reinforced unsaturated polyester composites," *Compos. Sci. Technol.*, **67**, pp. 1674-1683, 2007.

[17] Chinomomso M.E. and Isaac O.I." Properties of oil palm empty fruit bunch fibre filled high density polyethylene," *Int. J. Eng. Technol.*, **6**, pp. 458-471, 2012.

[18] Azman H., Arshad A.S., Farid N.A., and Aznizam A.B."A review on oil palm empty fruit bunch fibre-reinforced polymer composite materials," *Polym. Compos.*, **31**, pp. 2080-2101, 2010.

IOP Conf. Series: Materials Science and Engineering **368** (2018) 012025 doi:10.1088/1757-899X/368/1/012025

The physical properties of treated oil palm veneer used as face layer for laminated veneer lumber

K Masseat[1*], E S Bakar[2], I Kamal[3], H Husain[1] and P M Tahir[2]

[1]Forest Research Institute Malaysia (FRIM), 52109 Kepong, Selangor, Malaysia
[2]Laboratory of Biocomposite Technology, Faculty of Forestry and Institute for Tropical Forestry and Forest Products, Universiti Putra Malaysia, 43400 UPM Serdang, Selangor, Malaysia
[3]Johor Biotechnology and Biodiversity Corporation (J-Biotech), 81300 Skudai, Johor Bahru, Malaysia

*Corresponding author: khairulm@frim.gov.my

Abstract. The depletion of our tropical timber supply has encouraged the Malaysian wood-based industry to seek alternative resources such as fast-growing plantation trees and crops for wood-based products. Oil palm tree (*Elaeis guinensis*) has been identified as one of the potential alternative raw materials for the production of value-added products such as plywood, laminated veneer lumber (LVL) and even solid oil palm lumber. As we know, the properties of oil palm lumber or veneer is inferior to those of wood. Modification and treatment to the oil palm are needed to enhance the quality and dimensional stability. The main objective of this study is to determine the physical properties of LVLs made from oil palm veneers (OPVs) treated with low molecular weight phenol formaldehyde (LMWPF) resin with different resin solid content and different initial moisture content (MC) of veneer. OPVs were soaked into LMWPF and re-dried to semi gel condition before being hot-pressed. There was no additional resin was added to bond the veneers together into LVL. The physical properties such as density, moisture content, thickness swelling, water absorption and weight percent gain were determined in this study. The results showed that the LVLs made by treated OPVs exhibited better physical properties compared to LVLs made from untreated OPVs but still inferior with rubberwood LVL. It can be concluded that treating OPVs prior to their usage for LVL-making is essential and effective to produce LVLs with convincing physical properties.

1. Introduction

Oil palm veneer (OPV) is rapidly used for plywood manufacturing in Malaysia since years back and being used as core veneers for plywood production. This is due to the finding that oil palm trunk (OPT) can be transformed into veneer and can be used to produce high performance as engineered board such as plywood and laminated veneer lumber (LVL). The intention was to reduce usage of tropical hardwood wood veneer in the production. Recently, wood veneers were used only for the outer part while OPV were used as core of the board. Although, OPV can be used as material to produce high quality boards, it was found that the LVLs produced from OPVs exhibited inferior physical performance in comparison with those made of rubberwood [1]. Impregnating resins into the veneer was found great way to enhance its dimensional stability and strength. It had been confirmed by many researchers before [2]. According to [3], pre-treatment of veneer with phenolic or equivalent resin polymer can enhance the quality of OPV as well as final product such as plywood or LVL. The price of the final product can be reduced as the OPV is cheaper than normal hardwood veneer. This technology will also reduce amount of hardwood veneer used in the product because OPV has greater thickness that hardwood veneer.

Based on the research by [1], it was discovered that impregnating resins into veneers was a good way to enhance the LVL dimensional stability and their strength. Thus, this project was carried out to determine the effectiveness of low molecular weight phenol formaldehyde (LMWPF) resin in increasing the strength and dimensional stability of LVL boards made of OPV. Quality improvement studies on solid oil palm lumber revealed that initial moisture content of the material and the solid content of the resin used were significantly affected by the physical and mechanical properties [4], [5], durability [5], and machining properties of the treated oil palm lumber [6]. The veneers, with different initial MCs were dipped into LMWPF with different rate of solid content, to allow the resin penetration into the veneers before the veneers were hot-pressed to produce LVL boards. This research can help to promote the usage of OPT as an alternative material for the production of LVL. Consequently, it will reduce the usage of tropical hardwood and help to achieve more sustainable forest management.

2. Materials and methods

In this study, OPVs s were used as main raw material. The OPVs with a size of 127 cm (width) x 127 cm (length) were obtained from Kang Sem Enterprise, Johor. The average thickness of the OPVs was 4.5 mm (wet veneer after peeling) and only the outer/superior part of the OPVs was chosen for this study. Superior part was chosen to diversify this research as inferior part has been used in many previous researches. Four different target moisture content (MC) were used i.e. 100, 70, 30 and 15%. The veneers were trimmed to size 40 cm x 40 cm each before they were dried in a conventional oven separately to the four different targeted moisture content (MCs). All veneers were weighed by using digital balance before and after they were dried. The MC of the veneers was calculated by using reverse MC formula. The monitoring of the MC commenced until the targeted MC was recorded.

LMWPF resin was used as a treatment chemical and also as a binder. The resin impregnated into the veneers was left to semi gel/cure condition before being hot-pressed to form LVL boards. Two different resin solid contents of 30 and 45% solution were prepared. The solid content of the resin was reduced by diluting the resin with distilled water. The dried veneers, then, were soaked into the LMWPF for ten (10) minutes before they were pre-pressed to remove excessive resin. The soaked veneers were then dried by using conventional oven and were stacked with wood sticker to make sure each veneer was dried evenly until targeted MC (10%) was reached. The LMWPF resin in veneers was left to semi cured condition. Then, the veneers were arranged in parallel grain direction before they were being hot pressed with standard temperature and pressure to produce LVL boards. Five layers of OPV were used for the LVL fabrication process. The LVLs were trimmed and cut according to standard sizes before they were utilized for physical tests such as density and MC, thickness swelling (TS), water absorption (WA) and weight percent gain (WPG). Moisture content was determined in accordance with standard method for direct moisture content measurement of wood and wood-based materials, D4442 [7], water absorption and thickness swelling was conducted based on European Standard EN 317:1993 [8]. The WPG was conducted using a method described by [9] and [10].

3. Results and discussion

3.1 Weight Percent Gain (WPG)

Weight percent gain is to determine the amount of resin uptake by the OPVs after they were immersed into LMWPF resin. Table 1 shows highly significant difference of results between veneers at 15% MC (WPG value was 34.48% resin before re-dry) as compared veneer at 70% MC, where the WPG value was 11.97%. It shows that, the drier the samples, the more efficient they would be able to absorb the resin. It is because in dried samples, water in cell already removed and resin can easily filled up the empty space in cell. This is important to facilitate the resin to penetrate deeper and improve the physical and mechanical properties of the samples. From the table, it showed that both LMWPF SC gave similar result patterns. The lower OPV MC, the higher WPG value uptake was recorded. In addition, a comparison has been made using different LMWPF SC and the results showed that there was no significant difference of results between samples treated using 30 and 45% LMWPF SC. This finding was similar to [11], they was found that OPV treated with 41% SC of LMWPF was 32.5% WPG and a similar finding was reported by [12] using Japanese cedar (*Cryptomeria japonica*). These achievement of the highest level of WPG was attribute to the short chain of LMWPF resin, so that the resin could easily penetrate the parenchyma cells.

Table 1. Cross tabulation between LMWPF SC and initial OPV MC.

Properties	**LMWPF SC (%)**	**Initial OPV MC (%)**			
		100	**70**	**30**	**15**
Weight Percent Gain, WPG (%)	45	7.89^{c} (3.10)	11.97^{c} (9.55)	21.40^{b} (4.44)	34.48^{a} (5.08)
	30	11.51^{c} (7.51)	16.52^{c} (8.41)	24.70^{b} (3.97)	34.25^{a} (4.69)

Notes:
Values in parentheses are standard deviation
Means in the same row followed by the same letter are not significantly different at $p \leq 0.05$

3.2 Density and Moisture Content (MC)

MC determination of raw/fresh OPV is important to make sure the veneers were dried sufficiently and met the targeted MC. It is also important to calculate the oven dry weight of veneer using reversible MC formula. The study showed that the MC of the outer OPVs were in the range of 181.6 to 262.2% and the results were in line with [13], [14] and [15] who stated that the MC for oil palm was in the range of 76.5% up to more than 500%. Table 2 shows density and MC of OPVs treated using 30 and 45% of LMWPF SC at 100, 70, 30 and 15% initial veneer MC. From the result, it shows that there was no significant difference for density and MC for each LVL board using different initial MC of OPV for 45% LMWPF SC. But, significant difference ($p \leq 0.05$) was noted for density that using 30% LMWPF SC at 100 and 70% initial MC of OPV LVL. No significant difference was observed when 30 and 15% initial MC of OPV were used.

Table 2. Cross tabulation between LMWPF SC and initial OPV MC.

Properties	**LMWPF SC (%)**	**Initial OPV MC (%)**			
		100	**70**	**30**	**15**
Density (Kg/m^3)	45	532.01^{a} (12.76)	550.48^{a} (40.20)	517.44^{a} (37.71)	536.82^{a} (57.23)

Moisture Content (%)		7.89^{a} (0.28)	7.97^{a} (0.75)	8.01^{a} (0.89)	8.28^{a} (0.30)
Density (Kg/m^3)	30	473.11^{c} (19.09)	532.22^{b} (18.36)	571.83^{a} (46.38)	599.42^{a} (39.36)
Moisture Content (%)		7.92^{a} (0.48)	8.12^{a} (0.34)	8.08^{a} (0.34)	8.27^{a} (0.98)

Notes:
Values in parentheses are standard deviation
Means in the same row followed by the same letter are not significantly different at $p \leq 0.05$

Meanwhile, there was significant difference between 30 and 45% LMWPF SC when treated at 100, 30 and 15% of initial MC veneer before treatment as shown in Figure 1 and 2. But, when treated at 70% of initial MC of OPV, there was no significant difference observed when using either 30 or 45% LMWPF SC. Although, there was significant difference for 100, 30 and 15% initial MC by using 30 and 45 LMWPF SC, the mean density was still acceptable because the result almost similar to that of [16], [1] and [17]. On the other hand, the MC for all boards presented no significant difference because it was set at 10% before hot press.

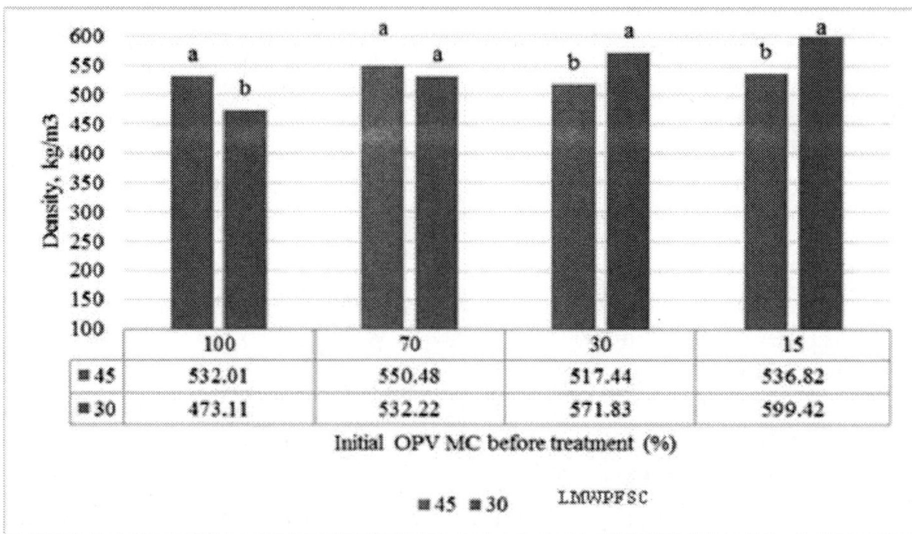

Note: Means in the same letter are not significantly different at $p \leq 0.05$

Figure 1. Bar chart of effect initial OPV MC before treatment with LMWPF SC for density.

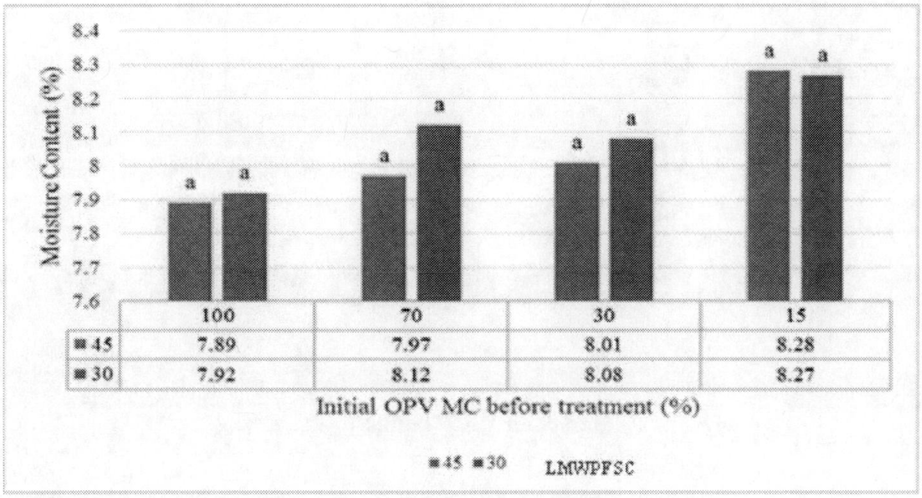

Note: Means in the same letter are not significantly different at p≤0.05

Figure 2. Bar chart of effect initial OPV MC before treatment with LMWPF SC for moisture content.

3.3 Thickness Swelling

The thickness swelling and water absorption were greatly influenced by the value of board density [1]. The higher the density, the lower the percentage of thickness swelling and water absorption. From Table 3, it shows that there was no significant difference when using 45% LMWPF SC at 70 and 30% initial OPV MC. Significant difference was noted at 15% MC compared to others. It showed that OPV MC at 15% has lowest thickness swelling results (0.76%). On the other hand, when 30% LMWPF SC was used, different results were obtained which there was no significant difference when treated at 70, 30 and 15% initial MC of OPV. These results were better than LVLs made from untreated OPV which is 2.99% as stated by [1]. In the other hand, the results of thickness swelling when treated using 30 and 45% LMWPF SC at 70, 30 and 15% initial OPV MC shows that there was no significant difference between 30 and 45% LMWPF SC when treated at 30 and 15% initial MC of OPV as shown in Figure 3. Although, there was no significant difference, samples treated with 45% LMWPF SC at 15% initial MC gave lesser thickness swelling value than the other samples. It was also observed that there was significant difference when the samples were treated at 70% initial MC of OPV.

Table 3. Cross tabulation between LMWPF SC and initial OPV MC.

Properties	LMWPF SC (%)	Initial OPV MC (%)			
		100	70	30	15
Thickness Swelling (%)	45	1.80^{a} (0.82)	2.19^{a} (1.01)	1.68^{a} (0.68)	0.76^{b} (0.59)
	30	5.24^{a} (1.72)	1.15^{b} (0.80)	1.20^{b} (0.72)	1.53^{b} (1.02)

Notes:
Values in parentheses are standard deviation
Means in the same row followed by the same letter are not significantly different at p≤0.05

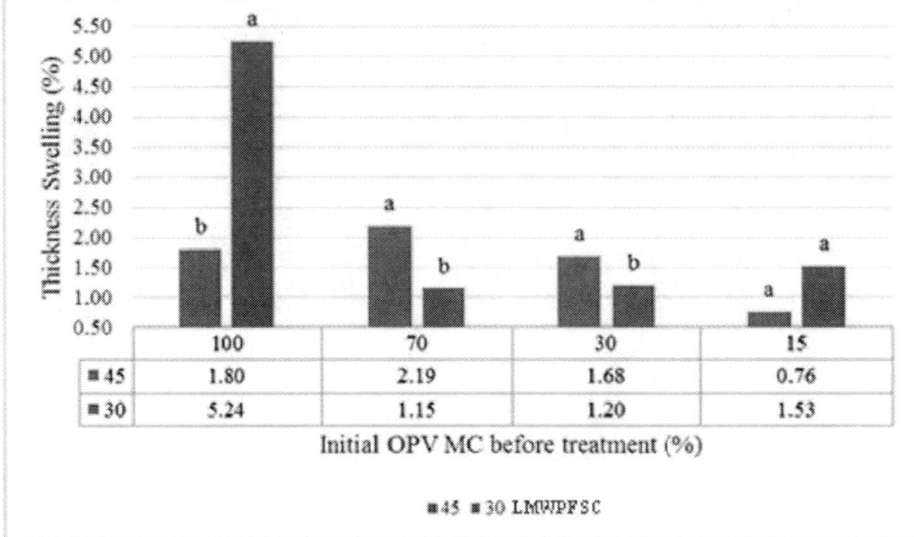

Note: Means in the same letter are not significantly different at $p \leq 0.05$

Figure 3. Bar chart of effect initial OPV MC before treatment with LMWPF SC for thickness swelling.

3.4 *Water Absorption*

Understanding water absorption in a great detail will help us to improve the dimensional stability of a wood-based product. It was expected that good dimensional stability contribute to good mechanical results. The higher the water absorption value, the weaker the wood-based product could be. From the Table 4, it showed, there was no significant difference when using 45% LMWPF SC treated at all initial OPV MC, but when using 15% initial MC, the water absorption value was lower than others. Samples treated with 30% LMWPF SC, also showed no significant difference at 70 and 15% initial MC but significant difference was recorded when the treatment was done at the MCs of 30% and (70, 15%). By looking at the results, it can be concluded that samples treated using 30% LMWPF SC at 30% initial MC gave the best results because the water absorption value was 17.60% which was the lowest in comparison with the others. According to [1], water absorption in LVL from untreated OPV is 63.03%, which was higher than treated OPV with LMWPF in this study. Figure 4 shows the results of water absorption for the samples treated using 30 and 45% LMWPF SC at 70, 30 and 15% MC before treatment are show that significant difference of results were recorded from the samples treated using 30% LMWPF SC and treated at 30% initial OPV MC. The treatment using that parameter was found effective to reduce water absorption compared to the other parameters. The reduction of water absorption was believed caused by the density of the samples.

Table 4. Cross tabulation between LMWPF SC and initial OPV MC.

Properties	LMWPF SC (%)	Initial OPV MC (%)			
		100	70	30	15
Water Absorption (%)	45	33.32^{a} (8.74)	35.49^{a} (15.87)	32.26^{a} (12.90)	24.40^{a} (9.54)
	30	88.29^{a} (13.85)	33.07^{b} (6.39)	17.60^{c} (13.26)	28.66^{b} (8.90)

Notes:
Values in parentheses are standard deviation
Means in the same column followed by the same letter are not significantly different at $p \leq 0.05$

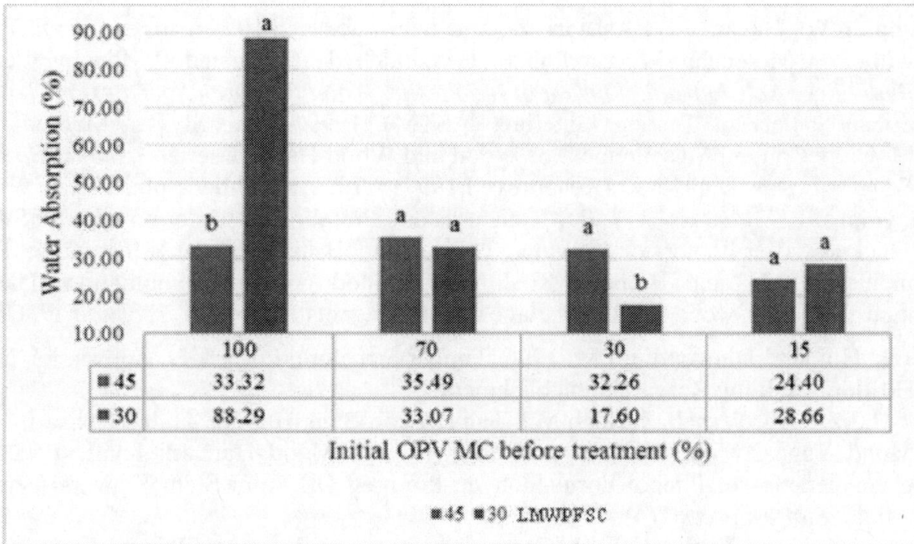

Note: Means in the same letter are not significantly different at $p \leq 0.05$

Figure 4. Bar chart of effect initial OPV MC before treatment with LMWPF SC for water absorption.

4. Conclusion

From this study, the density of oil palm LVL was almost similar to the untreated LVL. Although the density almost equal (LVL using outer veneer), the physical properties of LVLs using treated veneer with LMWPF shows better results than untreated LVL veneer. To get better properties of LVL board from OPV, the veneer need to do treatment especially using 45% LMWPF SC at 15% MC of OPV and re-dried to semi gel condition at 10% MC before the commencement of hot-pressing process.
As proven in this study, physical properties of oil palm LVLs was better than untreated oil palm LVLs.

5. References

[1] Abdul Razak, W., Hashim W.S., Azmy M. and Othman S. 2008. Utilization Potential of 30-Year Old Oil Palm Trunks Laminated Veneers for Non-Structural Purposes, *Journal of Sustainable Development*, Vol. **1(3)**: 109-113

[2] Loh, Y.F., Paridah, M. T., Hoong, Y. B. and Adrian, C. C. Y. 2011. Effects of treatment with low molecular weight phenol formaldehyde resin treatment on the surface characteristic of oil palm (*Elais quineensis*) stem veneer. *Material and Design* **32(4)**, 2277-2283.

[3] Paridah, M. T., Hoong, Y. B., Noryuziah, M. Y., Loh, Y. F. and Hashim W. S. 2014. Properties enhancement of oil palm plywood. *Handbook of oil palm trunk plywood manufacturing*: pp 89-108.

[4] Bakar, E.S., Hadi, Y.S. and Sunardi, I. 2001. Quality Improvement of Oil-Palm Wood: Impregnated with phenolic resin. *Indonesian Journal of Forest Products and Technology.* Vol. **XIV (2)**: 26-31.

[5] Bakar, E.S., Jun., H., Zaidon, A., Choo, A.C.Y. 2013. Durability of Phenolic Resin Treated Oil Palm Wood against Subterranean Termites and White Rot Fungi. *International Biodeterioration & Biodegradation* **85**: 126-130.

[6] Chong, Y.W., Bakar, E.S., Ashaari, Z., and Sahri, M.H. 2010. Treatment of Oil Palm Wood with Low-Molecular Weight Phenol Formaldehyde Resin and Its Planing Performance. *Wood Research Journal: Journal of Indonesian Wood Research Society* **1(1)**: 7-12.

[7] American Society of Testing Materials (ASTM). 1999. Standard Test Methods for Direct Moisture Content Measurement of Wood and Wood-Base Materials. *Annual Book of ASTM Standards Des.* D 4442-92 (Reapproved 1997). Vol. **4.10**. Philadelphia. PA.

[8] Anonymous. 1993. BS EN 317. Particleboards and Fibreboards. Determination of Swelling and Thickness after Immersion in Water. Test method. British Standards Institution (BSI).

[9] Schneider, M. H. and Brebner, K. I. 1985. Wood polymer combination: The chemical modification of wood by alkoxysilane coupling agents, *Wood Sci. Technol.* **19**, 67–73.

[10] Desch, H.E. and Dinwoodie, J.M. 1996. Timber Structure, Properties, Conversion and Use. **7th Edition**. Building Research Establishment.

[11] Nor Hafizah Ab. Wahab, Paridah Md. Tahir, Yeoh Beng Hoong, Zaidon Ashaari, Nor Yuziah Mohd Yunos, Mohd Khairun Anwar Uyup and Mohd Hamami Shari. 2012. Adhesion Characteristics of Phenol Formaldehyde Pre-preg Oil Palm Stem Veneers. Peer-reviewed article. *BioResources* 7(4), 4545-4562.

[12] Shams, M. I. and Yano, H. 2011. Comprehensive Deformation of Phenol Formaldehyde (PF) resin-impregnated Wood Related to the Molecular Weight of Resin. *Wood Science Technology* **45**, 73-81.

[13] Killmann, W. & Lim, S.C. 1985. Anatomy and properties of oil palm stem. *Proceeding of the national symposium of oil palm by-product for agro-based industries.* PORIM bulletin. **11**:18-42.

[14] Gan, K.S., Choo, K.T. and Lim, S.C. 2001. Basic density and moisture content distributions in 30-year-old oil palm (Elaeis guineensis) stems. *Journal of Tropical Forest Products* **7 (2)**: 184-191.

[15] Bakar, E. S., Mohd Hamami, S. and Paik, S. H. (2008). Anatomical characteristics and utilization of oil palm wood, In: Nobuchi, T. and Mohd Hamami, S. (eds.). *The Formation of Wood in Tropical Forest Trees, A Challenge from the Perspective of Functional Wood Anatomy*, Universiti Putra Malaysia, Selangor, 161-180.

[16] Abdul Hamid Saleh. 2005. Strength of Joints and Physical Performance of Chairs Made From LVL Oil Palm. *Master thesis*. University Teknologi MARA.

[17] Bakar, E.S., Tahir, P.M., Sahri, M.H., Mohd Noor, M.S., Zulkifli, F.F. 2013. Properties of Resin Impregnated Oil Palm Wood (*Elaeis guineensis* Jack). *Pertanika J. Trop. Agric. Sci.* **36(S)**: 93-100.

IOP Conf. Series: Materials Science and Engineering **368** (2018) 012026 doi:10.1088/1757-899X/368/1/012026

Properties of water-borne coating incorporate with nanoclay

U M K Anwar[1], S Y Alia[2], K Tumirah[1], H Hamdan[1] and M T Paridah[2]

[1] Forest Products Division, Forest Research Institute Malaysia, 52109, Kepong, Selangor.
[2] Institute of Tropical Forestry and Forest Products, Universiti Putra Malaysia, 43400 Serdang, Selangor

*Corresponding author: mkanwar@frim.gov.my

Abstract. Water-borne coatings, as well as other coating options, are having advantages and disadvantages besides having different mechanical properties from those of the conventional solvent based coatings. A nanoparticle such as nanoclay (NC) has found its applicability in coatings to meet the mechanical requirements. In this study, a special focus has been given in order to improve the properties of water-borne coating using NC. The optimum percentage of NC mixed into water-borne coating was determined. The samples were prepared by dispersing 2 % of NC in water-borne coating. Both of it were mixed by using homogenizer for a certain duration using high speed rotation. The dispersion process was then repeated for 4 %, 6 %, 8 % and 10 % of NC loading. Whilst water-based coating without NC was used as control for comparison purposed. Effect of using NC on the mechanical properties of coatings on wood such as adhesion (ISO 2409:2007 (E) (BS EN 2007)), impact test (BS3962:Part6 1980 (BS 1980) scratch resistant (ISO 1518-1:2011 (BS EN 2011)) and abrasion (ASTM D 4060-14) were investigated. For adhesion properties, no detectable failure was observed for nanocoating with 2 %, 4 % and 6 % of NC. However, NC 8 % and NC 10 % showed a little detachment, indicating poorer adhesion quality for these samples. There are significant differences for impact and abrasion tests but insignificant for scratch test. It was found that the increasing amount of NC percentage would improve the mechanical properties of water based coating until a certain limit.

1. Introduction

Coating is the process whereby a thin layer of material is applied on a substrate. It is primarily used for surface protection and provides a defense against chemical, mechanical and environmental threats by preventing water, oil or air penetration, chemical corrosion, UV penetration and electrical conduction. Water-based coatings was known as sorts of coating which have been around for decades but have demonstrated greatly improved performance characteristics in recent years to meet the high quality standards demanded by customers worldwide [1]. However, there are some issues on properties of water-based coating such as hardness, heat resistance, water and solvent resistance [2]. Thus, attempts have been made to enhance the mechanical properties of water-based coating by incorporating nanoparticles in the water-based coating [3].

Nanoparticles such as nanoclay found applicability in coatings to meet mechanical requirements and improvement in thermal stability of coatings [4]. Researchers have showed that the addition of nanoclay could provide excellent mechanical properties of nanocoatings due to their high surface area to volume ratio [5-6]. The objectives of this study are to determine the effect of nanoclay on adhesion and mechanical properties of water-based coating.

2. Materials and methods

Rubber wood timber was purchased from local company located in Selangor. Then it was cut and plane into panels. The panels were cut to size of 400 mm x 500 mm, having a moisture content of 12.

2.1 Preparation of nanocoatings

The water-based coatings, which was waterborne polyurethane contained 30 % clear finish, was purchased from local supplier located in Selangor, Malaysia. The typical characteristics of the coating are shown in Table 1. Nanoclay that have been chosen for this analysis contained octadecylamine and 3-aminopropyltriethoxysilane.

Table 1. Coating properties.

Characteristics	Properties
Appearance	Translucent liquid
Viscosity	Approximate 30-32 sec at 25°C
Solid content	33±2 %
Dilution	No required

The samples were prepared by dispersing 2 % v/v of nanoclay in 50 ml of water based coating. The solutions were mix by homogenization for 20 minutes at 10, 000 rpm. The dispersion process was then repeated for 4 %, 6 %,8 % of 10 % of nanoclay loading. The samples were compared with control (water-based coating) and referred to as NC 2%, NC 4%, NC 6%, NC 8% and NC 10%. Whilst water-based coating without nanoclay was used as control for comparison purposed. After finishing, the panels were kept at ambient room temperature in the laboratory.

2.2 Preparation of wood samples

The rubber wood panels were sanded with sandpaper with grit number of 80 followed by grit number of 360 along the grain to obtain a smooth surface. Panels were conditioned in a conditioning room at 22 ± 2 °C and relative humidity of 65 ± 5 % for one week. The dimensions of the panels were cut based on standard requirements of the test, in which the size for adhesion, impact and scratch test were 140 mm x 80 mm x 5 mm while for abrasion test was 100 mm x 100 mm x 5 mm. Each panel was coated for 3 layers using 75 µm wire bar coater. Three replicates were prepared for each test. After finishing, the panels were ambient at room temperature in the laboratory.

2.3 Adhesion test (cross-cut)

The adhesion test on the substrate was carried out using cross-cut tape method based on ISO 2409:2013 [7]. A single blade cutter was used to make lattice pattern. The number of cuts for each lattice pattern is six. The spacing between each cut at both directions is 2 mm. An adhesive tape was placed over the lattice area and removed after 5 minutes. The cut area percentage was examined.

2.4 Impact test (Resistance to mechanical damage)

The relative resistance of finishes was carried out based on BS3962:Part6 1980 [8]. whereby a steel ball was dropped from 2 m height on to the test panel. The area was examined and rated based on the assessment code stipulated in the standard.

2.5 *Scratch test*
The scratch test was carried out based on ISO 1518-1:2011 [9]. The finished panel was clamped onto the panel holder. A weight with a minimum load of 1000 g load was placed on the stylus. The load weight was increased by 100 g until a maximum load of 2000 g.

2.6 *Abrasion test*
The abrasion test was done on the coated rubber wood panels. The surface of the coated panels is abraded by rotating the panel under weighted abrasive wheels. Abrasion resistance is calculated as loss in weight at 1000 and 2000 of abrasion cycles based on ASTM D 4060-14 2014 [10].

3. Results and discussion

3.1 *Finishing properties of nanocoating*
Finishing properties of controlled and nanocoating were tabulated in Table 2. It showed that the adhesion and mechanical properties of coating with and without nanoclay. Coating adhesion can be evaluated in terms of the force at which the coating layers detached from the substrate surface or serious destruction occurs at the coated layer. No detectable failure was observed for nanocoating with 2 %, 4 % and 6 % of nanoclay loading. Thus, coating layer resided all along the indentation groove without any detachment or destruction. However, NC 8 % and NC 10 % showed a little detachment, showing poorer adhesion quality for these samples due to the agglomeration of the nanoclay. When the amount of nanoclay was increased with the same speed and time of homogenization, there was some nanoclay that was not well dispersed in the coating.

Table 2. Adhesion and mechanical properties of nanocoating.

	Adhesion	Impact	Scratch (N)	Abrasion (mg) at 2000 cycles
Control	0^a	2^c	19.70	274
NC 2%	0^a	3^d	19.70	241
NC 4%	0^a	3^d	19.70	290
NC 6%	0^a	3^d	19.70	294
NC 8%	1^b	2^c	19.70	396
NC 10%	1^b	2^c	19.70	561

[a]Notes are referenced for none of the lattice is detached.
[b]Detachment is <5 %.
[c]Slight flaking.
[d]Moderate cracking.

For impact test, the results show significant increase in resistance to mechanical damage for NC 2 %, NC 4% and NC 6 %. However, at higher percent of nanoclay, the impact strength decreased at NC 8 % and NC 10 %. However at higher percentage, reason for drop in impact strength is not yet addressed [11]. In this case, the decrement of impact strength was related to the not well dispersed of the nanoclay. There was some part of the panels was not smoothly covered with the nanocoating because of the nanoclay agglomeration.

It can be seen that scratch test results show for all samples of the coating with and without nanoclay can resist load maximum to 19.7 N. Increase of applied load being almost independent with nanoclay content. However, Figure 1 reveals a different penetration depth affected from the scratch test. There is a significant increase of penetration depth happens at NC 10 % compared to the NC 2 %. High of penetration depth as shown in Figure 4b was attributed with the bad adhesion properties. Hence, the penetration depth was highly related with the adhesion of the nanocoating.

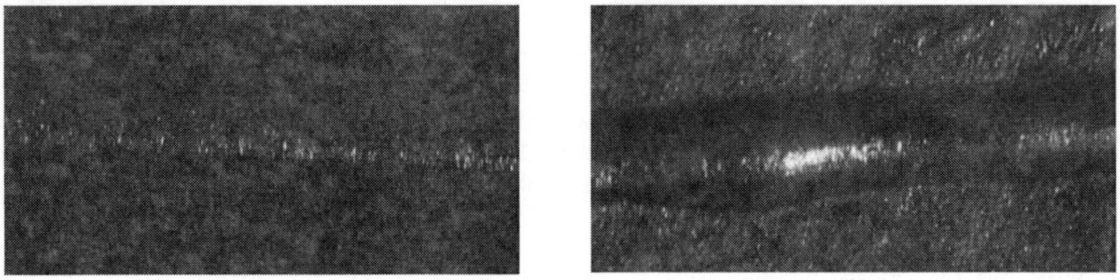

(a) (b)

Figure 1. Penetration depth of scratch on (a) NC 2% and (b) NC 10% under 18x magnification

Abrasion test was done up to 2000 cycles and the result obtained shows that there is a significant different of weight loss between control and NC 2 %. It reveals the addition of nanoclay in the coating can improve the abrasion resistant. However, the results give a bad effect while the weight loss increases after 2 % nanoclay. Thus, the scratch resistance decreases as the amount of nanoclay increase. The weight lost for control and nanocoating samples were 274 g and 241 g respectively. It reveals that by additional of nanoclay in the water-based coating can improve the abrasion resistant of this material. While having the optimum amount of nanoclay in water-based coating, a good abrasion resistance with smooth surface panels can be achived.

4. Conclusion

In the present study, it can be concluded that the addition of nanoclay into water-based coating resulted in an increase in adhesion and mechanical properties. The optimum percentage of nanoclay was found to be 2%, after which there was a decrease in properties due to the particle agglomeration due to the particle that was not well dispersed in the coating. The excessive amount of nanoclay was significantly affecting the adhesion, impact and abrasion resistance while it does not affect the scratch resistance. However, the penetration depth of the coating shows a significant different between the present of the nanoclay in the coating and coating without nanoclay.

References

[1] Shannon D 2001 *Water-borne Coatings* Jamestown Paint & Varnish Company, Jamestown, Pennsylvania pp 121-162
[2] Jhon, Cheong Y K, Kim I W, J H 2001 Advances in Urethane Science and Technology, Technom. Westport pp 121-162
[3] Katsikis N, Zahradnik F, Helmschrott A, Munstedt H and Vital A 2007 Thermal stability of poly(methylmethacrylate)/silica nano- and microcomposites as investigated by dynamic-mechanical experiments *Polym. Degrad. Stabil* **92** 1966-1976.
[4] Bertini F, Canetti M, Audisio G, Costa G and Falqui L 2006 Characterization and thermal degradation of polypropylene–montmorillonite nanocomposites *Polym Degrad Stab* **91** 600–5

[5] Haojie Y, Wang L, Shi Q, Jiang S and Jiang G 2006 Preparation of epoxy resin/$CaCO_3$ nanocomposites and performance of resultant powder coatings *J Appl Polym Sci* **101** 2656

[6] Bhanvase B and Sonawane S 2010 New approach for simultaneous enhancement of anticorrosive and mechanical properties of coatings:application of water repellant nano CaCO3–PANI emulsion nanocomposite in alkyd resin *Chem Eng J* **156** 177

[7] BS EN ISO 2409 2013. *Paints and varnishes – Cross-cut test.* BSI Standard Limited. London

[8] BS 3962: Part 6 1980. *Finishes for wooden furniture – Assessment of surface to mechanical damage*. BSI Standard Limited. London

[9] BS EN ISO 1518-1 2011. *Paints and varnishes. Determination of scratch resistance. Constant-loading method*. BSI Standard Limited. London

[10] ASTM D 4060-14 2014. *Standard test method for abrasion resistance of organic coatings by the taber abraser*. ASTM International, West Conshohocken, PA

[11] Svoboda P, Zeng C, Wang H, James L, David L and Tomasko L 2002 Journal of Applied Polymer Science **85** 1562-70

Biochemical methane potential (BMP) from anaerobic co-digestion of sewage sludge and decanter cake

N Khairul Anuar[1*], H Che Man[1], S Idrus[2], N N Nik Daud[2]

[1]Department of Biological and Agricultural Engineering, Faculty of Engineering, Universiti Putra Malaysia, 43400 UPM Serdang, Selangor, Malaysia
[2]Department of Civil Engineering, Faculty of Engineering, Universiti Putra Malaysia, 43400 UPM Serdang, Selangor, Malaysia

*Corresponding author: shafiqah.kn@gmail.com

Abstract. Nowadays, anaerobic co-digestion has become a focus of interest in generating methane gas for electricity supply. In this study, a biochemical methane potential (BMP) test was conducted to determine biogas production potential from anaerobic co-digestion of sewage sludge and decanter cake. Sewage sludge has high organic content and decanter cake has a high carbon content (42.70%). Batch fermentation were performed in 125mL serum bottle with 100 mL working volume and incubated for 10 days at mesophilic temperature (38°C) in a water bath. Each bottle containing sewage sludge and decanter cake at feedstocks to inoculums (F/I) ratios of 1:0, 1:1 and 2:1. Daily biogas productions were collected using water displacement method. The highest cumulative biogas produces 0.25 L per day at the F/I ratio of 1:1. However, it was observed that the anaerobic co-digestion of sewage sludge with decanter cake at F/I ratio 2:1 provides higher methane yield production than that provided by the sewage sludge alone (1:0) and F/I ratio of 1:1. The results obtained revealed that the co-digestion of sewage sludge and decanter cake has a potential to produce methane.

1. Introduction

Anaerobic co-digestion is a conversion or degradation process where microorganism breakdown multiple organic substrates/wastes in the absence of oxygen in the digester [1-2]. A wide range of wastewater types, solid waste and biomass have been used as the organic sources for co-digestion such as agricultural waste, agro-industrial wastes, animal wastes and industrial wastes [3-4]. The end product of anaerobic co-digestion is biogas; a mixture of methane, carbon dioxide and others. The typical biogas composition in an anaerobic co-digestion consists of 55-80% of methane, 20-45% of carbon dioxide and some other gases [5-6]. The biogas produced can be used for electricity generation, cooking, pipeline injection and heating. It is a valuable renewable energy source while it can reduce the greenhouse gas emissions from fossil fuel [7]. The benefits of anaerobic co-digestion include improving the nutrient balance, increasing digestion rate, increasing load of biodegradable organic matter and producing better biogas yield [8]. The anaerobic co-digestion also may improve the biogas production due to carbon, nitrogen and nutrient balance [9].

IOP Conf. Series: Materials Science and Engineering **368** (2018) 012027 doi:10.1088/1757-899X/368/1/012027

Recently, researches on decanter cake as a substrate in anaerobic co-digestion has been reported due to its high biodegradability [7]. Decanter cake is one type of waste produced from palm oil industry in the of oil purification [10]. Moreover, the decanter cake produced in each palm oil mill is sufficient enough to make a biogas plant in the mill cost-effective. In the palm oil mill, every ton of fresh fruit bunch (FFB) generates about 0.6 to 0.8 m^3 of palm oil mill effluent (POME), with an accumulation of 22 to 23% of empty fruit bunch (EFB), 3.5% of decanter cake (DC) and 13.5% of palm mesocarp fiber (PMF). Previous work has reported that decanter cake is suitable for animal feed [11], fertilizer and composting material [12], due to its high nutrient content. The pyrolyzed decanter cake also potential to be used as an adsorbent for the removal of metal ions from wastewater [13].

Anaerobic co-digestion between decanter cake with the POME gives practical use for the development of anaerobic process that providing hydrogen and methane co-production from palm oil residues [14]. The co-digestion between decanter cake and seafood wastewater also has a stimulatory effect on methane gas productions in BMP test [7].

Sludge is a precipitation yielded from wastewater treatment, sewage and excretion, and very difficult to process due to a high degree of organic constituents and moisture content. Sludge can be separated into organic and inorganic sludge, and organic sludge can be highly reduced by incineration [15]. Currently, the management of sewage sludge produced from wastewater treatment plants is one of the most serious challenges in biological wastewater treatment. Its amount is increasing worldwide as the quantity of wastewater being treated rises [16].

Sewage sludge is a valuable feedstock for anaerobic digestion as it contains a high percentage of nitrogen, which can provide a buffering capacity and an improved balance of nutrients [8]. Although convenient and feasible, it has been recognized that the individual digestion of sewage sludge did not achieve the most efficient production of biogas due to its inherent deficiency of carbon such as low carbon/nitrogen (C/N) ratio. In contrast, agro-wastes contain a high percentage of carbon and a low percentage of nitrogen (high C/N ratio). Different types of agro-wastes can be used in admixture with sewage sludge, such as rice straw [17], wheat straw [15] and municipal solid wastes [8]. Co-digestion of sewage sludge and decanter cake is considered the best solution for enhancing methane yield.

Biochemical methane potential (BMP) test is widely used to determine the anaerobic degradability of the wastes. It is also used to determine the quantity of methane that the waste can potentially produce in anaerobic condition [7]. BMP test is low cost and repeatable method for relative comparisons of the anaerobic digestibility and potential biogas production among various substrates. The BMP determine the amount of organic carbon in a given material that can be anaerobically converted to methane. The test is carried out by digesting nutritive materials and microorganisms into serum bottle under an optimal condition such as temperature and pH to identify the gas production amount and composition change. It was reported that it is easy to measure methane yield from organic waste and reliably compare to potential methane yield between different samples [13, 18].

Therefore, the aim of this study is to evaluate the biogas and methane potential of anaerobic co-digestion from sewage sludge and decanter cake using BMP test. The methane yield was measured at different ratio of inoculum to substrate. The biogas composition was also analyzed daily to determine the biogas produced in anaerobic co-digestion of sewage sludge and decanter cake.

2. Materials and Methods

2.1 Substrate and inoculum

Sewage sludge was obtained from wastewater treatment plant of Indah Water Konsortium. The sludge was sieved using a 1mm screen. The decanter cake was collected from Seri Ulu Langat Palm Oil Mill, Dengkil, Selangor. The sewage sludge and decanter cake were stored at 4°C for later use as the inoculum and substrate, respectively. The substrate and inoculum were characterized based on total solid (TS), volatile solid (VS), pH, carbon content (C) and nitrogen content (N). All analytical procedures were performed in accordance with standard method (APHA, 1998).

IOP Conf. Series: Materials Science and Engineering **368** (2018) 012027 doi:10.1088/1757-899X/368/1/012027

2.2 Experimental setup

The BMP test was conducted in 125mL serum bottles enclosed with rubber stoppers at working volume of 100mL (Figure 1). The samples were flushed with nitrogen gas for 2 min before seal to remove traces of oxygen and to ensure anaerobic condition. The cumulative biogas production was measured by water displacement method [9]. Batch fermentation was conducted in anaerobic mesophillic at 38±1°C temperature in a water bath (Memmert Waterbath WNB 45) for 10 days. The digesters containing sewage sludge and decanter cake were mixed at feedstocks to inoculums (F/I) ratios of 1:0, 1:1 and 2:1. The sewage sludge alone at F/I ratio 1:0 were used as control and incubated in the same water bath. The pH of samples were adjusted to 7±1.0 using sodium hydroxide (NaOH) to provide better growth conditions for methanogenic bacteria can produce biogas and methane effectively at a pH value from 6.5 to 8.0 [19]. The methane yield was expressed as the volume of methane produced based on the initial total VS of the feedstock [8]

2.3 Biogas measurement and analysis

The biogas composition was measured by gas chromatography (GC) (Agilent 6890) with Thermal Conductivity Detector (TCD). The quantity of hydrogen (H_2), methane (CH_4) and carbon dioxide (CO_2) in 1 mL of gas samples were calculated using Equation 1, 2 and 3 [17].

H_2 area / 3844.10 = qH_2 quantity in 1mL gas sample, (1)
CH_4 area / 2291.67 = qCH_4 quantity in 1mL gas sample, (2)
CO_2 area / 3667.21 = qCO_2 quantity in 1mL gas sample, (3)

In the next step, percentage of each individual gas in biogas was calculated using Equation 4, 5 and 6 [17]. Using the calculated percentage and the amount of accumulated biogas in the syringe, total quantity of each gas was measured. Daily amount of gas production volume per each trial was converted to STP conditions (standard temperature and pressure, 273 K, 1.01325 Pa).

$$100\ qH_2 / qH_2 + qCH_4 + qCO_2 = H_2\ \% \quad (4)$$
$$100\ qCH_4 / qH_2 + qCH_4 + qCO_2 = CH_4\ \% \quad (5)$$
$$100\ qCO_2 / qH_2 + qCH_4 + qCO_2 = CO_2\ \% \quad (6)$$

3. Results and discussion

3.1 Characteristic of sewage sludge and decanter cake

The characteristics of sewage sludge and decanter cake are shown in Table 1. The characteristic of sewage sludge was reported as; TS, VS and pH were 2.60 g/L, 12.60 g/L and 6.64, respectively. Alemahdi et al. (2015) reported that the characteristic of the sewage sludge of TS, VS and pH are 5.182 g/L, 3.734 g/L and 6.20 respectively. The characteristic of sewage sludge obtained were almost similar as compared to previous study by Alemahdi et al. (2015).

The characteristic of decanter cake was based on TS and VS were 19.40 g/L and 16.70 g/L, respectively. The previous study has reported that characteristic of decanter cake as TS and VS were 197.4 g/L and 169.7 g/L, respectively [10]. It was observed that the decanter cake was rich in the carbon contents with ratios of 42.70% relative to the dry weight. However, decanter cake has low nitrogen contents with percentages of 2.09%. In contrast, sewage sludge was characterized by a relatively low carbon content of 33.40% and a relatively high nitrogen content of 5.86% from its dry weight.

IOP Conf. Series: Materials Science and Engineering **368** (2018) 012027 doi:10.1088/1757-899X/368/1/012027

Table 1. Characteristic of sewage sludge and decanter cake

Parameters	Sewage sludge	Decanter cake
Total Solid (TS), g/L	2.60	19.40
Volatile solid (VS), g/L	12.60	16.70
Carbon content, %	33.40	42.70
Nitrogen content, %	5.86	2.09
C/N ratio	5.70	20.43
pH	6.64	5.52

Based on elemental analyses results, decanter cake indicated 20.43 of C/N ratio which is a suitable condition for anaerobic digestion while C/N ratio of sewage sludge was 5.7. Low C/N ratio of sewage sludge can affect the anaerobic digestion process that could be result in high total ammonia nitrogen (TAN) and high volatile fatty acids (VFAs) accumulated in the digester [9]. If C/N ratio was low, ammonia concentration can rise in the process of anaerobic digestion, and high ammonia concentration can hinder microorganism vitality serving as an obstacle to anaerobic digestion. Therefore, in the case of single digestion of sewage sludge, effective anaerobic digestion can be difficult due to low C/N ratio [13]. In consequence, co-digestion of sewage sludge and decanter cake seems to have positive effects to anaerobic digestion such as increasing in methane yield. Elsayed et al. (2016) indicated that the great variance in C/N values between sewage sludge and any agro-wastes in the mixture may increase the methane production as compared to the individual substrates.

3.2 Biochemical methane potential (BMP) test

3.2.1. Biogas production. The results of cumulative biogas production profiles for the three F/I ratios of 1:0, 1:1 and 2:1 are shown as in Figure 1. The biogas production was increase by 5 times in volume using co-digested samples as compared to sewage sludge alone. Approximately 250 mL, 140 mL and 30mL of biogas were produced in the samples that contain decanter cake and sewage sludge mixed at the F/I ratio of 1:1, 2:1 and 1:0, respectively. It was observed that biogas production was increased from day 2 until day 10. The F/I ratio of 1:1 produced more biogas than F/I ratio of 2:1. However, further biogas composition need to be investigated as to confirm the presence of methane and other component gases.

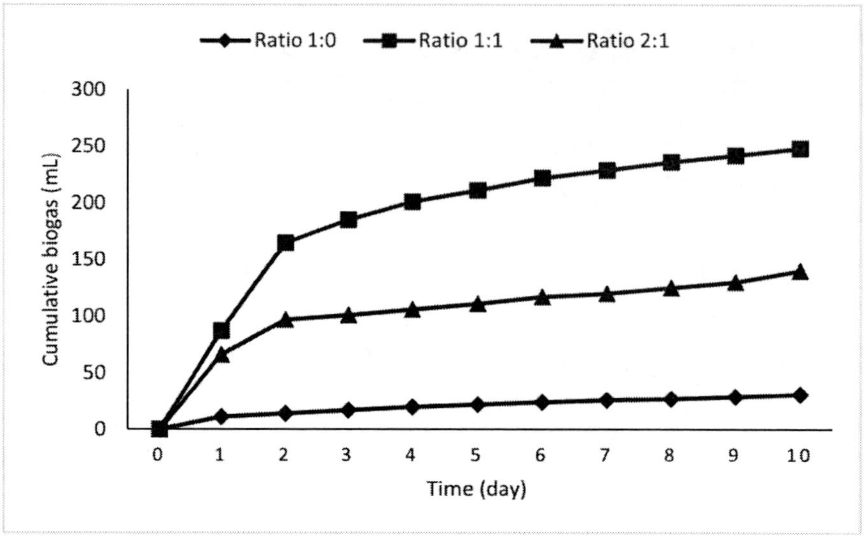

Figure 1. Cumulative biogas production

3.2.2. Biogas composition. The biogas composition in each gas samples for each day are shown in Figure 2. Based on the biogas composition, it was observed that high hydrogen yield and slightly some carbon dioxide were produced after two days of incubation. From day 3 until 10th, the CO_2 was produced and no hydrogen gas was detected. The methane gas was only observed at the 10th day. The hydrolysis and acidogenesis could happen from day 1 to 3 in converting carbohydrates, proteins and lipids to sugar, amino acid and fatty acid with the production of hydrogen, CO_2 and other volatile fatty acid (VFA). A sudden increase of methane at day 10th was observed with the highest methane yield of 3.3 mLCH4/gVS (Figure 3). This could be due to the methanogenic bacteria converting the VFA into methane. This shows that an acclimatization period of at least 10 days could be required when sewage sludge inoculum was employed for the anaerobic co-digestion with decanter cake at the F/I ratio of 2:1. Similarly, the acclimatization period for the F/I ratio of 1:1 was also observed at day 10th, but with much lower methane yield (0.6 mLCH4/gVS). Methane production decreased when increased F/I ratio [18].

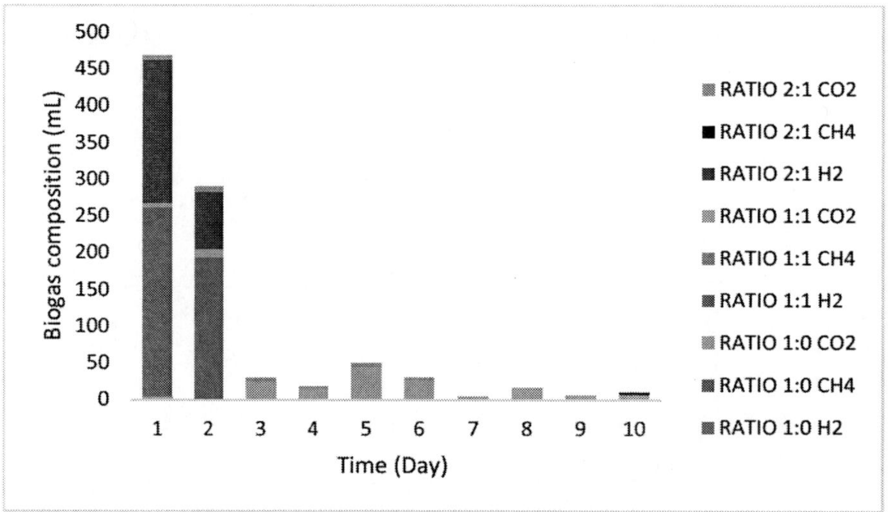

Figure 2. Biogas composition for each day

The period of acclimatization was independent upon the F/I ratio. In this study, the average of methane yield was found to be much lower with F/I ratio of 1:1 as compared to F/I ratio of 2:1. The finding was in line with a study reported by Suksong et al. (2015) showed that the methane production of decanter cake and POME mixture decreased with increasing proportion of decanter cake in mesophilic conditions. It has been reported by Suksong et al. (2015) that the lower methane yield could be due to difficulty of microorganism to degrade the decanter cake which has high amount of lignocellulose of 32.78%. Therefore, adding more lignocellulose material achieved longer acclimization period for methanogens.

Moreover, increase amount of inoculum could also adding more active microoorganisms for anaerobic degradation. It has been reported by Kaosol and Sohgrathok (2014) that the mixing ratio with frozen seafood wastewater (as inoculum) increased the methane gas production volume, and also co-digestion of wastewater and decanter cake at each different HRT produced larger amount of methane gas compared to single digestion, at ambient temperature (mesophilic).

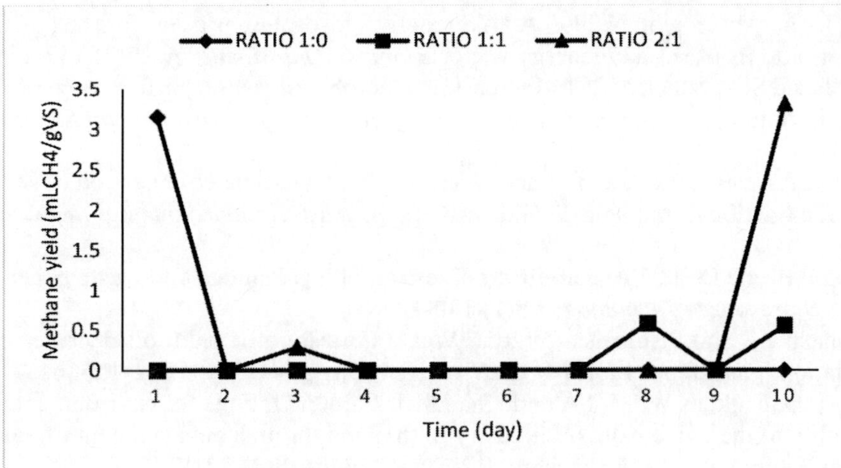

Figure 3. Methane gas production

The methane yield obtained from this co-digestion of sewage sludge and decanter cake increased with decrease F/I ratio. Therefore, F/I ratio of 2:1 produced more methane than F/I ratio of 1:1. Increasing in methane yield by co-digestion has been attributed to improve nutrient balance, mitigate toxicant levels or enhance microbial activity [20]. Solid waste in the palm oil mill such as decanter cake, empty fruit bunch and palm kernel shell are not utilize, due to its composition are difficult to degrade by methanogenic bacteria [18]. It seems necessary to introduce pre-treatment to decanter cake due to the lignocellulosic character of decanter cake. The pre-treatment techniques of the decanter cake could be with heat, ultrasonic or grinding in which to increase the degree of hydrolysis of feedstock and microbial stimulants.

4. Conclusion

The results indicated that the biochemical methane potential (BMP) tests could be used to assess the biogas and methane production from co-digestion process. It was found that the cumulative biogas production of the mixture of sewage sludge and decanter cake increased with increasing proportions of the decanter cake, however, the methane production was achieved much lower at higher load of decanter cake due to high degradability of the substrates since there is no pretreatment done to decanter cake. In consequences, still, the co-digestion of sewage sludge and decanter cake has a potential to produce methane. Therefore, preliminary study should be done to improve the digestion process. In this study, the suitable F/I ratio of anaerobic co-digestion between sewage sludge and decanter cake was F/I ratio of 2:1

IOP Conf. Series: Materials Science and Engineering **368** (2018) 012027 doi:10.1088/1757-899X/368/1/012027

References

[1] Rahmat 2014 Biogas production from tofu liquid waste on treated agricultural wastes. *American Journal Of Agricultural And Biological Sciences* **9(2)** 226-231.

[2] Saitawee 2014 Biogas production from anaerobic co-digestion of cow dung and organic wastes in thailand: temperature effect on biogas product *American Journal Of Environmental Sciences* **10(2)** 129-139.

[3] Pipatmanomai S, Kaewluan S and Vitidsant T 2009 Economic assessment of biogas-to-electricity generation system with H2S removal by activated carbon in small pig farm. *Applied Energy* **86(5)** 669-674.

[4] Roati C, Fiore S, Ruffino B, Marchese F, Novarino D and Zanetti M 2012 Preliminary evaluation of the potential biogas production of food-processing industrial wastes *American Journal Of Environmental Sciences* **8(3)** 291-296.

[5] Kolbitsch P, Pfeifer C and Hofbauer H 2008 Catalytic steam reforming of model biogas *Fuel* **87(6)** 701-706.

[6] Truong L and Abatzoglou N 2005 A H2S reactive adsorption process for the purification of biogas prior to its use as a bioenergy vector *Biomass And Bioenergy* **29(2)** 142-151.

[7] Kaosol T and Sohgrathok N 2014 Increasing anaerobic digestion performance of wastewater with co-digestion using decanter cake *American Journal Of Environmental Sciences* **10(5)** 469-479.

[8] Sosnowski P, Wieczorek A and Ledakowicz S 2003 Anaerobic co-digestion of sewage sludge and organic fraction of municipal solid wastes *Advances In Environmental Research* **7(3)** 609-616.

[9] Yen H and Brune D 2007 Anaerobic co-digestion of algal sludge and waste paper to produce methane *Bioresource Technology* **98(1)** 130-134.

[10] Kanchanasuta S and Pisutpaisal N 2016 Waste utilization of palm oil decanter cake on biogas fermentation *International Journal Of Hydrogen Energy* **41(35)** 15661-15666.

[11] Chavalparit O, Rulkens W, Mol A and Khaodhair S 2006 Options for environmental sustainability of the crude palm oil industry in thailand through enhancement of industrial ecosystems *Environment, Development And Sustainability* **8(2)** 271-287.

[12] Ramli A, Singh RP and Ibrahim MH 2012 Use decanter cake from palm oil mill as fertilizer supplement: the pattern of macronutrient accumulation in soil and plant with the amendment of decanter cake *UMT 11th International Annual Symposium on Sustainability Science and Management. Malaysia.*

[13] Dewayanto N, Husin MH, Yong LK and Nordin MR 2010 Waste to valuable by-product: kinetic and thermodynamics studie of Cd, Cu and Pb ion removal by decanter cake *Journal Engineering Technology* **1(1)** 85-98.

[14] Suksong W, Kongjan P and O-Thong S 2015 Biohythane production from co-digestion of palm oil mill effluent with solid residues by two-stage solid state anaerobic digestion process. *Energy Procedia* **79** 943-949.

[15] Yoon Y, Lee S, Kim K, Jeon T and Shin S 2017 Study of anaerobic co-digestion on wastewater treatment sludge and food waste leachate using BMP test *Journal Of Material Cycles And Waste Management*

[16] Elsayed M, Andres Y, Blel W, Gad A and Ahmed A 2016 Effect of VS organic loads and buckwheat husk on methane production by anaerobic co-digestion of primary sludge and wheat straw *Energy Conversion And Management* **117** 538-547.

[17] Alemahdi N, Che Man H, Abd Rahman N, Nasirian N and Yang Y 2015 Enhanced mesophilic bio-hydrogen production of raw rice straw and activated sewage sludge by co-digestion *International Journal Of Hydrogen Energy* **40(46)** 16033-16044.

[18] Owen W, Stuckey D, Healy J, Young L and McCarty P 1979 Bioassay for monitoring biochemical methane potential and anaerobic toxicity. *Water Research* **13(6)** 485-492.

[19] Sibiya N and Muzenda E 2014 A review of biogas production optimization from grass silage *Pretoria (South Africa): Int'l Conf. on Chemical Engineering & Advanced Computational Technologies (ICCEACT'2014)* 51-54.
[20] Chaikitkaew S, Kongjan P and O-Thong S 2015 Biogas production from biomass residues of palm oil mill by solid state anaerobic digestion *Energy Procedia* **79** 838-844.
[21] Saelor S, Kongjan P and O-Thong S 2017 Biogas production from anaerobic co-digestion of palm oil mill effluent and empty fruit bunches *Energy Procedia* **138** 717-722.

Effect of cellulose fiber from sago waste on properties of starch-based films

N Yacob[1,3*], M R Yusof[1,3], Z M A Ainun [2] and K H Badri[3*]

[1]Malaysian Nuclear Agency (Nuclear Malaysia) Bangi, 43000 Kajang, Selangor, Malaysia
[2]Program of Pulp and Paper, Institute of Tropical Forestry and Forest Products (INTROP), Universiti Putra Malaysia, 43400 UPM Serdang, Selangor, Malaysia
[3]School of Chemical Sciences and Food Technology, Faculty of Science and Technology, Universiti Kebangsaan Malaysia, 43600 Bangi, Selangor, Malaysia

*Corresponding author: kaybadri@ukm.edu.my; norzita@nuclearmalaysia.gov.my

Abstract. An increasing interest in developing biodegradable film using natural biopolymers such as polysaccharides is due to concerns on environmental problems caused by non-biodegradable petroleum-based materials. Though the film exhibits low mechanical properties, it can be improved by adding reinforcing agent, for instance, cellulose fiber. The objective of this study is to investigate the effect of adding cellulose fibers extracted from sago waste on the mechanical and thermal properties of sago starch-based films. Cellulose fibers extracted from sago waste showed reuction in size with cleaner surface as shown in FESEM micrographs. Thermal stability also improved indicated that the hemicelluloses and lignin were eliminated. Films were prepared by solution casting with 4 % sago starch and 3 % cellulose (from starch). These films were characterized using several techniques ie. tensile, x-ray diffractometry, TGA analysis, FESEM analysis and hazemeter. The films produced were highly transparent while film reinforced with cellulose fiber was more crystalline. The addition of fibers increased the tensile strength from 38.66 MPa to 40.70 MPa and Young's Modulus from 1710 MPa to 1713 MPa. The elongation at break also increased from 3.85 % to 4.62 %. On the other hand, the incorporation of cellulose fibers increased the thermal stability of the film. FE-SEM micrographs of the films showed homogenous distribution of the cellulose fibers. The inclusion of cellulose fiber has successfully increased the properties of the sago starch-based film.

1. Introduction

Malaysia has lavish natural resources contributed by it is being a tropical country. Other than being recognized as the main world supplier of palm oil and rubber, Malaysia is also producing sago starch. Sago starch is derived from the stem of palm (sago palm). Most of the sago palms in Malaysia are found in Sarawak and it grows well in swampy area. It is reported that sago plantation in Sarawak can produce up to 25 tons per hectar of starch in a year [1]. Sago palm becomes one of the most favourable plants because it is economically acceptable, relatively sustainable, environmentally friendly, and promotes socially stable agro-forestry system. It contains a large amount of starch in its trunk. Sago starch contains 27 % amylose and 73 % amylopectin. The molecular weight for amylose is in the range of 1.41×10^6 to 2.23×10^6, while for amylopectin is in the range of 6.70×10^6 to 9.23×10^6 [2]. Typically sago starches are processed for food, cosmetics, biodegradable plastics, adhesive and much more [3]. The residue left behind in the sago starch industry is called sago waste or *hampas sago*. Generally, sago waste is used for animal feed, compost for mushroom culture, waste adsorbent and fire retarding panel [4-5]. It is reported that 7 tons of sago waste are produced daily from a single sago starch processing mill. Currently, these residues which are mixed together with wastewater are either washed off into nearby streams or deposited in the factory's compound and may potentially lead to serious environmental problems [6]. Further, this sago waste also processed as a soil improver and natural fertilizer due to the small content of nutrient left in it [7]. Apart from that, this residue also contains fibers which can be extracted to cellulose. Cellulose from agricultural residues such as banana peel [8], sugarcane bagasse [9], potato peel waste [10] and soy hulls [11] have already been reported. The utilization of these residues can help in solving disposal problems. Therefore, development of other value added products such as cellulose as a reinforcing agent from these residues is a potential way to increase the value of agricultural wastes.

The use of cellulose fiber as a reinforcing agent in preparing starch film can improve the properties of the film. Starch can be considered as an alternative to synthetic films because it is cheap, renewable, biodegradable and widely available. However, films from starch have poor mechanical properties. Nevertheless, these limitations can be overcome by several approaches. The addition of plasticizers [12], blending the starch with synthetic polymers [13] and chemically modified the starch [14] are among the approaches to improve the properties of the starch film. The addition of fibers either synthetic or natural fibers as a reinforcing agent for a starch film also can overcome the problems.

In the present work, the effect of cellulose fibers on properties of starch films was examined. The fibers called treated sago waste (TSW) were extracted from sago waste through chemical treatment. Then, these TSW fibers were dispersed into sago starch films (SF) by using casting method. Characterization of the obtained films was made in order to study the effect of TSW on physical, mechanical and thermal properties of SF.

2. Methodology

2.1 Materials
Sago starch and sago waste were obtained from Ng Kia Heng Sago Industry Sdn. Bhd. (Batu Pahat, Johor). Sodium chlorite (80 %) and acetic acid (glacial, 99.85 %) were obtained from Sigma–Aldrich Inc. (St. Louis, MO, USA) and sodium hydroxide (98.9 %, ACS grade) were purchased from Fisher Scientific Inc. All reagents and solvents were used as received. Distilled water was used in the preparation of all solutions.

2.2 Extraction of cellulose
The extraction of cellulose fibers required the removal of other components such as lignin, hemicellulose and impurities from *sago waste* (SW). This step was achieved by chemical treatment. Sago waste which has not been treated (USW) was first soaked in hot water at 40 °C for 2 hours and oven dried at 40 °C for 48 hours to remove starch and impurities. The SW was then treated with 2 % NaOH aqueous solution at 60 °C for 1 h. The samples were then filtered and washed repeatedly with distilled water until a neutral pH was reached. The samples were dried at 40 °C for 48 hours to determine the cellulose yield and sample composition. Alkaline treated SW was then exposed to bleaching treatment. This was done by treating the alkaline treated SW with $NaClO_2$/glacial CH_3COOH mixture at 80 °C for 4 hours under mechanical stirring. The TSW was washed thoroughly with distilled water and dried at 60 °C. Each treatment was repeated three times in order to remove other constituents present in the waste.

2.3 Film preparation

The sago starch (SF) and sago starch with 3 wt% TSW cellulose fiber (SFSW) films were prepared by using casting method. Firstly, SFSW solution (3 wt%, starch basis) was prepared by dissolving the TSW in 10 ml of distilled water. The solution was then stirred for 1 hour at ambient temperature followed by ultrasonic treatment (Branson sonifier model 450) for 30 min. Meanwhile, sago starch 4 % (w/w) solution was prepared separately by mixing the starch with 15 ml of distilled water followed by heating to 70 °C while stirring for 30 min. During the heating process, the TSW solution was added to the starch solution. The mixture was poured into a 10 cm × 10 cm plastic mould. Then the films were dried in the oven at 40 °C for 48 hours, peeled off after drying and kept at 23 ± 2 °C and 50 ± 5 % relative humidity (RH) prior to testing. The same procedure was applied for SF films without the addition of TSW.

2.4 Morphological examination

Field Emission Scanning Electron Microscope (FESEM), Carl Zeiss model Gemini SEM 500 was used to investigate the morphology of the surface and cross-section for all samples. Images were acquired at 500× magnification for surface and 1000× for cross-section examination.

2.5 Light Transmittance

The light transmittance of films was measured using hazemeter spectrophotometer model BYK Gardner following standard method ASTM D1003. Films with the dimension of 10 cm × 10 cm were used and five replicates measurements were recorded.

2.6 X-ray diffraction (XRD) analysis

X-ray diffraction (XRD) measurement was carried out on a RÖntgen diffractometer system (PANalytical, Netherlands). The diffracted intensity of Cu-Ka radiation was measured at 40 kV, 30 mA at 25 °C. The diffraction patterns were determined over a range of diffraction angle 2θ from 5 ° to 40 ° at a scanning rate of 2 °C/min.

2.7 Thermogravimetric Analysis (TGA)

Thermogravimetric analysis was carried out using Perkin Elmer model Pyris 1 thermogravimetric analyzer to determine the thermal stability of sago waste at different stages of extraction. Thermal stability of SF and SFSW films was examined too. For analysis, the samples were heated from room temperature to 600 °C under nitrogen atmosphere at 10 °C/min.

2.8 Tensile properties

Tensile properties were measured according to ASTM D 882-97. The testing was conducted using 20 N load cell (Model UUK 5, Korea) equipped with Ezi Step micro stopper motor system (Fastec, Korea). The samples were run at crosshead speed of 0.5 mm/min.

3. Results and discussion

3.1 Morphology of fibers

The structural changes in the fibers were observed via FESEM and are presented in Figure 1. The micrographs displayed in Figure 1(a) and 1(b) showed that the diameter of the fiber decreased from 300 μm to 10-20 μm. The reduction in fiber size after treatment can be ascribed to elimination of extractable elements, other than defibrillation phenomena. The former caused damage to fiber structure resulted in finer and smaller fibers. The later segregated the fibers from bundles to single fibril. Figure 1(c) of USW displays rough surface cemented together with lignin, hemicellulose and wax [15]. Upon chemical treatment, the surface of the fiber became smoother because of the removal of non- cellulosic materials like lignin, hemicelluloses and wax as displayed in Figure 1 (d). This could result in an effective adhesion between the fiber surface and the matrix [16].

IOP Conf. Series: Materials Science and Engineering **368** (2018) 012028 doi:10.1088/1757-899X/368/1/012028

Figure 1. FESEM images of (a) USW and (b) TSW at 100× magnification while surfaces of (c) USW and (d) TSW at 500× magnification

3.2 Thermal properties of the fibers

Figure 2 and Figure 3 show the thermogravimetric (TG) and the differential thermogravimetric (DTG) curves obtained from the thermogravimetric analysis data for USW and TSW respectively. Table 1 lists the onset degradation temperature (T_0, ° C), maximum thermal degradation temperature (Tmax, ° C) and residue (%) at 600 °C for all samples.

Table 1. Onset degradation temperature (T_0, °C), maximum thermal degradation temperature (Tmax, °C) and residue (%) for USW and TSW

Sample	T_{Onset} (°C)	Tmax (°C)	Residue (%)
USW	250	310	9.97
TSW	297	343	7.90

The curves obtained for both fiber samples show that the initial weight loss started below 100 °C caused by water evaporation from the samples. USW shows an earlier weight loss at around 250 °C with maximum degradation temperature was 310 °C and residue at 9.97 %. For TSW, the degradation temperature was at 297 °C with maximum degradation temperature of 343 °C while the residue at 7.90 %. The increase in degradation temperature indicated that TSW has higher thermal stability than USW due to further removal of non-cellulosic impurities by bleaching process which is beneficial in improving the thermal stability of the cellulose fibers [17]. The percentage of residue for TSW fibers was lesser than USW due to the fact that hemicellulose and lignin were mostly removed from cellulose fibers. This finding supports the morphological observaton as discussed above.

The decomposition peak of USW shown in DTG curve in Figure 3 is wider than TSW due to the composition of hemicelluloses and lignin. Meanwhile, TSW shows narrow decomposition peak suggesting the component that degraded at lower temperature had been removed during treatments.

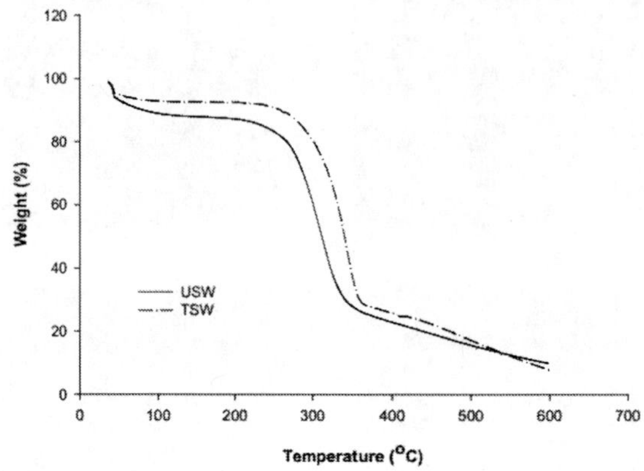

Figure 2. TG curves of USW and TSW

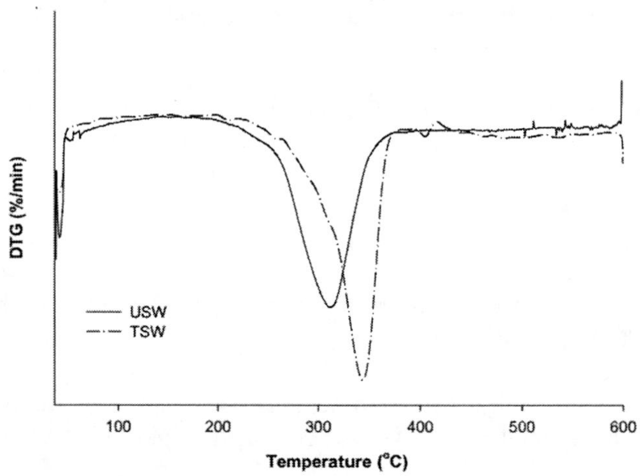

Figure 3. DTG curves of USW and TSW

3.3 X-ray diffraction analysis of the fibers
XRD analysis was performed in order to determine the crystallinity of USW and TSW as displayed in Figure 4. USW exhibited strong and broad peaks at both $2\theta = 17°$ and $23°$, medium peaks at around both $2\theta = 5.3°$ and $15.0°$ and a couple of weak peaks scattered around $10.2°$ and $11.7°$ representing the *C-Type* crystalline structure. This observation was consistent with previous studies by other researchers [18-19]. It is believed that these diffraction patterns belong to sago starch granule that occurred in sago waste. As can be seen, after chemical treatments, TSW displayed sharp diffraction peaks located at 2θ value of around 16° and 22.6°. This diffraction peaks were attributed to the diffraction planes of (110) and (002) representing typical cellulose I crystal form. The other peak observed in the XRD patterns of the TSW was at 2θ=34.5° which was attributed to the diffraction plane of (004), confirming that *Cellulose Type I* was formed [20]. The treatment of *sago hampas* was aimed to remove hemicellulose and lignin, which exist in the amorphous regions, leading to the realignment of cellulose molecules and

a better packing of cellulose chains. Consequently, the increase in crystallinity of the treated fiber has improved the thermal stability of TSW (Figure 2 and Figure 3).

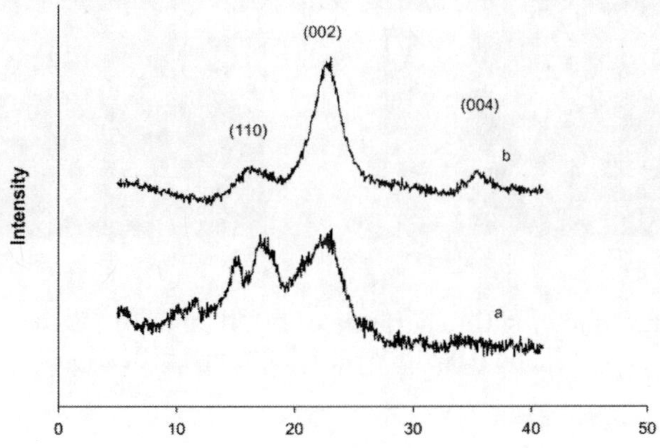

Figure 4. XRD patterns of (a) USW and (b) TSW

3.4 Morphological investigation of films

Figure 5 shows the FESEM observation onto the surface and cross-section of the films. Micrographs of the SF showed homogenous and smooth surface as can be seen in Figure 5(a). The surface of SFSW was rougher than SF with a very minimum presence of crack and more compact as displayed in figure 5(b). It shows that the TSW was well dispersed within the starch matrix. Meanwhile, the cross-section of SF and SFSW are depicted in Figure 5(c) and Figure 5(d) respectively. For SFSW, some holes and cracks were observed indicating the fiber pull out from the matrix. It shows that there is an interaction between the fiber and the matrix. Nevertheless, it can be clearly seen that presence holes indicate the adhesion between fiber-starch matrixes was less bonded. Thus, this can explain the reason of slightly increased in tensile properties. The treatment of the fiber or the addition of plasticizers is among the method that can be introduced in order to improve and enhance the adhesion.

(a)

(b)

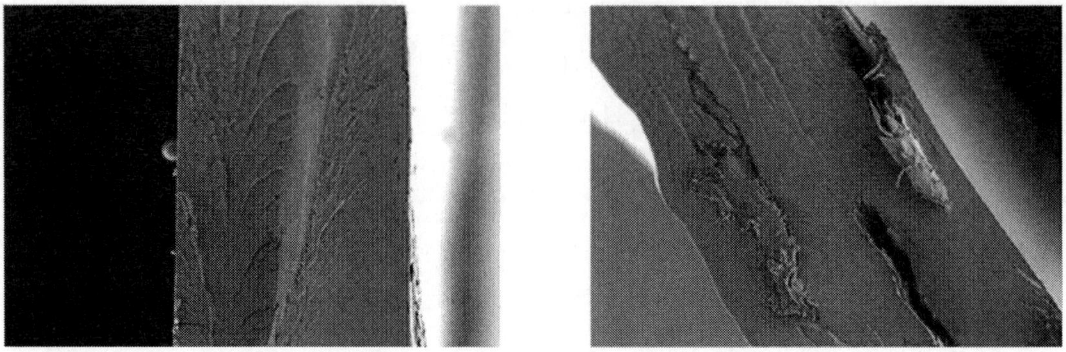

(c) (d)

Figure 5. FESEM images showing film surfaces of (a) SF and (b) SFSW at 500× magnification, and film cross-section of (c) SF and (d) SFSW at 1,000× magnification

3.5 Thermal properties of the films

The onset degradation temperature (T_0,°C), maximum thermal degradation temperature (Tmax, °C) and residue (%) of the films were summarized in table 2.

Table 2. The onset degradation temperature (T_0,°C), maximum thermal degradation temperature (Tmax, °C) and residue (%) of the films at 600°C.

Sample	T_{Onset} (°C)	Tmax (°C)	Residue (%)
Starch Film (SF)	291	323	2.32
Starch with cellulose (SFSW)	307	330	0.73

The thermal stability of the films was characterized using TGA analyzer. The TGA and DTGA curves were shown in Figure 6 and Figure 7. Both SF and SFSW showed two decomposition peaks. The first weight loss was observed from room temperature to 60 °C which was referred to evaporation of water. The second weight loss occurred from 270 °C to 400 °C with maximum degradation around 323 °C and 330 °C for SF and SFSW respectively. SFSW exhibited higher thermal stability than SF. This improvement is connected to the chemical interaction between starch matrix and fiber generated from chemical compatibility resulting in improved interfacial interactions between fiber and matrix. The amount of residue left after final decomposition of SFSW was less than that of SF. This is due to higher ash content of starch.

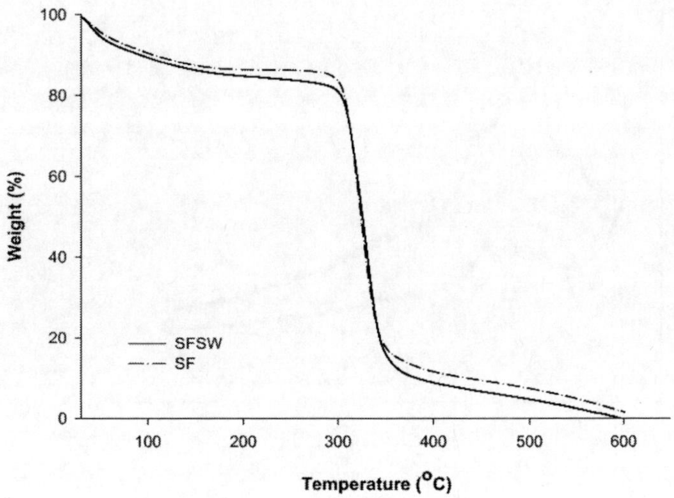

Figure 6. TGA curves of SF and SFSW

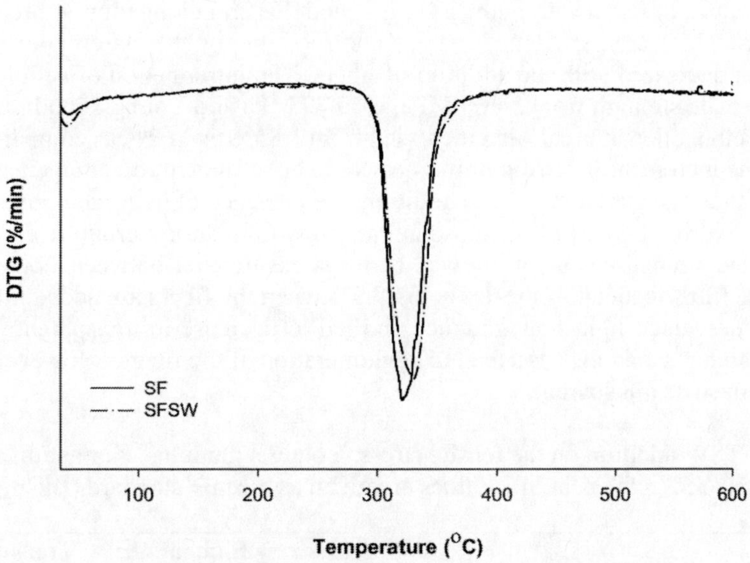

Figure 7. DTG curves for SF and SFSW films

3.6 X-ray diffraction analysis of the films

Figure 8 depicts the x-ray diffraction pattern of SF and SFSW. The sago film without cellulose displayed diffraction peaks around 2θ=5.9°, 15.2° and 17.3° representing the C-type crystalline structure. The addition of cellulose fiber decreased the intensity of peak at 2θ = 17.3° but increased the intensity peak at 2θ = 22.6°. The increasing value of 2θ showed that crystalline structure of cellulose I from cellulose fiber presents in the film. It is suggested that the incorporation of highly crystalline fibers enhances the crystallinity of the films.

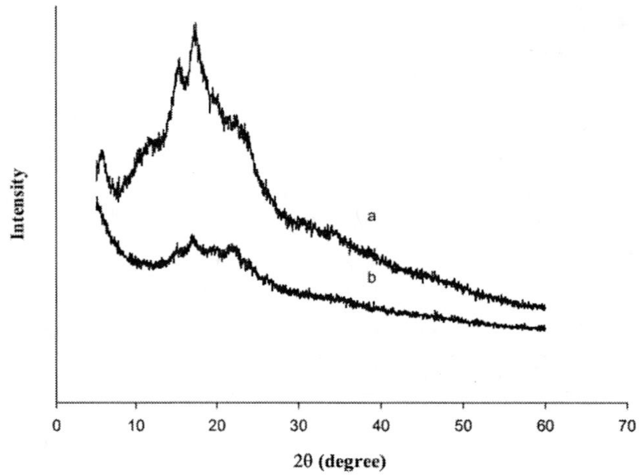

Figure 8. XRD patterns of (a) SF and (b) SFSW

3.7 Tensile properties of the films

Tensile properties of SF and SFSW are summarized in Table 3. Sago starch films reinforced with TSW presented higher values of tensile strength, Young's modulus and elongation at break as compared to film without fibers. Generally, the tensile stress and modulus of the starch film increased and the elongation at break decreased with the addition of fibers. The introduction of cellulose fibers into the SF increased the tensile strength from 38.66 MPa to 40.70 MPa and Young's modulus from 1710 MPa to 1713 MPa. The elongation at break also increased from 3.85% to 4.62% as compared to film without cellulose fiber. The increment in tensile properties could be attributed to homogeneous distribution of cellulose fibers in the sago starch matrix. In addition, the presence of hydrogen bonding between fiber and matrix, where hydroxyl group of cellulose interacted with hydroxyl group from starch matrix [21] further enhanced the strength. Thus, it allowed better stress transfer between fibers and matrix. The transparency of the films reduced from 91.4% to 89.9% when the fibers are added into the matrix. The addition of fibers prevented light transmittance and led to decrease in transparency of the film. The decrease in transparency could also ascribed to agglomeration of the fibers. However, the SFSW films still can be categorized as transparent.

Table 3. Effect of TSW addition on the tensile stress, Young's modulus, elongation and transparency of starch films compared to the neat SF. Values are given as mean ± standard deviation.

Sample	Stress, MPa	Young's Modulus, MPa	Elongation, %	Transmittance, %
SF	38.66 ± 1.37	1710 ± 0.07	3.85 ± 0.27	91.4
SFSW	40.70 ± 0.85	1713 ± 0.04	4.62 ± 0.35	89.9

Cellulose fibers and starch matrix are both hydrophilic and compatible with each other. Thus, the fiber has good adhesion with starch to allow a stress transfer from the matrix to the fiber [22]. This explained the increase in properties of starch films reinforced with cellulose fibers as compared to the properties of the starch film without cellulose fibers.

4. Conclusions

Sago waste cellulose fibers were successfully extracted using chemical treatment. Morphological analysis indicated a reduction in size from 300 µm to 20 µm, with cleaner and smoother surface of TSW observed. The thermal stability and crystallinity of TRW fibers increased upon alkaline hydrolysis. The hemicelluloses and lignin in sago waste have been successfully removed. Sago starch/treated sago waste fiber films were prepared via solution casting. The addition of TSW into SF has improved its performance. The thermal stability and tensile properties increased contributed by good interfacial adhesion between the fibers and the matrix. FESEM results indicated that chemical treatment on fibers plays an important role in enhancing the interfacial bonding between fiber and matrix. The treated fiber showed good adhesion between fiber and matrix. The present work demonstrates the cellulose extracted from sago waste are compatible with starch matrix and can be used as reinforcement in sago starch films.

Acknowledgements

The authors would like to thank the Malaysia Ministry of Higher education for the financial support on the analysis of data using grant allocation from ERGS/1/2013/STG01/UKM/02/2 and Center for Collaborative Innovation, Universiti Kebangsaan Malaysia through its Dana Inovasi INOVASI-2017-001.

References

[1] Sighal R S, Kennedy J F, Gopalakrishnan S M, Kaczmarek A, C J Knill and F A Putri 2008 Industrial production, processing and utilization of sago palm-derived products. *Carbohydrate Polym* **72** 1-20
[2] Karim A A, Tie A, Manan D M A, and Zaidul I S M, 2008 Starch from the sago (Metroxylon sagu) palm tree—properties, prospects, and challenges as a new industrial source for food and other uses *Compr Rev Food Sci F* **7** 215-228
[3] Bujang K B and Ahmad F B 2000 Country report production and utilization of sago starch in Malaysia Proceedings of the International Sago Seminar (Bogor, Indonesia, 22-23 March 2000) ed H M H Suwardi S, Kamal M, Setiawan K, Hadi pp 1–8
[4] Zainab N, Azlin A K A, Nazlina S, Hasnain H, Norhaizat S, Teng J X, and Lawai V 2013 Production of fire-retardant sound-absorbing panels from sago waste *J Trop For Sci* 510-515
[5] Amode J O, Santos J H, Alam Z M, Mirza A H, and Mei C C 2016 Adsorption of methylene blue from aqueous solution using untreated and treated (Metroxylon spp.) waste adsorbent: equilibrium and kinetics studies *International Journal of Industrial Chemistry* **7** 333-345
[6] Bujang K, Apun K, Salleh K M 1996 Astudy in the production and bioconversion of sago waste Sago-The Future Source of Food and Feed ed Jose C and Rasyad A (Riau, Indonesia:University Press) 195–201
[7] Singhal R S, Kennedy J F, Gopalakrishnan S M, Kaczmarek A, Knill C J, and Akmar P F 2008 Industrial production, processing, and utilization of sago palm-derived products *Carbohydrate Polym* **72** 1-20
[8] Elanthikkal S, Gopalakrishnapanicker U, Varghese S, and Guthrie J T 2010 Cellulose microfibres produced from banana plant wastes: Isolation and characterization *Carbohydrate Polym* **80** 852-859
[9] Kumar A, Negi Y S, Choudhary V, and Bhardwaj N K 2014 Characterization of cellulose nanocrystals produced by acid-hydrolysis from sugarcane bagasse as agro-waste *Journal of Materials Physics and Chemistry* **2** 1-8
[10] Chen D, Lawton D, Thompson M R, and Liu Q 2012 Biocomposites reinforced with cellulose nanocrystals derived from potato peel waste *Carbohydrate Polym* **90** 709-716
[11] Neto W P F, Silvério H A, Dantas N O, and Pasquini D 2013 Extraction and characterization of cellulose nanocrystals from agro-industrial residue–Soy hulls *Ind Crop Prod* **42** 480-488
[12] Sanyang M L, Sapuan S M, Jawaid M, Ishak M R, and Sahari J 2016 Effect of plasticizer type

and concentration on physical properties of biodegradable films based on sugar palm (Arenga pinnata) starch for food packaging *Journal of food science and technology* **53** 326-336

[13] Averous L U C and Fringant C 2001 Association between plasticized starch and polyesters: processing and performances of injected biodegradable systems *Polymer Engineering & Science* **41** 727-734

[14] Shah U, Naqash F, Gani A, and Masoodi FA 2016 Art and science behind modified starch edible films and coatings: a review *Compr Rev Food Sci F* **15** 568-580

[15] Fiore V, Scalici T, and Valenza A 2014 Characterization of a new natural fiber from Arundo donax L. as potential reinforcement of polymer composites *Carbohydrate Polym* **106** 77-83

[16] Narendar R and Dasan KP 2014 Chemical treatments of coir pith: Morphology, chemical composition, thermal and water retention behavior *Compos Part B-Eng* **56** 770-779

[17] Sonia A and Dasan K P 2013 Chemical, morphology and thermal evaluation of cellulose microfibers obtained from Hibiscus sabdariffa *Carbohydrate Polym* **92** 668-674

[18] Ahmad F B, Williams P A, Doublier J L, Durand S, and Buleon A 1999 Physico-chemical characterisation of sago starch *Carbohydrate Polym* **38** 361-370

[19] Polnaya F J, Marseno D W, and Cahyanto M N 2013 Effects of phosphorylation and cross-linking on the pasting properties and molecular structure of sago starch *International Food Research Journal* **20** 1609-15

[20] Obi R K, Ashok K and Reddy K R N 2014 *Carbohydrate Res.* **341** 2677-87

[21] Sabu T, Kuruvilla J, Malhotra S.K, Koichi G and Sreekala M.S 2012 Polymer Composites, Macro an Microcomposites (Weinheim, Germany :John Wiley & Sons) **1** 356-358

[22] Yu L, Liu X, Petinakis E, Dean K and Bateman S 2013 Advances in NaturalPolymers: Composites and Nanocomposites ed Thomas S, Visakh P M and Mathew A P (Berlin : Springer) p 121

The effect of lignin on processing and the properties of lignocellulose material recovered by ionic liquid

A Khosravani*, M Pourjafar and R Behrooz

Wood and Paper Science & Technology Department, Faculty of Natural Resources, Tarbiat Modares University, Noor, Mazandaran, Iran

*Corresponding author: khosravani@modares.ac.ir

Abstract. As a prospective raw material, lignin containing papermaking fibers can be easily recycled, dissolved with proper solvents and reused. However, in order to utilize this material, more eco-friendly green dissolution methods are also required. As a green solvent, ionic liquids (ILs) have been attracted too much interest in processing of lignocellulosic biopolymers, but the practical effect of lignin on this process was not clear. In this due, production of lignocellulosic bio-composite films from bleached (lignin-free) and unbleached (lignin-containing) soda pulps was studied. First, fibers were dissolved in 1-Buthyl 3-Methyl Imidazolium Chloride (BMIMCL) at 85-95 °C. Following fabrication and preparation of the lignocellulosic films, certain mechanical and physical properties of cellulose composite films were analyzed. FESEM and XRD analytical methods were applied to study the product morphology and structure. Microscopic studies showed both fibers were disappeared in the ionic liquid in less than 1 h, although the unbleached fibers took more time for complete dissolution. The results indicated that the films produced from unbleached pulp had less tensile strength and more dynamic contact angle with water droplet than the ones produced from bleached pulp. Moreover, the optical tests revealed that the lignin resulted in less transparency of the films, but increased absorption of UV radiations. This implies that even impure lignocellulose feedstock can have promising features.

1. Introduction

The fibers, films and composite materials made of biopolymers have attracted much attention recently, as they help decreasing dependence on oil resources, while they are biocompatible, biodegradable and also due to their high performance [1-2].

Other cellulose, which is the most abundant natural polymer in our environment, is a fibrous, tough, water-insoluble substance with unique crystalline structure that plays a substantial role in blend with different biopolymers to produce various bio-products [1–3]. Cellulose is mainly isolated from lignocellulosic resources, which are abundant renewable organic materials on earth. In order to extract cellulose from lignocellulosic feedstocks, especially lignocellulosic residues, for fabrication of various products, dissolution and further processing of cellulose is necessary [3-4].

However, usually dissolution of cellulose in common solvents is not readily achieved. Some features of this carbohydrate, such as: the strong inter and intera-molecular hydrogen bonding, high degrees of polymerization and crystallinity are some problematic factors in this respect. Although several solvent systems have been applied for preparation of regenerated cellulose materials,

IOP Conf. Series: Materials Science and Engineering **368** (2018) 012029 doi:10.1088/1757-899X/368/1/012029

generally, usual solvents are not industrially applicable due to high environmental toxicity, recovery problems or many adverse side reactions [5].

In this respect, as a solvent, ionic liquid can be named as one of the proper candidates. Ionic Liquids (ILs) have recently attracted too much attention as "green solvents" for various chemical processes. Owing to very small vapor pressure, ILs can be easily recovered and reused. They are mostly in the liquid form at relatively low temperatures (below 100 °C), while possessing thermal and chemical stability and non-flammability [6–8].

Pinkert et al. (2009), summarized a list of lignocellulosic materials which had been treated/dissolved with various ILs and different dissolution conditions [9]. Zhang et al. (2017), focused on fabricating methods of cellulose-based materials with ILs and its future trends [10]. In some studies, wood chips of different species were immersed in some of ionic liquids for even very long periods, at relatively high temperatures, but low solubility was observed [8]; while thermo-mechanical pulp (TMP) fibers or micro crystalline cellulose (MCC) particles, were readily dissolved in ionic liquids under similar conditions [4,9]. Therefore, in addition to lignocellulosic material particle size, the presence and structure of lignin may be important factors affecting the application of ILs for processing of lignocellulose materials [11], and moreover, the properties of the products.

2. Experimental

2.1 Materials

As raw fibrous material, bagasse was provided from Pars Paper mill (Khuzestan, Iran), and used for further delignification and bleaching. 1-Butyl-3-Methyl Imidazolium Chloride (BMIMCl), was purchased from Sigma-Aldrich chemical Co. (Steinheim, Germany). All other reagents were analytical grade.

2.2 Pulping and bleaching of lignocellulose material

As a main delignification process, pulping was performed in a rotating digester equipped with electrical power under controlled pressure and temperature. The liquor/bagasse ratio was 8:1 and the alkaline charge 18 % NaOH, based on oven-dry bagasse weight (22.5 g/L). The time-to-temperature, pulping time and final temperature were set 15 min, 45 min and 170 °C, respectively. Following cooking process, the pulp was washed on a sieve of 200 mesh size with tap water. The pulping yield was calculated about 52 %, with final Kappa number of 16.5, according to TAPPI test method T236 om-99, at least with three replications.

Then bleaching treatment was carried out to eliminate residual lignin according to Zobel and McElwee (1958) procedure [12]. Air-dried fiber (2g) was placed in an Erlenmeyer flask, and 160 ml preheated distilled water was added. Afterwards, 20 ml of buffer solution (made from 29 ml acetic acid, 10.5 g sodium hydroxide and 170 ml distilled water) and 20 ml of sodium chlorite (34 g/l) were added respectively. Following stirring the mixture, the flask was placed in a hot-water bath at 70 °C for 2 h. Then the content of the flask was neutralized by sodium sulfite. Finally, the fibers were filtered and washed thoroughly with distilled water. The final Kappa was about 1.5. Also, the degree of polymerization (DP) of cellulose was analyzed according to TAPPI T230 om-04 test method.

2.3 Treatment of lignocellulose material in IL and preparation of the film

Lignocellulosic materials were firstly oven dried at 103 ± 2 °C for 24h. Then, they were added calmly into BMIMCL at 85-95 °C, with 2 % concentration. Consequently, the IL-treated lignocellulosic solution was used for fabrication of the films. For this purpose, the solution was stirred on a hot plate at a constant stirring rate (about 300 rpm) under N2 gas atmosphere, for 1 h. The clear lignocellulosic solution was then casted on a glass plate and kept in the ambient temperature for 24 h. Afterward, the produced gel was washed periodically for 24 h with distilled water to remove the IL. Finally, the films were gently blotted between filter papers and oven-dried at 60 °C for 12 h.

Also, for visualization of the dissolution process, some pictures were taken at 250x magnification using MICROS N-200 optical microscope (Austria), at various stages of dissolution of lignocellulosic materials in the IL.

2.4 Characterization of the film samples

2.4.1 FESEM. To study the structure of the produced film samples, a field emission scanning electron microscope (FE-SEM) MIRA3 XMU (TESCAN, Czech Republic) was used, at accelerating voltages of 3 and 15 kV, at various magnifications. In this due, the regenerated film samples were frozen in liquid nitrogen, immediately broken and then the fractured surfaces were coated with a very thin layer of gold, prior to take micrographs.

2.4.2 X-ray diffraction. The crystalline structures of the produced cellulosic films were studied using a Philips X'Pert MPD diffractometer (Eindhoven, Netherlands) with Cu Kα radiation (40 kV and 40 mA). The diffracted intensities were measured in the range of $2\theta = 5°$ to $59°$ with the step size of 0.04° at 0.8 s per step.

2.4.3 Tensile strength of the films. To Tensile strength (TS) of the samples were tested on a Universal Testing Machine (Model TVT-300Xp, Perten, Sweden), according to ASTM standard method D 882-02. In this due, the film samples were cut into rectangular pieces (5 cm × 1 cm) and the thickness was determined at 5 points. The samples were carefully placed between the grips, vertically, with 30 mm of initial grip separation distance and constant elongation rate of 10 mm/min.

2.4.4 Contact angle measurements. The effect of lignin on affinity of the film surface toward water droplet was evaluated using a PG-X Goniometer, Switzerland, with a droplet volume of 60 μl, at 23 °C and 50 % RH.

2.4.5 UV–visible Spectroscopy (UV–vis). In order to analyze the optical transmittance (Tr) of the films, a UV–vis spectroscope (Lambda 25, PerkinElmer, Fremont, CA, USA) was used at wavelengths between 200 to 800 nm. The samples were cut into the shape of rectangular (9 mm × 30 mm) samples, directly placed in a spectrophotometer cell and a vacant quartz compartment was used as the reference.

3. Results and discussion

3.1 The effect of lignin on the dissolution process

Figure 1, visualizes the effect of lignin on the dissolution process of bagasse fibers in the IL. The figure demonstrates that raw bagasse fibers were not significantly dissolved in the IL, even after some hours (>12 hours) (Figure 1c), while unbleached and bleached bagasse soda pulp (UBSP and BBSP) fibers were readily disappeared in the IL within less than 1 hr. (Figure 1a and 1b). This revealed that by means of pulping and bleaching, in which lignin structure is broken up and kappa number decreases to 16.5 (after pulping) and finally to 1.5 (following harsh bleaching), the dissolution process was highly facilitated [4].

On the other hand, the DP of cellulose was analysed. The results showed the DP of cellulose in UBSP (about 400) was higher than that of BBSP (about 250). This explains that, in addition to diminishing lignin content and structure, pulping and bleaching processes may also facilitate the dissolution process due to shortening of the cellulose chain.

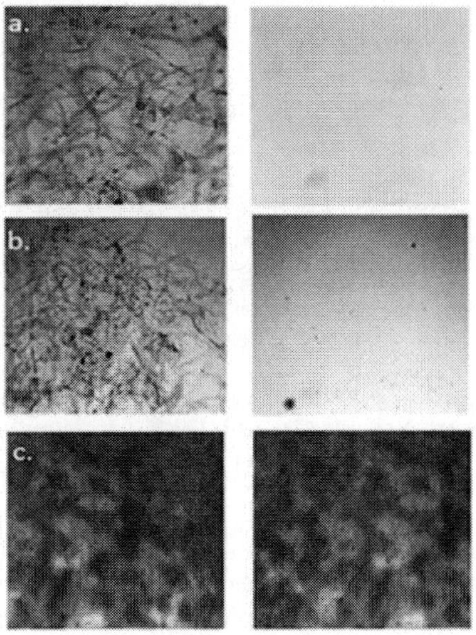

Figure 1: The dissolution stages of the lignocellulosic material in the ionic liquid (IL): a. Bleached bagasse soda pulp (BBSP) fibers, b. Unbleached bagasse soda pulp (UBSP) fibers, and c. Raw bagasse fibers (RBF), before and after 1 hr stirring in the IL (magnification: 250X).

3.2 FESEM micrographs
Figure 2 shows the FESEM micrographs form the cross sectional view of the produced film samples. The pictures demonstrate a uniform dense micro and nano structure for both BBSP and UBSP films, in which no border line (separation line) is observed, even in the lignin containing films.

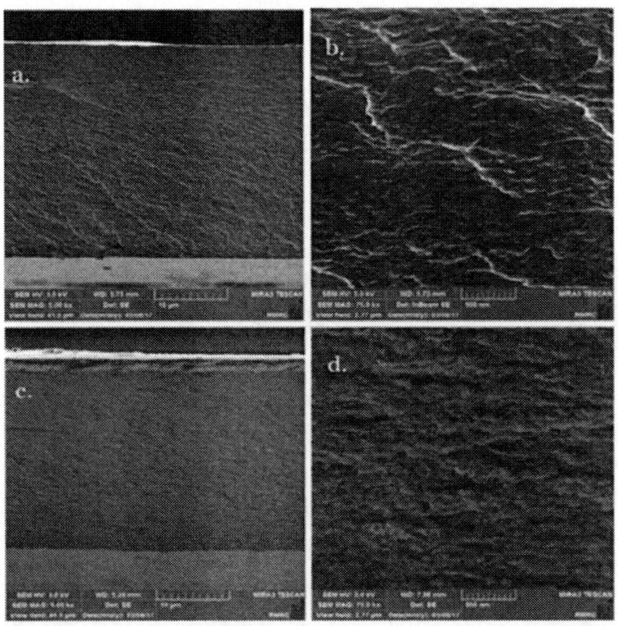

Figure 2: FE-SEM images from cross sectional view of the produced films (a., b. BBSP film, and c., d. UBSP film samples, at 5 kX and 75 kX magnification, respectively).

3.3 X-ray diffractograms
Figure 3 demonstrates the X-ray diffraction pattern for the initial lignocellulosic materials and the produced films. In the figure, the diffraction patterns for BBSP and UBSP fibers, both display the

clear peaks at 2θ: 15.9°, 22.21°, and 34.67°, with the most intensity at 2θ: 22.21°, which show the normal pattern of cellulose type-I (native cellulose). On the other hand, the diffraction pattern of regenerated films exhibit one broad scattering in a lower angle around 2θ: 20° with less intensity. This broad peak (more FWHM (full width at half maximum) of the peak), demonstrated the formation of cellulose structure with less crystallite sizes, less ordered domains in the cellulose film, or both [1–3,13]. In addition, less intensity of the main peak indicates that the regenerated films predominantly consist of amorphous regions.

Moreover, the X-ray pattern obtained for BBSP fibers and produced films mostly resembled to that of UBSP counterparts, indicating lignin did not significantly affect the material and product crystalline structure (Figure 3).

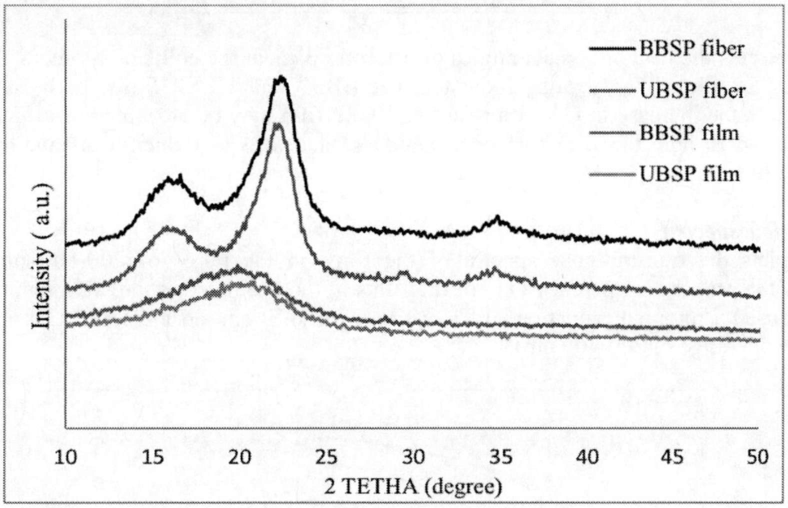

Figure 3: X-ray diffraction pattern of BBSP, UBSP fibers and their regenerated films.

3.4 Mechanical Strength

The tensile strength (TS) of the films was studied as shown in Figure 4. The figure shows that the TS of BBSP film samples (150.15 ± 5.6 MPa) was higher than that of UBSP ones, although the DP of cellulose was higher in UBSP. This may indicate that lignin, as an impurity in the cellulose matrix, had negative effect on hydrogen bonding during regeneration of cellulosic films [1, 14].

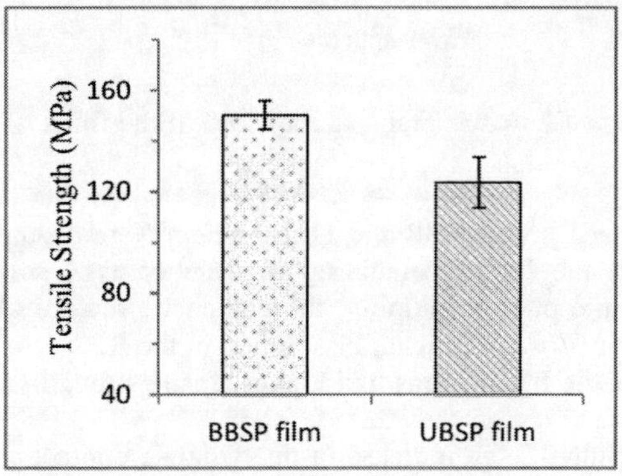

Figure 4: Tensile strength (TS) of the films.

3.5 Contact angle

Generally, the contact angle (CA) of the material surface with water droplets implies to the hydrophilicity of the samples. Furthermore, the morphology and roughness of the surface may affect the wettability. If the CA is smaller than 90 °, the film surface is considered hydrophilic [3].

Table 1. The contact angles of the films with water droplet

Samples	Contact angle at collision moment (°)
BBSP film	69.2 ± 4.6
UBSP film	82 ± 8.8

Table 1 presents the data of contact angles of water droplets at the collision moment with the films. BBSP and UBSP films. The results, indicates that BBSP and UBSP films, both had hydrophilic properties, while the increase in CA as a result of UBSP film may be attributed to effect of lignin, as lignin is reported to help the film surface smoothness and also to reduction of available hydroxyl groups on the film surface [15-16].

3.6 UV-visible spectra

Figure 5 depicts the transmittance spectra of the films in the range of 200-800 nm. The graph demonstrates the proper transparency (Tr) of the films in the evaluated range, although the data show that lignin caused a minimal reduction of Tr, while, instead, it caused increased UV absorption (for UBSP film in the range of 200-400 nm).

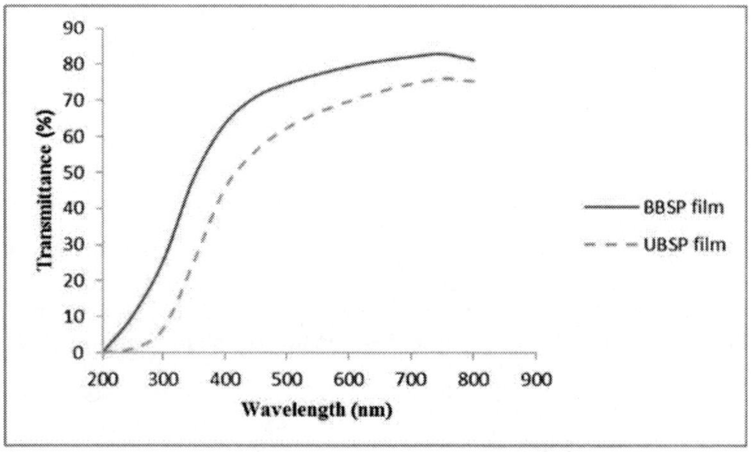

Figure 5: Transparency (Tr) of the films.

4. Conclusion

1. Microscopic studies showed both BBSP and UBSP fibers were disappeared in the ionic liquid in some minutes, although the lignin containing unbleached type of fibers took more time for complete dissolution, while prior to pulping, the compact structure of lignocellulose structure of raw bagasse fibers did not allow significant dissolution in the IL.
2. The results indicated that the BBSP films had higher tensile strength, although the DP of cellulose was less for it.
3. Presence of the lignin resulted in an increase in the dynamic contact angle of the films with water droplet.

4. The lignin resulted in less transparency of the films, but increased absorption of UV radiations.
5. The above mentioned results showed that even impure lignocellulose feedstock, specially following pulping process can have promising features.

References

[1] Abdulkhani A, Marvast E H, Ashori A and Karimi A N 2013 Effects of dissolution of some lignocellulosic materials with ionic liquids as green solvents on mechanical and physical properties of composite films *Carbohydrate Polym* **95** 57–63
[2] Soheilmoghaddam M, Wahit M U, Tuck Whye W, Ibrahim Akos N, Heidar Pour R and Ali Yussuf A 2014 Bionanocomposites of regenerated cellulose/zeolite prepared using environmentally benign ionic liquid solvent *Carbohydrate Polym* **106** 326–34
[3] Niroomand F, Khosravani A and Younesi H 2016 Fabrication and properties of cellulose-nanochitosan biocomposite film using ionic liquid *Cellulose* **23** 1311–24
[4] Lateef H, Grimes S, Kewcharoenwong P and Feinberg B 2009 Separation and recovery of cellulose and lignin using ionic liquids: a process for recovery from paper-based waste *Journal of Chemical Technology & Biotechnology* **84** 1818–27
[5] Zhang H, Wu J, Zhang J and He J 2005 1-Allyl-3-methylimidazolium Chloride Room Temperature Ionic Liquid: A New and Powerful Nonderivatizing Solvent for Cellulose *Macromolecules* **38** 8272–7
[6] Swatloski R P, Spear S K, Holbrey J D and Rogers R D 2002 Dissolution of Cellose with Ionic Liquids *J Am Chem Soc* **124** 4974–5
[7] Fort D A, Remsing R C, Swatloski R P, Moyna P, Moyna G, and Rogers R D 2007 Can ionic liquids dissolve wood? Processing and analysis of lignocellulosic materials with 1-n-butyl-3-methylimidazolium chloride *Green Chemistry* **9** 63–9
[8] Kilpeläinen I, Xie H, King A, Granstrom M, Heikkinen S and Argyropoulos D S 2007 Dissolution of Wood in Ionic Liquids *J Agr Food Chem* **55** 9142–8
[9] Pinkert A, Marsh K N, Pang S and Staiger M P 2009 Ionic Liquids and Their Interaction with Cellulose *Chem Rev* **109** 6712–28
[10] Zhang J, Wu J, Yu J, Zhang X, He J and Zhang J 2017 Application of ionic liquids for dissolving cellulose and fabricating cellulose-based materials: state of the art and future trends *Materials Chemistry Frontiers* **1** 1273–90
[11] Pu Y, Jiang N and Ragauskas A J 2007 Ionic Liquid as a Green Solvent for Lignin *Journal of Wood Chemistry and Technology* **27** 23–33
[12] Zobel B, Tappi R M- and 1958 U 1958 Variation of cellulose in loblolly pine *Tappi* **41** 167–70
[13] French A D 2014 Idealized powder diffraction patterns for cellulose polymorphs *Cellulose* **21** 885–96
[14] Wu R-L, Wang X-L, Li F, Li H-Z and Wang Y-Z 2009 Green composite films prepared from cellulose, starch and lignin in room-temperature ionic liquid *Bioresource Technol* **100** 2569–74
[15] Simmons T J, Lee S H, Miao J, Miyauchi M, Park T-J, Bale S S, Pangule R, Bult J, Martin J G, Dordick J S and Linhardt R J 2011 Preparation of synthetic wood composites using ionic liquids *Wood Sci Technol* **45** 719–33
[16] Mahmoudian S, Wahit M U, Ismail A F and Yussuf A A 2012 Preparation of regenerated cellulose/montmorillonite nanocomposite films via ionic liquids *Carbohydrate Polym* **88** 1251–7

IOP Conf. Series: Materials Science and Engineering **368** (2018) 012030 doi:10.1088/1757-899X/368/1/012030

Exploring Sudanese agricultural residues as alternative fibres for pulp and paper manufacturing

H A M Saeed[1,2*], Y Liu[1*] and H Chen[1]

[1]Key Lab of Pulp and Paper Science and Technology of Ministry of Education (Shandong Province) Qilu University of Technology, Jinan, 250353, Shandong, P.R. China.
[2] Center of Fibers, Papers and Recycling, Faculty of Textiles, University of Gezira, Box 20, Wad Medani, Sudan.

*E-mail: haroonsaeed75@gmail.com, leoliuyu@163.com

Abstract. In this work, the chemical composition, fibre dimensions and morphology, elemental analysis and paper characterisation of sesame, millet, karkadeh and sorghum stalks were studied. Soda and soda-anthraquinone pulping as sulfur-free processes were applied. It was found that, the cellulose content of karkadeh, sesame, millet and sorghum stalks were 45.80, 42.30, 40.99 and 35.40%, respectively. Klason lignin was found to be highest in sesame 20.85% while it was 19.32, 18.20 and 10.32% in karkadeh, millet and sorghum, respectively. Moreover, karkadeh showed highest pulp yield (46.60% and 43.50%) with viscosity of (655 ml/g and 640 ml/g) and kappa number of (18.00 and 19.40) for soda-AQ and soda pulps, followed by sesame, okra and sorghum, respectively. Overall strength properties of the handsheet made from karkadeh pulp were significantly higher than those made from sesame, millet, and sorghum pulps. Morphological analysis and chemical composition of the studied fibrous materials showed their suitability for producing paper of various grades including writing and printing paper as well as packaging applications.

1. Introduction

An increase in pulp and paperboard consumption, stricter environmental and sustainability regulations, and the increased use of wood materials for furniture production have prompted scientists and researchers to seek additional lignocellulosic material for pulp and papermaking [1]. However, agricultural residues have become one of the most important alternative sources of fibrous resources in the 21st century [2]. However, agricultural residues are not sometimes used eco-friendly as after harvesting they are directly burnt in opened land under not well-controlled conditions, causing destructive environmental impact due to the heavy smog [3-4]. Agricultural residues cellulosic materials can be applied as an effective alternative to constantly decreasing forest resources in different regions especially in the developing countries [5-6]. However, to make better utilization of these cheap and abundant agricultural residues, these stalks could be considered as promising cellulosic raw materials for nano-cellulose pulp and paper making [2, 7].

Sudan is rich in agricultural residues cellulose materials, which can be good candidate cellulosic materials for pulp and paper manufacturing [8]. However, this article investigates the application of Sudanese agricultural residues, namely; sesame, millet, karkadeh and sorghum in pulp and paper making.

Sesame is belongs to Pedaliaceae family, it is one of the most common traditional oilseed crops in the world. Sesame crop belongs to the tropical climatic zone; it is largely grown in India, China, Mynemar, and Sudan [9-10]. In Sudan, sesame grown on stored soil moisture with minimal irrigation or rainfall and can produce good yield under high temperature in many states. However, It represent the third important crop after sorghum and millet [11]. Ateş, F. *et al.* [12] investigated the pyrolysis of sesame stalk as biomass, they found that sesame stalk had 48 % cellulose content. This high cellulose content of sesame stalks make it a promising candidate for pulp and paper making.

Millet known as "Dukhun," is the most important cereal crop in Sudan (Kordofan and Darfur States). The average total area annually cultivated is approximately 2.5 million hectares [13] it is application in pulp and paper manufacturing can be more beneficial because it is abundant, inexpensive, and can provide economic and environmental benefits [14]. Harinarayana, G., et al [15] investigated millet stalks chemical composition, they found that the millet stalks were rich in cellulose content (39.4 %), hemicellulose (23.9 %), and relatively low in lignin (12.8 %), thus representing a promising feedstock.

Karkadeh is belongs to Malvaceae family, is commonly known as Roselle, hibiscus, Jamaica sorrel or red sorrel (English) and in Arabic, karkadeh [16]. It is an ideal crop for developing countries as it is relatively easy to grow, can be grown as part of multi-cropping systems and it has multi usages. It can grow annually and attain a height in range of 2 to 2.5 m [17-18]. In Sudan, karkadeh is grown in many different area, it is one of the cash crops cultivated by traditional farmers in Kordofan and Darfur states under rain-fed conditions, where large quantities are produced both for local consumption and for export [19]. Khristova, P. and M. Tissot [20] studied the chemical composition of the karkadeh, the results shown that it constituted of 45.56 % cellulose, lignin 20.11% and 2% ash. Sorghum is an important crop in Sudan, and the country ranked eighth in the world in sorghum grain production (3.5 million tonnes) in 2014 [21]. Jiménez *et al.* [22] studied the application of sorghum stalks in pulp and papermaking and determined they had better properties than other agricultural residues, such as olive tree fellings, wheat straw, sunflower stalks, vine shoots, and cotton plant stalks. Gençer *et al.* [23] identified conditions for producing pulp and paper from sorghum. Indeed, the length of sorghum stalks is sufficient to make them a viable alternative for pulp and paper production.

Agricultural residues materials are beneficial in pulp and paper industry in terms of environmental and socio-economic aspects. Due to this reason it was thought that these stalks could be utilised as an alternative resources for pulp and paper making. In his study, sesame, millet, karkadeh and sorghum stalks were investigated on their fibres dimensions and morphological properties, chemical composition, soda and soda-anthraquinone pulping and pulp and paper properties.

2. Experimental

2.1 Materials

Sesame, millet, karkadeh and sorghum stalks were collected from a farm in north Kordofan State (latitude 15° 40′ N and 3° 32′ E) Sudan in February 2016. The stalks were air dried for weeks then fragmented separately into small chip sized 25 x 25 x 2.5 mm using band and dick saw. All the chemicals were purchased from Sinopharm Chemical Reagent Co., Ltd. (Shanghai, China) and were used without further purification.

*2.2 **Chemical composition***

The chips were grinded in Wiley mill to produce sawdust on 40/60 mesh, which was analyzed for chemical composition. The amount of extractive in hot and cold water, 1% sodium hydroxide and benzene-ethanol mixture (20/80 v/v) were determined using the TAPPI standard methods; TAPPI T207 cm-08, TAPPI T212 om-07 and TAPPI T204 cm-07, respectively [24]. The mean chemical components; ash and klason lignin, were determined according to TAPPI T 211 om-02 [25] and

IOP Conf. Series: Materials Science and Engineering **368** (2018) 012030 doi:10.1088/1757-899X/368/1/012030

TAPPI T211-om-93, while holocellulose was determined by Wise's chloride method [26]. Moreover, acid-soluble lignin was determined based on the method of UV–vis spectro-metric at wavelength 205 nm TAPPI um 250 [27].

2.3 *Pulping*

The pulping of these agricultural residues was carried out using soda and soda-anthraquinone process. However, pulping were performed in electrically heated laboratory rotary a 5-L stationary stainless steel digester consisting of 4 vessels each with a capacity of 1 L (NAC Autoclave Co. Ltd., Japan) according to . A total of 100 g of each material were cooked in 25 % a freshly prepared caustic soda. The amount of anthraquinone was kept at 0.1% (w/w), with respect to the oven dried (o.d.) initial solid material for soda-AQ pulping. Liquor-to-material ratio of 5:1, time to maximum temperature of 60 min, maximum temperature of 160 °C, cooking time at maximum temperature of 120 min. At the end of pulping, the pulp was pressed to remove the liquor then washed with running water to neutral pH, then, disintegrated in a pulp mixer and screened on a 0.15 mm slotted plate. The pulp yield was determined by gravimetric method as dry matter obtained on the basis of oven dried raw material.

2.4 *Pulp Properties*

Kappa number, Canadian Standard Freeness (CSF) and viscosity, were determined according to the following methods TAPPI T236 om-99, TAPPI T227 om-99 and TAPPI T230 om-89, respectively [28]. Totally chlorine-free bleaching (ECF) was carried out for the screened pulp in polyethylene bags in a water bath at 10% pulp concentration using (D_0ED_1) bleaching sequence (D: chloride 3% ClO_2, Ep: alkaline-peroxide 1% NaOH) at 80 °C for 2 h for each stage [29].

2.5 *Paper making and testing*

The screened pulp was beaten in a Hollander beater for 5000 revolutions and 35 °SR freeness according to TAPPI T248 sp-00. Standard handsheets with basis weight of 60 g/m^2 were made using a laboratory hand-sheet former (PTI laboratory Equipment, Vorchdorf, Austria) according to the TAPPIT 205 sp-95 standard method.

2.6 *Paper Properties*

The handsheets properties were tested based on the following standards: apparent density TAPPI T 411 om-89, tear index TAPPI T 414 om-98, tensile strength TAPPI T 494 om-01, burst index TAPPI T 403 om-97, opacity TAPPI T1214 sp-98 and brightness TAPPIT1216 sp-98.

2.7 *The elemental analysis*

The elemental analysis (Carbon (C), oxygen (O), Aluminium (Al), Nitrogen (N), Silicon (Si), Magnesium (Mg), and Calcium (Ca)) were observed under a scanning electron microscope (SEM) (OCTANE 9.88/1114658 AMETEK®,Mahwah, USA).Before testing, paper samples were gold coated in a Sputtergerät SCD 005 sputter coater (England). A sputter current of 60 mA, sputter time 90 s.

2.8 *The fibre morphological characteristics*

For the measurements of fibre's dimensions, samples were macerated in a mixture of 30% hydrogen peroxide and acetic acid (1:1). Fibre length, width and kinked and curled indices were analyzed with an FQA device (LDA02128, OpTest Equipment Inc., Ontario, Canada), model LDA-02 according to TAPPI T271 om-07. Lumen diameter and wall thickness were measured by Leica DMLB (Leica Microsystems GmbH, Wetzlar, Germany) connected to a video camera Leica DFC490 (Leica Microsystems GmbH, Wetzlar, Germany) at 1000× magnifications.

3. Results and discussion

IOP Conf. Series: Materials Science and Engineering **368** (2018) 012030 doi:10.1088/1757-899X/368/1/012030

3.1 Chemical composition of residues material

Table 1 shows the percentages of chemical contents and some solubility values in sesame, millet, karkadeh and sorghum stalks. Karkadeh shown the highest cellulose content 45.80 % with low lignin 19.32 % , while cellulose and lignin contents are (42.30 and 20.85 %), (40.99 and 18.20 %) and (35.4 and 10.35 %) of sesame, millet and sorghum stalks respectively. Compared to other non-woods, karkadeh millet, and sesame shown same cellulose content as date palm 45.04 % and wheat straw 47.14 % [30], pineapple leaves 41.15 % [31] and higher than cotton and corn stalks [32]. In pulp and paper, a high cellulose content is preferable because it leads to produce high pulp yield as well as the quality paper. However, less lignin content is desirable in the pulp and paper because less chemicals and energy are required during the pulping process. Furthermore, low lignin will also produce paper with good optical properties and therefore requires less bleaching [33-34]. These raw fibrous materials content less lignin than rapeseed straw, rice straw and corn straw [35].

Table 1: Chemical composition of sesame, millet, karkadeh and sorghum stalks

Chemical Composition (%)	Sesame	Millet	Karkadeh	Sorghum
Cellulose	42.30±2.2	40.99±2.5	45.80±2.5	35.40±1.8
Hemicellulose	22.10±2.1	20.90±2.0	22.82±2.4	19.41±1.3
Klason lignin	20.85±1.3	18.20±1.8	19.32±1.4	10.27±0.8
Acid Insoluble Lignin	0.05±0.01	0.06±0.01	0.06±0.01	0.05±0.01
Ash	4.50±1.0	5.96±0.5	3.20±1.1	5.31±0.6
Hot water extractives	10.70±1.2	10.20±1.4	8.60±1.3	13.20±1.2
Cold water extractives	5.80±1.1	4.50±0.7	4.60±1.0	11.60±1.1
Alcohol-benzene	5.90±1.1	4.60±1.1	3.40±1.2	5.20±0.7
1% NaOH	20.30±2.5	21.30±2.0	19.55±2.3	16.20±1.6

Hemicellulose content of karkadeh (22.82 %) is the highest compared to sesame (22.10 %), millet (20.90 %) and sorghum (19.41 %) as shown in Table 1. High hemicellulose content is granulated for pulp and paper making because it enhance the mechanical properties of the paper produced particularly in tensile and burst index [36]. However, paper produced from karkadeh are expected to have high tensile strength and burst index. The highest ash was observed in millet stalk 5.90%, while karkadeh shown the lowest 3.20 %. However, the lower ash content is desirable, because, it expected to decrease chemicals required during pulping and bleaching.

Solubility in water is the measure of the extractive compositions of the ligncellulosic material. However, the high extractives values were indicating high presence of tannins, alkaloids and starch [37]. However, sorghum shown the highest amount of extractives in hot and cold water (24.80 %) compared to that (16.50, 14.70 and 13.20 %) for sesame, millet and karkadeh, respectively. These values are very high compared with that found in hardwood (2–8 %) [38] but similar to that of some non-wood materials such as sunflower and corn stalks [5], and less than that of rapeseed straw reported by 18 % [35]. Moreover, the solubility in different solvents indicated the extractive contents, which were not cell wall components [33].

The solubility predicts the mass loss of the carbohydrates and other alkali soluble matters in raw fibrous materials [4]. On the other hand, high solubility suggests high mass loss during the pulping process and low pulp yield [39]. The high extractives contents of okra led to decrease in pulp yield, thus, increasing in chemicals needed in cooking and bleaching [5].

3.2 *Pulp characterization*

Table 2: Properties of pulp from sesame, millet, karkadeh and sorghum stalks

Material	Pulp yield (%)		Kappa number		Viscosity (ml/g)		CSF (ml)	
	Soda	So-AQ	Soda	So-AQ	Soda	So-AQ	Soda	So-AQ
Sesame	40.60	43.30	20.31	18.17	620	635	410	380
Millet	40.10	42.04	14.50	12.43	650	665	560	555
Karkadeh	43.50	46.60	19.40	18.00	640	655	425	405
Sorghum	33.00	33.15	16.30	15.12	670	744	440	430

The properties of the pulps obtained from the sesame, millet, karkadeh and sorghum stalks are listed in Table 2. The results reveal that there were some differences in pulps properties of the investigated agricultural residues produced by the two pulping methods. However, pulp yield for the studied materials was in the range of (43.50 to 33.00 %) when cooked with soda process while the yield was increased by 3 % with the use of 0.1 % AQ similar results were reported previously [40]. Among the studied materials, karkadeh shown the highest pulp yield regardless the pulping methods, this is attributed to the high cellulose content in karkadeh stalks. However, a higher pulp yield with soda-AQ pulps, compared to the soda method, was due to higher hemicellulose retention [41]. Moreover, the combination of AQ to soda may accelerates the delignification rate and protect cellulose degradation, which was illustrated by the quantity of pulp yield [42]. Similar results have been observed by several authors [43]. On the other hand, sesame pulp shown the highest kappa numbers of 20.13 and 18.17 for soda and soda-AQ pulps, respectively, while millet shown the lowest kappa numbers 14.50 and 12.34 for soda and soda-AQ pulps, respectively. It is noted that pulp yield improved and kappa number reduced when soda method was modified by AQ, this result could be attributed to the acceleration of delignification rate with the addition of AQ [43-44]. Moreover, low kappa number also led to less consumption of bleaching chemicals, which directly led to decreasing the effluent load [33]. It was reported that the viscosity decreased with decreasing yield percentages for both types of pulping [45]. However, among the studied materials, sorghum shown highest viscosity (670 and 744 ml/g) while sesame shown the lowest (635 and 620 ml/g) for soda-AQ and soda pulp, respectively.

Table 3: Fibres properties of sesame, millet, karkadeh and sorghum stalks

Material	Sesame	Millet	Karkadeh	Sorghum
Fibre length (mm)	0.80±0.2	0.41±0.3	0.92±0.3	0.52±0.3
Fibre diameter (µm)	16.30±1.2	27.10±1.5	14.00±1.5	26.80±1.4
Lumen diameter (µm)	14.20±1.3	13.31±1.4	15.20±1.4	14.10±1.2
Cell wall thickness (µm)	6.80±0.3	7.40±0.4	6.20±0.2	6.10±0.3

As shown in Table 3, the results reveal that these materials contained short fibre. However, karkadeh shown the highest fibre length 0.92 mm [46] compared to that of 0.80 mm of sesame, 0.65mm of sorghum and 0.41mm of millet. These results are similar to pea [47], *C. orientalis* and *C. tataria* [48]. However, based on the results of the fibre length, karkadeh produced paper is expected to show best tea strength. On the other hand, higher fibre length improve tensile properties of paper but causes distortion of paper [49]. The average fibre diameter of karkadeh 14.00 µm is relatively smaller than that of sesame 16.30 µm, 26.80 sorghum and 27.10 µm millet and. A small diameter of fibre indicates that less fibre flexibility. Lumen width of the fibre has remarkable influence on the beating of the pulp. The wider the lumen width, better responses to the beating of the pulp can be observed because of the permeation of liquids into free spaces in and between the fibres. The thickness of the fibres cell wall play important role on most paper properties, with thick-walled fibres forming bulky sheets with less quality of tensile and burst indices and folding endurance but with a high tearing index [33]. It was reported that increase in fibre length and decrease in cell wall thickness had noticeable effects on the physical properties of paper [48]. However, millet stalks shown highest cell wall thickness 7.40 µm compared to sesame 6.40 µm and followed by karkadeh 6.20 µm and sorghum 6.10 µm. These are comparatively similar to kenaf 6.4 µm [42] and sunflower stalk 5.9 µm [7].

3.3 Elemental analysis

Table 4: Elemental analysis of sesame, millet, karkadeh and sorghum stalks

Material	C (%)	O (%)	N (%)	Mg (%)	Al (%)	Si (%)	Ca (%)
Sesame	52.60	41.20	4.20	0.34	0.54	2.80	0.45
Millet	45.21	20.90	2.09	0.38	0.17	0.48	0.50
Karkadeh	50.80	39.30	2.40	0.24	0.07	3.05	0.55
Sorghum	43.54	51.14	2.02	0.53	0.23	0.57	0.75

The SEM-EDS elemental analysis is shown in Table 4. However, the results reveal that the predominant elements in paper obtained from the studied materials were C and O and lesser amounts of mineral elements such as N, Mg, Al, Si and Ca were also discovered. Among the studied material sesame stalks shown highest C and O content (52.60 and 41.20%), respectively [50] compared to (50.8 and 39.30%) of karkadeh, (43.54 and 51.14%) of sorghum and (45.21 and 20.90%) of millet. However, similar results were observed in wheat straw [51], corn hull [52] and rice husk and bagasse [53]. However, the mineral substances appear in these agricultural residues do not present any counter indication for chemical pulping.

3.4 Paper Properties

Table 5: Physical properties of papers from of sesame, millet, karkadeh and sorghum stalk.

Properties	Sesame		Karkadeh		Sorghum		Millet	
	Soda	So-AQ	Soda	So-AQ	Soda	So-AQ	Soda	So-AQ
A. D (g/cm^3)	0.50	0.54	0.55	0.58	0.51	0.53	0.40	0.41
T. I (N m/g)	45.90	46.80	51.70	52.30	47.10	50.70	32.15	33.95
Te. I (mNm^2/g)	3.80	4.10	4.60	5.50	2.40	2.61	2.28	2.31
B. I (KPa m^2/g)	3.40	3.10	3.50	3.90	2.15	2.55	1.90	1.96
Opacity (%)	85.40	86.50	86.60	87.90	81.40	82.60	88.00	88.50
B (%)	65.90	67.25	75.50	77.64	67.10	71.15	65.10	66.30

AD: Apparent Density, T I: Tensile Index; Te.I: Tearing Index; B.I: Bursting Index; B: Brightness.

The paper properties tested were those pertinent for assessment the quality of liner and fluting papers [54]. Among the studied materials papers produced from karkadeh pulp shown the highest tensile strength (52.30 and 51.70N m/g), tearing indices (13.50 and 12.60 mNm^2/g) and bursting indices (3.90 and 3.50 KPa m^2/g) for soda-AQ and soda pulp, respectively Table 5. Similar result were reported by [55-56]. This result may be due to the higher fibre flexibility, higher bonding ability, higher slenderness ratio [7] and lower Runkel ratio [43] of the karkadeh pulp fibres. On the other hand, millet hand sheet shown the lowest tensile strength (33.95 and 32.15 N m/g), tearing indices (2.31 and 2.28 mNm^2/g) and bursting indices (1.96 and 1.90 KPa m^2/g) for soda-AQ and soda pulp, respectively. Comparing the paper properties obtained from soda and soda-AQ pulps, the results reveal that the mechanical properties of the paper samples were influenced by addition of AQ as well as the delignification degree through the reaction with quinine methide intermediates of lignin [42, 43]. Furthermore, the additional removing of lignin and extractives and formation of new hydrogen bonds between molecules of cellulose and hemicelluloses may also led to increasing in paper physical properties [5]. Moreover, the soda pulp of the studied materials shown a lower brightness than that soda-AQ.

Hand sheet produced from karkadeh fibres shown highest apparent density (0.58 and 0.55 %) compared to that (0.54 and 0.50%) of sesame followed by (0.53 and 0.51 %) of sorghum and (0.41 and 0.40 %) millet for soda-AQ and soda pulp, respectively. It was reported that the apparent density of the hand sheet increased with the increased of short fibres content in the pulp that led to increases the fibre packing efficiency . The apparent density of soda -AQ pulps were generally higher than soda pulps, this is could be due to highly refined flexible fibres obtained by the soda- AQ pulping which increase the fibre packing efficiency [57].

The best bleachability and highest brightness (77.64 and 75.50 %) was achieved by karkadeh for soda-AQ and soda pulps, respectively, while it was (71.15 and 67.10 %), (67.00 and 65.50 %) and (66.30 and 65.10 %) of soda-AQ and soda pulps of sorghum, sesame and millet respectively. However, brightness represent the kappa number, which in turn to evaluates the level of delignification. Opacity is an important aspect of paper appearance [58]. Millet showed the highest opacity of (88.50 and 88.00 %) followed by karkadeh (87.90 and 86.60 %) for soda-AQ and soda pulps, respectively, while it was (86.50 and 85.40 %) and (85.60 and 84.40 %) of soda-AQ and soda pulps of sesame and sorghum respectively. This could be attributed to the extraction of lignin during pulping which led to replace hemicellulose on the fibre surfaces, therefore, hemicellulose may enhances fibre-to-fibre bonding, this will led to less voids between the fibres then reduces the light scattering at fibre-air and pigment-air interfaces [58].

3.5 Surface Morphology Analysis

Scanning electron microscopy (SEM) analysis of the paper produced from sesame, millet, karkadeh and sorghum fibres were magnified at 1000X are shown in Figure1 (A, B, C &D). However, the strength of the fibre can be understood based on the order and formation of the fibre pattern [59]. From Figure 1 (B), karkadeh handsheet fibres are parallel, aligned and closed to each other rather than that of millet Figure 1 (A), sesame Figure 1 (C) and sorghum Figure 1 (D). This is attributed to the higher fibre content as well as long fibres in karkadeh compared to the other species. Moreover, in Figure 1 (B) as can be seen that the structure of the karkadeh fibres are closely compacted and formed many more bundles of fibre matrix and fibres are quite homogeneous and resembled more than that of millet, sorghum and sesame handsheets.

Figure 1. SEM micrographs of the surface of millet (A), karkadeh (B), sesame (C) and sorghum (D) handsheet (1000X)

4. Conclusion

Agricultural residues could be considered an interesting and promising candidate as cellulosic fibre source for pulp and paper industry since they are abundant, inexpensive and renewable. Moreover, the high cellulose contents (45.80, 42.30, 40.99 and 35.40 %) and low lignin contents (19.32, 20.85, 18.20 and 10.35 %) of karkadeh, sesame, millet and sorghum stalks, respectively, led to high-quality pulp and paper with good physical properties. The use of soda-AQ pulping process for these agricultural residues helped delignification, increased pulp yield and enhanced the paper mechanical properties. However, the results showed that a lower kappa number, higher screened pulp yield, and better paper properties were obtained for the pulp produced via soda-AQ pulping than the soda process. As paper is usually a mixture of short fibres and long fibres, the morphological analysis shows that the studied fibres can be as promising short fibres for various paper grades.

Acknowledgments

The authors are grateful for the financial support from the National Science Foundation of China (Grant Nos. 31470602, 31600472, and 31570566,), the Foundation for Outstanding Young Scientist in Shandong Province (BS2015SW011), the Project of Shandong Province Higher Education Science and Technology Program (No. J13LD03), the Foundation of Key Laboratory of Pulp and Paper Science and Technology of Ministry of Education (Shandong Province) of China.

References

[1] Xing L, Xu M and Pu J 2016 *BioResources* **12** 871
[2] Jonoobi M, Oladi R, Davoudpour Y, Oksman K, Dufresne A, Hamzeh Y and Davoodi R 2015 *Cellulose* **22** 935
[3] Fortunati E, Luzi F, Jiménez A, Gopakumar D, Puglia D, Thomas S, Kenny J, Chiralt A and 2016 *Carbohydr* Torre L . *Polym* **149** 357
[4] Espinosa E, Tarrés Q, Delgado-Aguilar M, González I, Mutjé P and Rodríguez A 2016 *Cellulose* **23** 837
[5] Barbash V, Trembus I, Alushkin S and Yashchenko O 2016 *ScienceRise* **3** 71
[6] Fahmy Y, Fahmy T Y, Mobarak F, El-Sakhawy M and Fadl M 2017 *International Journal of ChemTech Research* **2** 424

[7] Rudi H, Resalati H, Eshkiki R B and Kermanian H 2016 *J. Cleaner Prod* **127** 562
[8] Saeed H A, Liu Y, Lucia L A and Chen H 2017 *BioResources* **12** 4166
[9] Sehr E M, Okello-Anyanga W, Hasel-Hohl K, Burg A, Gaubitzer S, Rubaihayo P R, Okori P, Vollmann J, Gibson P and Fluch S 2016 *J. Crop Sci. Biotechnol* **19** 117
[10] Warra A 2011 *Bayero Journal of Pure and Applied Sciences* **4** 164
[11] Abdellatef E, Sirelkhatem R, Ahmed M M, Radwan K and Khalafalla M 2008 *African Journal of Biotechnology* **7** 24
[12] Ateş F, Pütün E and Pütün A 2004 *J. Anal. Appl. Pyrolysis* **71** 779
[13] Abuelgasim E H 2011 *Sudanese Encyclopedia of Agricultural Sciences*
[14] Elzaki O T, Khider T O and Omer S H 2012 *Cellulose* **46** 47
[15] Harinarayana G, Melkania N, Reddy B, Gupta S, Rai K and Kumar P S 2005 *Forage potential of sorghum and pearl millet* 292
[16] Shruthi V, Ramachandra C, Nidoni U, Hiregoudar S, Naik N and Kurubar A 2016 *Plant Archives* **16** 515
[17] Da-Costa-Rocha I, Bonnlaender B, Sievers H, Pischel I and Heinrich M 2014 *Food Chem* **165** 424
[18] Nadlene R, Sapuan S, Jawaid M, Ishak M R and Yusriah L 2016 *J. Nat. Fibers* **13** 10
[19] Mohamed B B, Sulaiman A A and Dahab A A 2012 *Bull. Environ. Pharmacol. Life Sci* **1** 48
[20] Khristova P and Tissot M 1995 *Bioresour Technol* **53** 67
[21] Hamed A H, Abbas S O, Ali K A and Elimam M E 2015 *Animal Review* **2** 68
[22] Jiménez L, López F and Martínez C 1993 *Holzforschung-International Journal of the Biology, Chemistry, Physics and Technology of Wood* **47** 529
[23] Gençer A and Şahin M 2015 *BioResources* **10** 2850
[24] Tappi T 2004 TAPPI test methods
[25] Standard T 2007 Tappi Standard
[26] Wise L E 1946 *Paper Trade Journal* **122** 35
[27] Tappi T 2002 2002–2003 TAPPI Test Methods
[28] Sarker M, Sutradhar S, Sarwar A G, Uddin M N, Chanda S C and Jahan M S 2017 *Journal of Bioresources and Bioproducts* **2** 24
[29] ATEŞ S, Deniz I, KIRCI H, Atik C and Okan O T 2015 *Turk. J. Agric. For.* **39** 144
[30] Nasser R A, Hiziroglu S, Abdel-Aal M A, Al-Mefarrej H A, Shetta N D and Aref I M 2015 *Measurement* **62** 179
[31] Banerjee R, Chintagunta A D and Ray S 2017 *J. Cleaner Prod* **149** 387
[32] Huang L, Ding T, Liu R and Cai J 2015 *J. Therm. Anal. Calorim* **120** 1473
[33] Shakhes J, Marandi M A, Zeinaly F, Saraian A and Saghafi T 2011 *BioResources* **6** 4481
[34] Syed N F N, Zakaria M H and Bujang J S 2016 *BioResources* **11** 5358
[35] Chaker A, Mutjé P, Vilar M R and Boufi S 2014 *Cellulose* **21** 4247
[36] Kassim A, Aripin A, Hatta Z and Daud Z, *2014S pringer* 19
[37] Tutuş A, Kazaskeroğlu Y and Çiçekler M 2015 *BioResources* **10** 5407
[38] Ferhi F, Das S, Moussaoui Y, Elaloui E and Yanez J G 2014 *Ind. Crops Prod* **59** 109
[39] Van Tran A 2006 *Ind. Crops Prod* **24** 66
[40] Akgül M and Tozluoglu A 2009 *African Journal of Biotechnology* **8** 6127
[41] Jahan M S, Rahman M and Rahman M M 2012 J-FOR **2** 12
[42] Shakhes J, Zeinaly F, Marandi M A and Saghafi T 2011 *BioResources* **6** 4626
[43] GÜLSOY S K, PEKGÖZLÜ A K and AKTAŞ A C 2015*Turk. J. Agric. For* **39** 295
[44] Akguuml M and Tozluoğlu A 2010 *Scientific Research and essays* **5** 1068
[45] Das A K, Nakagawa-izumi A and Ohi H 2016 *Carbohydrate polymers* 147 133
[46] Khristova P, Gabir S, Bentcheva S and Dafalla S 1998 *Ind. Crops Prod* **9** 9
[47] Gonzalo A, Bimbela F, Sánchez J L, Labidi J, Marín F and Arauzo J 2017 *J. Cleaner Prod* **156** 184

[48] Tutus A, Comlekcioglu N, Karaman S and Alma M H 2010 *International Journal of Agriculture and Biology* **12** 286
[49] Omotoso M A and Owolabi A W 2015 *International Journal of Chemistry* **7** 113
[50] Ateş F, Pütün A E and Pütün E 2006 Fuel **85** 1851
[51] Taherymoosavi S, Joseph S and Munroe P 2016 *J. Anal. Appl. Pyrolysis* **120** 441
[52] Zhang T, Walawender W P, Fan L, Fan M, Daugaard D and Brown R 2004 *Chem. Eng. J.* **105** 53
[53] Munir S, Daood S, Nimmo W, Cunliffe A and Gibbs B 2009 *Bioresour. Technol* **100** 1413
[54] Gonzalo A, Bimbela F, Sánchez J, Labidi J, Marín F and Arauzo J 2017 *J. Cleaner Prod*
[55] Sharma A K, Anupam K, Swaroop V, Lal P S and Bist V 2015 *J. Cleaner Prod* **106** 422
[56] de Andrés M Á, Sequeiros A, Sánchez R, Requejo A, Rodríguez A and Serrano L 2016 *Environ. Eng. Manage. J.* **15** 2479
[57] Kamoga O L, Kirabira J, Byaruhanga J, Godiyal R and Anupam K 2016 *Cellul. Chem. Technol* **50** 275
[58] Moradbak A, Tahir P M, Mohamed A Z, Peng L C and Halis R 2016 *BioResources* **11** 5994
[59] Daud Z, Hatta M Z M, Kassim A S M, Awang H and Aripin A M 2013 *BioResources* **9** 872

Activated carbons from oil palm shell for hydrogen storage

W Zhao[1], L Luo[1], T Chen[1], Z Li[1], Z Zhang[1] and M Fan[1, 2*]

[1] College of Material Engineering, Fujian Agriculture and Forestry University, NO 15 Shangxiadian Road, Fuzhou 350002, PR China

[2] College of Engineering, Design, and Physical Sciences, Brunel University, Uxbridge UB8 3PH, United Kingdom

*Corresponding author: weigang-zhao@fafu.edu.cn

Abstract. Activated carbons (ACs) with controlled microporosity have been developed from oil palm shell and their H_2 storage performances been have tested at 77 K. Such adsorbents are the natural agricultural by-products of chemically activated with KOH. N_2 adsorption-desorption at 77 K was used to investigate the pore structure of the ACs produced with different weight ratio of KOH/oil palm shell charcoal. The results showed that the developed oil palm shell ACs achieved surface areas (S_{BET}) as high as 3508 m^2/g and micropores volumes (V_{DR}) as high as 1.10 cm^3/g. Outstanding storage capacities of hydrogen, as high as 6.8 wt.% have been obtained at 4 MPa and 2.86 wt.% have been obtained at 1 bar for some of these adsorbents at 77 K. These values of hydrogen adsorption are among the best that ever published so far, which well above those of some reference commercial materials, e.g. Maxsorb-3, and also the ACs from the open literature.

1. Introduction

Hydrogen is a renewable energy, which holds the potential to be a means of providing clean, safe, affordable, and secure energy in the future because it is a renewable energy source that generates no carbon emissions [1]. It is frequently proposed as an alternative to the current fossil fuel-based energy system to solve problems pertaining to energy distribution and environmental issues. However, one of the greatest problems of hydrogen arises from its supercritical nature under standard conditions (Tc = 33.19 K, Pc = 1.296 MPa), giving it a very low density: only 90g m^{-3} at 1 bar and 273 K. The US Department of Energy (DOE) set new targets for hydrogen storage systems in light-duty vehicles in 2009, which correspond to reversibility, a gravimetric density of approximately 5.5 wt.%, and consideration of the entire storage system [2]. The methods for hydrogen storage include traditional compressing and liquefying hydrogen, but also with different materials by physisorption or chemisorption [3-6]. However, there is no method can reach the new targets at room temperature, even at 77 K. Since stationary applications are not limited by their weight and therefore technical solutions could be available in a shorter time.

The success of any future hydrogen economy depends on the development of inexpensive materials with sufficiently high hydrogen-storage capacity. Among all the materials, porous and high-surface area activated carbons (ACs) are among the most important candidates to store efficiently hydrogen due to their advantages of being reversible, having a high specific surface area and pore microstructure, being light weight (low density), being stable for large scale production, and having fast kinetics. The other important reason is for the stationary applications, in which the weight and volume are not limiting factors. ACs can be prepared from the precursors of carbonaceous materials having a high carbon content and moderate inorganic contents, such as coal, peat, polymeric materials, wood, bamboo, and various agricultural by-products [5-8]. Among them, previous research reported that oil palm shell consisting of 26.4 % fix carbon which indicated as a good source as raw material for preparation activated carbon [9]. Meanwhile, oil palm shell is a common low-cost agricultural waste in Malaysia, more than 4.7 million tonnes produced in 2007 [10]. All above mentioned make that the oil palm shell is a suitable precursor for activated carbons production. Some studies have reported that using the activated carbon from oil palm shell for CO_2, chromium (VI), phenol, and methylene blue dye adsorption [9-11]. Efficient utilization of oil palm shell, in particular for high value-added applications, has yet to be explored.

This research presents the development of activated carbons using KOH as the activation agent from oil palm shell, which combine low cost, low ash content and an ideal microtexture for getting a tunable nanoporosity, for hydrogen storage application. The impact of the weight ratio of KOH/oil palm shell charcoal (OPC) on pore structure and hydrogen storage performance has been investigated. The best lab made AC was also compared with well-known commercial activated carbon Maxsorb-3 and also the ACs from open literatures [23].

2. Experimental

2.1 *Materials*

The oil palm shell was selected as a char precursor was obtained from Malaysia by Chinalight Guangzhou Co., Ltd. (China). All reagents were of analytic grade and used without further purification or treatment. The commercial AC Maxsorb-3 (formerly AX21) from Kansai Coke and Chemicals Co., Ltd. (Japan) was used for the sake of comparison.

2.2 *AC's synthesis*

The precursor of the ACs investigated here was oil palm shell. It was first washed with water to remove oil and dirt attached on the sample then it was dried under sunlight for few days. The dried oil palm shell were put into a horizontal tubular furnace under a pure nitrogen environment and subjected to a gradual increase in temperature (5 °C/min) from room temperature to 650 °C, at which they were kept for 1 h. The flow rate of the nitrogen was 500 mL/min. After the furnace had been cooled down to room temperature under nitrogen atmosphere, the oil palm shell charcoal (OPC) was finally produced. The OPC was then ground and sieved in order to collect the grains having an average size within the range 100 – 200 μm. The OPC powder was mixed with KOH with different weight ratio of (W) of KOH/OPC. The resultant mixture was then introduced into a nickel crucible and heat-treated in a muffle furnace (KDF, 80-plus) at a constant heating rate (3 °C/min) up to the final activation temperature (T=800 °C), which was maintained for 2 h. The crucible was then allowed to cool down to room temperature. The whole process was conducted under a stream of nitrogen with a flow rate of 300 mL/min. Finally, the activated product was washed, first with distilled water (for the recovery of the rudimentary KOH), then with HCl (1 M), and finally, with distilled water again until the pH of the rinse had stabilized. After being dried in an oven for 12 h, a very pure activated carbon (AC) material was obtained. The developed AC was stored in a vacuum desiccator for subsequent pore texture characterization and hydrogen adsorption experiments.

2.3 *AC's characterisation*

2.3.1 Pore structure. N_2 adsorption isotherms were obtained at 77K using an automatic adsorption apparatus (Micromeritics ASAP 2020, USA). The samples were outgassed for 48 h under vacuum at 523 K prior to N_2 adsorption. Surface areas, SBET, were determined by the BET calculation method [12] applied to the adsorption branch of the isotherms. SBET and pore volumes were estimated with a typical, maximum, uncertainty of 3 %.The pore-size distributions (PSD) were calculated by application of the DFT model [13] supplied by Micromeritics software, considering slit-shaped pores.

2.3.2 H_2 adsorption. The H_2 storage properties both at low pressure and high pressure at 77 K for the ACs were determined. The ASAP 2020 HD88 from Micromeritics in USA was used for the low-pressure measurements (up to 1 bar) and the gravimetric analyzer HPVA-100 (VTI Corporation of TA Instruments, USA) was used for the high-pressure measurements (up to 8 MPa). For both sets of measurements, the samples were outgassed under a primary vacuum at 350 °C for 24 h before any sorption measurements were taken. Repeatability was always found to be very satisfactory, leading to errors lower than 3% on each measurement.

3. Results and discussion

3.1 Pore structure

All the ACs prepared were labelled as AC_x. In this nomenclature, the "x" at the end corresponds to the different weight ratios. Table 1 listed the textural characteristics that reflected the N_2 adsorption-desorption at 77 K for the OPC and all ACs. The result shown that the oil palm shell after carbonization presented a very low S_{BET} (198 m^2/g). During the activation process, the redox reactions between the potassium compounds and carbon can result in an increase in the value of the specific surface area [1]. After the activation with different weight ratio, W, from 2 to 6, the S_{BET} first increased from 1600 to 3503 m^2/g and then decreased to 3183 m^2/g, the V_{DR} increased from 0.63 to 1.1 cm^3/g, and the $V_{0.99}$ increased from 0.7 to 1.68 cm^3/g. As the previous studies reported, the weight ratio is the most important parameter in the KOH chemical activation process [1, 14]. It is interesting that the surface area and micropore volume were not increased with the W linearly. AC4 sample with weight ratio 4 achieved the highest S_{BET} and V_{DR}, then both the S_{BET} and V_{DR} decreased when W further increased to 6. The proportion of microporosity to total porosity was expressed as $V_{DR}/V_{0.99}$, which varied from 60 to 90 %, means these developed ACs were essentially microporous. The average micropore diameter (L_0) was in the range of 0.74 to 1.41 nm.

Table 1. Textural characterization of all the ACs.

Sample	W	T (℃)	S_{BET} (m^2/g)	$V_{0.99}$ (cm^3/g)	V_{DR} (cm^3/g)	EA (kJ/mol)	Lo (nm)	$V_{DR}/V_{0.99}$	Vmeso (cm^3/g)
OPC	N	N	198	0.91	N	N	N	N	N
AC2	2	800	1600	0.70	0.63	25.98	0.74	0.90	0.07
AC3	3	800	3201	1.51	1.01	20.01	1.25	0.67	0.50
AC4	4	800	3503	1.68	1.10	20.24	1.22	0.66	0.58
AC6	6	800	3183	1.71	1.03	19.05	1.41	0.60	0.64

N_2 adsorption and desorption isotherms at 77 K consisting of both the linear and logarithmic patterns were shown in Figure 1. According to the IUPAC classification, all AC samples presented were type I isotherms with no hysteresis cycle [15]. When the W increased, the amount of nitrogen adsorbed increased and showed a more extensive widening isotherm 'knee' at a low relative pressure. It is well known that the sorption behavior at low pressures corresponds to nitrogen adsorption in the micropores, where a much steeper increase means more micropores. Accordingly, as the W increased from 2 to 6, the knee of the isotherm widened, indicating both the broadening of the micropore width and development of the mesoporosity.

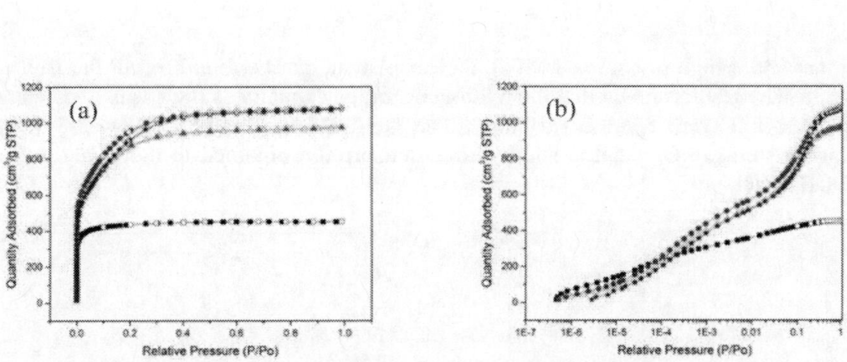

Figure 1. N_2 adsorption (solid symbols) and desorption (hollow symbols) isotherms on both (a) linear and (b) logarithmic scales (■ W=2, ▲ W=3, ● W=4, and ✱ W=6).

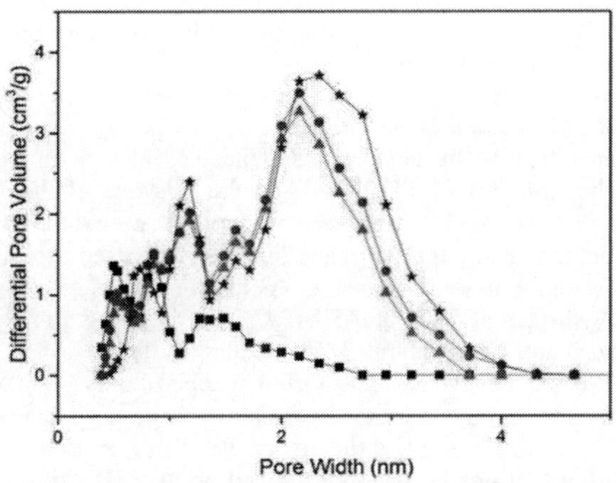

Figure 2. Pore size distributions from N_2 adsorption-desorption for all the ACs calculated by DFT (■ W=2, ▲ W=3, ● W=4, and ✱ W=6).

The PSD was calculated using computer software by application of density functional theory (DFT) (Figure 2). The development of surface area by KOH activation is always accompanied of pore widening. The results shown that PSDs shifted to wider pores with higher W. The result also reinforced and confirmed the aforementioned finding that the micropores were the major component of the AC structure, coexisting with a few mesopores.

In one word, the weight ratio plays the most important role, the optimum W may change with the precursors used. The reason why the surface area increased with W and then dropped could be explained by the fact that during KOH activation, the formation of mesopores and pore widening are two main mechanisms of the development of porosity, and in general the process of pore widening occurred at high Ws, which may damage the pores structure.

3.2 Hydrogen Storage Performance

Based on the previous experiments that the hydrogen storage capacities at 77K has linear relationship with surface area and micropore volume [1, 5], only AC4, which has the highest S_{BET} and V_{DR} were considered for hydrogen storage analysis. The hydrogen adsorption isotherms up to 1 bar and up to 8 MPa of the BC and ACs are given in Figure 3a and 3b. There was no hysteresis cycle. For low pressures up to 1 bar, the isotherms increased continuously, as shown in Figure 3a. Meanwhile, it is

IOP Conf. Series: Materials Science and Engineering **368** (2018) 012031 doi:10.1088/1757-899X/368/1/012031

clearly shown that under high pressure (8 MPa), a clear plateau can be found for all the isotherms, as show in Figure 3b. The results shown that the hydrogen storage capacity as high as 6.7 wt.% at 4 MPa, and 2.86 wt.% at 1 bar were obtained at 77 K. The US DOE target for 2010 was 5.5 wt.% for hydrogen storage systems (AC + tank). The hydrogen adsorption obtained in this study was still far from the US DOE target.

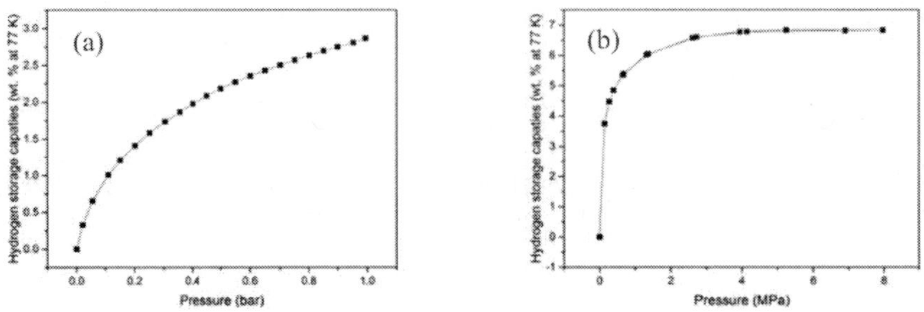

Figure 3. H_2 storage isotherms (wt. %), (a) low pressure up to 1 bar; (b) high pressure up to 8 MPa

3.3 Comparison with Maxsorb-3 and Open Literature

The hydrogen storage capacities of the lab made sample was also compared with the well-known commercial activated carbon Maxsorb-3 (3203 m^2/g), which is one of the ACs frequently used for hydrogen storage. Both the lab made ACs and Maxsorb-3 prepared by activation with alkali hydroxide, but different precursor that the lab made sample AC4 used the biomass material oil palm shell, and the commercial sample maxsorb-3 used the petroleum coke. Maxsorb-3 has been claimed to adsorb about 6 wt. % of hydrogen at 77 K and 5 MPa, 5 wt. % being already reached at 2MPa [16]. Other results on the same material are about 3 wt. % at 80K and 10MPa [17]. Our lab-made AC reached 6.8 wt. % and 2.86 wt. % whereas Maxsorb-3 reached "only" 5.8 wt. % and 2.1 wt.% at 4 MPa and 1 bar. Our lab-made AC has a better hydrogen storage performance because it has higher S_{BET} than Maxsorb-3: 3503 and 3203 $m^2\ g^{-1}$, respectively. But this is not the only reason. The reason of the higher performances of our lab-made AC is also based on its PSD shifted towards narrower pores, as shown in Figure 4.

Table 2 summarize the hydrogen adsorption data at 77 K but different pressure for AC produced from different precursors in the open literature and compared with the ACs prepared in this study for H_2 adsorption. However, the performances reached in this study are among the best reported by the latest reviews [18, 19] and recent literature in general, e.g. [1, 20-21]. Therefore, it can be seen that the activated carbon produced by the oil palm shell shows that our experimental results were among the best of the activated carbons from others.

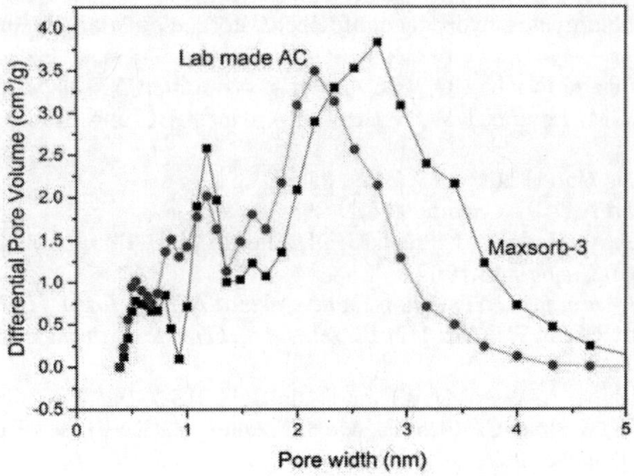

Figure 4. Pore size distributions from N_2 adsorption-desorption calculated by DFT for Mawsorb-3 and for our best lab-made AC.

Table 2. Comparison of activated carbons by KOH activation for H2 storage

Feedstock	Surface area (m^2/g)	H_2 uptake (wt.%)	Ref.
Oil palm shell	3503	6.7 (77K, 40bar) 2.86 (77K, 1bar)	This work
Bamboo	3208	6.60 (77K, 40bar) 2.74 (77K, 1bar)	1
Carbon precursor	2386	2.94 (77K, 1bar)	20
Chitosan	3100	5.6 (77K, 40bar)	21
Corncob	3708	5.8 (77K, 40bar)	22
Anthracites	3220	6.6 (77K, 40bar)	23
Carbon aerogels	1980	4.3 (77K, 20bar)	24
Bark	3170	4.08 (77K, 10bar)	21

4. Conclusions

We have prepared a series highly microporous ACs from oil palm shell by activation with KOH due to the uniform pore structure and higher surface area from chemical activation. ACs with S_{BET} values as high as 3503 m^2/g and V_{DR} values as high as 1.10 cm^3/g were developed, representing a very promising material for hydrogen storage, which showed outstanding performances for hydrogen storage at 77K. The hydrogen storage capacity as high as 6.7 wt.% at 4 MPa, and 2.86 wt.% at 1 bar were obtained. These performances are well above those of the well-known commercial AC Maxsorb-3 and also the ACs from open literatures. Such exceptional hydrogen storage capacities may be ascribed to a higher surface area and volume of micropores of our ACs.

Acknowledgements

The authors are grateful for the support of the National Natural Science Foundation of China (31300488) and the Fujian Agriculture and Forestry University Fund for Distinguished Young Scholars (xjq201420).

References

[1] Zhao W, Luo L, Wang H and Fan M 2017 *Bioresources.* **12** 1246
[2] http://www1.eere.energy.gov/hydrogenandfuelcells/storage/pdfs/targets_onboard_hydro_storage_explanation.pdf
[3] Rango PD, Marty P and Fruchart D 2016 *Appl. Phys. A Mater. Sci. Process.* **122** 1
[4] Ren J, Musyoka NM, Langmi HW, Swartbooi A, North BC and Mathe M 2015 J. *Hydrogen Energy.* **40** 4617
[5] Zhao W, Fan M and Gao H 2016 *RSC Adv.* **6** 46977
[6] Sethia G and Sayari A 2016 *Carbon* **99** 289-294
[7] Fierro V, Szczurek A, Zlotea C, Marêché JF, Izquierdo MT, Albiniak A, Latroche M, Furdin G and Celzard A 2010 *Carbon* **48** 1902
[8] Fierro V, Zhao W, Izquierdo MT, Aylon E and Celzard A 2010 *Int. J. Hydrogen Energ.* **35** 9038
[9] Allwar A, Hartati R and Fatimah I 2017 *Int. Con. Chemical Process and Engineering* **1823** 2017
[10] Hesas RH, Arami-Niya A, Wan MAWD and Sahu JN 2015 *J. Ind. Eng. Chem.* **24** 196
[11] Montoya-Suarez S, Colpas-Castillo F, Meza-Fuentes E, Rodríguez-Ruiz J.and Fernandez-Maestre R 2016 *Water Science & Technology A Journal of the International Association on Water Pollution Research.* **73** 21
[12] Brunauer S, Emmet PH and Teller E 1938 *J. Am. Chem. Soc.* **60** 309
[13] Tarazona P 1995 Surf. Sci. **331-333** 989
[14] Lillo-Ródenas MA, Cazorla-Amorós D and Linares-Solano A 2003 *Carbon* **41** 267
[15] Findenegg GH *EngineeringFoundation* **1984** 207
[16] Ansón A, Lafuente E, Urriolabeitia E, Navarro R, Benito AM, Maser WK and Martínez MT 2007 *J. Alloys Compd.* **436** 294
[17] Khalil YF 2013 *J. Loss Prevent. Proc.* **26** 96
[18] Dutta S 2014 *J. Ind. Eng. Chem.* **20** 1148
[19] Durbin DJ and Malardier-Jugroot C 2013 *Int. J. Hydrogen Energy.* **38** 14595
[20] Sethia G and Sayari A 2016 Carbon **99** 289
[21] Wróbel-Iwaniec I, Díez N and Gryglewicz G 2015 *Int. J. Hydrogen Energy.* **40** 5788
[22] Wang D, Geng Z, Zhang C, Zhou X and Liu X 2014 *J. Energ. Chem.* **23** 601
[23] Zhao W, Fierro V, Fernández-Huerta N, Izquierdo MT and Celzard A 2013 *Int. J. Hydrogen Energy* **38** 10453
[24] Robertson C and Mokaya R 2013 *Micropor. Mesopor. Mat.* **179** 151
[25] Xiao Y, Dong H, Long C, Zheng M, Lei B and Zhang H 2014 *Int. J. Hydrogen Energy.* **39** 11661

IOP Conf. Series: Materials Science and Engineering **368** (2018) 012032 doi:10.1088/1757-899X/368/1/012032

The application of 2D woven kenaf reinforced unsaturated polyester composite in automotive interiors

Z Othman*, M P Saiman, M F A Hamid, M N Talibin

Department of Mechanical Engineering, Seberang Perai Polytechnic, 13500 Permatang Pauh, Pulau Pinang, Malaysia

*E-mail: zulizudin@psp.edu.my

Abstract. A 2D woven kenaf reinforced Unsaturated Polyester (UPE) has been successfully developed for pillar A and door panel for interior automotive parts. This study of woven kenaf fibers reinforced unsaturated polyester focused on application of 2D woven kenaf UPE for the purpose of automotive interiors parts. The reinforced preform selection was done based on the design with fixed and identified parameters. Plain weave design was selected as a reinforcement due to the stability and superior flexural and impact strength. Vacuum Infusion Process (VIP) was selected as a processing method to prepare the interior composite parts. Two different parts were selected in this study, which are pillar A and upper door panel, and the products show similar physical properties. The result indicates that the 2D woven kenaf UPE composite was proven capable to be applied as automotive interior parts.

1. Introduction

In recent decades, global automotive industries have been under pressure due to environmental issues where various regulation was introduced for the sake of environmental protection. One of the issues is recycling automotive interior parts, which has become difficult because most petro based polymer materials are mixed with composite structures. Current disposal methods for these auto interior parts are either garbage dumps or burning, both causing environmental concerns [1]. A substitution of common raw materials, which are currently largely produced from fossil (petrochemical) or mineral resources, to products produced from renewable (plant and animal based) resources is required.

Interior parts in automotive consists of door panels, pillar, cup holder, glove box cover, and dashboard. Traditionally, it used petroleum based material such as polypropylene (PP), Acrylonitrile Butadiene Styrene (ABS), polyethylene terephthalate (PET), polyoxymethylene (POM) and polyvinyl chloride (PVC) [2]. This is because automotive material use lighter material to reduce the weight of the vehicle but pure plastic is low in strength. Parts that require strength use reinforced plastic, which consist mostly of glass fibers due to its strength and ease of works. Pradeep [3] has listed some of the properties and the use of composite in door panel by automakers as shown in table 1.

Table 1. Properties and Use of composite in Door Panels by Various Automakers

Polymer	Tensile Strength (MPa)	Flexural Modulus (GPa)	Automaker
Glass fiber–PUR	253.8	0.4–0.5	VW
Long glass fiber reinforced–PP	151	9.36	Ford

However, due to environmental issues, the replacement with biodegradable material such as natural fibers reinforce composite is more preferable. At present, the use of natural fibers reinforced plastics for interior parts can be classified as low. Among the wide range of natural fiber that is being used as reinforcement, kenaf bast fiber is known to have the potential as a reinforcing fiber in plastic composites. This is because of its superior toughness and high aspect ratio in comparison to other fibers [4]. The use of kenaf fibers reinforced plastics for interior parts require several major technical considerations before it is accepted as a substitute for existing material. Most interiors parts require a high standard of mechanical performance but it is difficult to find an optimum material that can fulfil all requirements [5].

In general to improve the mechanical performance of kenaf fibers as reinforced material, it is necessary to transform it into more stable materials. Among natural polymer composite for long fiber is two dimension (2D) woven preform consisting of two yarns, a warp yarn in vertical direction and weft yarn in horizontal direction that are interlaced together. The advantages of 2D woven preform are good dimensional stability in warp and weft direction, high yarn packing density, high out-of-plane strength, low shear rigidity and a good formability [6]. To obtain the best woven preform for the reinforced in composite, consideration must be given to the strength, stiffness and lightweight of the material as shown in Figure 1.

The weave design plays an important role as it covers all the properties for the composite and that less crimp will lead to superior properties. Weave design with less yarn crimp has great influence on the mechanical properties of the textile composite. A good design with low tightness of pattern will contribute less crimp and increase the preform properties. The lightweight and stiffness properties have opposite parameters but when considering the composite, the stiffness becomes the priority. However, a drawback on linear density of the yarn may appear due to the use of bast staple yarn which is linked to irregularity. The unevenness, or coefficient of variation of linear density for yarns made of short and stiff fibres such as flax, have an unevenness that can be as high as 15-20% [7]. However, it can be overcome by selecting suitable processing method.

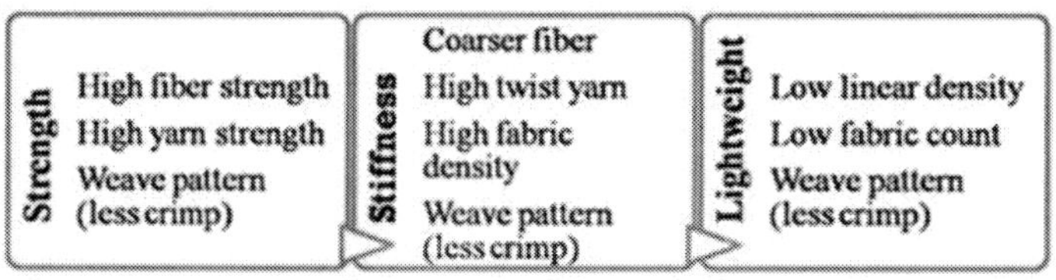

Figure 1. Diagram of the parameters affecting 2D woven preform

Typically producers of natural fiber-based used mat or other material forms with processing method of compression molding, injection molding, thermoforming, and structural reaction injection molding, which are all processes utilized to process natural-fiber composites [8]. In general, using injection molding to process 2D woven preform using thermoset resin and to transform it into composite is not suitable. Therefore, most studies have used resin infusion process which covers the use of RTM, SCRIMP, VARTM and VIP. An open mould application such as VIP has advantages, as it produced one side moulded with uniform surface while the uneven surface can be hidden. Other than that, it produces low voids composite, low cost and complex parts. In addition, VIP technology, which uses vacuum pressure to force resin into a composite, gets more attention due to its capability of large structures with excellent mechanical properties [9]. However, for small parts and mass production, it is less appropriate since it cannot be manufactured repeatedly, but for the purpose of study it is an option. Other than that, the main drawback to use VIP is that the resin must be very low in viscosity.

Unsaturated polyester (UPE) resin is a low cost thermoset resin which is preferred by manufacturers because of its capability to be processed, good mechanical properties and curing in ambient temperature. Thermoset UPE have many applications as in the production of components for automotive, aircraft, building and marine. The processing viscosity of UPE is controllable with the presence of styrene which make it preferable to process fiber reinforced composite. The purpose of styrene is to dilute the polyester which reduces the viscosity to facilitate the flow and it is compatible with the use of equipment for resin infusion process.

Therefore, in this work, a study on the structure of the composite has been done by fixing a certain parameters and focusing on the weave design as the independent variable. The best results will be selected to be used as a material for automotive interior parts.

2. Methodology

2.1 Materials.
Considering that various factors effecting 2D woven preform have been described in Figure 1, related information from the kenaf yarn as fixed parameters was taken through testing and applied to each sample.

2.1.1 Kenaf yarn. The kenaf yarn was supplied by Juteko Bangladesh Pvt. Ltd, Dhaka, Bangladesh with linear density of 759tex or 1mm in diameter with Z twist direction. The testing was conducted on the kenaf yarn to gain its properties. It consists of twists per meter using SDL Electric Twist Tester Y220B according to standard test ASTM D1422. The single yarn strength test used LLYOD, LR30K Universal Tensile Machine according to ASTM D2256.

2.1.2 Fabrication of Dry Preform and Composite. In this work, the selection of the reinforced material begins with the fabrication of 2D woven preform consisting of familiar weave design used in composite industries for man-made fibres. The design consists of Plain 1/1, Twill 4/4, Satin 8/3 and Basket as shown in Figure 2.

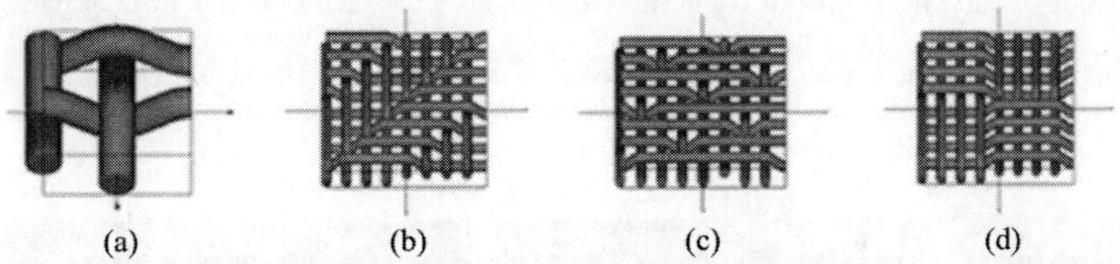

Figure 2. Selected design of a) Plain 1/1 b) Twill 4/4 c) Satin 8/3 d) Basket 4/4

The fabrication of the preform as shown in Figure 3 begins by weaving the 759 tex of kenaf yarn into selected design using a floor loom. The density of the woven preform is 5 ends x 4 picks per cm with the length and the width according to the area of the sample for mechanical testing.

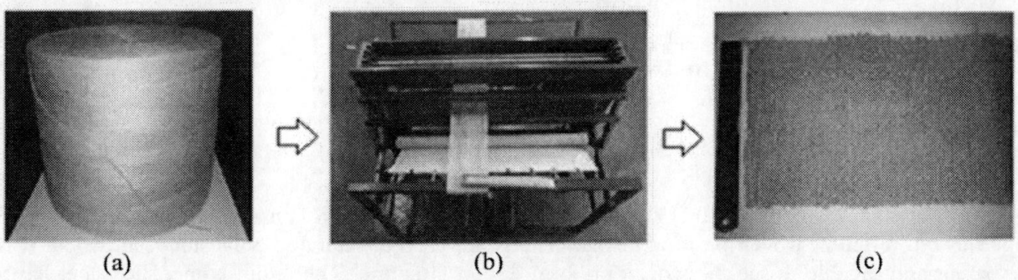

Figure 3. Fabrication of 2D woven preform (a) Kenaf yarn (b) Floor loom (c) 2D woven preform.

2.2 Fabrication of Composite Panel.
The composite panels were fabricated using vacuum infusion process. The matrix material used unsaturated polyester resin (2597P-I), and methyl ethyl ketone peroxide (MEKP) as the catalyst was provided by Wee Tee Tong Chemical Pvt. Ltd., Singapore.

2.3 Testing
2.3.1 Preform testing. The testing was conducted on 2D woven preform and was conducted only on warp direction. The woven preform thickness was tested using fabric thickness gauge. The testing for woven preform breaking strength used standard test ASTM D5035 and the fabric cover factor was calculated according to (1):

$$Cover\ factor\ (S.I) = \left(Yarns\ per\ cm\ \times\ \sqrt{Tex}\right) \div 10 \quad (1)$$

The crimp percentage is determined according to the formula below (2):

$$Percentage\ of\ Crimp = \frac{P-L}{L} \times 100 \qquad (2)$$

where L = distance between two ends of a yarn onto the plane of the preform
P = actual length of the yarn

2.3.2 Mechanical testing on composite. The woven kenaf composites panel of 759tex of yarn were tested on the tensile and flexural using LLYOD, LR30K Universal Tensile Machine. The test sample was conducted on the warp direction. The tensile test for the composite was carried out according to ASTM D3039 at a crosshead speed of 1mm/min and the flexural test used ASTM D790. The impact test used ASTM D3763 using SHIMADZU Hydroshot Impact Test Machine HITS-P10 using test speed of 10 m/s. The high speed impact data processing program generated the impact energy value in Joule (J) at maximum load applied.

2.4 Fabrication of automotive interior component.
The fabrication of automotive interior composite parts begins with the preparation of fiber glass mould for pillar A and upper part door panel from *Proton Persona* as shown in Figure (a) & (b). The mould was made using hand lay-up method and used actual component as a pattern. The mould was placed in a flat surface with each empty space covered. A preform was then carefully placed on the fiber glass mould and ready for full set of VIP as shown in figure 4 (c).

(a) (b) (c)

Figure 4. The fabrication of automotive interior parts (a) door panel (b) fabrication of mould (c) fabrication of door panel using VIP.

3. Results and Discussion

3.1 Composite panel
Figure 5 shows the composite panel from 759 tex of kenaf yarn with constant specification of yarn properties but different woven preform design. The panel used unsaturated polyester resin where it is transparent and capable to see the woven preform design structure. The entire composite panel has a stable structure after being processed using VIP, but basket 4/4 shows that the structure has changed. The adjacent yarns close between each other for each group of interlacement leaves a big space without any reinforcement. In this situation, two possible things might happen, as the space without reinforcement will become the weakest point while the tight yarns become jam structure making it difficult for resins to penetrate.

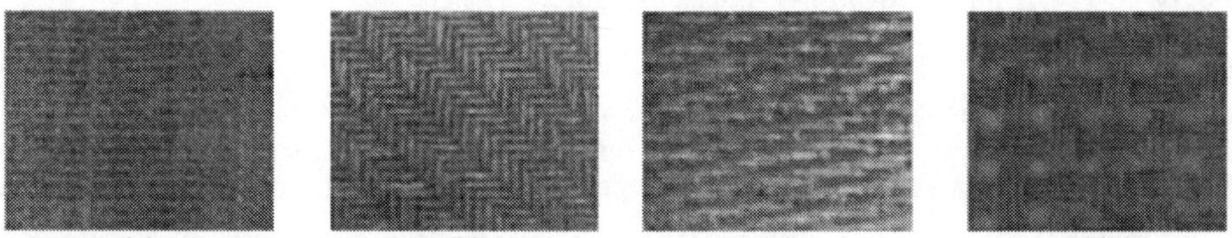

(a) (b) (c) (d)

Figure 5. The composite panel of (a) Plain 1/1 (b) Basket 4/4 (c) Satin 8/3 (d) Basket 4/4.

3.2 *The properties of preform and composite*

Table 2 shows the specification of woven preform taken in consideration following the parameters in Figure 1. The yarn twist is 153tpm with yarn strength of 97N obtained from the test on kenaf yarn supplied by JUTEKO. The value for cover factor shows that the structure is suitable to be woven into a preform with acceptable properties for composite processing. It is supported by Sondhelm [10] that when cover factor reached 16, it is a tight structure and starts to jam. This will affect the penetration of the resin between intra yarns and inter yarn during fabrication process.

As stated earlier, crimp percentage plays an important role on woven preform properties. The result shows that the plain weave preform has the highest crimp percentage which cause a significant result on the thickness of the preform. This is due to the design which made the preform more compact due to compression between yarns. It is different compared to the other preforms where it has more float causing the yarns to become loose. It is also found that a significant result on the areal density of the preform as the higher the crimp percentage, the heavier the preform is. This is because the higher the crimp, the longer the yarn being used and more fibers are inside the preform or the higher the fiber volume fraction is. However, between twill 4/4, satin 8/3 and basket 4/4, it can be seen that the value for thickness and areal density for the preforms are more likely to be influenced by external factors during the fabrication, while for all the composites, due to the VIP process which is an open mould, the result does not show a significant trend. Nevertheless, the values for thickness and areal density may serve as a guide during the fabrication for the automotive interior parts.

Table 2. The specification of preform and composite in warp direction according to suggested parameters.

			Thickness		Areal Density		Tensile Test		
							Preform	Composite	
Type of weave	Cover factor	Crimp %	Preform (mm)	Composite (mm)	Preform (g/m^2)	Composite (g/m^2)	Tensile strength (N/mm)	Tensile Strength (MPa)	Tensile Modulus (GPa)
Plain 1/1	13.77	16	2.77	2.50	868	2518	38	32	2.6
Twill 4/4	13.77	3	3.36	2.42	848	2613	42	35	2.9
Satin 8/3	13.77	4	3.37	2.35	825	2268	44	39	3.7
Basket 4/4	13.77	3	3.17	2.25	819	2703	49	33	5.8

The crimp percentage of plain woven preform also effect the preform strength where it has the lowest value. This is due to the behavior of a preform where yarns with crimp will use the initial load to consume straightening bent yarns and then take up the load, subsequently leading to low strength materials [11]. The result is also proportional with the tensile strength of the composite; the higher the woven preform strength, the higher the strength of the composite. However, it is different for Basket 4/4 as it decreases when it is transformed into a composite. This has already been explained in previous study as the structure of basket 4/4 changes after been process, which leads to a jammed structure and produces large empty spaces without any reinforcement [12]. The tensile modulus determines the stiffness of the composite and it is found that the strength of the woven preform is proportional to the stiffness of the composite. Therefore, in terms of tensile strength and the stiffness of the composite, it shows that satin 8/3 has the greatest strength while basket 4/4 has the highest stiffness properties.

Figure 6 shows the results on the flexural strength and the impact strength of the composite for different woven preform design. The results shows that the composite from plain weave has the highest strength either for flexural or on the impact strength. This means that the plain weave is superior to resist deformation and capable to withstand a sudden force compared to other designs. Both properties have been mentioned by Faris and Sapuan [13] as among properties which are important in the making of automotive parts from natural fiber composite.

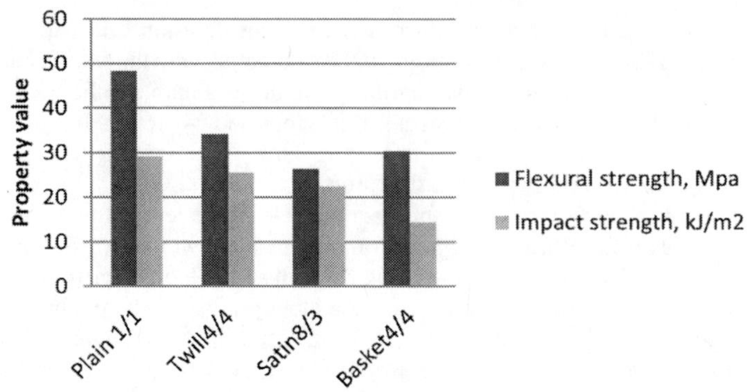

Figure 6. Flexural and impact strength of composite for different woven preform design.

Overall, the results have shown that the use of 2D woven plain composite is low in tensile strength and tensile modulus compared to the other designs. However, the structure of plain weave is more stable while the thickness and the areal density is still within acceptable limits. Moreover, the result from the flexural and impact strength shows that the plain weave is more superior. Therefore, the 2D plain woven preform has been selected as a material for the fabrication of pillar A and door panel.

3.3 Automotive Interior from 2D Woven Kenaf.

Figure 7 shows the result of 2D woven kenaf preform transformed into automotive material. As highlighted, the selected 2D woven plain preform was fabricated into pillar A (Figure 7 (a)) and upper part of the door panel (Figure 7 (b)). It shows that VIP can be produced for each part with dimensions as accurate as the original. However, there are areas that have voids which will effect the mechanical properties and the surface of the parts, especially when painted. Most of the voids are at the back, edges and end of the parts where it most probably occurs as a result of suction of vacuum infusion process.

(a) (b)

Figure 7. Automotive interior parts (a) actual pillar A from Proton Persona, pillar A composite before painting and pillar A after painting (b) parts of door panel composite painted green.

Figure 8 shows the pillar A and door panel made from 2D woven kenaf reinforced UPE composite was successfully assembled on the Proton Persona car. Each parts indicates that the dimension is like the original parts and it fits with the other parts when it is assembled. However, each part requires finishing process either surface finish or to obtain dimensional accuracy.

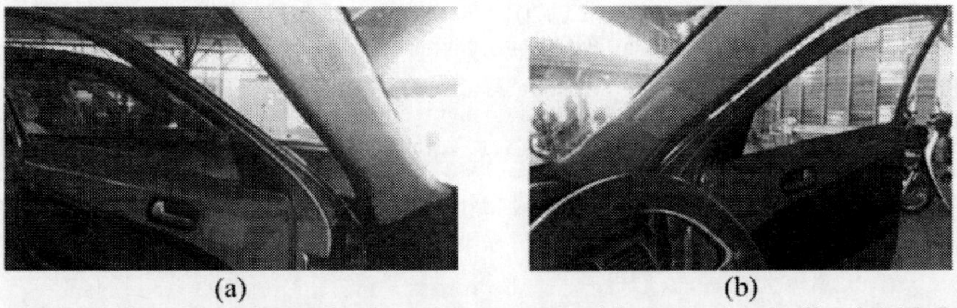

Figure 8. Composite parts consists of upper parts of door panel and pillar A (a) left side of car door (b) right side of car door.

A review by Ashori [14] shows that the weight of automotive interior components made from plant fibers by many vehicles manufacturers is between 0.4kg to 2.5kg. It can reach 5kg depending on the component parts. The work has resulted in pillar A with weight of 245g and the thickness of 2.34mm, which is in the range of the panel samples. Compared with the actual pillar A for Proton Persona, the weight is 250g with the thickness of 2.55mm, which is almost the same. However, the thickness of the weight of the component depends on the kenaf yarn size, which will affect the other properties.

4. Conclusion

The 2D Woven Kenaf Reinforced UPE composite were successfully fabricated into automotive interior components. The selected plain weave design is low in tensile properties but it is structurally stable with superior flexural and impact strength compared to other weave designs suitable to be fabricated into composite for interior parts. Based on the output of this work, the 2D woven kenaf preform is capable to be used as interior parts but requires a process that is more suitable for large-scale production. The results also indicate that the composite has the same physical properties with the existing parts. In addition, natural long fibers reinforced composite has been considered a very interesting alternative to overcome the problems related to land pollution due to automotive interior parts. It is recommended for future study, to include on the processing technique such as sandwich structures as it may increase the strength on the natural fiber composite.

References

[1] Yan C, Liangfeng S, Ovidiu C, Ioan N, Val Y, and Mary W 2005 Kenaf/Ramie Composite forAutomotive Headliner *J Polym Environ* **13** 107-114

[2] Akshat P, Arun P, Rajesh P 2017 An overview of Polymeric Materials for Automotive Applications *Materials Today: Proc.* **4** 3807-3815

[3] Pradeep S A, Iyer R K, Kazan H, Pilla S 2017 Applied Plastic Engineering Handbook (Second Edition), Pages 651-673

[4] Akil H M, Omar M F, Mazuki A A M, Safiee S, Ishak Z A M and Abu Bakar A 2011 Kenaf Fiber Reinforced Composite: A Review *Materials and Design* **32** 4107-4121

[5] Albrecht N, Udo B and Wulf T 2002 Improved Acoustic Behavior of Interior Parts of Renewable Resources in the Automotive Industry *J Polym Environ* **10** 115–11

[6] Yeoung S S, Jung T L, Dong S J, Myung W K, Seung H L and Jae Rn Y 2012 Viscoelastic and thermal behavior of woven hemp fiber reinforced poly(lactic acid) composites *Compos Part B-Eng* **43** 856-860.

[7] Lomov S and Verpoest I 2005 Manufacturing and Internal Geometry of Textiles in Design and Manufacturing of Textile Composite *Design and Manufacture of Textile Composite* ed Long A C (Cambridge: Woodhead Publishing Limited) pp 1-61

[8] James H and Dan H 2006 Natural-fiber reinforced polymer composites in automotive applications. *JOM* **58** 80-86

[9] Kim Y H, Eum S H, Choi B K, Kim J C, Kwon S C and Dutta P K 2004 A study on the dimensional stability of interior composite panel made by vacuum bagging and hand lay-up fabrication method *Key Engineering Materials* **261–263** 1493-1498

[10] Sondhelm W S 2004 Technical fabric structures - Woven fabrics *Handbook of technical textiles* ed Horrocks A R and Anand S C (Abington Cambridge: Woodhead publishing limited)

[11] Peerzada M H, Abbasi S A and Khatri A 2012 Effect of weave strucrure on tensile strength and yarn crimp of three-dimensional fibre woven fabric. *Sci.Int.(Lahore)* **24** 47-50
[12] Saiman M P, Wahab M S and Wahit M U 2014 The Effect of Fabric Weave on Tensile Strength of Woven Kenaf Reinforced Unsaturated Polyester Composite *Int. Colloquium on Textile Engineering, Fashion, Apparel & Design*
[13] Faris M A and Sapuan S M 2014 Natural fiber reinforced polymer composites in industrial applications: feasibility of date palm fibers for sustainable automotive industry *J Clean Prod* **66** 347-354
[14] Alireza A 2008 Wood–plastic composites as promising green-composites for automotive industries! *Bioresource Technol* **99** 4661–4667

IOP Conf. Series: Materials Science and Engineering **368** (2018) 012033 doi:10.1088/1757-899X/368/1/012033

Characterization of cellulose nanofiber from oil palm mesocarp fiber produced by ultrasonication

T A T Yasim-Anuar[1], H Ariffin[1,2]*, M A Hassan[2]

[1]Laboratory of Biopolymer and Derivatives, Institute of Tropical Forestry and Forest Products, Universiti Putra Malaysia, 43400 UPM Serdang, Selangor
[2]Faculty of Biotechnology and Biomolecular Sciences, Universiti Putra Malaysia, 43400 UPM Serdang, Selangor

*Corresponding author: hidayah@upm.edu.my

Abstract. Oil palm mesocarp fibre (OPMF) is one of the major biomass generated from palm oil industry and it is currently inefficiently burnt at palm oil mill for energy and as a mean of disposal. With the growing concern on environmental sustainability and finite supply of non-renewable resources, particular attention has been given to produce value-added materials from OPMF, such as biocomposite and biocompost. Since OPMF contains mainly cellulose, there is high potential for the production of cellulose nanofiber (CNF) from OPMF. In this study, cellulose was obtained from OPMF by chemical extraction. OPMF-cellulose obtained was then soaked in distilled water [0.2 - 1% (w/v)] prior to nanofibrillation using an ultrasonicator at 125 W and 36 kHz for 3 to 12 hours. Scanning electron microscopy (SEM) micrographs of fiber undergone ultrasonication confirmed the occurrence of CNF having diameter in the range of 40-200 nm. CNF produced had crystallinity value in the range of 31-43%, and was thermally stable up to 340 °C.

1. Introduction

Palm oil industry is currently ranked fourth as the major economic contributor to Malaysia, as it is contributing approximately 8% to the country's GNI per capita and 6% of GDP [1-2]. As one of the major economic backbone to the country, this industry covers nearly three quarters of the Malaysia's agricultural land which is about 5.39 million hectares and this consequently cause an environmental issue, especially in managing wastes contributed from oil palm plantations and mills. Umar et al. (2013) [3] stated that there is about 80 million tons of oil palm biomass generated by the Malaysia's oil palm industry annually, and the value is expected to rise to 100 million tons by the year 2020. Given its abundant availability, various industries had started to use oil palm biomass as an alternative raw material for various product manufacturing such as biocomposite [4], biofuel [5], biochar [6], biocompost [7] and many more. This is due to the abundant and infinite supply of oil palm biomass that can be collected anytime due to huge plantation area and rapid fruits production.

Focus is given mainly on the utilization of oil palm empty fruit bunch (OPEFB) and palm kernel shell (PKS), leaving behind the OPMF. OPMF is the solid residue obtained from oil palm fruits after oil extraction and it is usually being left in palm oil mills to be used as a boiler fuel or burnt as a mean of disposal, which is wasteful and this indirectly contributes to various environmental pollution [8,9]. As a lignocellulosic material, OPMF has potential to be used as a raw material for various high value bioproducts manufacturing including cellulose nanofiber (CNF). CNF which is a term referring to nano-structured cellulose is a tiny fiber with light weight, yet possesses great strength, stiffness, biodegradability and renewability [10]. Owing to their good mechanical properties, CNF can be used for several applications ranging from consumer products to high-tech industrial applications. The opportunity to utilize OPMF for the production of CNF would be a new approach in utilizing OPMF for the production of value-added products.

Nevertheless, there are few major issues in the production of CNF especially in delaminating the fibers cell wall and separating the nano-sized fibrils without affecting their properties. Due to this matter, selection of nanofibrillation method need to be carried out thoroughly to ensure the chosen materials are well fibrillated. Electrospinning is the most common and conventional method used for nanofibrillation of cellulose. Previous study by Yasim-Anuar and colleagues (2017) [11] demonstrated that OPMF can be converted to CNF by electrospinning. Nevertheless, the diameter of the CNF was larger than 200 nm. It has been reported that other nanofibrillation such as ultrasonication can produce CNF with small diameter and relatively high length fiber [12]. Ultrasonication is able to produce nano-size fibers due to its acoustic frequency cavitation which involves formation, expansion and implosion of microbubbles in aqueous solution. This process will cause erosion, which then lead to the breakage of relatively weak Van der Waals forces resulting in the separation of the microfibrils into individual fibers [13].

Even though ultrasonication has been widely studied and reported before, yet there have been lack of reports on CNF production from oil palm biomass using this method. Since each natural fiber has different physical and chemical properties, this may influence the morphology, crystallinity as well as thermal properties of the CNF. This was supported by Zhao et al. (2015) [14] that stated each natural fibers which were constituted by microfibrils defibrillated the fibers in different degree, even though similar mechanical process was conducted. This was highly attributed to the physical and chemical properties of the fibers.

Therefore, the goal of this study was to investigate the characteristics of CNF produced from OPMF using ultrasonication. The cellulose extracted from OPMF were subjected to ultrasonication at different ultrasonication duration, to identify the optimum duration to produce CNF with smallest diameter. Subsequent ultrasonication was then carried out using different cellulose concentration, to identify the effect of ultrasonication on different cellulose concentration.

2. Experimental

2.1 Materials

Samples of oil palm mesocarp fiber (OPMF) obtained from Seri Hulu Langat Palm Oil Mill (Dengkil, Selangor, Malaysia) was used to prepare cellulose nanofiber (CNF). The OPMF samples were cleaned using tap water to remove any residuals and sun dried prior to further study [15]. Cellulose extraction was conducted by treating the OPMF using sodium chlorite ($NaClO_2$) and potassium hydroxide (KOH) according to the method by Yasim-Anuar et al. (2017) [11]. OPMF-cellulose obtained after extraction had purity of 99 %.

2.2 Methods

2.2.1 Preparation of cellulose nanofiber. In order to produce cellulose nanofiber (CNF) by ultrasonication, the cellulose extracted from OPMF was first soaked in distilled water at 0.2, 0.4 and 0.8 % (w/v) concentration. Each cellulose suspension was then placed in an ultrasonicator (Q125, Serial No: 63863T-07-11, QSonica, Connecticut, USA) at 125 W, 20 kHz in frequency and 100 amp. In order to determine the effect of ultrasonication time on the CNF formation, ultrasonication was conducted at 3, 6, 9 and 12 hours. The ultrasonication was carried out in ice and this condition was maintained throughout the ultrasonication process to prevent suspensions from overheating as cavitation during ultrasonication may rise the temperature of the cellulose suspension (method modified from Chen et al., 2011 [16]). After completing this process, the CNF samples were subjected to freeze drying process prior to further analysis.

2.2.2 Characterization

2.2.2.1 Scanning electron microscopy analysis (SEM). The morphology of the CNF was observed using a field-emission scanning electron microscopy (FE-SEM) (Sirion 200, FEI, Eindhoven, Netherlands) with accelerating voltage of 5 kV. The freeze-dried CNF were coated with platinum using a vacuum sputter coater prior to FE-SEM observation.

2.2.2.2 X-ray diffraction analysis. The crystallinity index of CNF was analysed using X-ray diffractometer (XRD) which was performed on a Philips PW3040/00 X'Pert MPD system (Eindhoven, Netherlands). The X-ray diffractogram was recorded in the 2θ range of 5 to 50° and the crystallinity index, *CrI* was calculated using the following equation:

$$\mathrm{CrI} = \frac{I_{002} - I_{am}}{I_{002}} \times 100\% \quad (1)$$

where I_{002} is equal to $2\theta = 23^0$ and it refers to the peak of the crystalline portion of biomass (*i.e.*, cellulose), while I_{am} refers to the peak at about $2\theta = 18^0$, which corresponds to the amorphous region [15].

2.2.2.3 Thermal stability analysis. Thermal stability and decomposition temperature of the cellulose and CNF was determined by using a thermogravimetric analyzer (TG 400, Perkin Elmer, Waltham, MA, USA). The samples (approximately 10 mg) were placed in aluminum cups, and the analysis was carried out at a heating rate 10 °C/min under nitrogen flow of 20 mL/min from 50 to 550 °C [11].

3. Results and Discussion

3.1 Morphology of the ultrasonicated CNF

The morphology of 0.2% (w/v) ultrasonicated CNF produced at 3, 6, 9 and 12 hours was examined to observe the effect of ultrasonication time towards CNF formation. The average diameter of the CNF was determined by measuring it using Image J software. The SEM micrograph shows that the OPMF-cellulose gradually reduced in its size by time after 3 hours of ultrasonication, CNF with average size of approximately 150 to 1000 nm was produced. Prolonged ultrasonication for 6, 9 and 12 hours were able to produce CNF with diameter around 80 to 600 nm, 40 to 200 nm and 100 to 500 nm, respectively. This result indicates that the hydrogen bonding between the CNF was lower after ultrasonication and smallest diameter of CNF can be obtained after 9 hours of treatment. This proved that ultrasonication can break the relatively weak interfaces among cellulose fibers, which are bonded to each other by H-bonds. Thus, it can disintegrate the micro-size cellulose fibers into nano-size cellulose fibers [16-17].

Theoretically, it was expected that the longer the processing time, the smaller the diameter of CNF can be obtained. However, when ultrasonication was prolonged to 12 hours, the diameter of CNF was observed to be larger than CNF produced at 9 hours of processing time. From the SEM micrographs, ultrasonication did helped in breaking fibers' agglomeration. However, subsequent ultrasonication can lead to re-agglomeration of the CNF and similar results were observed from other studies by Cengiz et al., (2009), Gokce et al., (2014) and Mandzy et al., (2005) [18–20]. This phenomenon occurred probably due to the attraction of van der Walls forces and it also may be due to the reformation of H-bonds between the CNF which may be due to stress on the CNF surface because of longer processing time or because of the adhesion of CNF to one another. Based on this SEM micrograph, the smallest diameter of CNF can be produced after 9 hours of ultrasonication.

Subsequent ultrasonication was carried out using 0.4 and 0.8% (w/v) concentration to test the effect of concentration on CNF formation.

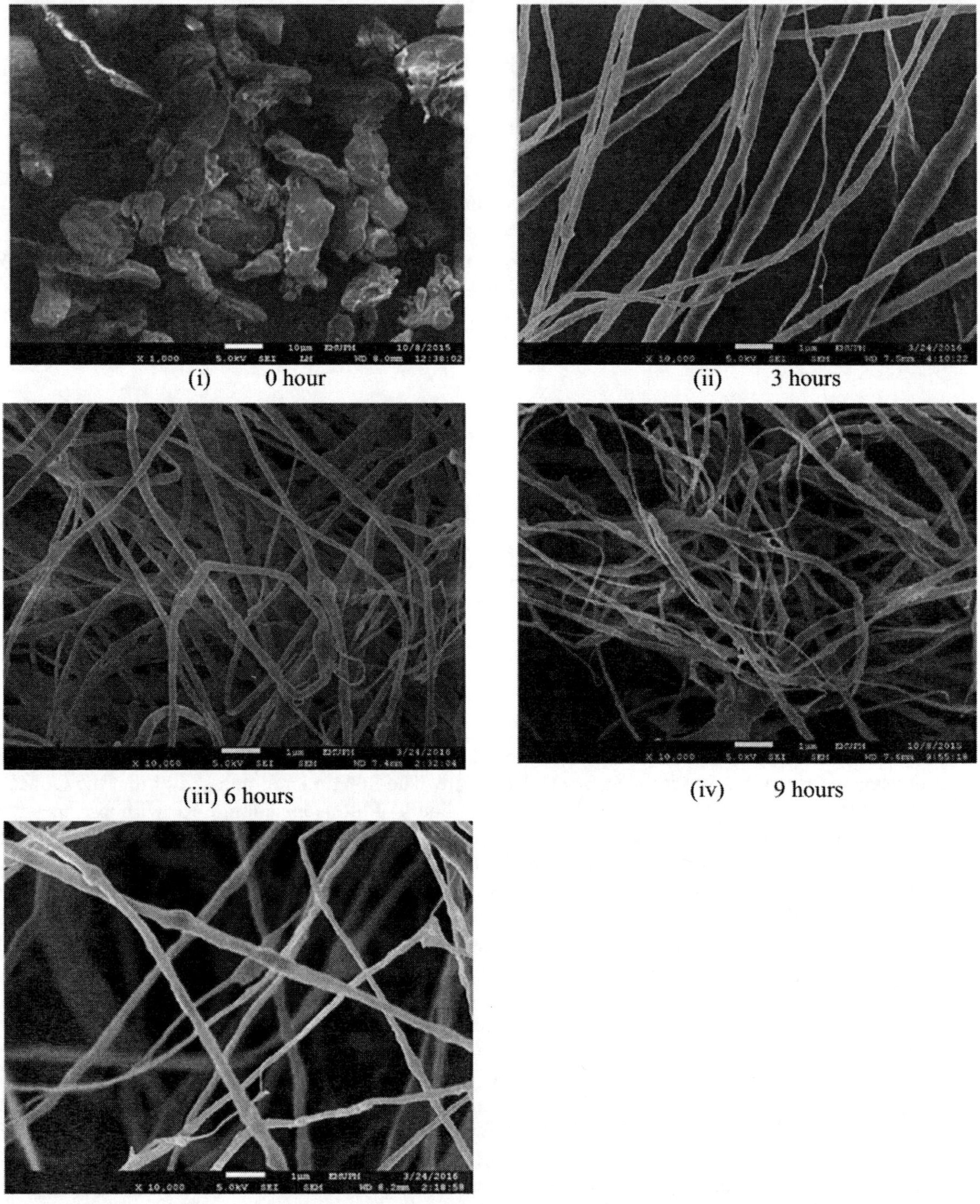

(i) 0 hour (ii) 3 hours

(iii) 6 hours (iv) 9 hours

(v) 12 hours

Figure 1. The SEM micrographs of (i) OPMF-cellulose, and CNF after (ii) 3 hours (iii) 6 hours, (iv) 9 hours and (v) 12 hours of ultrasonication (10,000x magnification).

In terms of the effect of concentration on the formation of CNF, the SEM micrograph shows that larger diameter of CNF was recorded when using both 0.4 and 0.8% (w/v) OPMF-cellulose concentration. The average diameter was approximately around 110 to 560 nm and 150 to 800 nm for 0.4 and 0.8% (w/v) cellulose concentration respectively. This shows that high cellulose concentration can increase the intra-molecular interaction, which consequently raised the agglomeration effect due to van der Waals forces. This was also supported by Li et al. (2014) [21] which stated that cavitation of ultrasonication requires more energy to break cellulose chains for higher cellulose concentration due to higher viscous system of the solution. Hence, it can be concluded that when the OPMF-cellulose solution was more than 0.2% (w/v), the viscosity of the solution increase hence, cellulose were not able to disperse uniformly. Therefore, it was discovered that CNF with smallest diameter can be produced using 0.2% (w/v) cellulose concentration.

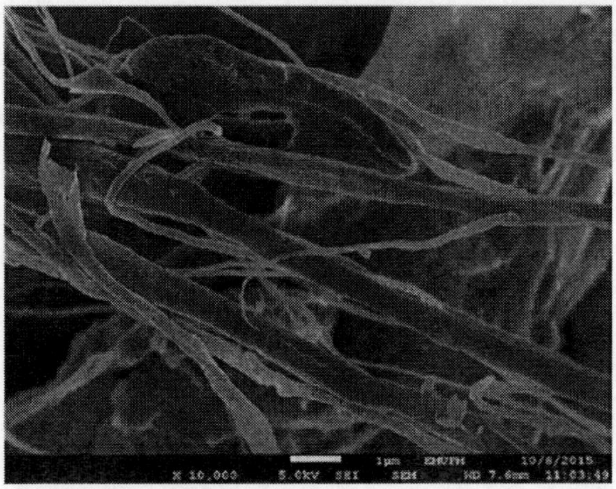

(i) 0.4% (w/v)

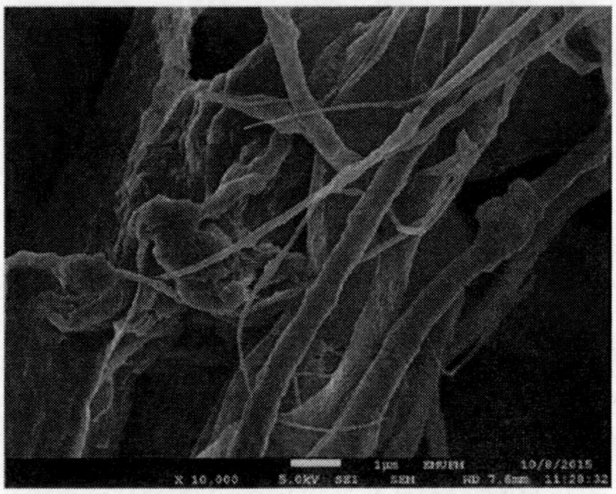

(ii) 0.8% (w/v)

Figure 2. The SEM micrographs of CNF using (i) 0.4% and (ii) 0.8% (w/v) OPMF-cellulose for 9 hours ultrasonication (10,000x magnification).

3.2 Crystallinity by X-ray diffraction

X-ray diffraction analysis was conducted to investigate the crystallinity of the OPMF-cellulose and CNF obtained after ultrasonication. The XRD diffractograms of OPMF-cellulose and CNF produced after ultrasonication as shown in Figure 3 exhibited an intense reflection at $2\theta = 21$ to $22°$, indicates the cellulose crystallographic plane (002, Bragg reflection), and another small peak at $2\theta = 18$ to $19°$ which corresponds to the amorphous region (101, Bragg reflection). The crystallinity index of OPMF-cellulose was 51 % and the crystallinity index of CNF after 3, 6, 9 and 12 hours of ultrasonication were 31, 37, 41 and 32 %, respectively. It is apparent that after ultrasonication, the crystallinity index of CNF decreased significantly. This result shows that the exposure to ultrasonication may disrupt the CNF structure thus, led to reduction in crystallinity. This was supported by Xie et al. (2016) [22] that stated ultrasonication may break down the hydrogen bonds of the cellulose fibers, which then consequently caused damage to the crystalline part of the CNF.

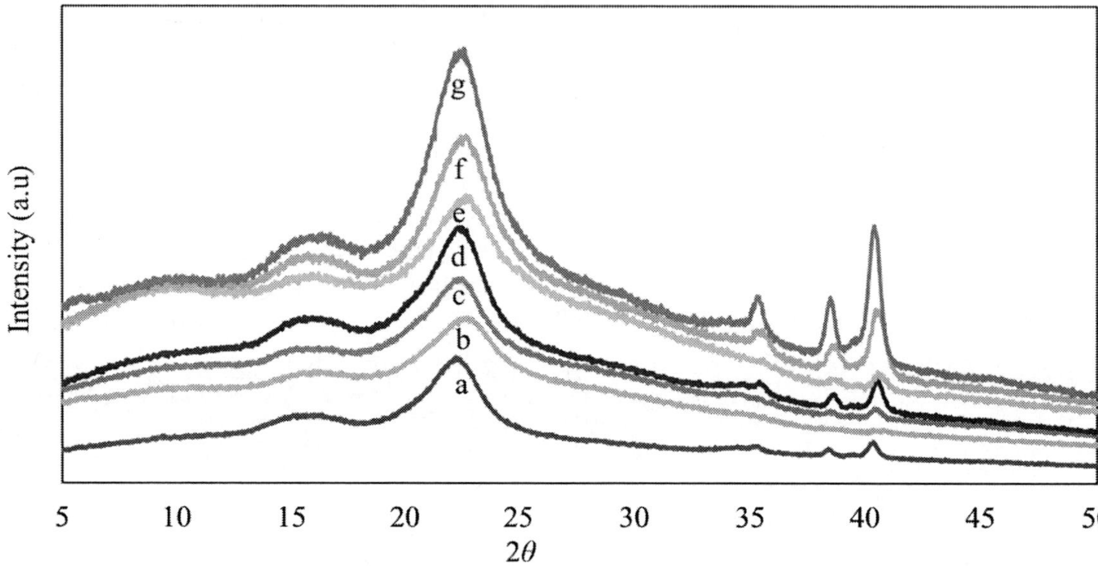

Figure 3. XRD diffractograms of (a) OPMF-cellulose, CNF from 0.2% (w/v) OPMF-cellulose ultrasonicated for (b) 3 hours, (c) 6 hours, (d) 9 hours and (e) 12 hours. While (f) CNF from 0.4% (w/v) OPMF-cellulose ultrasonicated for 9 hours and (g) CNF from 0.8% (w/v) OPMF-cellulose ultrasonicated for 9 hours.

The reduction in crystallinity after ultrasonication might be due to the impact of hot spots generated during the collapse of cavitation [23]. In this study, water which acted as a solvent and hydrolysis reagent produced bubbles throughout ultrasonication process. Those bubbles oscillated and by time, they violently collapsed, thus able to generate high pressure and temperature on the surface of the cellulose and caused damage to the crystalline parts of the CNF. Hence, it can be concluded that despite of able to broken down the microfibrils, acoustic cavitation of ultrasonication however may reduce the crystallinity index of obtained CNF.

Subsequent 9 hours ultrasonication using different OPMF-cellulose concentration; 0.4 and 0.8% were tested to determine the effect of concentration on the crystallinity of CNF. The crystallinity for CNF produced at 9 hours ultrasonication from 0.2, 0.4 and 0.8 % (w/v) OPMF-cellulose was almost similar which was in the range of 41 - 43% as shown in Table 2. This indicates that there was no significant different in terms of crystallinity between different OPMF-cellulose concentration.

Table 1. Crystallinity index of the OPMF-cellulose and CNF obtained after ultrasonication.

Samples	Crystallinity index (%)
OPMF-cellulose	51
CNF from 0.2% (w/v) OPMF-cellulose at 3 hours	31
CNF from 0.2% (w/v) OPMF-cellulose at 6 hours	37
CNF from 0.2% (w/v) OPMF-cellulose at 9 hours	41
CNF from 0.2% (w/v) OPMF-cellulose at 12 hours	32
CNF from 0.4% (w/v) OPMF-cellulose at 9 hours	42
CNF from 0.8% (w/v) OPMF-cellulose at 9 hours	43

3.3 Thermal stability of ultrasonicated CNF

The thermogram of the OPMF-cellulose and CNF samples are illustrated in Figure 4 and Figure 5 respectively. A slight weight loss at around 50 to 100 °C indicates the evaporation of moisture and initial thermal decomposition, while the broad peak between 220 to 400 °C is attributed to cellulose degradation [13]. The TG curve of CNF shows a slight reduction at $Td_{50\%}$ which was from 343°C (OPMF-cellulose) to 342, 341, 338 and 336 °C after 3, 6, 9 and 12 hours of ultrasonication respectively. A slight decrease in thermal stability for CNF might be due to their high surface area, which was due to their nanoscale. It must be noted from FE-SEM images that the surface areas of CNF are much higher than OPMF-cellulose, hence the CNF may initiate more active sites and accelerate the decomposition [24]. This similar behavior has also been reported by Oliveira et al. (2013) [25] . This TG result is also significantly related to the reduction in crystallinity after ultrasonication and as according to Pacaphol and Aht-Ong, (2017) [26], this is due to the greater number of the chain ends in CNF, which was able to move independently when undergoing ultrasonication. In contrast to the CNF, OPMF-cellulose which appeared in microscale contained less amount of the chain ends and as a result, the degradation occurred late.

Besides that, it was also observed that by prolonged the ultrasonication duration may resulted in low thermal degradation temperature. This clearly showed that longer ultrasonication duration may disrupt the structure of the CNF, thus negatively affect the thermal stability. This was corresponds to the crystallinity index as well. On the other hand, the thermal stability of CNF from different OPMF-cellulose concentration is presented in Figure 5. The degradation temperature was almost similar for CNF produced at 9 hours ultrasonication from 0.2, 0.4 and 0.8 % (w/v) OPMF-cellulose suspension, which was in a range of 338 - 340°C. This indicated that no significant different in terms of thermal degradation temperature between different OPMF-cellulose concentration.

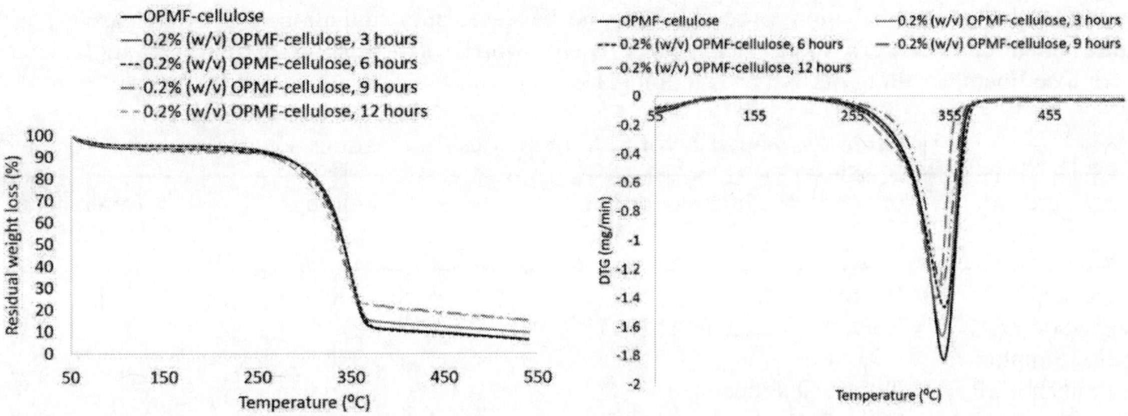

Figure 4. The TG and DTG thermograms of CNF obtained using 3, 6, 9 and 12 hours of ultrasonication from 0.2 % (w/v) OPMF-cellulose suspension.

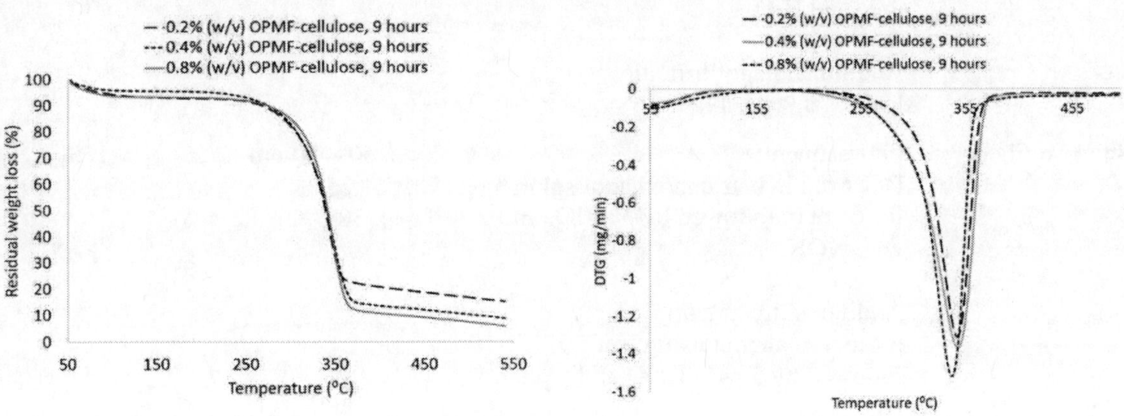

Figure 5. The TG and DTG thermograms of CNF obtained using 0.2, 0.4 and 0.8 % (w/v) OPMF-cellulose suspension at 9 hours of ultrasonication.

3.4 Comparison of CNF

Based on the morphology, average diameter, crystallinity and thermal stability of CNF produced from OPMF, it was revealed that ultrasonication exhibited excellent CNF formability using water as the main solvent throughout nanofibrillation process. The CNF produced from this method has almost similar properties to those reported in previous studies as shown in Table 2. In this study, CNF with diameter less than 200 nm, high crystallinity and thermally stable was produced without having to rely on chemicals as a solvent during ultrasonication. This is the advantage of this method as compared to other reports. For example, Wang et al. (2015) [27] suspended and ultrasonicated the cellulose fibers in several solvents such as hydrochloric acid, formic acid, and mixture of formic acid and hydrochloric acid in order to produce CNF. Li et al., (2014) [21] on the other hand used a mixture of NaOH, thiourea, urea and H_2O.

In terms of power usage and duration of ultrasonication, this study only conducted ultrasonication for 9 hours at 125 W, which was lowered than the other studies. For example, Chen et al. (2011) [16] applied ultrasonication at 1200 W for 30 minutes, despite of conducting additional chemical treatment prior to ultrasonication. Wang et al. (2015) [27] also applied high power during ultrasonication which was approximately 1080 – 1620 W for 4 – 6 hours. Li et al. (2014) [21] also conducted homogenization process after ultrasonication at 100 – 500 W to produce smaller diameter size of CNF. The comparison of the processing parameter indicates that the use of high ultrasonication power and additional nanofibrillation were not the ultimate condition to produce CNF. This study prove that ultrasonication solely can be able to produce CNF, without having to proceed with another nanofibrillation process. This is because, the cavitation energy released during ultrasonication created a mechanical shock wave, which then able to disrupt the agglomerated cellulose fibers, resulting in CNF production [21].

Table 2. Comparison of CNF properties from various sources.

Source of cellulose	Methods condition	Findings	References
Bleached softwood kraft pulp (Supplied by Suzano, SP, Brazil)	Pretreatment: NaOH: thiourea: urea: H_2O = 7:9:9:75 Ultrasonication: Time: 30 min Power: 100 – 500 W Solution: Mixture of NaOH, thiourea, urea and H_2O Additional nanofibrillation: Homogenization	Size: > 30 nm Crl: 43.7 % $T_{d50\%}$: 370 °C	[21]
Bamboo fibers	Pretreatment: Dewaxed in benzene/ethanol solution for 6 hours followed by $NaClO_2$ and 2 wt% KOH. Additional treatment: Further treatment using $NaClO_2$ followed by 5 wt% KOH and 1 wt% HCl.	Size: 30 – 80 nm Crl: 61.25 % $T_{d50\%}$: 309 °C	[16]

	Ultrasonication: Time: 30 min Power: 1200 W Solution: Distilled water		
Silk derived from *Bombyx mori* silkworm	Degummed in 0.2% (w/v) Na_2CO_3 solution Ultrasonication: Time: 4 - 6 hours Power: 1080 – 1620 W Solution: hydrochloric acid, formic acid, and a mixture of formic acid and hydrochloric acid.	Size: 30 – 120 nm CrI: - $T_{d50\%}$: -	[27]
Oil palm mesocarp fiber	Chemical pretreatment: 6% (w/v) $NaClO_2$ and KOH Ultrasonication : Time: 9 hours Power: 125 W Solution: Distilled water	Size: 40 – 200 nm CrI: 41% $T_{d50\%}$: 338 °C	This study

4. Conclusions

The present study showed that ultrasonication was successfully used to convert OPMF-cellulose into CNF. The applied acoustic cavitation reduced the crystallinity index and thermal degradation temperature after subsequent ultrasonication. The resulting CNF produced from OPMF by ultrasonication had diameter in the range of 40 to 200 nm, crystallinity index of 41% and was thermally stable with $T_{d50\%}$ of approximately 338 °C. It was also revealed that the optimum condition for CNF production by ultrasonication was at 125 W for 9 hours, with 0.2% (w/v) cellulose concentration. Overall, CNF produced from OPMF-cellulose can be considered as a promising nanomaterial as it offers high renewability, high thermal stability and high crystallinity. Due to their small diameter size and high thermal stability properties, the CNF produced in this study has potential to be used for the manufacturing of various applications such as transparent paper.

Acknowledgements

The authors would like to thank the Ministry of Higher Education (MOHE), Malaysia through SATREPS research grant (Vote No.: 6300156) for the research funding. The first author of this paper is sponsored by MOHE under MyBrain15 during her study at Universiti Putra Malaysia and Japan Student Services Organization (JASSO) during research attachment at Kyushu Institute of Technology, Japan.

References

[1] Tang J P, Lam H L, Abdul Aziz M K and Morad N A 2017 Palm biomass strategic resource managment – A competitive game analysis *Energy* **118** 456–63
[2] Abdul-Manan A F, Baharuddin A and Wei Chang L 2014 A detailed survey of the palm and biodiesel industry landscape in Malaysia *Energy* **76** 931–41
[3] Umar M S, Jennings P and Urmee T 2013 Strengthening the palm oil biomass Renewable Energy industry in Malaysia *Renew. Energy* **60** 107–15
[4] Beg M D H, Mina M F, Yunus R M and Moshiul Alam A K M 2015 *Biofiber Reinforcements*

in Composite Materials (Elsevier)
[5] Johari A, Nyakuma B B, Mohd Nor S H, Mat R, Hashim H, Ahmad A, Yamani Zakaria Z and Tuan Abdullah T A 2015 The challenges and prospects of palm oil based biodiesel in Malaysia *Energy* **81** 255–61
[6] Idris J, Shirai Y, Anduo Y, Ali A A M, Othman M R, Ibrahim I, Husen R and Hassan M A 2015 Improved yield and higher heating value of biochar from oil palm biomass at low retention time under self-sustained carbonization *J. Clean. Prod.* **104** 475–9
[7] Chin C F S, Furuya Y, Zainudin M H M, Ramli N, Hassan M A, Tashiro Y and Sakai K 2017 Novel multifunctional plant growth–promoting bacteria in co-compost of palm oil industry waste *J. Biosci. Bioeng.* **124** 506–13
[8] Warid M N M, Ariffin H, Hassan M A and Shirai Y 2016 Optimization of Superheated Steam Treatment to Improve Surface Modification of Oil Palm Biomass Fiber *Bioresources* **11** 5780–96
[9] Ali A A M, Othman M R, Shirai Y and Hassan M A 2015 Sustainable and integrated palm oil biorefinery concept with value-addition of biomass and zero emission system *J. Clean. Prod.* **91** 96–9
[10] Zhu J, Wei S, Patil R, Rutman D, Kucknoor A S, Wang A and Guo Z 2011 Ionic liquid assisted electrospinning of quantum dots/elastomer composite nanofibers *Polymer (Guildf).* **52** 1954–62
[11] Yasim-Anuar T A T, Ariffin H, Norrrahim M N F and Hassan M A 2017 Factors Affecting Spinnability of Oil Palm Mesocarp Fiber Cellulose Solution for the Production of Microfiber *Bioresources* **12** 715–34
[12] Chen W, Yu H and Liu Y 2011 Preparation of millimeter-long cellulose I nanofibers with diameters of 30–80nm from bamboo fibers *Carbohydr. Polym.* **86** 453–61
[13] Khawas P and Deka S C 2015 Isolation and characterization of cellulose nanofibers from culinary banana peel using high-intensity ultrasonication combined with chemical treatment *Carbohydr. Polym.* **137** 608–16
[14] Zhao Y, Xu C, Xing C, Shi X, Matuana L M, Zhou H and Ma X 2015 Fabrication and characteristics of cellulose nanofibril films from coconut palm petiole prepared by different mechanical processing *Ind. Crops Prod.* **65** 96–101
[15] Nordin N I A A, Ariffin H, Andou Y, Hassan M A, Shirai Y, Nishida H, Yunus W M Z W, Karuppuchamy S and Ibrahim N A 2013 Modification of oil palm mesocarp fiber characteristics using superheated steam treatment *Molecules* **18** 9132–46
[16] Chen W, Yu H and Liu Y 2011 Preparation of millimeter-long cellulose i nanofibers with diameters of 30-80 nm from bamboo fibers *Carbohydr. Polym.* **86** 453–61
[17] Zhao H-P, Feng X-Q and Gao H 2007 Ultrasonic technique for extracting nanofibers from nature materials *Appl. Phys. Lett.* **90** 73112
[18] Gokce Y, Cengiz B, Yildiz N, Calimli A and Aktas Z 2014 Ultrasonication of chitosan nanoparticle suspension: Influence on particle size *Colloids Surfaces A Physicochem. Eng. Asp.* **462** 75–81
[19] Cengiz F, Krucińska I, Gliścińska E, Chrzanowski M and Göktepe F 2009 Comparative analysis of various electrospinning methods of nanofibre formation *Fibres Text. East. Eur.* **72** 13–9
[20] Mandzy N, Grulke E and Druffel T 2005 Breakage of TiO2 agglomerates in electrostatically stabilized aqueous dispersions *Powder Technol.* **160** 121–6
[21] Li Y, Zhu H, Xu M, Zhuang Z, Xu M and Dai H 2014 High yield preparation method of thermally stable cellulose nanofibers *BioResources* **9** 1986–97
[22] Xie J, Hse C, Hoop C F De, Hu T, Qi J and Shupe T F 2016 Isolation and characterization of cellulose nanofibers from bamboo using microwave liquefaction combined with chemical treatment and ultrasonication *Carbohydr. Polym.* **151** 725–34
[23] Sumari S, Roesyadi A and Sumarno S 2013 Effects of ultrasound on the morphology, particle

size, crystallinity and crystallite size of cellulose *Sci. Study Res.* **14** 229–39

[24] Xiea J, Chung-Yun Hse, Hoopc C F De, Hua T, Qi J and Shupec T F 2016 Isolation and characterization of cellulose nanofibers from bamboo using microwave liquefaction combined with chemical treatment and ultrasonication *Carbohydr. Polym.* **151** 725–34

[25] Oliveira J E, Mattoso L H C, Orts W J and Medeiros E S 2013 Structural and morphological characterization of micro and nanofibers produced by electrospinning and solution blow spinning: A comparative study *Adv. Mater. Sci. Eng.* **2013**

[26] Pacaphol K and Aht-Ong D 2017 Preparation of hemp nanofibers from agricultural waste by mechanical defibrillation in water *J. Clean. Prod.* **142** 1283–95

[27] Wang H Y, Chen Y Y and Zhang Y Q 2015 Processing and characterization of powdered silk micro- and nanofibers by ultrasonication *Mater. Sci. Eng. C* **48** 444–52

IOP Conf. Series: Materials Science and Engineering **368** (2018) 012034 doi:10.1088/1757-899X/368/1/012034

One-pot nanofibrillation of cellulose and nanocomposite production in a twin-screw extruder

M N F Norrahim[1], H Ariffin[1,2*], T A T Yasim-Anuar[2], M A Hassan[1], H Nishida[3*], T Tsukegi[4]

[1]Department of Bioprocess Technology, Faculty of Biotechnology and Biomolecular Sciences, Universiti Putra Malaysia, 43400 UPM Serdang, Selangor, Malaysia.
[2]Laboratory of Biopolymer and Derivatives, Institute of Tropical Forestry and Forest Products (INTROP), Universiti Putra Malaysia, 43400 UPM Serdang, Selangor, Malaysia.
[3]Department of Biological Functions and Engineering, Graduate School of Life Science and Systems Engineering, Kyushu Institute of Technology, 2-4 Hibikino, Wakamatsu-ku, Kitakyushu, Fukuoka, 808-0196, Japan.
[4]Innovative Composite Materials Research and Development Center (ICC), Kanazawa Institute of Technology, Hakusan, Ishikawa 924-0838, Japan.

*Corresponding authors: hidayah@upm.edu.my (H Ariffin), nishida@lsse.kyutech.ac.jp (H Nishida)

Abstract. Oil palm mesocarp fiber (OPMF) is rich in cellulose and suitable to be used as raw material for the production of cellulose nanofiber (CNF) and biocomposite. Recently, there have been reports on the use of CNF as filler in polypropylene (PP) for improving the mechanical properties of PP, however the process requires two steps: (i) nanofibrillation for CNF production and (ii) biocomposite compounding. In this study, a one-pot process was developed whereby nanofibrillation of cellulose and subsequently melt-blending of the CNF with PP were conducted at once, in a twin-screw extruder. Morphological analysis of the biocomposites by SEM showed that the cellulose was successfully fibrillated into CNF and compounded homogeneously with PP. The highest tensile strength, Young's modulus, flexural strength, and flexural modulus recorded were 34.9 ± 0.5 MPa, 12.1 ± 0.1 GPa, 59.3 ± 1.3 MPa, and 2.3 ± 0.05 GPa, respectively when 3 % CNF was used in the biocomposite. The reinforcement of CNF-OPMF in PP improved the mechanical properties of the biocomposite by 33.4 % compared to neat PP. It is interesting to note that the addition of CNF managed to improve the crystallinity of the biocomposite (54.6 %) compared to neat PP (50.1 %), despite of the lower crystallinity of CNF compared to PP. This observation can be attributed to the high density of covalent bonds per cross-sectional area and the large number of hydrogen bonding sites. Additionally, the observation can be explained by the role of CNF in composite which acted as nucleation agent, which eventually increased the crystallinity of the biocomposite.

1. Introduction

Cellulose nanofiber (CNF) from renewable biomass has attracted much interest as an alternative to micro-sized reinforcement materials in composite plastics. There have been significant advances in the development of plant-plastic composites [1]. Owing to its high stiffness, it can be used to increase the mechanical strength of general-purpose thermoplastics such as polypropylene (PP), polyethylene (PE),

IOP Conf. Series: Materials Science and Engineering **368** (2018) 012034 doi:10.1088/1757-899X/368/1/012034

polyvinyl chloride, polylactic acid and nylon. PP is the most important and widely used polyolefin [2]. Its low density, low production cost, design flexibility, and recyclability make it a popular choice as a matrix for composite making. PP is hydrophobic, leading to compatibility issues when fillers with polar surfaces such as cellulose, are used.

A challenge when using CNF in biocomposites is due to its non-uniform dispersion. This is contributed by their polar surfaces which is difficult to disperse uniformly in a non-polar medium. This might be the reason why the processing of CNF biocomposites was first limited to solvent casting, where water soluble or dispersive polymers were the most common matrices [3]. However, the development of other, more flexible and industrially viable processing techniques is necessary to promote commercialization of these materials. Therefore, several interesting processing methods have been recently reported for these materials. One of the developed processing methods is melt compounding [4]. Melt compounding of cellulose nanocomposites presents several challenges [3]. The major difficulties in melt compounding are to feed the CNF into the extruder and achieve uniform dispersion in the polymer matrix. The CNF has high surface area and has tendency to aggregate when dried. This is expected to be avoided by firstly mixing them in a suitable medium, which is then fed into the extruder. Another method to provide uniform dispersion of nanocellulose in polymer matrix is by using solvent casting, nevertheless the process is not industrially applicable.

Reinforcement of CNF in PP by extrusion was discovered by Hassan et al. (2014) [5]. However, the fibrillation of cellulose into CNF was conducted separately from the extrusion. Based on results obtained, the mechanical properties of PP/CNF composites were improved as compared to neat PP. Furthermore, the reinforcing effects of CNF was better in the injection molded composites as compared to compression molded composites.

One-pot nanofibrillation of cellulose and compounding process for biocomposite production could be done by extrusion. The advantage of this kind of processing lies on its efficient production of CNF biocomposites due to fewer processing steps involved. The ability to blend CNF with PP by melt compounding using an extruder would be advantageous for large scale biocomposites production since extrusion will allow for high solid content processing and provide a continuous process. In this study, nanofibrillation of cellulose for the production of CNF and subsequently melt compounding of the CNF with PP was taken place in the same vessel, *i.e.* a twin-screw extruder. The biocomposites produced were characterized for their mechanical, morphological, crystallinity and thermal properties.

2. Materials and methods

2.1 Materials
OPMF was collected from Seri Ulu Langat Palm Oil Mill, Dengkil, Selangor. The fibers were first disintegrated, washed and dried. Sodium chlorite ($NaClO_2$) and potassium hydroxide (KOH) were obtained from ACROS ORGANICS and J.T Baker Neutracit, respectively.

2.2 Fiber treatment
In order to disintegrate the OPMF fibers physically, they were washed, cleaned, sorted and sun dried. Thereafter, pretreatment of fibers was conducted to recover the cellulose by removing hemicellulose and lignin in a method as explained below.

Lignin removal: Delignification was conducted by treating the OPMF with 5 wt% of sodium chlorite ($NaClO_2$) aqueous solution. pH of the solution was adjusted to pH 5 and the fibers were soaked in this solution at 70 °C for 1.5 hour. Then the fibers were washed and filtered with deionized water until the pH became neutral [6]. Following that, fibers were oven dried overnight.

Hemicellulose removal: KOH pretreatment was conducted by chemically treating the fibers with KOH to completely remove hemicellulose. Initially, fibers were dissolved in 6 wt % of KOH for 24 hours with constant stirring rate [6]. Sample was then filtered and rinsed with deionized water until to reach the pH=7.

Chemical composition of untreated and treated OPMF was shown in Table 1. More than 99% of cellulose was isolated from treated OPMF.

Table 1. Chemical composition of untreated and cellulose-OPMF

	Chemical composition * (%)		
Oil palm mesocarp fiber	**Cellulose**	**Hemicellulose**	**Lignin**
Untreated	38.38 ± 1.21	36.74 ± 1.59	24.88 ± 0.59
Cellulose-OPMF	99.46 ± 0.21	0.00 ± 0.00	0.54 ± 0.69

*Data provided is mean of triplicate samples

2.3 One-pot nanofibrillation and nanocomposite production in a twin-screw extruder

Fibrillation and melt compounding were carried out using a twin-screw extruder (Imoto machinery Co. Limited: Model IMO-160B). The extruder consisted of two major parts: fibrillation and mixing. Fibrillation of cellulose was conducted with a specially designed screw element: tooth-mixing elements (TME) for nanofibrillation at 120 °C, with screw speed of 100 rpm for 1 hour [7]. After nanofibrillation was completed, the rotational speed was reduced to 25 rpm. This is to allow the movement of the fibrillated cellulose into mixing part. PP was fed through a hopper in between fibrillation and mixing section with the ratio of PP to cellulose of 97:3, 98:4 and 95:5. For melt compounding, temperature used was 180 °C and screw speed was set at 25 rpm for 30 minutes. PP/CNF-OPMF composites were then pressed at the same temperature in an electrically heated platen press for 5 min under a pressure of 10 atm to form composite sheets.

2.4 Analysis

2.4.1 Visual examination. The visual appearance analysis of composites was conducted by using digital camera. Composites film were placed on the black fabric in order to avoid reflections. Therefore, a clear dispersion of CNF in composites can be observed. Pictures were taken in order to investigate the distribution of CNF in PP matrix.

2.4.2 Morphological analysis. The morphology of the CNF and the fracture surfaces of the composite samples were observed using SEM (FE-SEM S-3400N, Hitachi, Japan). Composites fracture surfaces were obtained through flexural tests. The dried CNF and fracture surfaces were coated with Pt-Pd for 100 s (Ion sputter: Hitachi, Japan) before the observations.

2.4.3 Mechanical properties. Tensile and flexural tests were performed to determine the mechanical properties of the composites using a mechanical tester (AGS-5kNG, Autograph, Shimadzu Co, Kyoto, Japan). Three-point flexural tests were performed on the bar-shaped composites samples at a crosshead speed of 5 mm/min; the span length for the tests was 50 mm. Flexural tensile (MPa) and flexural modulus (MPa) were obtained in this test. The dumbbell-like composites samples were subjected to tensile tests. The tensile tests were performed at a crosshead speed of 10 mm/min. Results of tensile strength (MPa), young's modulus (MPa), and elongation at break (%) were recorded.

2.4.4 Crystallinity analysis. X-ray diffraction (XRD) pattern was recorded using X-ray powder diffractometer (Rigaku Corporation, Tokyo, Japan) equipped with nickel filtered CuKα radiation (k= 0.1542 nm) at 40 kV and 25 mA. The diffractograms were detected in the range $2\theta = 2 - 50^{\circ}$ at a scan rate of 2°/min.

2.4.5 Differential scanning calorimetry analysis. Differential scanning calorimetry (DSC): measurements were performed using a Pyris 1 DSC calorimeter (Perkin–Elmer Co., Waltham, MA, USA). The samples were first heated from 30 to 200 °C at a rate of 10 °C/min and held at 200 °C for 1 min. They were then cooled to 50 °C at a rate of 10 °C/min and held at 50 °C for 1 min. Next, they were again heated to 200 °C at a rate of 10 °C/min and held at 200 °C for 1 min. Finally, they were quenched to 135 °C at a rate of 50 °C/min to determine their $\tau 1/2$ values. The temperatures corresponding to the exothermic and endothermic peaks in the first heating step and the cooling step were called T_m and T_c, respectively. ΔH_m values were determined from the areas of the melting and crystallization peaks, respectively. The ΔH_m values was converted on the basis of the PP weight ratio of the composites.

2.4.6 Thermal stability. Thermal stability analysis was determined by using Thermo Gravimetry Analyser (TGA) (TGA – 9, Perkin Elmer, USA) under nitrogen flow. Sample (5-11 mg) was placed on a ceramic pan and set on the TGA. The sample was heated at heating rate of 10 °C min^{-1} within the temperature range of 50–550 °C.

3. Results and Discussion

3.1 Degree of fibrillation by extrusion

The effectiveness of nanofibrillation of cellulose-OPMF into CNF-OPMF by extrusion was evaluated and depicted in Figure 1. The SEM micrograph shows that the fibrillation did not occur when less than 3 % of fiber used. Small amount of fiber used gives a low force in the fibrillation part of extrusion which reduced the degree of fibrillation. In spite of this, only small amount of fiber was managed to move into nanocomposite part after fibrillation process completed when less than 3 % of fiber used. The fibrillated fiber was moved into the composite production part after the RPM of the extrusion screws was slow down to 25 RPM. However, as the amount of fiber increased, the fibrillation was increased. Fibrillation was occurred very well at 3% of cellulose as shown in Figure 1 (c). At higher magnification, the size of CNF was identified as <50 nm (Figure 1 (d)). Therefore, in this study, composites were produced from 3 % to 5 % of CNF to PP.

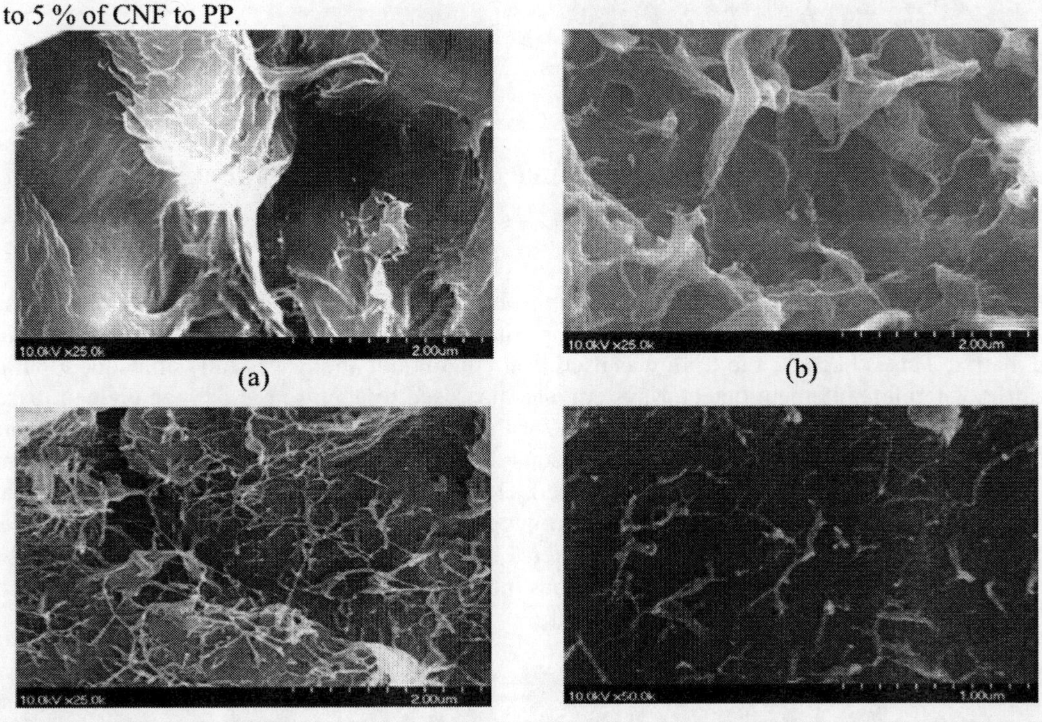

Figure 1. Morphology of fibrillated cellulose-OPMF by extrusion, (a) cellulose-OPMF (1%), (b) cellulose-OPMF (2%), (c) cellulose-OPMF (3%), (d) cellulose-OPMF (3%) (50,000 magnification).

3.2 Visual appearance

Visual appearance of the composites samples was observed to determine dispersion characteristic of CNF-OPMF in the PP matrix. It was seen that CNF in PP/CNF-OPMF (3%), PP/CNF-OPMF (4%), and PP/CNF-OPMF (5%) was well dispersed as shown in Figures 2 (a), (b) and (c). However, aggregation of CNF was observed in some parts of PP/CNF-OPMF (5%) composite (Figure 2 (c)). The aggregated CNF in PP matrix could be correlated to the decrement of the mechanical properties when CNF amount was increased. In order to support this finding, morphological analysis of composites by SEM was also evaluated.

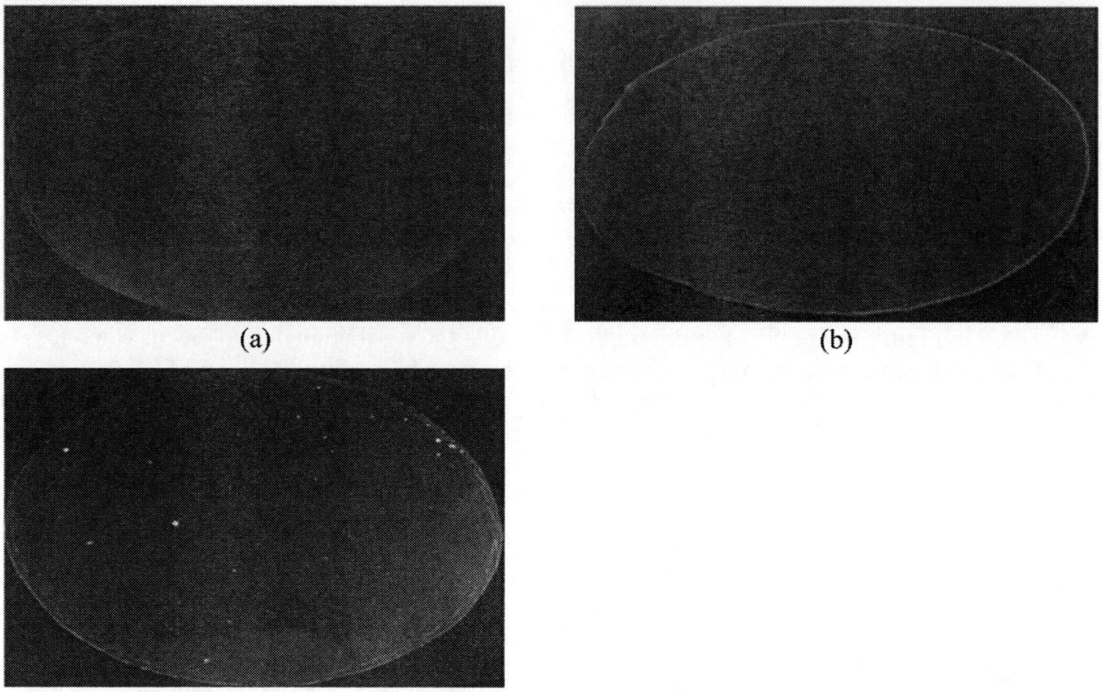

(a) (b)

(c)

Figure 2. Visual appearance of composites, (a) PP/CNF-OPMF (3%), (b) PP/CNF-OPMF (4%), (c) PP/CNF-OPMF (5%)

3.3 Morphological analysis

In correlation to the visual appearance analysis, morphology of the fractured sample of composites were analysed and depicted in Figure 3. For nanocomposite, it is difficult to identify the dispersion of CNF in the matrix. This is because the CNF was used in small amount and it was fully embedded within the PP matrix. Therefore, element detector system was also used to analyse the oxygen element based on chemical structure of celluloses as an indication for the dispersion of CNF in the matrix. Addition of 3 % CNF-OPMF in PP matrix results a homogeneous dispersion of CNF. There was only small amount of aggregations of oxygen element were observed. However, at 4 % and 5 % of CNF in PP matrix as shown in Figure 3 (b) and (c), aggregations of oxygen element were increased and clearly observed. This is concurrent with the visual appearance, where clear white spots of aggregated CNF were observed in the film. As discussed before, these aggregations indicated a limited chemical interaction between PP and CNF-OPMF which decrease the compatibility.

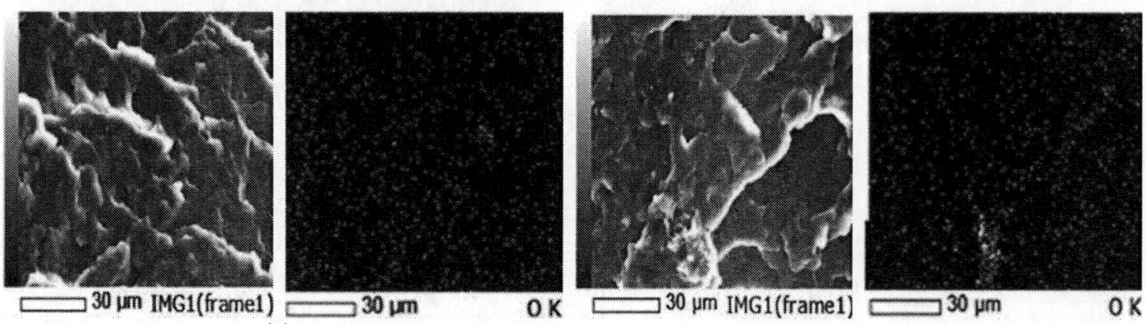

(a) (b)

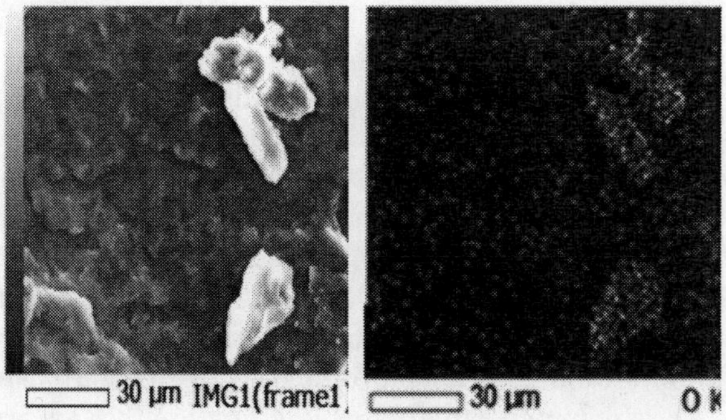

(c)

Figure 3. Micrograph of composites, (a) PP/CNF-OPMF (3%), (b) PP/CNF-OPMF (4%), (c) PP/CNF-OPMF (5%)

3.4 Mechanical properties

Mechanical properties of PP/CNF-OPMF composites and neat PP are shown in Table 2. The addition of CNF-OPMF to PP were markedly increased the tensile and flexural properties as compared to neat PP. For an example, tensile strength of PP/CNF-OPMF (3%) is 34.88 ± 0.5 MPa, as compared to neat PP 25.99 ± 0.5 MPa. Moreover, flexural strength of PP/CNF-OPMF (3%) was 59.32 ± 1.3 MPa as compared to neat PP 46.43 ± 0.8 MPa.

Table 2. Tensile and flexural properties of composites

Composites	Tensile strength (MPa)	Young's Modulus (MPa)	Elongation at break (%)	Flexural stress (MPa)	Flexural modulus (MPa)
Neat PP	25.99 ± 0.5	4394 ± 53	162 ± 13	46.43 ± 0.8	1237 ± 27
PP/CNF-OPMF (3 %)	34.88 ± 0.5	12121 ± 117	145 ± 8	59.32 ± 1.3	2337 ± 55
PP/CNF-OPMF (4 %)	31.49 ± 1.1	9386 ± 547	140 ± 7	53.06 ± 1.0	2201 ± 23
PP/CNF-OPMF (5 %)	29.01 ± 0.8	8315 ± 645	122 ± 5	42.75 ± 0.3	1609 ± 68

The Young's modulus and Flexural modulus of composites were significantly higher as compared to neat PP. The Young's modulus of PP/CNF-OPMF (3 %) was increased to 12121 ± 117 MPa as compared to neat PP 4394 ± 53 MPa. Meanwhile, the Flexural modulus of PP/OPMF-CNF (3 %) was 2337 ± 55 MPa as compared to neat PP 1237 ± 27 MPa. Since CNF is highly crystalline and tougher than neat PP, its influenced the increment on tensile and flexural properties of the composites than neat PP. In spite of this, these improvements can be attributed to the high density of covalent bonds per cross-sectional area and the large number of hydrogen bonding sites between CNF and PP.

Besides, as the amount of CNF increased in the composites, the mechanical strength was decreased. This is related to the dispersion of CNF in PP as discussed before. Aggregation of CNF could lead a poor stress transfer in the polymer matrix which resulted the decrement of the mechanical properties of PP/CNF-OPMF (4%) and PP/CNF-OPMF (5%).

3.5 Crystallinity properties

In order to support the mechanical properties analysis, crystallinity analysis was also conducted and depicted in Figure 4. High crystalline of CNF helps to improve the crystallinity of composites. This was due to the CNF acts as nucleation agent for crystallization occurs (Iwamoto *et al.,* 2014). As the amount of CNF increased, the crystallinity of the composites was also increased. The crystallinity of PP/CNF-OPMF (3 %), PP/CNF-OPMF (4 %), and PP/CNF-OPMF (5 %) were 54.64 %, 56.21 % and 57.00 % respectively as compared to neat PP 50.12 % (Table 3). Composites samples had higher crystallinity than neat PP which resulted in better mechanical properties as discussed before. Nevertheless, even though PP/CNF-OPMF (5 %) had the highest crystallinity, the mechanical properties of the composite sample were lower compared to PP/CNF-OPMF (3 %), which may explain that aggregation of CNF and eventually low stress transfer mainly affected the mechanical properties.

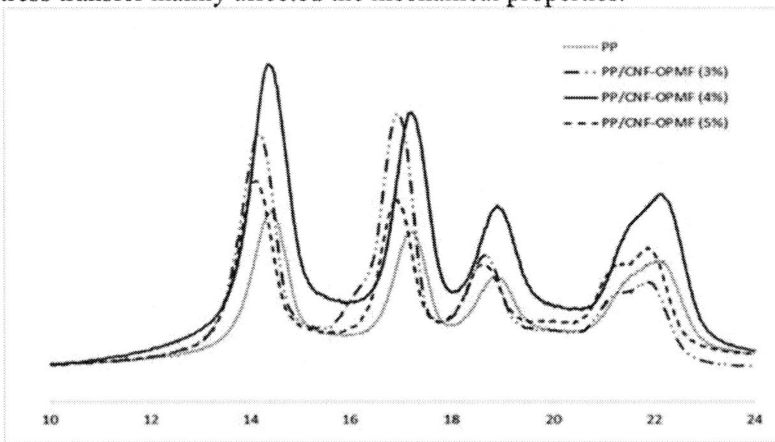

Figure 4. Crystallinity of composite samples.

Table 3. Crystallinity properties of composites.

Composition	**Crystallinity (%)**
Neat PP	50.12
PP/CNF-OPMF (3 %)	54.64
PP/CNF-OPMF (4 %)	56.21
PP/CNF-OPMF (5 %)	57.00

3.6 Thermal properties

Table 4 lists the thermal properties of PP/CNF-OPMF composites. The T_m values were almost similar in all samples, indicating that the CNF blended well with the polymer matrix. However, the T_c and ΔH_m of the composites were slightly higher than that of the neat PP, inferring that the crystallinity of the PP matrix in the composites was higher than that of the neat PP. This is correlate to the crystallinity analysis, where composites have higher crystallinity than neat PP. As the amount of CNF increased, T_c and ΔH_m were also increased. This is also indicated that the crystallization of the PP matrix in the composites started earlier in a non-isothermal cooling process than the neat PP.

Table 4. Thermal properties of composites.

Composition	***Tm* (°C)**	**Δ*Hm* (°C)**	***Tc* (°C)**
Neat PP	162.23	99.2	116.59
PP/CNF-OPMF (3 %)	162.21	105.4	117.89
PP/CNF-OPMF (4 %)	162.24	106.1	118.32
PP/CNF-OPMF (5 %)	161.27	107.2	118.55

3.7 Thermal stability analysis

Thermal stability analysis of composites was analysed and depicted in Figure 5. It has been reported that cellulose acts as a nucleating agent [8], suggesting the presence of an interaction between cellulose and PP sequence. The interaction with solid materials must limit the mobility of polymers even at higher temperatures than melting points. Thus, the thermal stability of composites must be influenced by the crystallinity of PP in the composites. As the crystallinity of the composites prepared by one-step extrusion method was increased as compared to neat PP, the thermal stability based on the $T_{d50\%}$ were also increased. the $T_{d50\%}$ for PP/CNF-OPMF (3 %), PP/CNF-OPMF (4 %) and PP/CNF-OPMF (5 %) were 436.9 °C, 444.1 °C, and 449.3 °C respectively, as compared to neat PP 426.8 °C (Table 5). These results are correlated to the crystallization of PP enhanced by CNF in the composites as discussed before. Looking at different angle, the composites had more amorphous regions, which are easy to degrade compared to crystalline regions. Therefore, it also caused the neat PP easier to be decomposed than the composites.

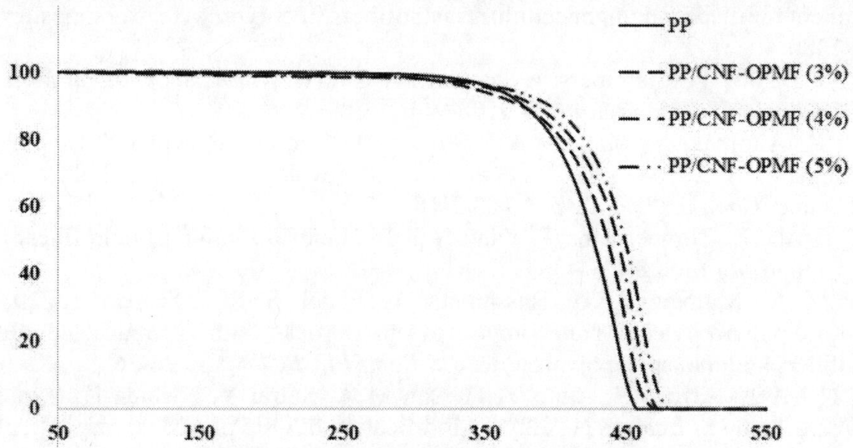

Figure 5. Thermal stability of composites.

Table 5. Thermal properties of composites.

Composition	$T_{d50\%}$ (°C)
Neat PP	426.8
PP/CNF-OPMF (3 %)	436.9
PP/CNF-OPMF (4 %)	444.1
PP/CNF-OPMF (5 %)	449.3

4. Conclusions

A one-pot nanofibrillation of cellulose-OPMF and PP/CNF-OPMF composites processing was successfully developed by using a twin-screw extruder. The advantage of using a twin-screw extrusion process is that it allows for the processing and fibrillation of cellulose at high solid content. To the best of our knowledge, there is lack of research on one-pot nanofibrillation of cellulose and nanocomposite production by extrusion, and this report provides extensive information on the potential use of the method for PP/CNF-OPMF composite production. From our results, the morphological observation of PP and its composites showed that a relatively good dispersion of CNF was achieved as no CNF aggregates was visible in the fractured surfaces of the PP/CNF-OPMF (3%). CNF started to aggregate at higher concentration (5 wt%). Evaluation of the mechanical properties of the PP/CNF-OPMF composites showed that the composite samples had higher tensile and flexural properties compared to neat PP. The improvement in mechanical properties of the composite samples could be related to the high crystallinity of PP/CNF-OPMF. Nevertheless, despite of steady increment in crystallinity with the increased amount of CNF, the mechanical properties of the composite samples with higher CNF content reduced slightly, mainly due to the CNF aggregation which resulted in low stress transfer. Generally speaking, the ability to fibrillate and produce the composite from CNF-OPMF by one-step extrusion method as proposed herewith, is possible to be commercialised due to its fast, simple and easy technique.

Acknowledgements
The authors gratefully acknowledge the Ministry of Higher Education Malaysia (MOHE) for financial support through SATREPS funding (vote no: 6300156) and the Ministry of Science, Technology and Innovation (MOSTI) for the provision of scholarship (MyPhD) to the first author. Authors also would like to acknowledge Japan International Cooperation Agency (JICA) for travel grant to Kyushu Institute of Technology, Japan.

References
[1] Iwamoto S, Yamamoto S, Lee S H, Endo T 2014 Solid-state shear pulverization as effective treatment for dispersing lignocelluloses nanofibers in polypropylene composites. *Cellulose* **21** 573-1580
[2] Hearle J W S 2001 Textile Fibers: A comparative Overview. *Encyclopedia of Materials: Science and Technology (Second Edition)* 9100-9116
[3] Oksman K, Aitomaki Y, Mathew A P Siqueira G Zhou Q Butylina S Tanpichai S Zhou X Hooshmand S 2016 Review of the recent developments in cellulose nanocomposite processing. *Compos Part A-Appl S* **83** 2-18
[4] Ho T T T, Abe K, Zimmermann T, Yano H 2015 Nanofibrillation of pulp fibers by twin-screw extrusion. *Cellulose* **22** 421-433
[5] Hassan M L, Mathew A P, Hassan E A, Fadel S M, Oksman K 2014 Improving cellulose/polypropylene nanocomposites properties with chemical modified bagasse nanofibers and maleated polypropylene. *J Reinf Plast Comp* **33** 26-36
[6] Nordin N I A A, Ariffin H, Ando Y, Hassan M A, Shirai Y, Nishida H, Yunus W M Z W, Karuppuchamy S, Ibrahim N A 2013 Modification of Oil Palm Mesocarp Fiber Characteristic Using Superheated Steam Treatment. *Molecules* **18** 9132-9146
[7] Nishida H, Yamashiro K, Tsukegi T 2017 Biomass Composites from Bamboo- Based Micro / Nanofibers, in *Handbook of Composite from Renewable Materials*. V. K. Thakur, M. K. Thakur, and M. R. Kessler, eds., Scrivener Publishing LLC, pp. 339–361
[8] Quillin D T, Caulfield D F, Koutsky J A 1993 Crystallinity in the polypropyrene/cellulose system. I. Nucleation and crystalline morphology. *J Appl Polym Sci* **50** 1187-1194
[9] Wang H W, Zhou H W, Gui L L, Ji H W, Zhang X C 2014 Analysis of effect of fiber orientation on Young's modulus for unidirectional fiber reinforced composites. *Compos Part B-Eng* **56** 733-739

IOP Conf. Series: Materials Science and Engineering **368** (2018) 012035 doi:10.1088/1757-899X/368/1/012035

The effects of alkali treatment on the mechanical and chemical properties of pineapple leaf fibres (PALF) and adhesion to epoxy resin

M H Zin[1,2]*, K Abdan[1,3], N Mazlan[1,4], E S Zainudin[5] and K E Liew[2]

[1]Institute of Tropical Forestry and Forest Products, Universiti Putra Malaysia, 43400 UPM Serdang, Selangor, Malaysia.
[2]Aerospace Malaysia Innovation Centre, No 3-6-01, German Malaysian Institute, Jalan Ilmiah, Taman Universiti, 43000 Bangi, Selangor, Malaysia.
[3]Department of Biological and Agricultural Engineering, Faculty of Engineering, Universiti Putra Malaysia, 43400 Serdang, Selangor, Malaysia.
[4]Department of Aerospace Engineering, Faculty of Engineering, Universiti Putra Malaysia, 43400 UPM Serdang. Selangor, Malaysia.
[5]Department of Mechanical and Manufacturing Engineering, Faculty of Engineering, Universiti Putra Malaysia, 43400 UPM Serdang. Selangor, Malaysia.

*Corresponding author: hanafeezin@gmail.com

Abstract. The focus of this study was to obtain the optimum alkaline treatment for pineapple leaf fibre and its effect on the mechanical and chemical properties, surface topography, heat resistivity, as well as its interfacial bonding with epoxy matrix. There were 6 different treatment conditions set for the fibre. The morphology of a single fibre observed under the Digital Image Analyzer indicates slight reduction in fibre diameter with increasing NaOH concentration. The Scanning Electron Microscope (SEM) results show the deteriorating effect of alkali, which can be seen from the removal of impurities and increase in surface roughness. The mechanical analysis indicates that 6% NaOH treatment with 3-hour immersion period yielded the highest tensile strength. The adhesion between single fibre and epoxy resin was analysed through the micro-droplet test. Alkaline treatment results in better mechanical bonding between fibre and epoxy resin. It was found that 6% NaOH treatment with 1-hour immersion yielded highest interfacial shear stress. However, as NaOH concentration went above 6 %, the fibre started to show reduction in mechanical properties, as well as fibre-matrix interlocking. The TGA analysis implies that alkaline treatment improved the thermal stability and heat resistivity of the fibre.

1. Introduction

Natural fibre has gained growing interest and usage over the recent years due to their potential capability as a substituent for synthetic fibres. Natural fibres possess valuable properties such as low density and high specific strength and stiffness [1]. Besides economical reason, natural fibres are a renewable resource, for which production requires little energy, with CO_2 absorption, and returning oxygen to the environment [2].

IOP Conf. Series: Materials Science and Engineering **368** (2018) 012035 doi:10.1088/1757-899X/368/1/012035

There are numerous research being carried out previously on natural fibre reinforced composite (NFRC) [3]. When talking about NFRC, it is crucial to understand the chemical composition and surface adhesive bonding between matrix and reinforcement that affect the mechanical properties of the composite. Natural fibres are known for their hydrophilic polar characteristic, which is against the non-polar hydrophobic matrix properties [4]. This condition lead to weak mechanical bonding between the matrix and reinforcement, resulting in poor mechanical properties of the composite produced.

Hydrophilic nature of natural fibre requires chemical modification to improve interfacial properties between fibre and resin [5]. One of the most economical and conservative chemical modification methods is alkali treatment. Alkaline treatment increases surface roughness, as well as the amount of exposed cellulose on the fibre surface by eliminating cellulosic content that covers the external surface of the fibre cell wall [6], resulting in better mechanical interlocking.

Previous works have recorded numerous trials involving alkali treatment for natural fibres. Atiqah et al. [7] treated kenaf fibre with 6% sodium hydroxide (NaOH) solution for 3 hours and showed optimum results for flexural, tensile and impact strengths. Claudia Merlini et al. [8] attempted alkaline treatment on banana short fibres with 10% NaOH solution for 1 hourPanyasart et. al [9] treated pineapple leaf fibre (PALF) with 5% NaOH solution and 5 hours immersion period at room temperature. Previous work by Asim et. al [4] on alkali treatment for pineapple fibre showed increase in mechanical properties for fibres treated with 6% NaOH. The alkaline treatments showed improved behaviour in mechanical properties as compared to untreated fibres. The objective of this study was to obtain the optimum alkaline treatment for pineapple leaf single fibre strand and its effect on the mechanical properties, surface micrography, heat resistivity, as well as interfacial bonding with epoxy matrix.

2. Materials

The Pineapple Leaf Fibre (PALF) was obtained from Johor, Malaysia. Properties of pineapple leaf fibre [10] is shown in Table 1:

Table 1. Properties of Pineapple Leaf Fibre (PALF)

Properties	PALF
Density (g/cm^3)	1.07
Tensile strength (MPa)	126.60
Elongation at break (%)	2.2
Young's modulus (MPa)	4405

The chemical used in this research is Sodium Hydroxide (NaOH) pellet. For resin system, epoxy DM15 made of Bisphenol A was selected, with low molecular weight and high mechanical strength. The resin was made of 2 parts; Part A (base) and Part B (hardener). The ratio between base and hardener is shown as follow:

$$\text{DM15 (Part A)}: \text{DM15 (Part B)} = 5:1 \qquad (1)$$

2.1 Alkali Treatment

The PALF fibres were immersed in NaOH with various treatment concentration; 4%, 6% and 8% (w/v) as illustrated in Table 2:

Table 2. NaOH treatment condition on PALF

Sample	NaOH %	Duration (Hours)	Abbreviation
Untreated	0	0	Untreated
4% + 1 hour	4	1	4%1h
4% + 3 hour	4	3	4%3h
6% + 1 hour	6	1	6%1h
6% + 3 hour	6	3	6%3h
8% + 1 hour	8	1	8%1h
8% + 3 hour	8	3	8%3h

The alkaline treatment involved dissolving NaOH pellet according to the designated concentration, as performed previously by [11]. For instance, to make 4% NaOH solution, 40g NaOH pellet was dissolved in 1 litre of distilled water. PALF fibres were cut to about 400mm, and immersed in the NaOH solution accordingly, followed by rinse off using distilled water. pH paper was used to check alkalinity by immersing it in the solution. The fibres were washed until pH 7 was obtained. Last step of the alkaline treatment was oven drying, at 60°C for 24 hours.

2.2 Characterizations

2.2.1 Physical Analysis. The PALF fibre's diameter was measured under Olympus SZX 12-CCD Digital Image Analyzer at 20 times magnification. The average diameter of 5 different points was taken for each fibre strand. For each type of concentration, 10 samples were prepared and averaged.

2.2.2 Morphology Analysis. The surface morphology of the fibre was observed under Hitachi S-3400N Scanning Electron Microscope (SEM) with the setting of 5.0 kV magnification between 300 to 800 times.

2.2.3 Single Fibre Tensile Test. Single fibre tensile test was conducted as per ASTM D3822 using Instron 3365 Dual Column Table Top Universal Testing Systems with 5kN maximum load at the rate of 2 mm/min. The single fibre was attached to a paper holder, as illustrated in Figure 1.

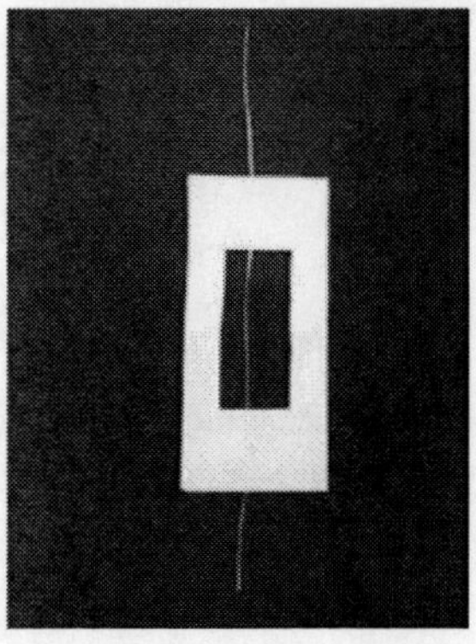

Figure 1. PALF single fibre attached to paper holder

The sample was attached to the gripper, and average of 10 samples were taken for each combination of NaOH treatment.

IOP Conf. Series: Materials Science and Engineering **368** (2018) 012035 doi:10.1088/1757-899X/368/1/012035

2.2.4 Adhesion to Epoxy Resin. The micro-droplet test was carried out to investigate the adhesive bonding between PALF fibre and epoxy resin. The technique was based on previous work of Dai et. al, (2011), whereby resin was wetted onto the single PALF fibre that was attached to a paper holder to form a micro-droplet surrounding the fibre diameter due to surface tension. The embedded length of the micro-droplet was observed under a digital image analyser. The schematic of the micro-droplet test is shown in Figure 2.

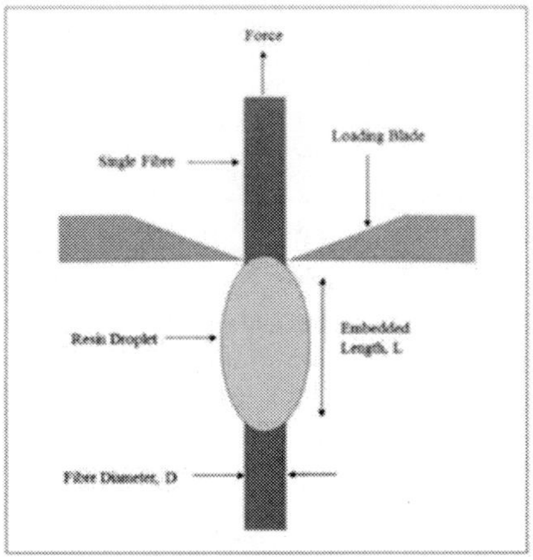

Figure 2. The schematic of the micro-droplet test

The figure illustrated vertical arrangement of the micro droplet test. The top part of the paper holder was clamped to the gripper. While the fibre being pulled upwards, the resin droplet was hold by the loading blade, opposing the pulling direction. The interfacial shear stress (IFSS) was calculated as follow:

$$\tau = \frac{F}{\pi DL} \quad (2)$$

Where F is the maximum load, D is the fibre diameter, and L is the embedded length. Average of 10 samples per NaOH combination was taken for record. The tensile test was conducted with 5kN maximum load and at a rate of 0.1 mm/min.

2.2.5 Thermal Stability Analysis. The thermogravimetric analysis (TGA) was conducted to investigate thermal behaviour of PALF fibre with a temperature setting between 30 °C up to 600 °C at a rate of 10 °C/min.

3.0 Results and Discussion

3.1 Physical Analysis
The effect of alkaline treatment on single fibre diameter is shown in Table 3:

Table 3. Effect of alkalization on PALF single fibre diameter

Treatment conditions	Diameter (mm)	Std Dev
Untreated	0.070	0.024
4%1h	0.107	0.029
4%3h	0.070	0.009
6%1h	0.102	0.023
6%3h	0.121	0.014
8%1h	0.115	0.047
8%3h	0.096	0.016

In general, there is a variation in fibre diameter throughout the range of different NaOH concentration and period treatment. Average diameter of untreated PALF fibre is 0.070 ± 0.024 mm, which is in accordance with the finding by previous researcher [12]. From the result, the diameter of single strand fibre increased after the alkaline treatment. This is because, the treatment caused the fibres to clumped together, made it difficult to be segregated into single fibre strand, unlike the untreated fibre. In terms of the aspect ratio, the result will be normalised for tensile test as the calculation will consider the fibre diameter with respect to the force needed to break the fibre. Looking at the trend, treatment beyond 6% of NaOH concentration yielded in reduction of fibre diameter. This could correspond to removal of waxy epidermal tissue, as also reported by previous researcher [2].

3.2 Morphology Analysis
Figure 3, 4 and 5 showed the effect of alkaline treatment on surface morphology of the PALF fibre.

Figure 3. SEM micro-graph for untreated fibre.

Figure 4.. SEM micro-graph for fibre treated with 6% NaOH for 4 hours

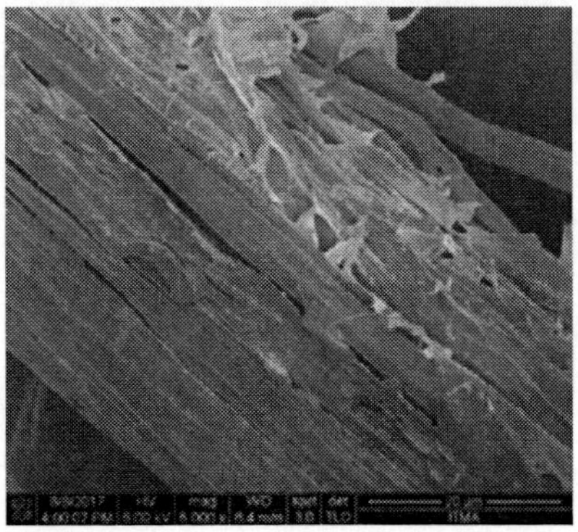

Figure 5. SEM micro-graph for fibre treated with 8% NaOH for 4 hours.

Figure 3 illustrated the untreated PALF whereby considerable deposits of impurities could be seen on the surface. As also mentioned by previous researcher, the impurities existed on fibre surface caused poor fibre-matrix interfacial adhesion due to limited area of contact between fibre and matrix [13]. Figure x showed the surface of PALF fibre treated with 4% NaOH. It can be seen that impurities have been removed, making the surface cleaner and rougher. Damages on the fibre are observable on a few spots. The removal of waxy layer, impurities and elements such as lignin and hemicellulose due to alkaline treatment contribute to better surface adhesion between the fibre and epoxy resin, as also depicted by previous study [11]. This is illustrated in Figure 4 whereby the fibre surface is cleaner due to removal of impurities.

However, as NaOH concentration further increased, damages on the surface became more severe, due to corrosive effect of the alkaline solution, as shown in Figure 5. Excessive delignification caused fibre deterioration, hence making fibre weak and reduce its mechanical properties. The micrograph result is in line with the tensile test finding explained in section 3.3.

3.3 Single Fibre Tensile Test

Figure 6 presents the average Ultimate Tensile Strength of PALF single fibre.

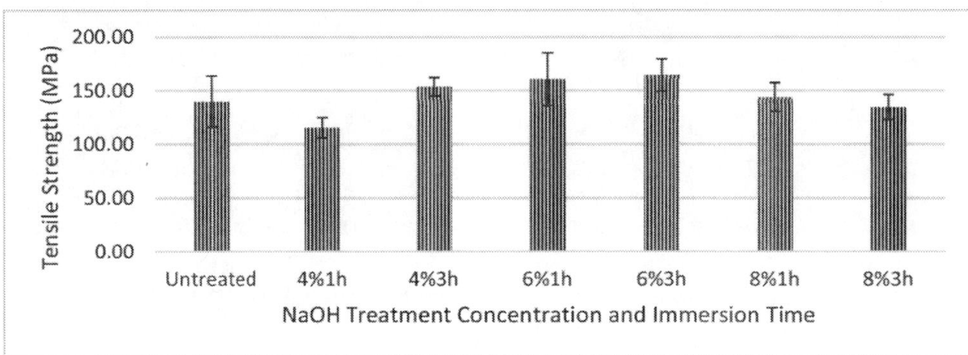

Figure 6. PALF single fibre tensile strength with different NaOH concentration

The untreated fibre recorded tensile strength of 139.90 MPa, which is in accordance with previous findings [10]. The alkali treatment resulted in increasing trend of the tensile strength properties, until the optimum point is reached. The highest tensile strength recorded was 164.55 MPa, corresponding to NaOH treatment with 6% concentration and 3 hours immersion time. This is equivalent to 18% increment as compared to the untreated fibre. As the NaOH concentration went higher to 8%, the tensile strength of PALF single fibre started to show decreasing pattern. This could be due to excessive removal of waxy layers and lignin of the fibre in high alkaline concentration, resulting in weaker or damaged fibre. Similar finding was also reported by previous researcher [13]. This can be proven from morphology presented in Figure 5.

3.4 Adhesion Analysis to Epoxy Resin

The interfacial shear strength for untreated and treated PALF with epoxy resin is presented in Figure 7.

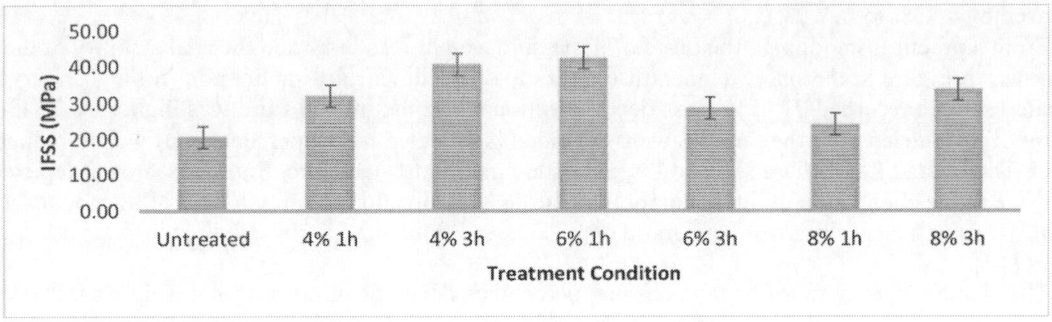

Figure 7. The interfacial shear strength for untreated and treated PALF with epoxy resin

The interfacial shear strength (IFSS) denoted the adhesive interlocking between single fibre and epoxy resin droplet. Higher IFSS means greater force is needed to detach the resin droplet from fibre surface. The untreated fibre recorded the IFSS of 20.64 MPa. The IFSS improved with alkali treatment. The peak was recorded at 42.67 MPa, corresponded with 6% NaOH concentration and 1 hour soaking period. This is equivalent to 106% improvement compared to untreated fibre. Alkali treatment removed natural and artificial impurities, which resulted in increase in surface roughness, hence leading to better interlocking adhesion and greater amount of exposed cellulose on the fibre surface. Similar finding was obtained by previous researcher [14]. Rougher surface enabled better fibre wetting through increased number of possible reaction sites. The effect of rough surface is evident in morphology analysis in Figure 4.

3.5 Thermal Stability Analysis

Figure 8 illustrate the thermogravimetric analysis (TGA) for both treated and untreated PALF single fibre.

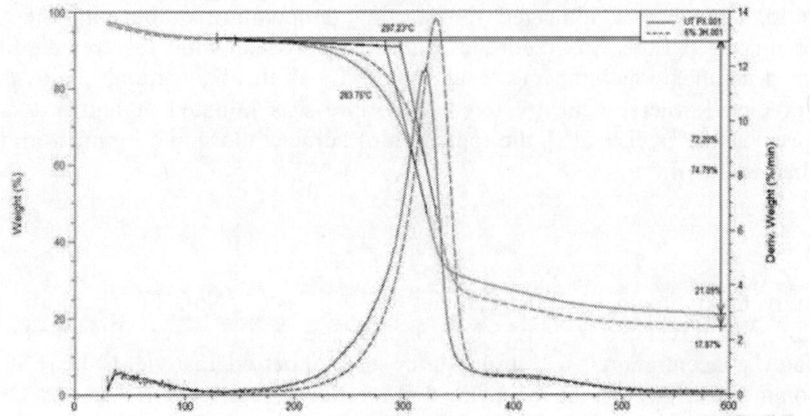

Figure 8. Thermal stability behaviour for untreated and treated PALF

From figure 8, untreated PALF fibre started to lose weight earlier than treated fibre. This is corresponding to removal of water content of untreated fibre, whereby the presence of hemicellulose lead to higher moisture absorption of the fibre [15].

At temperature 80°C, moisture started to evaporate from the sample. [16] depicted that due to hydrophilic property of natural fibre, it is impossible to completely remove moisture from the sample. The decomposition temperature for untreated fibre started at 283.75°C, whilst for the treated fibre, temperature increased by 4.75% to 297.23°C.

From a practical standpoint, the onset of a massive weight loss indicated thermal stability of the fibre. As shown in Figure 8, the onset temperature refers to sharp downward inclination in the TG curve. The untreated sample recorded 71.75% mass decomposition while the treated fibre was higher by 2.71%. The amount of residue left after the samples were degraded is reflected by the percentage of weight reduction at 500°C. The treated PALF fibre showed lower residue than untreated fibre. Some researcher depicted that this is due to the elimination of lignin component through alkalization. Lignin in natural fibre contributes to charring thus untreated fibre will have more char. The result obtained in this study is in line with previous studies [17].

The change in decomposition temperature, percentage decomposition and final weight are shown in Table 4.

Table 4. Thermal stability comparison between untreated and treated PALF

Treatment	Decomposition temperature (°C)	% Decomposition	% Final weight after decomposition
Untreated	282.25	71.75	21.05
4% 1h	296.52	76.79	16.49
4% 3h	298.1	74.68	17.14
6% 1h	308.34	80.52	12.83
6% 3h	296.57	74.46	17.83
8% 1h	302.16	81.01	13.22
8% 3h	300.54	76.85	15.51

The decomposition temperature indicated the starting temperature of burning the sample. The peak temperature for untreated fibre was recorded at 282.25°C. Meanwhile the treated fibre of 6% NaOH recorded higher decomposition temperature at 296.57°C. With alkalization, removal of hemicellulose component is predicted whereby the treated PALF fibre was initiated at higher temperature. As also mentioned by previous researcher [18], the reduction of hemicellulose and lignin from fibre treatment led to increase in thermal stability.

4. Conclusion

Based on the study, the optimum alkaline treatment for PALF is 6% NaOH concentration with 3-hour-hour immersion period, which resulted in 164.55 MPa tensile strength. However, for interfacial shear strength, the optimum NaOH concentration is 6 % with 1-hour soaking period, that yields the IFSS of 42.67 MPa. As the concentration increases to 8 %, the mechanical properties of PALF deteriorates by 18% down to 134.75 MPa. This could be due to excess delignification on fibre surface that contributed to the deteriorating effect. The SEM micrographs show that alkali treatment increases the surface roughness of the fibre, while the TGA results show that the treatment enhances the thermal stability of the fibre by an increase of 14.32 °C for the burning temperature.

Acknowledgement
This study was conducted in a collaborative laboratory known as the "Sustainable Aviation" - joint lab with Aerospace Malaysia Innovation Centre, Universiti Putra Malaysia and University of Nottingham Malaysia Campus, under Project no. BF-P06.

Reference

[1] K. L. Pickering, M. G. A. Efendy, and T. M. Le, "A review of recent developments in natural fibre composites and their mechanical performance," *Compos. Part A Appl. Sci. Manuf.*, vol. 83, pp. 98–112, 2015.

[2] Y. Cao, F. Chan, Y. H. Chui, and H. Xiao, "Characterization of flax fibres modified by alkaline, enzyme, and steam-heat treatments," *BioResources*, vol. 7, no. 3, pp. 4109–4121, 2012.

[3] A. Alavudeen, N. Rajini, S. Karthikeyan, M. Thiruchitrambalam, and N. Venkateshwaren, "Mechanical properties of banana/kenaf fiber-reinforced hybrid polyester composites: Effect of woven fabric and random orientation," *Mater. Des.*, vol. 66, pp. 246–257, 2015.

[4] M. Asim, M. Jawaid, K. Abdan, and M. R. Ishak, "Effect of Alkali and Silane Treatments on Mechanical and Fibre-matrix Bond Strength of Kenaf and Pineapple Leaf Fibres," *J. Bionic Eng.*, vol. 13, no. 3, pp. 426–435, 2016.

[5] X. Li, L. G. Tabil, and S. Panigrahi, "Chemical treatments of natural fiber for use in natural fiber-reinforced composites: A review," *J. Polym. Environ.*, vol. 15, no. 1, pp. 25–33, 2007.

[6] T. Gurunathan, S. Mohanty, and S. K. Nayak, "A Review of the Recent Developments in Biocomposites Based on Natural Fibres and Their Application Perspectives," *Compos. Part A Appl. Sci. Manuf.*, vol. 77, pp. 1–25, 2015.

[7] A. Atiqah, M. A. Maleque, M. Jawaid, and M. Iqbal, "Development of kenaf-glass reinforced unsaturated polyester hybrid composite for structural applications," *Compos. Part B Eng.*, vol. 56, pp. 68–73, 2014.

[8] C. Merlini, V. Soldi, and G. M. O. Barra, "Influence of fiber surface treatment and length on physico-chemical properties of short random banana fiber-reinforced castor oil polyurethane composites," *Polym. Test.*, vol. 30, no. 8, pp. 833–840, 2011.

[9] K. Panyasart, N. Chaiyut, T. Amornsakchai, and O. Santawitee, "Effect of surface treatment on the properties of pineapple leaf fibers reinforced polyamide 6 composites," *Energy Procedia*, vol. 56, no. C, pp. 406–413, 2014.

[10] R. M. N. Arib, S. M. Sapuan, M. M. H. M. Ahmad, M. T. Paridah, and H. M. D. Khairul Zaman, "Mechanical properties of pineapple leaf fibre reinforced polypropylene composites," *Mater. Des.*, vol. 27, no. 5, pp. 391–396, 2006.

[11] R. Mahjoub, J. M. Yatim, A. R. Mohd Sam, and S. H. Hashemi, "Tensile properties of kenaf fiber due to various conditions of chemical fiber surface modifications," *Constr. Build. Mater.*, vol. 55, pp. 103–113, 2014.

[12] M. Jawaid and H. P. S. Abdul Khalil, "Cellulosic/synthetic fibre reinforced polymer hybrid composites: A review," *Carbohydr. Polym.*, vol. 86, no. 1, pp. 1–18, 2011.

[13] O. M. L. Asumani, R. G. Reid, and R. Paskaramoorthy, "The effects of alkali-silane treatment on the tensile and flexural properties of short fibre non-woven kenaf reinforced polypropylene composites," *Compos. Part A Appl. Sci. Manuf.*, vol. 43, no. 9, pp. 1431–1440, 2012.

[14] M. Mostafa and N. Uddin, "Effect of Banana Fibers on the Compressive and Flexural Strength of Compressed Earth Blocks," *Buildings*, vol. 5, no. 1, pp. 282–296, 2015.

[15] P. Methacanon, U. Weerawatsophon, N. Sumransin, C. Prahsarn, and D. T. Bergado, "Properties and potential application of the selected natural fibers as limited life geotextiles," *Carbohydr. Polym.*, vol. 82, no. 4, pp. 1090–1096, 2010.

[16] S. Rodriguez, F. M. Margem, S. N. Monteiro, and V. Calado, "Materials Science & Engineering A

Thermogravimetric behavior of natural fibers reinforced polymer composites — An overview," *Mater. Sci. Eng. A*, vol. 557, pp. 17–28, 2012.
[17] S. H. Lee and S. Wang, "Biodegradable polymers/bamboo fiber biocomposite with bio-based coupling agent," *Compos. Part A Appl. Sci. Manuf.*, vol. 37, no. 1, pp. 80–91, 2006.
[18] M. D. H. Beg and K. L. Pickering, "Accelerated weathering of unbleached and bleached Kraft wood fibre reinforced polypropylene composites," *Polym. Degrad. Stab.*, vol. 93, no. 10, pp. 1939–1946, 2008.

IOP Conf. Series: Materials Science and Engineering **368** (2018) 012036 doi:10.1088/1757-899X/368/1/012036

Physical changes and FTIR analysis of kenaf core fiber heat treated in air

H W Hussin[1*], K Abdan[1,2], M S Salit[1,3] and E S Bakar[1,4]

[1]Laboratory of Biocomposite Technology, Institute of Tropical Forestry and Forest Product (INTROP), Universiti Putra Malaysia, 43400 UPM Serdang, Selangor, Malaysia.
[2]Department of Biological and Agricultural Engineering, Faculty of Engineering, Universiti Putra Malaysia, 43400 UPM Serdang, Selangor, Malaysia.
[3]Department of Mechanical and Manufacturing Engineering, Faculty of Engineering, Universiti Putra Malaysia, 43400 UPM Serdang, Selangor, Malaysia.
[4]Department of Forest Production, Faculty of Forestry, Universiti Putra Malaysia, 43400 UPM Serdang, Selangor, Malaysia.

*Corresponding author: wmh76wh@gmail.com

Abstract. In this paper physical changes (hygroscopic and color properties) and FTIR analyses of kenaf core fiber heat treated in air were studied. The fibers were heated in air at temperature of 103, 165, 175, 185, 195 and 205 °C for 15 minutes. For physical changes hygroscopic and color analysis were carried out where moisture absorption and kenaf core color lightness were measured. FTIR of treated and untreated fibers were analyses to find possible changes in the spectrum and also relation between moisture absorption and OH groups of the fibers. Ratio between OH band height (3332 - 3346 cm^{-1}) and absorbed moisture band height (1630 - 1640 cm^{-1}) was used to find relation between absorbed moisture and OH groups. The results showed moisture absorption decreased as treatment temperature increased. However for fiber treated at 165 and 175 °C, moisture absorption was not significantly different. Fiber darkness only increased for fibers treated above 175 °C. FTIR results showed no substantial chemical changes between treatments; however, through deduction, heat treatment cause degradation of OH groups. As a conclusion, chemical modification that changes hygroscopic and color properties of fiber only minimum and possibly occurred at surface of fiber, thus cannot be detected directly by FTIR analysis.

1. Introduction

Heat treatment has been known to improve dimensional stability and durability of wood. Commercially, heat treated wood has been introduced as early as 1930's with process call Lignostone and Staybwood. In more recent times, several heat treatment processes for wood emerges in Europe such as Thermowood, Plato Wood, OHT-Oil Heat Treatment, Bois Perdure, Rectification, WTT and Huber Holz. These process treats wood at temperatures between 160 °C and 260 °C with different process conditions and duration. Normally, duration of treatment would involves hours of exposure to heat [1]. At around 250 °C, even minimum exposure of heat flux can cause wood ignition [2]. Apart from temperature, duration of exposure at particular heat flux can also lead to wood ignition[3]. Therefore, to prevent wood ignition, heat treatment with high temperature and/or long duration require

IOP Conf. Series: Materials Science and Engineering **368** (2018) 012036 doi:10.1088/1757-899X/368/1/012036

deprived of oxygen condition.

Dimensional stability that achieved through thermal treatments on wood relates to reduction of available hydroxyl site in wood cell wall to form hydrogen bond water [4]. The other physical change experienced by wood is color changes where longer duration and higher treatment temperature produce darker color wood [5-6]. Degradation of lignin and non-cellulosic polysaccharides (hemicellulose, pectin, gum, etc.) cause changes in wood color. Heated acetylated hemicelluloses produce acetic acid which catalyzed hydrolysis of hemicellulose into soluble sugars. Heat caused sugars to caramelized into brown color which affect the wood color [7].

Fourier-transform infrared spectroscopy (FTIR) has been used in many study to qualitatively confirm chemical changes caused by chemical reactions, heating or decaying wood [8-10]. Other researchers use FTIR as quantitative tools to measure the ratio of lignocellulose components and composites material [11-12]. FTIR analysis was used in this study to detect changes on chemical groups through qualitative and quantitative analysis.

Although woods and kenaf core material are both lignocelluloses materials, smaller size of kenaf core fiber would influence drying, heating and reaction of heat treatment process [13]. As this study was carried out in air with the presence of oxygen, fiber degradation is expected to be accelerated. To prevent kenaf core fiber ignition, treatment parameter in this study was limited to 205 °C and 15 minutes of exposure.

Few studies were carried out on kenaf bast fiber. Heat treatment of kenaf bast shows an improvement on tensile strength for fiber treated at 140 °C for 2.5 to 10 hours[14-15]. While kenaf bast fiber heated for 1 hour at 160 °C produce higher tensile modulus compare to fiber heated at 140°C for same duration [16]. Due to the morphological difference between kenaf bast and core and short duration of heating (15 minutes), this study chose 165 °C to 205 °C as parameter of studies. Treatment temperature of 165 °C is within range where of extractives and hemicellulose begin to degrade [17].

The objectives of this paper are to study the effect heat treatment on moisture absorption and color of kenaf core fiber. Furthermore, because moisture absorption relates to formation of hydrogen bond between hydroxyl groups of fiber and water molecules from environment, FTIR is used to detect changes of -OH composition. Reduction in -OH composition can reduce the hydrophilicity of kenaf core fibre thus increase the compatibility between kenaf core fiber and hydrophobic polymer matrix in natural fiber composite.

2. Materials and Methods

Kenaf core fiber size 40 μm was supplied by LKTN (Kota Bharu, Malaysia). Fiber was heated in gravimetric oven for 15 minutes at 103 °C (drying temperature), 165, 175, 185, 195 and 205 °C.

For moisture absorption, heat treated fiber were directly stored in desiccator for 10 days and then weighted. Then samples were heated for 3 days at 103 °C to achieve bone dry condition. Then the fiber weighted again. Moisture content was determined using equation 1.

$$Moisture\ content\ (\%) = \frac{Weight_{after\ dessicator} - Weight_{Bone\ dry}}{Weight_{Bone\ dry}}\ x\ 100\% \quad (1)$$

For color changes, samples were arranged and a photo was taken in a single shot as shown in **Figure 1**. The colors from each specimen were cropped and analyzed using open source software, GIMP (GNU image manipulating program). The peak for each color histogram (Top menu: Colors > Curve) were acquired and compared between the treatments.

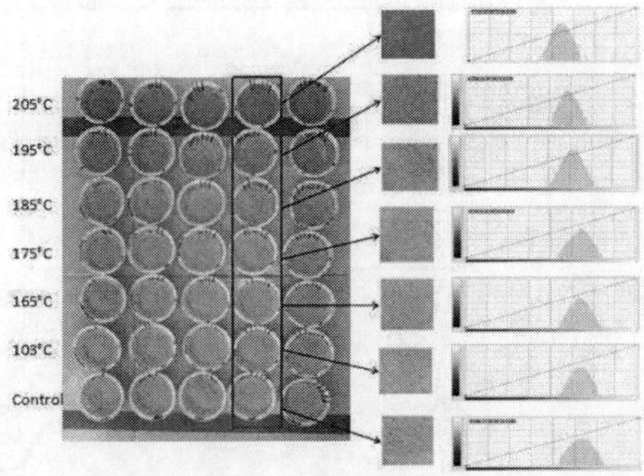

Figure 1 Color analysis of heat treated kenaf core fiber.

In this study, color changes were analyzed using lightness data (also referred as intensity or value) which were transformed RGB (red, green and blue) system to HSL (hue, saturation and lightness) based on equation 2.

$$L = \frac{\max(r,g,b) + \min(r,g,b)}{2} \quad (2)$$

Where L is lightness (in GIMP color histogram; lightness referred as Value channel), max and min is maximum and minimum value among red, green and blue channel respectively [18].

FTIR analysis was carried out using Nicolat Spectroscopy machine. The samples were scanned (4 replications for each treatment) from wavelength of 650 to 4000 cm^{-1}. For qualitative analysis, bands between treatments were compared. While for qualitative analysis, the ratio of bands within the treatments was determined and then the ratio is compared between the treatments. The results for both qualitative and quantitative analysis were modified from transmittance to absorbance throughout the study for determination of area under and band height using equation 3.

$$A = 2 - \text{Log}_{10}(T\,\%) \quad (3)$$

Where A is absorbance and T % is transmission in percentage [19]. The absorbance results, as shown in Figure 2, were analyzed using OriginPro 8 SR4 v8.0951 (B951) for baseline correction, band height, area and position.

Samples thickness of powdery samples like kenaf core fiber cannot be precisely controlled. Even for same samples, amount and size of sample will produce variation in band because height of band is related to amount of sample thickness or cencentration as mention in Lambert-Beer law [20]. Therefore this study use bands ratio to empirically analyze the changes caused by heat treatment.

The ratio of Band A and B was introduced in this study to show changes in C-OH in plane stretching based on OH stretching of absorbed water (refer to Figure 2). Band labelled A (3332 - 3346 cm^{-1}) were identified as hydrogen bonded OH group [21-24]. While band labelled B (1630 - 1640 cm^{-1}) represents OH stretching from absorbed water [8], [12], [24-27].

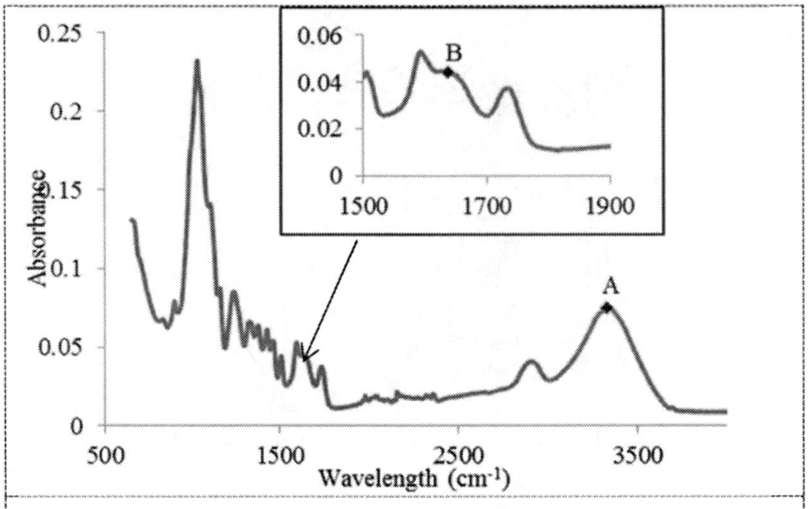

Figure 2 Identified bands for analysis

Two approaches were used to acquire the data for ratio; ratio based on band height and ratio based area under graph.

3. Results and discussions

3.1 Moisture absorption and color changes

Figure 3 shows the ability to absorb water from environment decrease as treatment temperature increase. Even drying process (treatment at 103 °C) has reduce the ability of fiber to absorbed moisture, while fiber treated at 165 and 175 °C does not significantly differ in amount of moisture absorbed. After treatment temperature increased more than 175 °C, the ability to absorb moisture gradually reduced as the temperature increase. Similar trend is also reported for heat treated bamboo scrimber [28]. There is a possibility that the reduction of moisture absorption relates to increase in crystallinity of cellulose and loss of -OH groups due to degradation of hemicellulose.

Figure 3 also shows the similarities between moisture content and color changes for heat treated kenaf core fiber. Compare to untreated fiber; the ability to absorb moisture decrease by 0.4 % for fiber treated at 103 °C (dried fiber), 1.0 % for 165 and 175 °C, 1.5 % for 185 °C, 2.0 % for 195 °C and 2.6 % for 205 °C. However, lightness value when compare to untreated fiber, the value increase by 4.4 for dried fiber (103 °C), 2.4 for fiber treated at 165 °C and 0.4 for fiber treated at 175 °C. Then , compared to untreated fiber, the value decrease by 6.6 for fiber treated at 185 °C, significantly reduce further by 13.2 and 28.6 for fiber treated at 195 and 205 °C respectively.

Both properties show that at temperature of 185 °C and above, ability to absorb moisture and maintain color lightness decrease. The darker color indicates changes in extractives and structural component such as lignin and/or hemicellulose. The other possibility is reaction between sugars and sugar/protein in presence of heat, such as caramelization and Maillard reaction, may produce colored compound of kenaf core fiber [7], [29]. Therefore the chemical changes that affect the color changes begin at 175 °C. This is also in agreement with bamboo scimber heated at 170 °C where the L value begin to drop significantly [28]. This is an indicator that heat treatment begin to significantly alter the composition of the fiber.

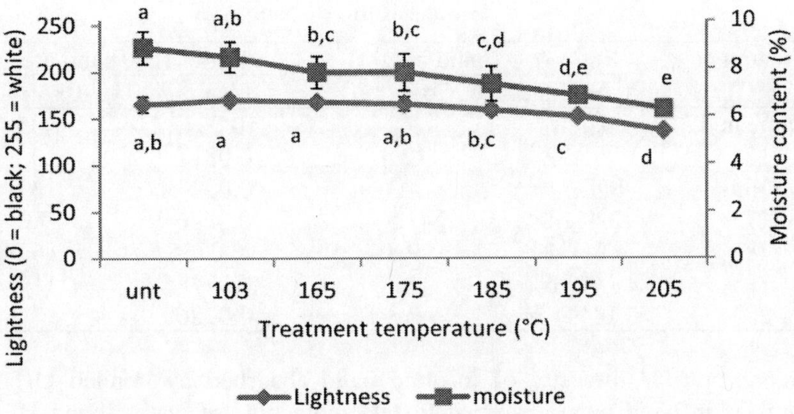

Figure 3 Color and moisture content of heat treated fiber.

3.2 FTIR analysis

Figure 4 shows the bands for qualitative analysis using FTIR. It is worth to mention that the presented figures are carefully selected among treatment replication so that the bands do not overlapped with each other. The bands position and shapes is almost similar among all treatments. This results is in agreement with untreated kenaf fibre analyzed using FTIR reported by other researcher [30-31]. This indicates the treatment does not cause major chemical changes to kenaf core fiber.

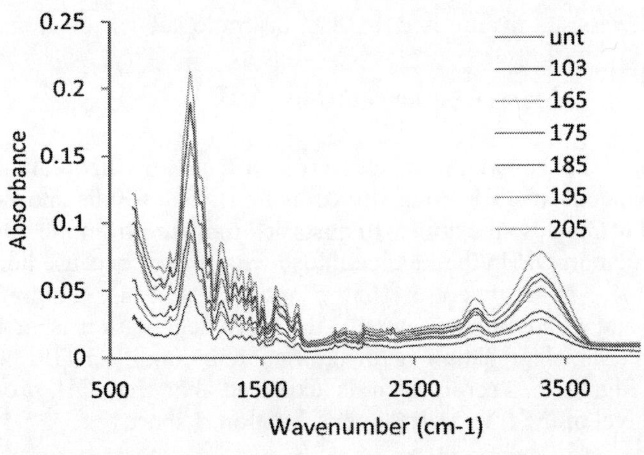

Figure 4 FTIR spectroscopy of untreated and heat treated fiber.

For qualitative analysis, Table 1 shows the A:B ratio of FTIR bands. Using band height ratio produce significant reduction trend for fiber heated at 103 until 185 °C. On contrary, ratio based on band area show insignificant reduction between fibers treated at 103 and 175 °C; significant reduction only begin at 185 °C. After 185 °C, both ratios either calculated by height or area show a significant increase for fiber treated at 195 and 205 °C. Based on dispersion of data, as shown by coefficient of variance (CoV) also known as relative standard deviation (RSD), the dispersion for ratio based on band height is smaller compare to ratio based on integrated area thus produce more significant results for ratio based on height. Therefore ratios based on band height were selected in further analysis. Other researchers also use band height due to better value provided RSD [32].

Table 1 Ratio of band A:B

Treatment	Ratio A:B (band height)		Ratio A:B (band area)	
(°C)	Average	CoV (%)	Average	CoV(%)
untreated	0.488 ^{B,A}	9.2	0.0912 ^{A}	7.5
103	0.499 ^{A}	12.4	0.0817 ^{B,A}	12.6
165	0.477 ^{B,A,C}	4.3	0.0824 ^{B,A}	13.6
175	0.451 ^{B,A,C}	3.5	0.0826 ^{B,A}	7.2
185	0.431 ^{C}	9.8	0.0735 ^{B}	16.0
195	0.438 ^{B,C}	1.5	0.0778 ^{B,A}	11.9
205	0.444 ^{B,C}	7.3	0.0786 ^{B,A}	17.2

Ratio A:B represents the intensity of infrared band absorbed by bonded OH group (band A) compare to 1 unit of infrared band absorbed to OH molecules of water (band B). Generally band height and/or integrated areas are proportional to the quantity of a specific component present in a sample [33]. Therefore, actual unit of OH group affected by the treatment did not presented by Table 1 because the amount of water within the fiber affected by the treatment temperature is lower as shown by Figure 3. Because of bands are directly proportion quantity and ratio between bands must be same; mass and band height of band A equal to mass and height of band B. Therefore equation 4 can be applied.

$$\text{Band}_{(A)} \text{ x Mass}_{(A)} = \text{Band}_{(B)} \text{ x Mass}_{(B)} \tag{4}$$

$$\text{Mass}_{(A)} \text{ x [Band}_{(A)} / \text{Band}_{(B)}] = \text{Mass}_{(B)} \tag{5}$$

Therefore,

$$\text{Mass}_{(A)} \text{ x (ratio A:B)} = \text{Mass}_{(B)}/\text{(ratio A:B)} \tag{6}$$

$$\text{Mass}_{(A)} = \text{Mass}_{(B)} /\text{(ratio A:B)} \tag{7}$$

By applying Equation 7, the actual unit of -OH group affected by the treatment were deducted and presented in Figure 5 Amount of -OH group (by deduction). The results shows that although drying process (treatment at 103 °C) cause the ability to absorbed moisture from the environment decrease, it does not cause the degradation OH in the hemicellulose within fiber because hemicellulose only starts to degrade at 160 °C [17]. The treatment at 165 °C and 175 °C cause a significant decrease of OH compared to untreated and dried fiber. Generally the -OH degradation is gradual as the treatment temperature increase but the degradation is insignificant to modify FTIR bands as inspected by qualitative analysis in Figure 4. Therefore, it is expected that the -OH group modification only occurred at the surface level of the fiber as treatment duration is short.

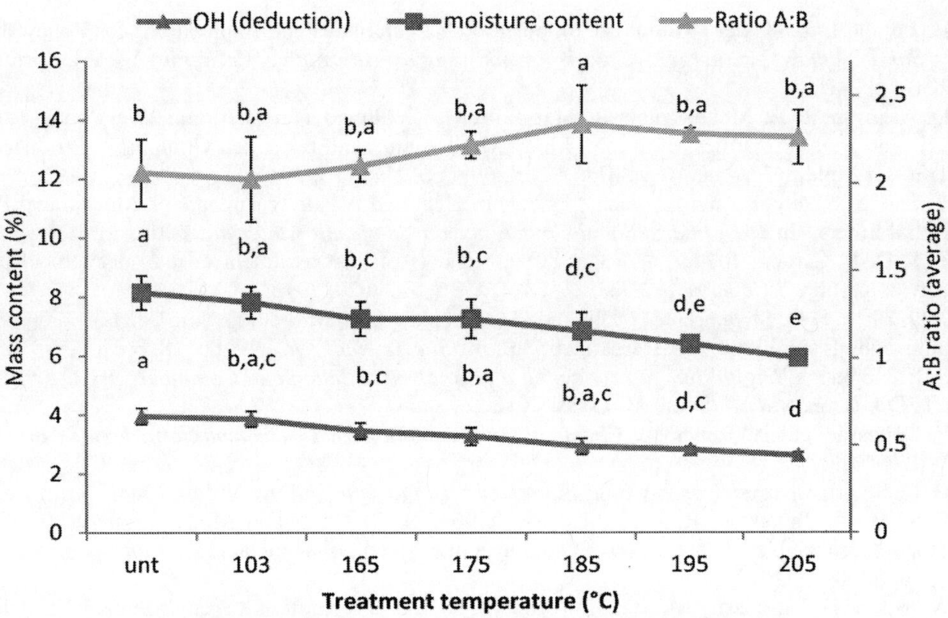

Figure 5 Amount of -OH group (by deduction)

4. Conclusion

This study concludes that heat treatment on kenaf core fiber causes physical modification in term of moisture absorption and color of kenaf core fiber. Chemical modification is not observable in qualitative FTIR band analysis but can be quantitatively deduced using band of absorbed water in FTIR and actual moisture absorption. The results show the reduction of OH group within kenaf core fiber due to heat treatment only begins after 175 °C.

References

[1] B. M. Esteves and H. M. Pereira, "Wood modification by heat treatment: A review," *BioResources*, vol. 4, no. 1, pp. 370–404, 2009.

[2] V. (Vyto) Babrauskas, "Ignition of Wood: A Review of the State of the Art," *J. Fire Prot. Eng.*, vol. 12, no. 3, pp. 163–189, 2002.

[3] W. Jaskółowski and P. Ogrodnik, "The study of time to ignition of woods under external heat flux by piloted ignition and autoignition," *For. Wood Technol.*, no. 86, pp. 133–137, 2014.

[4] R. M. Rowell, "Chemical Modification of Wood," *For. Prod. Abstr.*, vol. 6, no. 12, pp. 363–382, 1983.

[5] M. Sehlstedt-Persson, "Colour Responses to Heat-Treatment of Extractives and Sap from Pine and Spruce," in *8th International IUFRO Wood Drying Conference*, 2003, pp. 459–464.

[6] A. Chen, "Comparisons of Heat Treated Wood to Chemically Treated and Untreated Wood in Commercial Usages Abstract :," 2013.

[7] E. Baysal, S. Kart, H. Toker, S. Degirmentepe, and H. Militz, "Some physical characteristics of thermally modified oriental- beech wood," *Maderas. Cienc. y Tecnol.*, vol. 16, no. 3, pp. 291–298, 2014.

[8] H. S. Sulaiman, C. H. Chan, C. H. Chia, S. Zakaria, and S. N. S. Jaafar, "Isolation and fractionation of cellulose nanocrystals from Kenaf Core," *Sains Malaysiana*, vol. 44, no. 11, pp. 1635–16432, 2015.

[9] K. K. Pandey and A. J. Pitman, "FTIR studies of the changes in wood chemistry following decay by brown-rot and white-rot fungi," *Int. Biodeterior. Biodegrad.*, vol. 52, no. 3, pp. 151–160, 2003.

[10] S. Borysiak, "The thermo-oxidative stability and flammability of wood/polypropylene composites," *J. Therm. Anal. Calorim.*, vol. 119, no. 3, pp. 1955–1962, Dec. 2014.

[11] W. Lao, G. Li, Q. Zhou, and T. Qin, "Quantitative analysis of biomass in three types of wood-plastic composites by FTIR spectroscopy," *BioResources*, vol. 9, no. 4, pp. 6073–6086, 2014.

[12] A. Emandi, C. I. Vasiliu, P. Budrugeac, and I. Stamatin, "Quantitative investigation of wood composition by integrated FT-IR and thermogravimetric methods," *Cellul. Chem. Technol.*, vol. 45, no. 9–10, pp.

579–584, 2011.

[13] H. Lu and L. L. Baxter, "Biomass Combustion Characteristica and Implications for Renewable Energy," in *Solid Biofuels for Energy: A Lower Greenhouse Gas Alternative*, P. Grammelis, Ed. Springer London, 2010, p. 242.

[14] D. Ariawan, Z. A. Mohd. Ishak, R. Mat Taib, M. Z. Ahmad Thirmizir, and Y. J. Phua, "Effect of Heat Treatment on Properties of Kenaf Fiber Mat/Unsaturated Polyester Composite Produced by Resin Transfer Molding," *Appl. Mech. Mater.*, vol. 699, pp. 118–123, 2014.

[15] Y. Cao, S. Sakamoto, and K. Goda, "Effects of Heat and Alkali Treatments on Mechanical Properties of Kenaf Fibers," in *16Th International Conference on Composite Materials*, 2007, pp. 1–4.

[16] P. T. D. L. Carada, T. Fuji, and K. Okubo, "Effects of heat treatment on the mechanical properties of Kevlar-29 fibre," *Compos. Sci. Technol.*, vol. 1736, p. 20029, 2016.

[17] J. O. Brito, F. G. Silva, M. M. Leão, and G. Almeida, "Chemical composition changes in eucalyptus and pinus woods submitted to heat treatment," *Bioresour. Technol.*, vol. 99, no. 18, pp. 8545–8548, 2008.

[18] S. E. Umbaugh, *Digital Image Processing and Analysis: Human and Computer Vision Applications with CVIPtools, Second Edition*. CRC Press, 2016.

[19] F. Rouessac and A. Rouessac, *Chemical Analysis: Modern Instrumentation Methods and Techniques*. Wiley, 2013.

[20] B. H. Stuart, *Infrared Spectroscopy: Fundamentals and Applications*. Wiley, 2004.

[21] I. S. M. A. Tawakkal, R. A. Talib, A. Khalina, N. L. Chin, and M. N. Ibrahim, "Optimisation of processing variables of kenaf derived cellulose reinforced polylactic acid," *Asian J. Chem.*, vol. 22, no. 9, pp. 6652–6662, 2010.

[22] A. H. Faris, M. Nasir, M. Ibrahim, and A. Iqbal, "Preparation of environmental friendly phenol-formaldehyde wood adhesive modified with kenaf lignin," *Beni-Suef Univ. J. Basic Appl. Sci.*, pp. 0–9, 2017.

[23] N. I. A. Razak, N. A. Ibrahim, N. Zainuddin, M. Rayung, and W. Z. Saad, "The influence of chemical surface modification of kenaf fiber using hydrogen peroxide on the mechanical properties of biodegradable kenaf fiber/poly(Lactic Acid) composites," *Molecules*, vol. 19, no. 3, pp. 2957–2968, 2014.

[24] S. Y. Oh, D. Il Yoo, Y. Shin, and G. Seo, "FTIR analysis of cellulose treated with sodium hydroxide and carbon dioxide," *Carbohydr. Res.*, vol. 340, no. 3, pp. 417–428, 2005.

[25] C. H. Chan, C. H. Chia, S. Zakaria, I. Ahmad, and A. Dufresne, "Production and characterisation of cellulose and nano- crystalline cellulose from kenaf core wood," *BioResources*, vol. 8, no. 1, pp. 785–794, 2013.

[26] F. T. G. Dias, R. C. R. Souza, and C. T. Andrade, "Development of maleated starches using an internal mixer," *Macromol. Symp.*, vol. 299–300, no. 1, pp. 139–146, 2011.

[27] P. L. King, M. S. Ramsey, P. F. McMillan, and G. Swayze, "Laboratory fourier transform infrared spectroscopy methods for geologic samples," in *Infrared Spectroscopy in Geochemistry, Exploration Geochemistry and Remote Sensing*, Mineral Association of Canada, 2004, pp. 57–91.

[28] W. Shangguan, Y. Gong, R. Zhao, and H. Ren, "Effects of heat treatment on the properties of bamboo scrimber," *J. Wood Sci.*, vol. 62, no. 5, pp. 383–391, 2016.

[29] B. Sundqvist, "Colour changes and acid formation in wood during heating," Luleå University of Technology, 2004.

[30] C. Pappas, P. A. Tarantilis, I. Daliani, T. Mavromoustakos, and M. Polissiou, "Comparison of classical and ultrasound-assisted isolation procedures of cellulose from kenaf (Hibiscus cannabinus L.) and eucalyptus (Eucalyptus rodustrus Sm.)," *Ultrason. Sonochem.*, vol. 9, no. 1, pp. 19–23, 2002.

[31] N. Dehbari, N. Moazeni, and W. A. Wan Abdul Rahman, "Effects of Kenaf Core on Properties of Poly(lactic acid) Bio-Composite," *Polym. Compos.*, vol. 35, no. 6, pp. 1220–1227, 2014.

[32] E. Konoz and B. Amir Hossein Mohsen Sarraffi Marjaneh, Samadizadeh Samaneh, "Quantitative analysis of Lorazepam in pharmaceutical formulations through FTIR spectroscopy," *E-Journal Chem.*, vol. 9, no. 4, pp. 2232–2238, 2012.

[33] P. Ducheyne, K. Healy, D. W. Hutmacher, D. W. Grainger, and C. J. Kirkpatrick, *Comprehensive Biomaterials II*. Elsevier Science, 2017.

Effectiveness of acetic, propionic and butyric anhydrides to protect rubberwood from decayed by white rot *Trametes versicolor*

N H Hamid[1], **N Hassan**[1], **P M Tahir**[2], **S Ujang**[3]
[1]Faculty of Forestry, Universiti Putra Malaysia,43400 UPM, Serdang, Selangor, Malaysia
[2]Institute of Tropical Forestry and Forest Product, Universiti Putra Malaysia,43400 UPM, Serdang, Selangor, Malaysia
[3]Mycology Unit, Product Development Division, Forest Research Institute Malaysia (FRIM), Kepong, 52109, Kuala Lumpur

*Corresponding author: nuraishah_23@yahoo.com

Abstract. Fresh rubber tree was harvested, cut into sized (100 cm x 14 cm x 25 mm) (l x w x t) and kiln dried (10-12 %) MC. The specimen sized of 5 mm x 20 mm x 20 mm (l x w x t) were prepared, undergone soxhlet extraction with toluene/methanol/acetone (4:1:1 by volume) for 8 hours, followed by oven dried at 103 °C for 24 hours and cooled in silica gel. The specimens were vacuum impregnated and then reacted with anhydrides for times ranging from 0.25 to 24 hours at 120 °C. The weight percent gain (WPG) was calculated and the chemical bonding was analysed with FTIR. All the specimens were leached in dionised water according to EN 84 (1997) and exposed to white rot in incubation room set at 22 °C (16 weeks). The study was found that modification only managed to reduce the rate of decay; without totally protected the rubberwood. Modified rubberwood was still classified as durability class 1 when compared to scott pine and beech woods. The final MC had a positive correlation with the weight loss following decay. The SEM showed that the hyphae were penetrated in the cells in both rubberwood.

1. Introduction

Rubberwood is the most famous wood for manufacturing both local and export oriented furniture as a result of its inexpensive, availability and adaptable processing technology. Rubberwood sawn timber is the top raw material used in the furniture industry, which make up 85% of total wood furniture exports for Malaysia in 2013 [1]. Rubberwood emerged as an alternative source of timber for the wood industry, when restriction in logging activities was implemented by the Malaysian government. Rubberwood is recognized as the most ecologically friendly lumber used in today's furniture industry. Generally, rubber trees are fell and replanted after 26-30 years for latex tapping. The wood is then harvested for use as sawn timber, moldings, furniture, medium density fibreboard, chipboard, builders, joinery and carpentry and wooden frames [2].

Despite the success story of rubberwood products, the fact that low dimensional stability and biodegradation problems are the main reasons for rubberwood to be less attractive for building materials. Products from rubberwood were generally less susceptible to bio deterioration agents than solid wood, unless they were used in situations where exposure to moisture or risk of deterioration was likely [3].

Rubberwood appeared to be the most susceptible to soft rot decay compared to other non-durable such as punggai (*Coelostegia griffi thii*) and jelutong (*Dyera costulata*); which this was caused by a variety of micro fungi [4)] . However, rubberwood was found to be more susceptible than non-durable jelutong and ramin (*G. bancanus*) against the white rot *Trametes versicolor* [5-6].

Wood modification is getting attention in this new millennium which offers less or no toxic solution to extend the wood service life, as alternative for toxic copper chromium and aresenic (CCA). The CCA was reported to be able to leach out into soils from wood during its service. As a consequence, the CCA treated wood was banned for toys at playground directly contact with children skin [7]. Alternatively, chemical modification offers the best solution to improve wood dimensional stability and decay resistance, without impairing the mechanical properties especially the load bearing [8-9].

One of the most excellent chemical modification treatments is acetylation [9]. The increased resistance of acetylated wood against fungi caused by the bulking effect of the chemical in wood cell walls and the reduction of hydroxyl (OH) groups, which reduced the cell wall moisture content [7, 9]. In acetylated rattan (*C. manan*), the reduction of moisture content was as a result of decreasing in cell wall void volume following with modification maybe the factor which reducing the fungal attack. The moisture content of acetylated rattan was lower than the untreated rattan [8, 10].

Despite its supreme decay resistance, the acetylation process produced a byproduct of acetic acid needs further separation process to recycle the acetic anhydride [7, 9]. Further studies using a longer chain of carboxylic acid anhydrides such as propionic and butyric anhydrides reported a good decay protection. The performance varies either the source is dicotyledon or monocotyledon plants, derived by variation in the amount of structural organic chemical and microstructure.

The aim of this study was to investigate the effectiveness of acetic, propionic and butyric anhydrides to protect rubberwood against *Trametes versicolor.*

2. Materials and Method

2.1 Source of material

Rubberwood (*Hevea brasiliensis*), aged 25 to 30 years were obtained by local plantation. They were processed at Forest Research Institute Malaysia (FRIM), Kuala Lumpur; cut into sized of 25 mm x 14 cm x 100 cm (r *x* t *x* l) and kiln dried to 10-12 % moisture content. The specimens with sized of 20 mm x 20 mm x 5 mm (r *x* t *x* l) were cut using small band saw.

2.2 Preparation of the specimens

The rubberwood specimens were sanded from any loosely adhering fibers. Then it was placed in a soxhlet extractor for solvent extraction using toluene/methanol/acetone (4:1:1 by volume) for 8 hours. Specimens were then oven dried in an oven for 24 hours at 103 °C, weighed to three decimal places and its volume was measured using Mitutoyo digimatic height gage, transferred to a desiccator and allowed to cool to ambient temperature over silica gel. The rubberwood specimens were arranged in container by reaction time, weighed under metal wrapped with aluminium foil and submerged in absolute acetic, butyric and propionic anhydrides (Table 1). The containers were vacuum impregnation at 0.7 kPa and the specimens were kept submerged under the anhydride for 3 days to maintain in wet and swollen state before transferred into reaction flask.

Table 1. The molecular weight of anhydrides used in this study.

Anhydride	Molecular weight	Molecular formula
Acetic anhydride	102.09 g/mol	$(CH_3CO)_2O$
Propionic anhydride	130.14 g/mol	$(CH_3CH_2CO)_2O$
Butyric anhydride	158.19 g/mol	$(CH_3CH_2CH_2CO)_2O$

2.3 Reaction procedures

The reaction procedures were conducted according to Hisham *et al*, 2014 [11]. The impregnated specimens were heated with oil in bath for the time ranging from 0.25, 4, 10, and 24 hours at 120 °C to obtain weight percent gain (WPGs) ranging from lowest (4.7 to 8.3), low (7.2 to 12.4), medium (10.2 to 14.0) to the highest (13.3 to 15.5) as shown in Figure 1. At the end of the reaction period, the reaction was quenched in ice until liquid temperature reached 20 °C and washed with cold acetone at the interval time of 1 hour for three times. The modified rubberwood was finally extracted using soxhlet consist of toluene/methanol/acetone solution mixture (4:1:1) for 8 hours and oven dried at 103 °C for 24 hours. Dry specimens were cooled, weighed and measured followed by air dried in conditioning room at 20 °C and 65 % RH to constant weight. The WPG was calculated as the following formula:

$$\text{Weight percent gain (\%)} = [(W_m - W_{um})/W_{um}] \times 100 \quad (1)$$

Where: W_m = mass of modified wood, W_{um} = mass of unmodified wood.

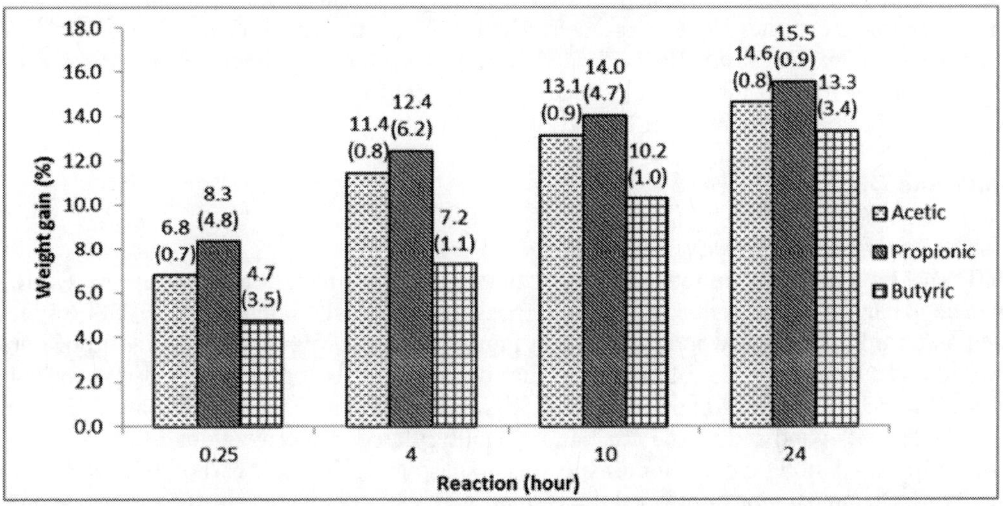

Figure 1. The averages WPG of acetylated rubberwood, propionylated rubberwood and butyrylated rubberwood.

2.4 FTIR analysis

The FTIR analysis was performed using Perkin Elmer Spectrum 100, FT-IR spectrometer to confirm the chemical bonding of rubberwood and anhydrides. The specimens were shaved prior to the FTIR test. The infrared spectra of each specimen which represented unmodified and modified rubberwood were obtained and evaluated with FTIR.

2.5 Decay test

All rubberwood specimens were leached in water following EN 84 (1997) [12] standard, and oven dried at 103 °C for 24 hours above. The specimens were then sterilised with propylene oxide for 5 days and exposed to decay over 4% malt extract agar in vented 500 mL squat jars following procedures in EN 113 (1996) [13]. For these purposes, 60 ml of 4% malt agar (40 g L21 Oxoid powdered malt extract, 20 g L21 Oxoid no. 3 agar, deionised water) was dispensed into 500 mL squat jars. These were sealed with lids vented with non-absorbent cotton wool plugs and the jars were autoclave sterilised. *Trametes versicolor* (CTB863A) was allowed to grow on the medium at 22 °C, 65% relative humidity for 2 weeks before exposure of the blocks. The blocks, one reacted and one untreated, were exposed over a sterilised polypropylene mesh in each jar. Twelve replicates were used for each reaction period and anhydride. Similarly sized of Scots pine (*Pinus sylvestris*) sapwood and European beech (*Fagus sylvatica*) blocks

IOP Conf. Series: Materials Science and Engineering **368** (2018) 012037 doi:10.1088/1757-899X/368/1/012037

were exposed, six jars each with two blocks as reference specimens. All blocks were incubated for 16 weeks. At the end of the test, excessive mycelium was removed and the moisture content and mass loss due to decay were determined. The durability classification was then determined following the guidance given in EN 350-1 (1994) [14)] where a ratio, expressed as the 'x value' is determined in comparison to the reference species, i.e. Scots pine or beech, i.e. x= average mass loss of test specimens/average mass loss of reference species. In this case, x value using both reference species was calculated. Durability classes were assigned according to EN 350-1 [14)], where class 1 (very durable) with an x value of $x \leq 0.15$, 2 (durable) is $0.15 < x \leq 0.30$, 3 (moderately durable) is $0.30 < x \leq 0.60$, 4 (slightly durable) is $0.60 < x \leq 0.90$ and 5 (not durable) is $x > 0.90$.

2.6 Scanning electron microscopy (SEM)
A thin layer surface of decayed specimens was cut using a sharp razor-blade for a smooth surface. The specimen was then separated into vials and fixed in 4% glutaraldehyde for two days at 4 °C, washed with 0.1 m sodium cacodylate buffer for 30 minutes (three times), and post fixed in 1% osimium tetroxide for 2 hours at 4 °C. The specimens were washed again with 0.1 M sodium cacodylate buffer for 30 minutes for three times. They were dehydrated with acetone at concentration of 35 %, 50 %, 75 %, 95 % for 30 minutes and lastly 100 % for 1 hour (3 times). The specimens were critically dried with critical point drying for 30 minutes and coated with a gold using sputter coater. Scanning electron microscope was performed using PHILIPS XL30 ESEM.

3. Results and Discussion

3.1 Fourier Transform Infra-Red (FTIR) analysis
The FTIR spectra were differentiated by large and small peak of the line graph in the FTIR spectra. Larger peak means high intensity and smaller peak means lower intensity. The FTIR spectrum of untreated and modified rubberwood is shown in Figure 2. It showed that the spectrum of unmodified rubberwood had a lower intensity of intermolecular bonded hydroxyl groups at 3339.94 cm^{-1}. The peak intensity of bonded hydroxyl groups was found higher in acetylated rubberwood at 3338.45 cm^{-1}, for propionylated rubberwood at 3338.65 cm^{-1} and, for butyrylated rubberwood at 3338.97 cm^{-1}. A weak C-H stretch appeared at 2900 cm^{-1} for all spectra. A sharp intensity ester carbonyl vibration appears at 1720 cm^{-1} to 1730 cm^{-1} region confirmed the formation of ester bonds or esterification. The unmodified rubberwood spectrums had a lower intensity in this region of 1729.70 cm^{-1}. The increased in the intensity of C=O bonds occurred for all modified wood with 1729.46 cm^{-1} for butyrylated rubberwood, 1728.81 cm^{-1} for acetylated rubberwood and 1728.51 for propionylated rubberwood.

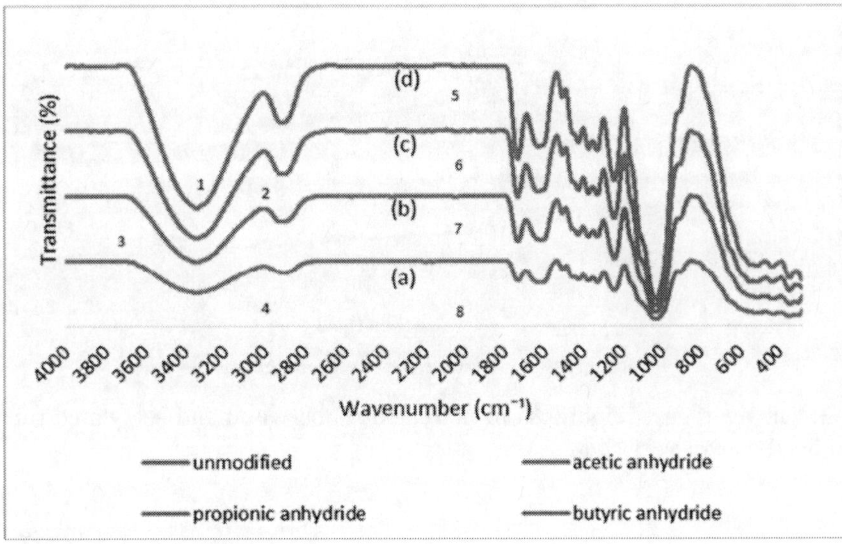

Figure 2. The FTIR spectra of unmodified and modified rubberwood. Band position - 1: 3338.97; 2: 3338.65; 3: 3338.45; 4: 3339.94; 5: 1728.51; 6: 1728.81; 7: 1729.46; 8: 1729.79.

3.2 Decay resistance

The percent weight loss of untreated and modified rubberwood decayed by *Trametes versicolor* are shown in Figures 3 to 5.The weight loss of untreated rubberwood was almost identical when exposed together with each set of acetylated rubberwood (26.07 % to 36.65 %), propionylated rubberwood (23.07 % to 37.57 %) and butyrylated rubberwood (17.48 % to 28.78 %). The weight loss was gradually declined with increasing the reaction times. The weight loss was gradually decline from 16.13% to 4.31%, 18.88 to 4.41 % and 25.41 to 5.08 % for acetylated rubberwood, propionylated rubberwood and butyrylated rubberwood, respectively. The lowest percent weight losses were obtained by the highest WPGs of 14.6 (acetylated rubberwood), 15.5 (propionylated rubberwood) and 13.3 (butyrylated rubberwood) after reacted for 24 hours. In all the cases, the percent weight losses of modified rubberwood in this study were above 3 % and the test was valid according to EN113 [13]. In terms of effectiveness, these anhydrides treatment were not effective enough to protect rubberwood against *T. versicolor*. In case of chemical modifications, the decay protection of wood is referred as decay protection threshold or the percent weight gain above which no microbial degradation occurs [7]. The decay protection threshold of 10 WPG in acetylated Corsican pine (*Pinus nigra*) gave 0 % weight loss after 16 weeks exposed to *T. versicolor* [15]. In acetylated beech (*Fagus sylvatica*), threshold of 20 WPG gave 0 % weight loss. In acetylated rattan (*Calamus manan*) aged 10 and 13 years, threshold 13.4 and 9 WPGs gave 0 % weight loss after exposed for 16 weeks [8]. A less effectiveness of anhydrides to protect rubberwood was not only due to its low WPG. Another reason is that rubberwood was difficult to react with anhydrides. Acetylated rubberwood obtained 13 WPG after reacted for 27 hours and the extension of reaction time for another 94 and 144 hours only gave 15 and 16.6 WPGs, respectively [16]. This probably caused by a lower kinetic reaction in rubberwood. The kinetics reaction was explained by the access reagent to the hydroxyl groups (OH) to promote the reaction or the real chemical reaction occurred between anhydrides and the OH groups in cellulose, hemicellulose and lignin [17]. The lower WPG resulted from lower reaction rate made anhydride treatment was not favourable to totally protect rubberwood from *T. versicolar*.

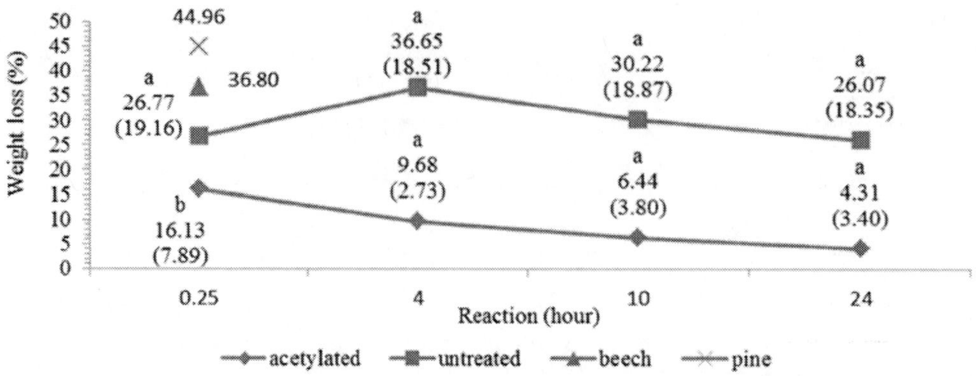

Figure 3. The averages weight loss of untreated rubberwood and acetylated rubberwood decayed by *Trametes versicolor*.

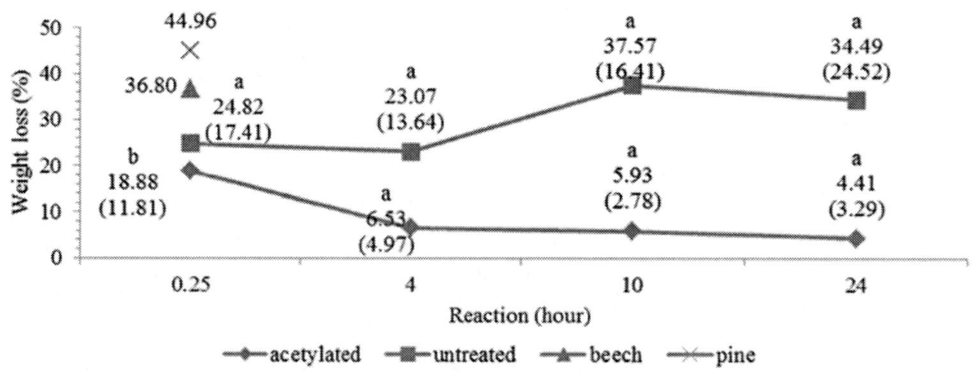

Figure 4. The averages weight loss of untreated rubberwood and propionated rubberwood decayed by *Trametes versicolor*.

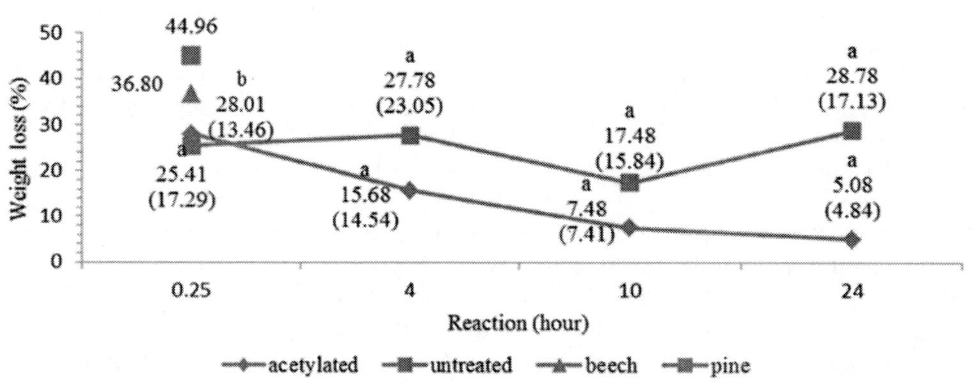

Figure 5. The averages weight loss of untreated rubberwood and butryrated rubberwood decayed by *Trametes versicolor*.

The moisture content of untreated and modified rubberwood is shown in Figures 6 to 8. The untreated rubberwood took up more than 100 % moisture regardless of anhydrides, except for untreated rubberwood exposed together with butyrylated rubberwood. They were ranged from 132.8% to 159.4%, 107.6% to 161.3% and 84.9% to 143.4% when exposed together with acetylated rubberwood, propionylated rubberwood and butyrylated rubberwood, respectively. In all cases, the moisture content of modified rubberwood was greatly reduced with increasing reaction times. This was consistent with the percent weight loss after decayed by fungus. The moisture content was ranged from 17.8 % to 84.3 %, 15.8 % to 70.9 % and 23.54 % to 111.5 % for acetylated rubberwood, propionylated rubberwood and butyrylated rubberwood respectively.

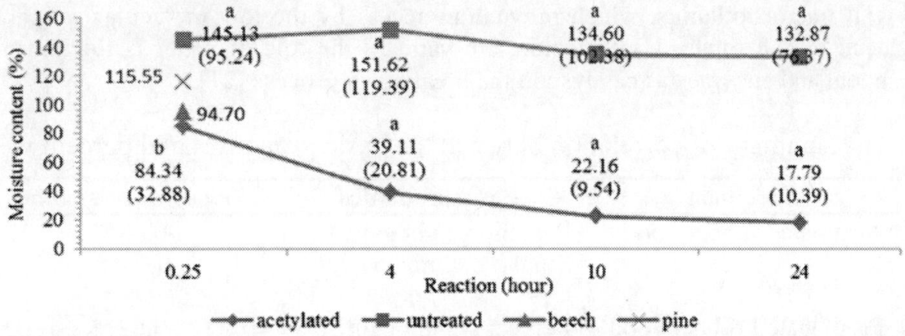

Figure 6. The averages final moisture content of untreated rubberwood and acetylated rubberwood decayed by *T. versicolor.*

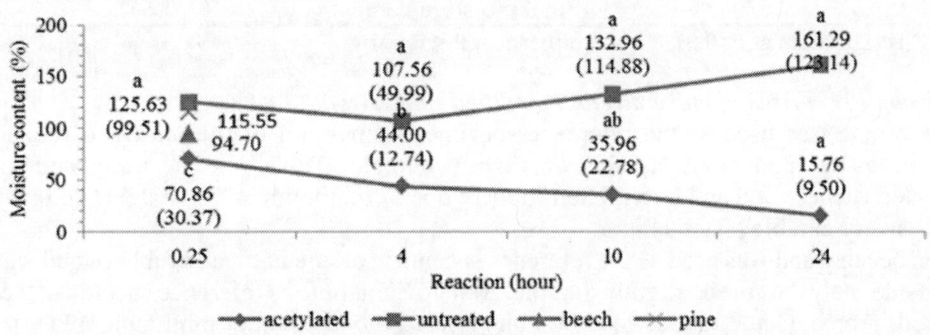

Figure 7. The averages final moisture content of untreated rubberwood and propionated rubberwood decayed by *T. versicolor.*

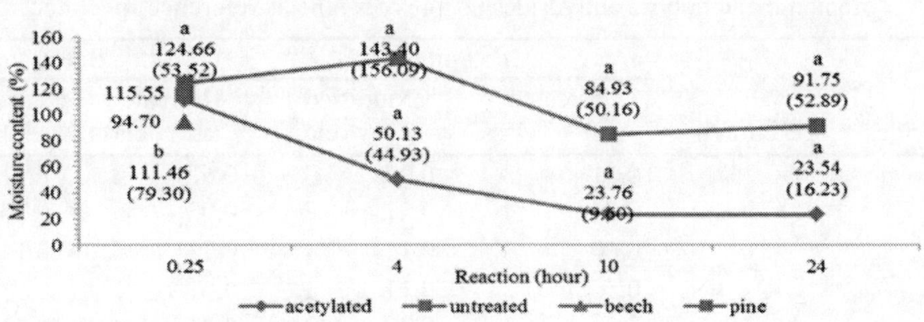

Figure 8. The averages final moisture content of untreated rubberwood and butyrated rubberwood decayed by *T. versicolor.*

3.3 The correlation between the basic properties and percent weight loss of decayed untreated and modified rubberwood.

The correlation between the basic properties and percent weight loss of modified rubberwood is shown in Table 2. The final moisture content following decay was the only property positively correlated with percent weight loss. Overall, the strongest correlation was obtained by butyrylated rubberwood (P=0.84) followed by acetylated rubberwood (P=0.69) and propionylated rubberwood (P=0.65). The WPG was transversely correlated with percent weight loss. This correlation trend was consistent with acetylated rattan (*Calamus manan*) decayed by *T. versicolor* [8]. Hill et al. (2005) [19] concluded that acetylation reduced decay by reducing cell wall moisture content. Papodopoulus & Hill (2003) [19] and Papodopoulus et al. (2010) [20] proposed that the mechanism for decay protection was due to blocking of the cell wall microcapillaries, which preventing access by the low molecular weight degradative agents produced by the fungus. Both factors are valid as the role of water is for transporting food, degradative agent and enzymes as catalyse to the breakdown process [21].

Table 2. The summary of correlation between the basic properties and percent weight loss.

Specimen	Basic properties	*Trametes versicolor*
Acetylated rubberwood	Percent weight gain	-0.68**
	Final moisture content	0.69**
Propionated rubberwood	Percent weight gain	-0.62**
	Final moisture content	0.84**
Butryalated rubberwood	Percent weight gain	-0.26
	Final moisture content	0.65*

* is significant at P<0.01, ** is significant at P < 0.001

3.4 The durability classes of untreated and modified rubberwood

When Scott pine was used as the reference specimen, almost all of the untreated rubberwood was classified as either moderately durable or slightly durable (Table 3). The acetylated rubberwood, propionylated rubberwood and butyrylated rubberwood at minimum WPG of 13.1, 14.0 and 10.2 was classified as very durable (class 1).

When beech wood was used as the reference specimen, all the untreated rubberwood was classified either as moderately durable or slightly durable. Akin to Scott pine as reference specimen, the acetylated rubberwood, propionylated rubberwood and butyrylated rubberwood at minimum WPG of 13.1, 14.0 and 10.2 was classified as very durable (class 1).

Table 3. X values and durability classes of untreated and modified rubber woods with acetic, propionic and butyric anhydrides using scots pine as reference specimen.

		Scott pine				Beech			
		Untreated		Modified		Untreated		Modified	
Anhydrides	Reaction/h	x	class	x	class	x	class	x	class
Acetic	0.25	0.51	3	0.35	3	0.62	4	0.41	3
	4	0.74	4	0.21	2	0.90	5	0.26	2
	10	0.60	3	0.15	1	0.73	4	0.14	1
	24	0.50	3	0.11	1	0.61	4	0.13	1
	Average	0.59	3	0.20	2	0.72	4	0.24	2
Propionic	0.25	0.48	3	0.37	3	0.58	3	0.45	3

IOP Conf. Series: Materials Science and Engineering **368** (2018) 012037 doi:10.1088/1757-899X/368/1/012037

	4	0.46	3	0.13	1	0.56	3	0.17	2
	10	0.77	4	0.12	1	0.93	5	0.14	1
	24	0.73	4	0.10	1	0.89	4	0.12	1
	Average	0.61	4	0.18	2	0.74	4	0.22	2
Butyric	0.25	0.49	3	0.58	3	0.6	4	0.71	4
	4	0.53	3	0.36	3	0.65	4	0.44	4
	10	0.38	3	0.13	1	0.57	3	0.15	1
	24	0.62	4	0.12	1	0.75	4	0.14	1
	Average	0.51	3	0.32	3	0.64	4	0.39	3

X values and durability classes of untreated and acetylated rubber wood using Scots pine and beech as reference specimen

3.5 Microstructure of decayed untreated rubberwood and rubberwood modified with a linear chain carboxylic acid anhydrides.

The microstructures of untreated rubberwood, acetylated rubberwood, propionylated rubberwood and butyrylated rubberwood decayed by fungi are shown in Figures 9 to 11. In almost all cases the hyphae penetration was more prominant at longitudinal direction than those from transverse direction. All the untreated and modified rubberwood were penetrated by hyphae even at the highest WPG. This indicated that the fungi hyphae can penetrated the wood cells but the access was limited due to slightly lower moisture content as shown in Figures 6 to 8. This study was consistent with others finding [8, 18]

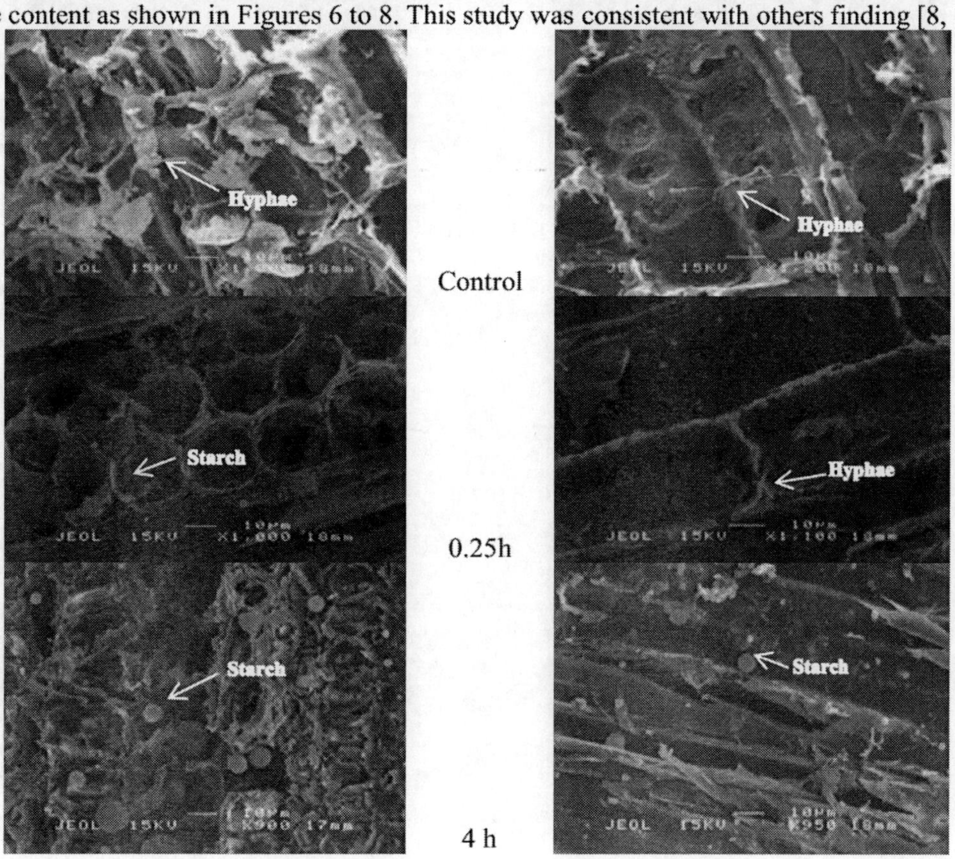

IOP Conf. Series: Materials Science and Engineering **368** (2018) 012037 doi:10.1088/1757-899X/368/1/012037

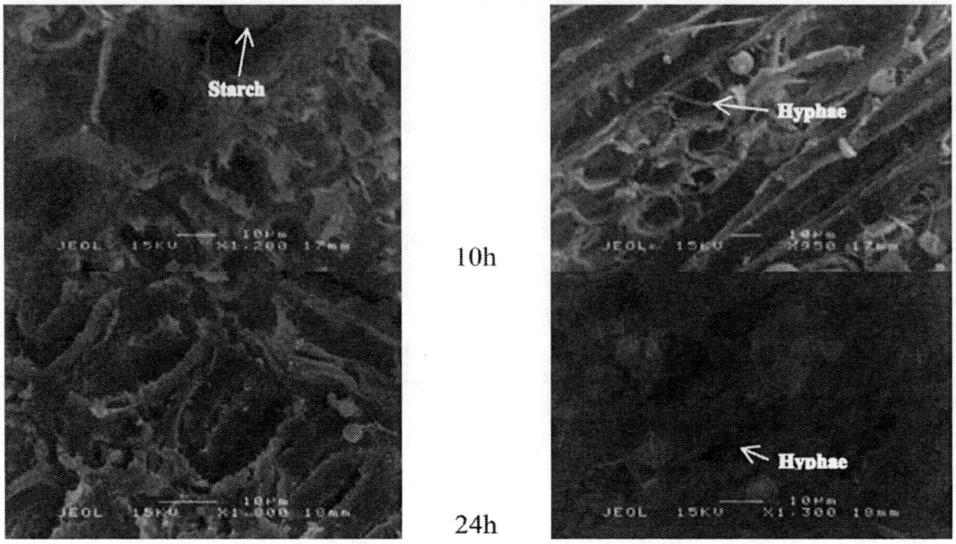

Fig 9. The micostructure of untreated rubberwood and acetylated rubberwood decayed by *Trametes versicolor* (left side: transverse, right side: longitudinal)

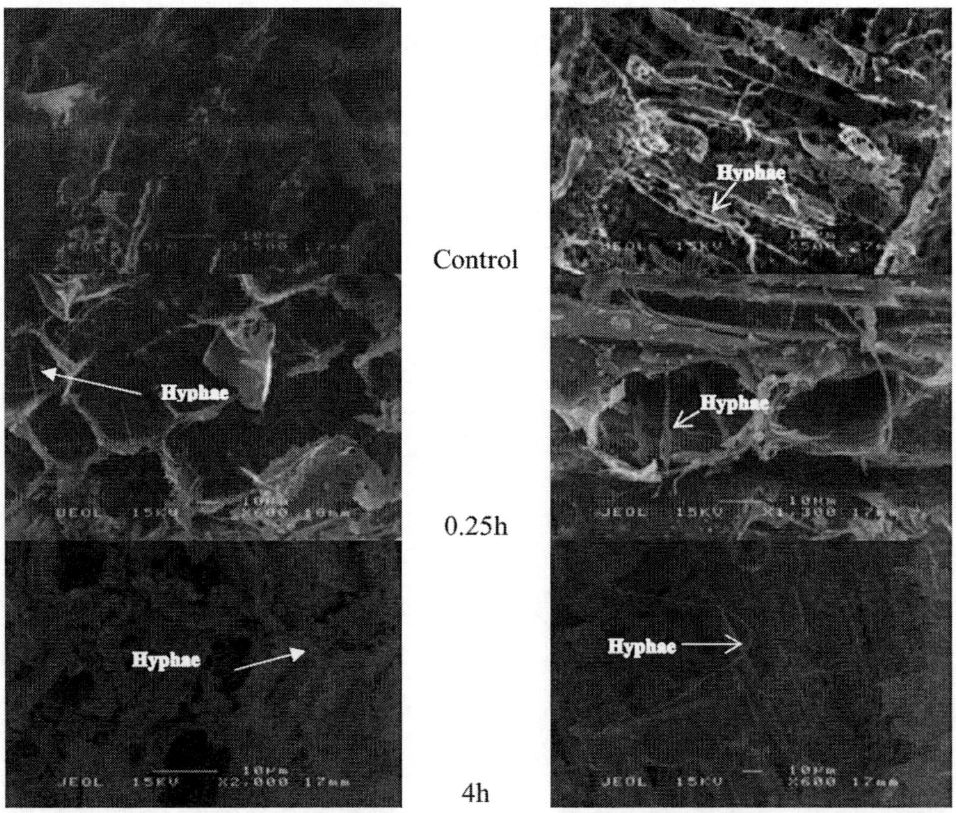

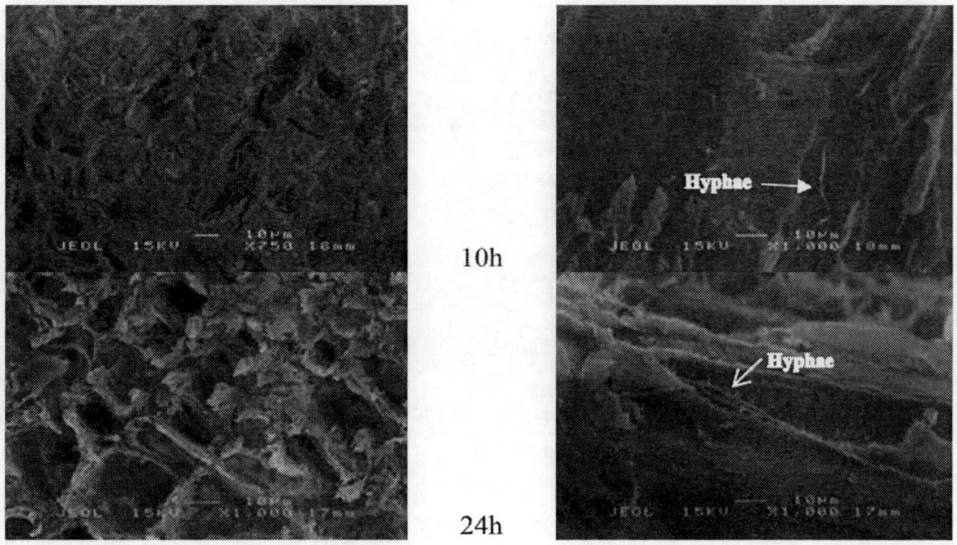

Fig 10. The micostructure of untreated rubberwood and propionylated rubberwood decayed by *Trametes versicolor* (left side: transverse, right side: longitudinal)

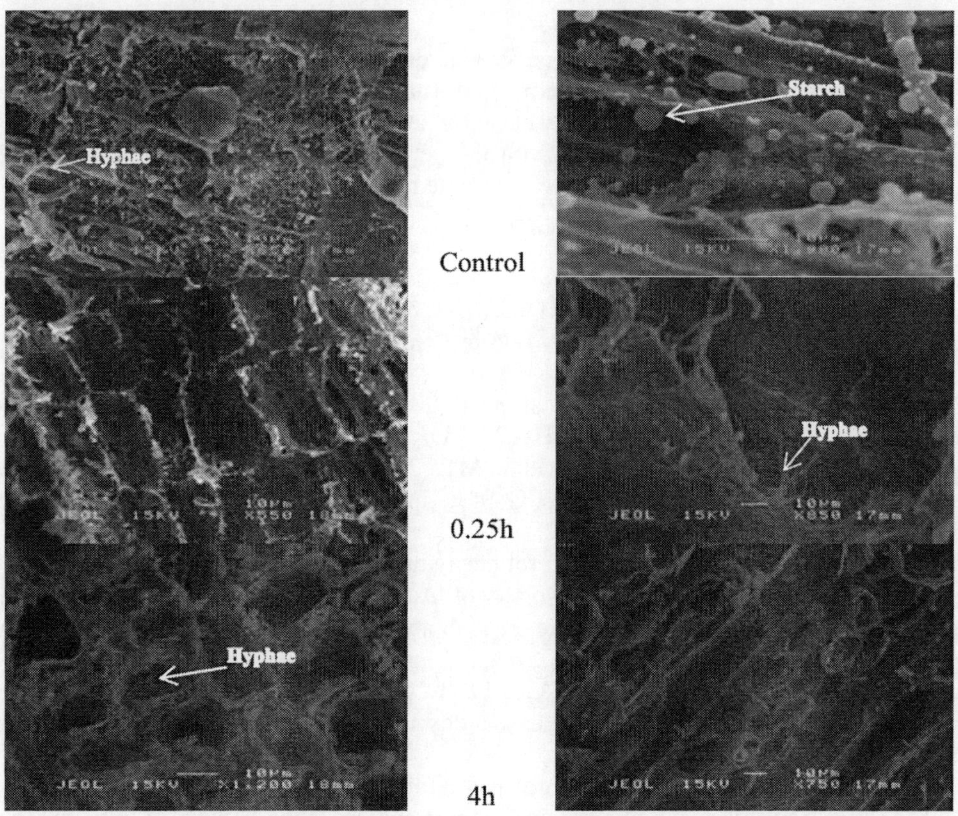

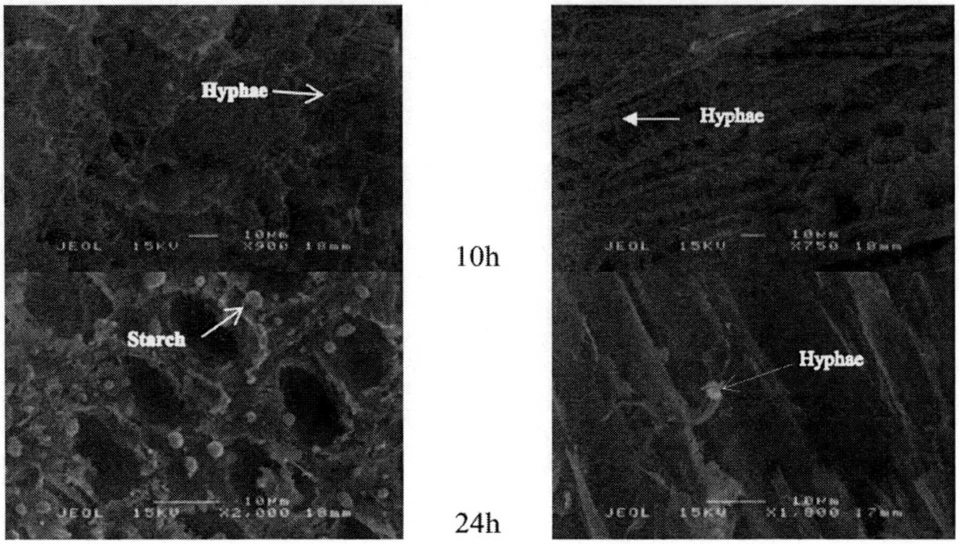

Fig 11. The micostructure of untreated rubberwood and butyrylated rubberwood decayed by *Trametes versicolor* (left side: transverse, right side: longitudinal).

4. Conclusion

Modification of rubberwood with anhydrides at certain degree, only improve the decay resistance to *T. versicolor* but the treatment was not totally protected from deterioration. All the modified rubberwood at maximum WPG was classified as durability class 1 when compared to Scott pine and beech. The final moisture content had a positive correlation with the percent weight loss following decay. The SEM observation confirmed that the hyphae were penetrated in the cells either in untreated or modified rubberwood.

References

[1] Merous N, Rahman R, Mohamed N, Ahmad IHL & Othman M. 2015. Factor determination on availability and consumption of rubberwood in Peninsular Malaysia. *Journal Tropical Resource Sustainable Science* **3**:191-209.

[2] Shigematsu A, Mizoue N, Kajisa T & Yoshida M. 2011. Importance of rubberwood in wood export of Malaysia and Thailand. *New Forest* **41**:179-189.

[3] Zaidon A, Moy CS, Sajap AS & Paridah MT. 2003. Resistance of CCA and boron-treated rubberwood composites against termites, Coptotermes curvignathus Holmgren. Pertanika Journal of Science and Technology **11(1)**: 65-72.

[4] Wong AHH. 1992. Susceptibility to soft rot decay in copper-chrome-arsenic treated and untreated Malaysian hardwoods. PhD thesis, University of Oxford, **341 pp**.

[5] Hong, LT & Yamamoto, K. 1989. A note on a laboratory method for estimating durability of some tropical hardwoods. *Journal of Tropical Forest Science* **2(2):** 167-170.

[6] Teoh YP, Don MM & Ujang S. 2011. Assessment of the properties, utilization, and preservation of rubberwood (*Hevea Brasiliensis*): A case study in Malaysia. *Journal of Wood Science.* **57(4)**: 255-266.

[7] Hill CAS. 2006. Chemical modification of wood (1): acetic anhydride modification. **Pp 45-75** in Wood Modification: chemical, thermal and other processes, John Wiley & Sons Ltd, 45–76.

[8] Hisham HN & Hale M. 2012. Decay threshold of acetylated rattan against white and brown rot fungi. International Wood Product Journal **3(2)**: 96-106

[9] Rowell RM. 2014. Acetylation of wood-a review. *International Journal of Lignocellulosic Products* **1(1)**:1-27

[10] Hisham HN & Hale M. 2013. Decay threshold of acetylated rattan (*Calamus manan*) against soft rot. *Journal of Forestry Research* **24(2)**: 375-380.

[11] Hisham HN, Ahmad Ludin N & Hale M. 2014. Equlibrium moisture content and moisture exclusion efficiency of acetylated rattan (Calamus manan). Journal of Tropical Forest Science **26(1)**: 32-34.

[12] EN 84. 1997. Wood preservatives. Accelerated ageing of treated wood prior to biological testing. Leaching procedure.

[13] EN 350-1. 1994. Durability of wood and wood based products-natural durability of solid wood – Part 1: guide to the principles of testing a classification of the natural durability of wood.

[14] EN 113. 1996. Wood preservatives. Test method for determining the protective effectiveness against wood destroying basidiomycetes. Determination of the toxic values.

[15] Hill CAS, Hale MD, Farahani MR, Forster S, Suttie ED, Jones D, & Papodopoulus, AN. 2003. Decay of anhydride modified wood. **Pp 143-152** in (Hill CAS) *Proc. 1st Eur. Conf. on Wood Modification*, Ghent University, Ghent, Belgium.

[16] Karim SR, Hill CAS & Ormondroyd GA. 2006. Dimensional stabilization of rubberwood (*Hevea brasiliensis*) with acetic or hexanoic anhydride. *Journal of Tropical forest Science* **18 (4)**: 261–268.

[17] Minato K & Ogura K. 2003. Dependence of reaction kinetics and physical and mechanical properties on the reaction system of acetylation 1: Reaction kinetic aspects. *Journal of Wood Science.* **49(5)**:418-422.

[18] Hill CAS, Foster SC, Farahani MRM, Hale MDC, Ormondroyd GA & Williams GR. 2005. An investigation of cell wall micropore blocking as a possible mechanism for the decay resistance of anhydride modified wood. *International Biodeterioration & Biodegradation* 55(1), 69–76.

[19] Papadopoulos AN & Hill CAS. 2003. The sorption of water vapour by anhydride modified softwood. Wood Science and Technology **37 (3–4)**: 221–223.

[20] Papadopoulos AN, Militz H & Pfeffer A. 2010. The biological behaviour of pine wood modified with linear chain carboxylic acid anhydrides against soft rot fungi. International Biodeterioration & Biodegredation **64**:409-412.

[21] Bowyer JL, Shmulsky R and Haygreen, JG. 2003. Wood durability and protection. **Pp 61-286** in Forest products and wood science: an introduction. 4th ed. New York: Wiley-Blackwell, 61–286.

IOP Conf. Series: Materials Science and Engineering **368** (2018) 012038 doi:10.1088/1757-899X/368/1/012038

Thermal and physicochemical properties of sugar palm fibre treated with borax

I Mukhtar[1,4*], Z Leman[1,2], M R Ishak[2,3] and E S Zainudin[1,2]

[1] Department of Mechanical and Manufacturing Engineering, Faculty of Engineering, Universiti Putra Malaysia, 43400 Serdang, Selangor, Malaysia
[2] Laboratory of Biocomposites Technology, Institute of Tropical Forestry & Forest Product (INTROP) Universiti Putra Malaysia, 43400 Serdang, Selangor, Malaysia
[3] Department of Aerospace Engineering, Faculty of Engineering, Universiti Putra Malaysia, 43400 Serdang, Selangor, Malaysia
[4] Department of Mechanical Engineering, Faculty of Engineering, Kano University of Science and Technology Wudil, P.M.B. 3244 Kano State, Nigeria

*Corresponding author: ismailmuk@gmail.com

Abstract. Natural fibres are biodegradable, low-cost, and low-density materials found in many parts of the world in large quantities. The comparative properties and competitive advantages make it be a preferred material in many engineering applications. The major challenges associated with natural fibres are its low thermal stability, lack of interfacial adhesion, poor resistance to moisture and inferior mechanical properties when compared with synthetic fibres. The solution is to treat it with either alkali, silane, or other treatment methods. Specifically, alkaline treatment comes with an additional problem such as fibre degradation. Therefore, this work employed the use of borax as a treatment chemical rather than its usage as a fire retardant. This could give a treated natural fibre with mild degradation and acceptable properties. Therefore, in this research sugar palm fibre was treated with 5 w/w% borax, washed several times and dried in an oven. The treated and untreated fibres were characterized based on X-Ray Diffraction, thermal stability, and morphological analysis. The experimental results showed that the crystallinity index increased from 44.02 % to 49.56 %, while the thermal stability increased from 246.76 °C to 255.5 °C with borax treatment. The morphological investigation revealed that the surface of the treated fibre is rough with no silica deposit when compared with the untreated fibre. In conclusion, the investigation proved that treatment with the borax has a significant effect on the physicochemical properties of sugar palm fibre.

1. Introduction

The increase in stringent regulation on the level of carbon footprint and other environmental concerns like global warming, disposal of components at the end of their useful life, necessitates the use of environmentally friendly materials in industries. In the recent years, industries like furniture, automobile, packaging, aerospace, building construction are all in search of new materials that will replace the conventional glass fibre and other synthetic fibres through research and development. The most promising and adopted materials nowadays are natural fibre reinforced polymer composites, they are evolving as the realistic alternatives to glass reinforced polymer composites in many applications since the 1990s [1]. These natural fibres like sisal, hemp, kenaf, jute, coir and sugar palm fibre among others offer numerous advantages when compared with synthetic fibres. Some of these advantages are light in weight, biodegradability, non-toxic, low-cost, non-abrasive, availability, less pollutant emission and low environmental impact [1, 2]. Unlike natural fibre, glass fibres can cause acute irritation to the skin, eyes, and upper respiratory tract. The fibrous glass and other synthetic vitreous fibres, when disturbed, release fibres that can become airborne, inhaled and retained in the respiratory tract [3].

Just like other natural fibres, sugar palm fibre that is abundantly found in South East Asia is of research interest in the field of material science. The sugar palm fibre is obtained from sugar palm tree, mostly from its four parts, namely trunk, frond, bunch, and trunk's surface [4]. It has been characterized in terms of physical, mechanical, chemical, thermal, and its morphology in previous research [5-8]. A research was done by Ticoalu, et al. [6] and revealed that sugar palm fibre diameter is within the range of 81 to 313 μm. This shows that the diameter of sugar palm fibre is within the range of natural fibres as compared with sisal 100 to 500 μm [9] and kenaf 40 to 200 μm [10]. In terms of moisture absorption, the bagasse, coir, sisal, and jute have a moisture absorption of 8.8, 10, 11, and 12% respectively [11]. This is much high than sugar palm fibre with 5.36 to 8.7% as reported in the literature by Ishak, et al. [8]. In this case, natural fibres are the obvious choice to substitute glass fibre which will normally improve the environmental performance of the components.

Despite the comparative properties of sugar palm fibre with other natural fibres, this fibre is also having some drawbacks. These drawbacks include but not limited to interfacial adhesion, low thermal stability, low durability, quality variation and poor resistance to environment among others [12]. One of the drawbacks of the natural fibres is moisture absorption, sugar palm fibre is not an exception. Moisture absorption test on sugar palm fibre reinforced epoxy conducted by Leman, et al. [13]. The findings showed that composite plate with 20% fibre loading has the highest moisture content (0.93%) prior to the absorption test. It is shown that a sample with high fibre content possesses the highest rate of moisture absorption. The moisture absorption of a fibre makes it degrade faster and will result in poor adhesion between fibre-matrix interfaces, this phenomenon is as a result of hydroxyl group presents in the fibre. Steps were taken by researchers to treat the fibre surface either by chemical, biological or plasma treatments. This problem was looked into by Bachtiar, et al. [14], in order to reduce the level of water uptake, an alkaline treatment was carried out to the fibre prior to the composite fabrication. The results showed an improvement in both strength and modulus after treatment with NaOH, but the effect is more pronounced for the modulus. These researchers obtained inconsistent results at high concentration and long soaking time and this was presumed to be attributed to the damage of the fibre. In another effort to reduce the moisture absorption, a cost-effective and environmentally friendly method was adopted by Leman, et al. [3]. In this research, the fibres were treated with a seawater and fresh water. The tensile results indicated that the fibre treated with seawater for 30 days had the highest stress value of 23.04 MPa i.e. improvement by 67.26% when compared with untreated fibre. While an improvement of 54.37% was recorded for fresh water treatment. Another experiment by Ishak, et al. [15], shows that with a 30 days seawater treatment the outer layer of hemicellulose and pectin were removed thereby improving interfacial bonding between fibre and matrix. In their results, both impact and flexural results had a higher value than untreated sugar palm fibre.

In this research, the sugar palm fibre was treated with borax. The treated and untreated fibres were characterized in terms of thermal stability (TG), crystallinity (XRD), physical (density and moisture absorption), and morphological analysis. This study was conducted to study the effect of the borax treatments for effective utilisation of this fibre in many fields.

2. Experimental Procedure

2.1. Materials

The sugar palm fibre used in this research was obtained from sugar palm tree in Jempol, Negeri Sembilan, Malaysia. It was collected from the surface of the trunk and it has a cellulose and hemicellulose contents of 43.87% and 5.57% respectively [16]. The treatment chemical used in this research is di-sodium tetraborate ($Na_2B_4O_7.10H_2O$) obtained from R & M chemicals Malaysia.

2.2. Chemical Treatment

The treatment was carried out at room temperature with an aqueous solution of di-sodium tetraborate at 5 % (w/v) [17]. The fibre was soaked for 24 hours and was thoroughly washed to remove excess borax. It was later dried in an oven at 60 °C for 24 hours and kept in a zip bag to avoid moisture ingress.

2.3. Density Measurement

Low density is one of the important properties of natural fibres that make them more attractive to substitute synthetic fibres. To verify the effect of the borax treatment on sugar palm fibre, helium pycnometer was used for the density measurement. This test was done in accordance with ASTM D3800M-10 [18] which is similar except that helium gas is used in placed of liquid. The measurement was done at room temperature and was repeated 10 times to ensure accurate results. To obtain the final value of the density, the weight of the fibre was measured with Mettler Toledo weighing balance while the volume was measured by the pycnometer.

2.4. XRD

The XRD analysis help to obtain important parameters that will verify the effectiveness of the borax treatment on sugar palm fibre. Most importantly the amorphous and crystalline nature of treated and untreated fibre can be determined through this analysis. This analysis was done in the XRD laboratory at the faculty of science, University Putra Malaysia. The machine used is PANalytical (Philips) X'Pert Pro PW3050/60 with scanning done at angle 2θ within the range of 5° to 40°. The tube was operated at 40kV and 20mA with Cu Kα X-ray source.

2.5. Moisture Absorption

To determine the moisture absorption of the samples, both untreated and borax treated sugar palm were allowed to absorbed moisture in humidity chamber as described by Ridzuan, et al. [19]. The humidity chamber was set at 70% and 90% relative humidity at a temperature of 23 °C. The samples were left inside the chamber for a period of 8 days for each of the conditions. Prior to the moisture absorption test, the samples were dried at 60 °C for 24 hours. The weight of the samples were measured before and after exposure to the humidity chamber using Mettler Toledo analytical weighing balance. While the moisture absorbed by the samples was computed using equation 1 as follows:

$$Moisture\ Absorption = \frac{W_{wf} - W_{df}}{W_{df}} \times 100 \tag{1}$$

Where W_{mf} is the weight of the moist fibre and W_{df} is the weight of the dry fibre

2.6. Morphological Analysis

Micrographs of the untreated and borax treated fibres were obtained in a scanning electron microscope SEM (S-3400N SEM HITACHI). The longitudinal image of the fibre was captured at 1000 X magnification. The aim of the analysis was to examine the surface of the fibre for a better understanding of the effectiveness of the borax treatment on the sugar palm fibre.

2.7. Thermal Analysis

TGA analysis of the untreated and borax treated sugar palm fibre were performed to study the degradation behaviour of the sugar palm fibre with an increase in temperature. Thermo-Gravimetric Analyser (TGA/DSC 1 Mettler Toledo) was used to obtain the TG curves for both treated and control sugar palm fibre. The analysis was done at a heating rate of 10 K/min under nitrogen atmosphere.

3. Result and Discussion

3.1. Density

The densities of the untreated and borax treated fibre were determined with the aid of helium pycnometer. The values were found to be 1.3304 g/cm^3 and 1.3220 g/cm^3 for untreated and borax treated sugar palm fibre respectively. It can be seen from the values that the borax treatment has a slight effect on the density of the treated sugar palm fibre. The density of treated fibre (1.3220 g/cm^3) is slightly lower than the density of the untreated sugar palm fibre (1.3304 g/cm^3). This density of sugar palm fibre is similar and in some cases lower than other natural fibres like coir 1.25 g/cm^3, cotton 1.51 g/cm^3, flax 1.4 g/cm^3, hemp 1.48 g/cm^3, jute 1.46 g/cm^3, ramie 1.5 g/cm^3, and sisal 1.33 g/cm^3 [20]. This change in density can be associated with the partial removal of hemicellulose and lignin from the fibre. A similar reduction in density was reported by Birnin-Yauri, et al. [21] due to borax treatment.

3.2. *XRD*

The results of the XRD analysis are shown in figure 1(a) and 1(b) for untreated and borax treated sugar palm fibre respectively. Both XRD of the untreated and borax treated patterns appeared with four (4) major peaks as expected. These peaks were at 15.4°, 16.2°, 22.5, and 34.6° 2θ, they correspond to Miller indices 1-10, 110, 200, and 004 respectively. These patterns are associated to cellulose found in natural fibres [22]. The crystallinity index of both treated and the treated fibres was computed using empirical formula proposed by Segal, et al. [23] given in equation 2.

$$CrI = (I_{200} - I_{am}) / I_{200} \times 100 \quad (2)$$

Based on the formula, the untreated sugar palm fibre has a crystallinity index of 44.02%, while the borax treated fibre has an index of 49.56%. It's clear that the treatment had improved the crystallinity index which is linked to improved mechanical properties. This index is low compared to other natural fibres like flax (72%) [24] and hemp (66%) [25], but it is still within the range of natural fibres.

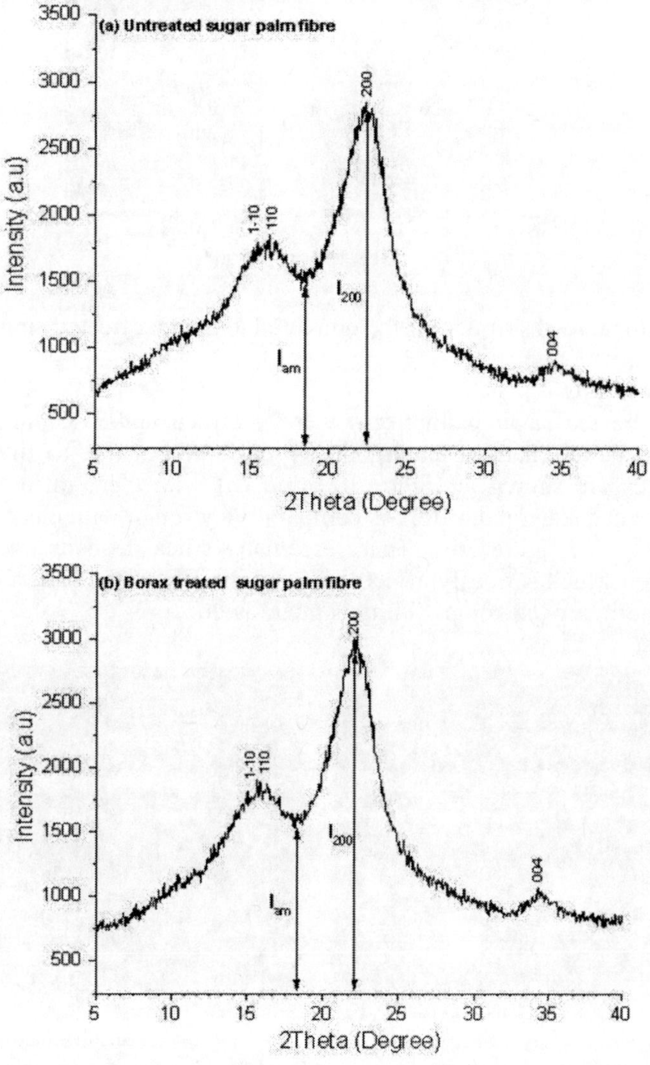

Figure 1. XRD patterns of (a) untreated and (b) borax treated sugar palm fibre.

IOP Conf. Series: Materials Science and Engineering **368** (2018) 012038 doi:10.1088/1757-899X/368/1/012038

3.3. Moisture Absorption

The moisture uptake of the untreated and borax treated sugar palm fibre when exposed to humidity at 70% and 90% RH is shown in figure 2. The moisture absorption test was conducted for a period of 8 days, i.e. not up to equilibrium stage. The untreated sugar palm fibre showed high moisture sensitivity when compared with modified sugar palm fibre with borax. This means that the hydrophilic nature of the sugar palm fibre was decreased with the treatment. The results showed that at 70% relative humidity, the untreated fibre absorbed 1.15% of moisture as compared to 0.86% for the borax treated sugar palm fibre. Furthermore, the moisture absorption rate increased with the increase in the percentage of the relative humidity. This result was in agreement with the previous study on moisture absorption for alkaline treated abaca fibre [26]. The hydroxyl group in the fibre which is responsible for the moisture absorption had been partially removed by treating the fibre with borax. This analysis has further confirmed the effectiveness of this treatment as found in the earlier XRD analysis.

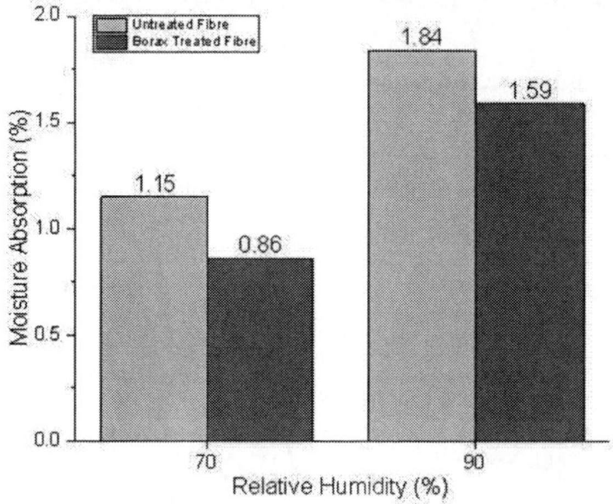

Figure 2. Moisture absorption of the untreated and borax treated sugar palm fibre.

3.4. Morphological Analysis

Untreated and borax treated sugar palm fibres were analysed under scanning electron microscope SEM in order to reveal the changes undergone by the surfaces due to the borax treatment. The morphological surfaces are shown in figure 3(a) and (b) with clear differences between the two samples. The borax treated sugar palm fibre is comparatively rough with clear pits due to the removal of silica deposit as seen in figure 3(b). There are many silica deposits and hemicellulose on the surface of the untreated which seriously affect the compatibility of the natural fibres with the matrix. But with this treatment, improved compatibility is guaranteed.

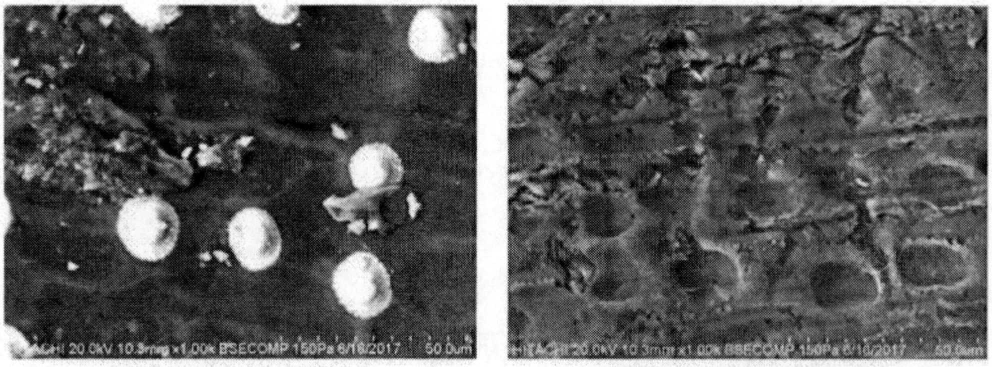

(a) Untreated sugar palm fibre (b) Borax treated sugar palm fibre

Figure 3. SEM images of (a) untreated and (b) borax treated sugar palm fibre.

3.5. Thermal analysis

The curves of weight loss (TG) and its derivatives (DTG) for untreated and borax treated sugar palm fibres are shown in figure 4 and 5 respectively. The curves are characterized with three (3) main stages as shown in the figures. In all the samples, these stages are associated with water loss, degradation of hemicellulose and cellulose, and lastly degradation of lignin. The untreated fibre has a thermal stability of 246.76 °C which is lower than the borax treated fibre with 255.5 °C as its thermal stability. This showed that the thermal stability increased with this treatment. The residue left for treated fibre was high at 31.90% as compared with untreated which is 28.08 %. This further confirmed the high thermal stability of borax treated sugar palm fibre.

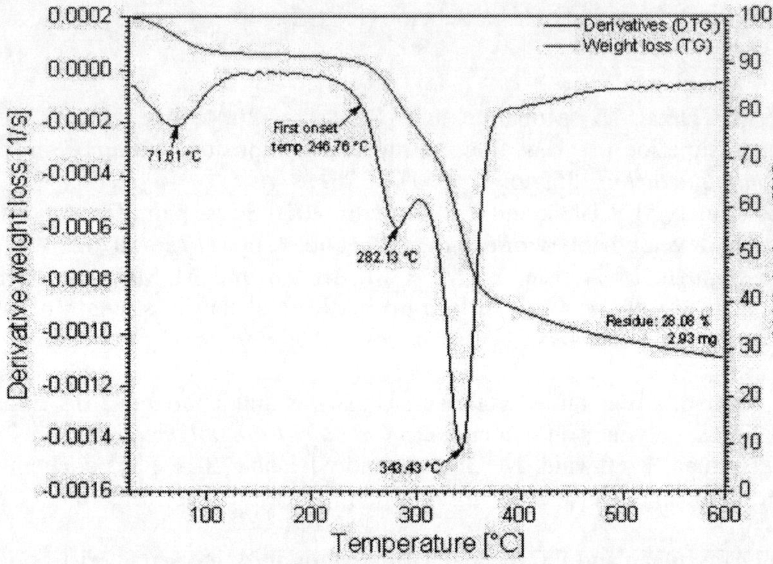

Figure 4. Thermogravimetric and its derivative curve for untreated sugar palm fibre.

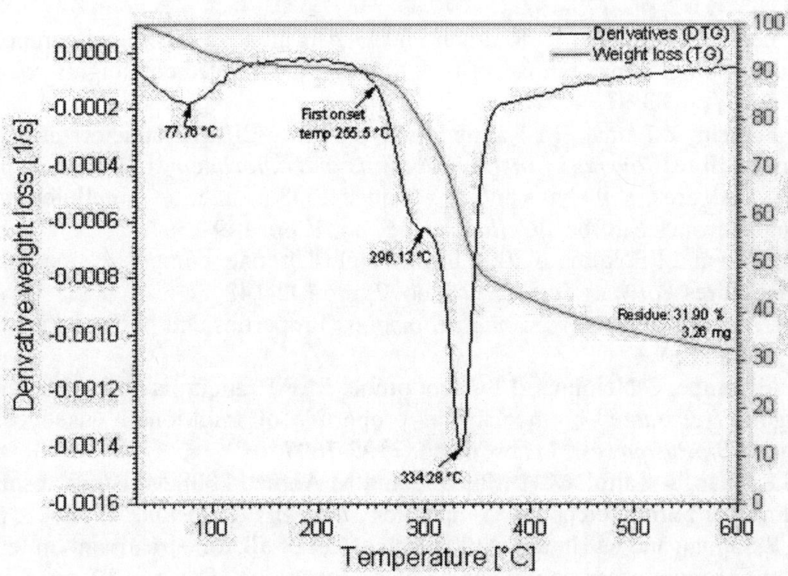

Figure 5. Thermogravimetric and its derivative curve for borax treated sugar palm fibre.

4. Conclusion

In this research, density, XRD, moisture absorption, morphological and thermal analysis were used to characterize untreated and borax treated sugar palm fibre. The sugar palm fibre was treated with 5% (w/v) of borax for the purpose of improving the adhesiveness between the fibre and the matrix. With this modification, the crystallinity index was increased from 44.02% to 49.56% and also reduced moisture absorption was noticed when compared with the untreated sugar palm fibre. The borax treatment had proved that a thermally stable lignocellulosic sugar palm fibre with an onset temperature of 255.5 °C can be obtained. As clearly revealed by SEM analysis the silica deposit had been removed leaving only the pits which are term as the roughness of the fibre. It is concluded in this research that borax can effectively treat sugar palm fibre and other natural fibres not necessarily as fire retardant additives as shown in many research.

Acknowledgment
The authors would like to thank the University Putra Malaysia (UPM) for financially supporting this research through Research University Grant Scheme (RUGS).

References

[1] Joshi S V, L Drzal, A Mohanty and S Arora 2004 Are natural fiber composites environmentallysuperior to glass fiber reinforced composites? *Composites Part A: Applied science and manufacturing.* **35**, no. 3, pp. 371-376.
[2] Mukhtar I, Z Leman, M R Ishak and E S Zainudin 2016 Sugar palm fibre and its composites: A review of recent developments *BioResources.* **11**, no. 4, pp. 10756-10782.
[3] Leman Z, S Sapuan, M Azwan, M M H M Ahmad and M Maleque 2008 The effect of environmental treatments on fiber surface properties and tensile strength of sugar palm fiber-reinforced epoxy composites *Polymer-Plastics Technology and Engineering.* **47**, no. 6, pp. 606-612.
[4] Ishak M, S Sapuan, Z Leman, M Rahman, U Anwar and J Siregar 2013 Sugar palm (arenga pinnata): Its fibres, polymers and composites *Carbohydrate polymers.* **91**, no. 2, pp. 699-710.
[5] Rashid B, Z Leman, M Jawaid, M Ghazali and M Ishak 2016 Physicochemical and thermal properties of lignocellulosic fiber from sugar palm fibers: Effect of treatment *Cellulose.* **23**, no. 5, pp. 2905-2916.
[6] Ticoalu A, T Aravinthan and F Cardona 2014 A study into the characteristics of gomuti (arenga pinnata) fibre for usage as natural fibre composites *Journal of Reinforced Plastics and Composites.* **33**, no. 2, pp. 179-192.
[7] Ishak M, Z Leman, S Sapuan, M Rahman and U Anwar 2013 Chemical composition and ft-ir spectra of sugar palm (arenga pinnata) fibers obtained from different heights *Journal of Natural Fibers.* **10**, no. 2, pp. 83-97.
[8] Ishak M, S Sapuan, Z Leman, M Rahman and U Anwar 2012 Characterization of sugar palm (arenga pinnata) fibres *Journal of thermal analysis and calorimetry.* **109**, no. 2, pp. 981-989.
[9] Morán J I, V A Alvarez, V P Cyras and A Vázquez 2008 Extraction of cellulose and preparation of nanocellulose from sisal fibers *Cellulose.* **15**, no. 1, pp. 149-159.
[10] Shibata S, Y Cao and I Fukumoto 2006 Lightweight laminate composites made from kenaf and polypropylene fibres *Polymer Testing.* **25**, no. 2, pp. 142-148.
[11] Rowell R M, *Natural fibres: Types and properties* (Properties and performance of natural-fibre composites). 2008, p. 1.
[12] Senwitz C, A Kempe, C Neinhuis, J L Mandombe, M F Branquima and T Lautenschläger 2016 Almost forgotten resources–biomechanical properties of traditionally used bast fibers from northern angola *BioResources.* **11**, no. 3, pp. 7595-7607.
[13] Leman Z, S Sapuan, A Saifol, M A Maleque and M Ahmad 2008 Moisture absorption behavior of sugar palm fiber reinforced epoxy composites *Materials & Design.* **29**, no. 8, pp. 1666-1670.
[14] Bachtiar D, S Sapuan and M Hamdan 2008 The effect of alkaline treatment on tensile properties of sugar palm fibre reinforced epoxy composites *Materials & Design.* **29**, no. 7, pp. 1285-1290.
[15] Ishak M, Z Leman, S Sapuan, M Salleh and S Misri 2009 The effect of sea water treatment on the impact and flexural strength of sugar palm fibre reinforced epoxy composites *International Journal of Mechanical and Materials Engineering (IJMME).* **4**, no. 3, pp. 316-320.
[16] Edhirej A, S Sapuan, M Jawaid and N I Zahari 2017 Cassava/sugar palm fiber reinforced cassava starch hybrid composites: Physical, thermal and structural properties *International Journal of Biological Macromolecules.* **101**, pp. 75-83.

[17] Cavdar A D, F Mengeloğlu and K Karakus 2015 Effect of boric acid and borax on mechanical, fire and thermal properties of wood flour filled high density polyethylene composites *Measurement.* **60**, pp. 6-12.
[18] ASTM-D3800M 2010 Standard test methods for density of high-modulus fibers *ASTM International West Conshohocken.*
[19] Ridzuan M, M A Majid, M Afendi, S A Kanafiah, J Zahri and A Gibson 2016 Characterisation of natural cellulosic fibre from pennisetum purpureum stem as potential reinforcement of polymer composites *Materials & Design.* **89**, pp. 839-847.
[20] Saba N, P M Tahir and M Jawaid 2014 A review on potentiality of nano filler/natural fiber filled polymer hybrid composites *Polymers.* **6**, no. 8, pp. 2247-2273.
[21] Birnin-Yauri A U, N A Ibrahim, N Zainuddin, K Abdan, Y Y Then and B W Chieng 2016 Enhancement of the mechanical properties and dimensional stability of oil palm empty fruit bunch-kenaf core and oil palm mesocarp-kenaf core hybrid fiber-reinforced poly (lactic acid) biocomposites by borax decahydrate modification of fibers *BioResources.* **11**, no. 2, pp. 4865-4884.
[22] Morales-Cepeda A B, M E Ponce-Medina, H Salas-Papayanopolos, T Lozano, M Zamudio and P G Lafleur 2015 Preparation and characterization of candelilla fiber (euphorbia antisyphilitica) and its reinforcing effect in polypropylene composites *Cellulose.* **22**, no. 6, pp. 3839-3849.
[23] Segal L, J Creely, A Martin Jr and C Conrad 1959 An empirical method for estimating the degree of crystallinity of native cellulose using the x-ray diffractometer *Textile Research Journal.* **29**, no. 10, pp. 786-794.
[24] Cao Y, F Chan, Y-H Chui and H Xiao 2012 Characterization of flax fibres modified by alkaline, enzyme, and steam-heat treatments *BioResources.* **7**, no. 3, pp. 4109-4121.
[25] Li Y and K L Pickering 2008 Hemp fibre reinforced composites using chelator and enzyme treatments *Composites Science and Technology.* **68**, no. 15, pp. 3293-3298.
[26] Punyamurthy R, D Sampathkumar, C V Srinivasa and B Bennehalli 2012 Effect of alkali treatment on water absorption of single cellulosic abaca fiber *BioResources.* **7**, no. 3, pp. 3515-3524.

IOP Conf. Series: Materials Science and Engineering **368** (2018) 012039 doi:10.1088/1757-899X/368/1/012039

Effect of lignin on the thermal properties of nanocrystalline prepared from kenaf core

F A Sabaruddin* **and M T Paridah***

Institute of Tropical Forestry and Forest Products, Universiti Putra Malaysia, 43400 Seri Kembangan, Selangor, Malaysia.

*Corresponding authors: atiyah88@gmail.com, parida@upm.edu.my

Abstract. Lignin and cellulose work together to provide a structural function in plants. While cellulose is the primary load-bearing element, lignin acts as the matrix that provides stiffness and rigidity to the fibre. In the production of nanocellulose, the removal of lignin is essential which is done through bleaching. Nevertheless, studies have shown that lignin can impart positive effects on thermal stability besides being a natural compatibilizer due to the presence of both aliphatic and polar groups in the chain network. The objective of this study is to evaluate the effect of lignin on the thermal properties of nanocrystalline cellulose prepared from kenaf core. Nanocellulose of different lignin content was prepared by undergoing first, pulping using 20 % NaOH and 0.1 % anthraquinone (AQ) followed by a bleaching sequence. The study shows that, sample having the least amount of lignin had good thermal stability during the first and second stage of decomposition by shifting the temperature higher as compared to sample with higher lignin content. However, when the temperature reached more than 300 °C up to the end of the degradation process, the latter gave a higher amount of residue suggesting that lignin can help the NCC to withstand heat better.

1. Introduction

Cellulose is the most abundant renewable natural biopolymer. Rapid developments of nanotechnologies have led to the emergence of the next generation of cellulose based products and one of these is nanocrystalline cellulose (NCC). NCC can be produced by a sequel series of isolation processes, including complete or partial removal of matrix materials and isolation of the cellulosic fibres and followed by controlling chemical treatment to remove amorphous regions of cellulose polymer [1].

Other than cellulose and hemicellulose, lignin is one of the polymeric components in a plant which interpenetrate with each other to form complex higher order structure. Over hundred years, lignin has been considered as un-welcomed by product and it is excluded from wood in order to extract cellulose especially in the pulping process [2]. However, it is very important to keep the lignin component as it has been investigated that lignin can be a natural compatibilizer due to the presence of both aliphatic and polar groups which can provide compatibility between nonpolar polymers and lignocellulosic fibres [3-6]. In fact, cellulose and lignin is analogous to that of epoxy resin and glass fibres where the fibrous components, cellulose or glass fibres, are the primary load-bearing elements while the matrix, lignin or epoxy resin, provides stiffness and rigidity.

Recently, there has been an interesting finding about lignin on its positive effect on thermal performance other than in mechanical properties [7]. However, most of the studies had reported the application of isolated and industrial lignins. For example, Hatakeyama and Hatakeyama [2], in their study reported the application of two types of industrial lignins to prepare a polyurethane matrix which to be applied together with various types of organic and inorganic fillers to produce densely filled sandwich composites with the objective to improve its thermal stability for practical used. These lignin PU composites can be used as panels in the field of construction including civil construction and housing materials. Another finding was done by Morandim-Gianetti et al. [6] in their studies on isolated lignin through acetosolv process from coir which incorporated in polypropylene reinforced coir fiber (PP/CF) also reported significant improvement in both initial decomposition temperature and oxidation induction times may due to role of lignin as an antioxidant and its action as a barrier against thermal degradation process.

Kenaf is another source of plant fibre that comprises two different parts, bast fibre and woody core. Kenaf bast fibres, an aggregate of long fibres, are well known for its high strength, which is widely used in natural fibre composites as reinforcement. On the contrary, kenaf core has short fibres and resembles that of low density wood, thus its application in the industries is limited. For this reason, and due to relatively high cellulose content, 31–33 %, kenaf core is a preferred source for the production of nanocellulose. Many studies have reported on the properties of NCC produced from kenaf core [8-11]. However, all these studies used NCC that has undergone the bleaching process, with almost zero lignin. Therefore, the aim of this study is to evaluate the effect of lignin content on the thermal decomposition and thermal stability of the NCC prepared from kenaf core.

2. Experimental Procedure

2.1 Materials

Raw kenaf core fibres (*Hibiscus cannibinus)* were supplied by National Kenaf and Tobacco Board (LKTN), Kota Bharu, Kelantan, Malaysia.

2.2 Methods

2.2.1 Pulping of kenaf

Kenaf core chips (200 g) of 2-3 cm long were placed in in a twin digester with 20% sodium hydroxide (NaOH) and 0.1% anthraquinone (AQ). The ratio of cooking liquid was 1:10 and was cooked at temperature of 170°C for 90 minutes. The condition of pulping process was listed in Table 1 below.

Table 1. The condition of kenaf core pulping process

Condition	
Maximum Temperature	170 °C
Time	1 hour 30 minutes
Chemical	20% NaOH, 1% AQ
Ratio	1:10

2.2.2 Bleaching

The kenaf pulp, then was later purified through three-stage bleaching. Table 2 shows the bleaching conditions employed in the study. Three types of pulp were prepared: (1) Unbleached pulps, denoted as UB, (2) pulps that have undergone bleaching process up to stage D_1, denoted as B1, and (3) pulps that were subjected to all bleaching stages (D1 followed by E_P then D2) and denoted as B2.

Table 2. Bleaching conditions used in the study

Bleaching stage	Chemical charge	Reaction time (min)	Temperature (°C)	Consistency (%)
D_1	2% Sodium Chlorite + 3% Acetic Acid	120	70	10
E_P	1.5% NaOH + 1% H_2O_2	90	70	10
D_2	1.5% Sodium Chlorite + 3% Acetic Acid	90	60	10

2.2.3 Preparation of NCC via acid hydrolysis

All unbleached and bleached samples were treated with sulfuric acid for producing the NCC. High content of lignin tends to impede the acid hydrolysis process, thus requires higher percentage of sulfuric acid and *vice versa* [10]. Hence the hydrolysis were conducted using the same conditions except for the concentration of the acid depend on the lignin content (Table 3).

The pulps were stirred vigorously in sulfuric acid solution for 60 minutes at 45 °C. After that, the pulps were rinsed using multiple centrifugations until the solution become a cloudy suspension. The process was continued with dialysis using cellulose membrane until the pH changed from acidic to neutral. The suspensions were then sonicated for 30 minutes before they were freeze dried. Later, the pulp was subjected to sonification followed by freeze drying to produce NCC. The kenaf core NCC powder was kept in a cool and dry place for further characterization.

Table 3. Acid Hydrolysis conditions employed for the production of NCC from kenaf core

Sample	Acid concentration (%)	Temperature (°C)	Time (sec)
NCC-UB[1]	64% H_2SO_4	45	60
NCC-B1[2]	62% H_2SO_4	45	60
NCC-B2[3]	60% H_2SO_4	45	60

[1] NCC produced from unbleached pulp
[2] NCC produced from D1-stage bleached pulp
[3] NCC produced from D, E_P, D2-stage bleached pulp

2.3 Characterization

2.3.1 Chemical analysis

The chemical compositions of kenaf core fibres (raw, bleached and unbleached) were investigated. The content of extractive was carried out according to TAPPI T204 cm-97 standards. The acid insoluble lignin in kenaf core, then were prepared according to TAPPI T222 om-02 using 72% sulfuric acid. Then the cellulose content of the kenaf pulps was carried out according to the TAPPI standard 203 using 8.3% sodium hydroxide, 17.5 % sodium hydroxide and 2N acetic acid. Next, the percentage of holocellulose is based on a study done by Wise et al., [12] using 1.5g sodium chlorite, 10% acetic acid and acetone.

IOP Conf. Series: Materials Science and Engineering **368** (2018) 012039 doi:10.1088/1757-899X/368/1/012039

2.3.2 Zeta potential and particle size distribution

The zeta potential and nano particle size distribution for kenaf core NCC suspensions were analyzed using Zetasizer Nano equipment (Malvern Instruments, UK). The NCC suspension was sonicated for 15 minutes before the process.

2.3.3 Thermal analysis

The thermal composition and thermal stability were obtained using TGA/SDTA 851 (Mettler Toledo) thermogravimetric analyzer under a linear temperature condition with temperature 50 – 700 °C with heating rate of 10 °C/min under nitrogen atmosphere. The DSC test was done using Perkin Elmer Instrument (Pyris Diamond). The temperature range is from 50-300 °C with a heating rate of 10 °C/min.

3. Results and Discussion

3.1 Chemical properties

The chemical compositions of kenaf core fibre before and after bleaching treatments are presented in Table 4.

Table 4. Chemical components of the NCC prepared from kenaf core fibres

Sample	Lignin (%)	α-cellulose (%)	Hemicellulose (%)	Extractive (%)
RAW kenaf core	33.7	43.7	17.8	4.3
UB[1]	11.5	67.2	15.3	3.4
B1[2]	9.2	77.4	8.9	3.1
B2[3]	2.2	89.9	5.8	2.7

[1] Unbleached pulp
[2] D1-stage bleached pulp
[3] D1, E_P, and D2-stage bleached pulp

Literally, the amount of lignin from woody materials varies from 12 % to 39 %, according to acid insoluble Klason lignin analysis [2, 13]. The lignin content of raw kenaf core is 33.7 %, which falls within this range. Alkaline treatment appears to reduce the amount of lignin content in the kenaf core; from 33.7 % to 11.5 % indicating the effectiveness of NaOH in removing the lignin content. Jonoobi et al. [14] also reported the same effect for kenaf core. The content of cellulose also was seen to correspondingly increase from 43.7 % to 67.2 % after the alkali treatment.

As the kenaf pulp was subjected to first stage bleaching (B1), the content of lignin dropped to 9.2 % (from 11.5 %) whilst the content of cellulose increased from 67.2 % to 77.4 %. At this stage, sodium chlorite and acetic acid appear to partially remove the lignin, as well as some amounts of hemicellulose [14]. On further bleaching using E_P followed by D2 bleaching sequence (B2), almost all the lignin and hemicellulose were substantially extracted out as shown by the relatively low percentage of lignin (2.2 %) and hemicellulose (5.8 %) contents in NCC-B2 whilst the cellulose content increased up to 89.9%. This result indicates the effectiveness of bleaching process used in removing hemicellulose and lignin, producing high amounts of pure cellulose. According to Jonoobi et al. [14], hydrogen peroxide that was used at stage E_p helps to remove the residual chlorite in the pulp and ease the removal of lignin. The last stage of bleaching, D_2, removed the remainder of hemicellulose and lignin. The significant increment in cellulose content proved that the alkaline and bleaching process are an effective sequence for kenaf core.

A similar trend was reported by Jonoobi et al., [8] and Jonoobi et al., [14] showing increment of cellulose content of kenaf bast and kenaf core after pulping and bleaching process while content of hemicellulose and lignin decrease after this chemical treatment. They agree that those effects attributed to the effect of NaOH as well other chemical employed during chemical treatment process. In other study done by Chan et al. [10], kenaf core powder was treated using NaOH and $NaClO_2$ for pulping and bleaching process show increase percentage of holocellulose and marked decrement of Klason lignin as the kenaf core fibre chemically treated. Also show a similar trend was study done by Shi et al., [15] on kenaf bast fibre which treated using slightly different methods of pulping and bleaching process using NaOH and H_2O_2.

3.2 Physical properties (Zeta Potential and Particle Size Analysis)

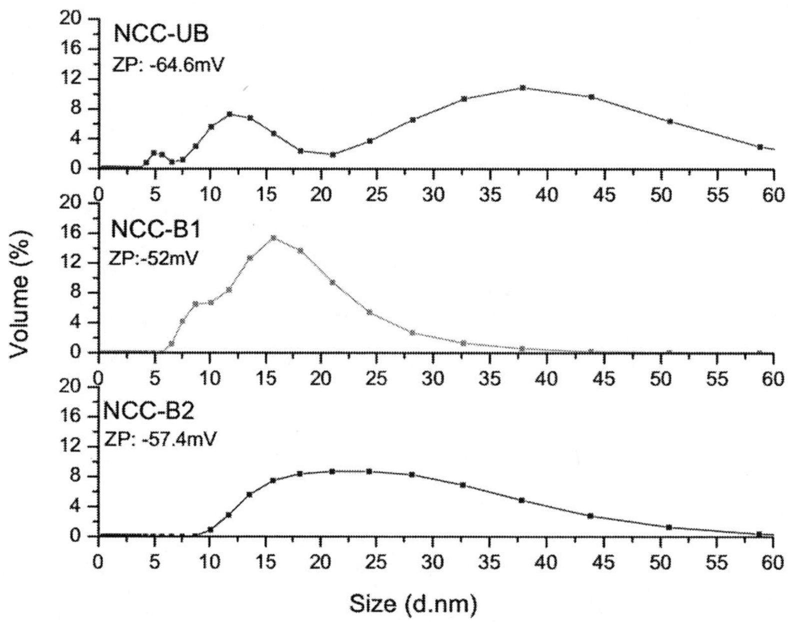

Figure 1. Particle size distribution of NCCs and zeta potential value

Figure 1 shows the diameter of NCCs suspension corresponding to the volume percentage. During the acid hydrolysis process, the presence of hydrogen ions (H^+) helps to remove the amorphous materials and effect the degree of removal is depending on the availability of these hydrogen ions. The effect of acid hydrolysis occurred in the amorphous parts continued by the erosion of crystalline regions of the cellulose fibre [16, 17]. Unbleached sample (NCC-UB) was treat with 64 % of sulfuric acid show two peaks considering two highest mean of diameter size; ~13 nm (volume of 8 %) and ~37 nm (volume of 12 %) respectively. Sample NCC-B1 have a means of diameter value of ~15 nm (volume of 16 %) whilst sample NCC-B2 show broad mean peak around15 – 30 nm (volume 8 %). The major portion of the NCC-UB have higher diameter size which may be attributed to insufficient hydrogen ions (H^+) to remove the amorphous part of the cellulose chains thus, insufficient to penetrate to the other parts of cellulose. Apparently, the unbleached pulps still contain a lot of hemicellulose and lignin that may have prevented the hydrogen ions to reach the amorphous region in the cellulose chains resulting in a substantial amount of the cellulose chains are still intact with diameter of more than 30 nm. Chan et al. [10], Ng et al. [16] and Kumar et al. [18] associated this behaviour with the impediment by the presence of lignin and hemicellulose for the hydrolysis to occur. This result also suggests that the acid hydrolysis condition for unbleached pulp need to be intensified by increasing the acid concentration.

IOP Conf. Series: Materials Science and Engineering **368** (2018) 012039 doi:10.1088/1757-899X/368/1/012039

On the other hand, both NCC-B1 and NCC-B2 show an almost perfect nano size cellulose distribution which mainly has < 30 nm in diameter.

Figure 1 also shows the values of zeta potential for NCC-UB, NCC-B1 and NCC-B2. Zeta potential measurement has been used to investigate the colloidal stability of NCC dispersion in an aqueous media. The ideal value for zeta potential is higher than -25.0 mV [19]. Higher zeta-potential value show higher capacity for dispersion in water with no tendency to flocculate [17, 20]. As shown in Figure 1, the values of zeta potential for NCC-UB, NCC-B1 and NCC-B2 samples are respectively -64.6 mV, -52.0 mV and -57.4 mV implying a stable NCC suspension due to electrostatic repulsion between sulfate groups. According to Roman and Winter [21] and Brinchi et al. [1], during acid hydrolysis process, especially using sulfuric acid, there is a certain degree of grafting of sulfate group onto the surface of NCC. These groups impart negative surface charge to NCC and stabilize the suspension against flocculation; thus give better zeta potential values.

3.3 Thermal properties

TG and DTG graphs for unbleached NCC (NCC-UB), sample bleached at the first stage (NCC-B1) and bleached at the second stage (NCC-B2) are shown in Figures 2(a) and 2(b) respectively. The initial weight loss of NCCs at first stage occurred around 90 - 100°C were indicated the vaporization of water. Similar findings were reported by Rosa et al. [7] and Chan et al. [10].

The main decomposition happens at the second stage. Based on the value listed in Table 5, the onset temperature at the second stage shifted from lower to higher temperature from sample NCC-UB, NCC-B1 and NCC-B2 respectively may due to removal of hemicellulose and partial removal of lignin. The decomposition value for this study is lower compared to other study involving NCC may due to the presence of acid sulfate groups which leads to decrement of thermal stability [1, 7, 10, 21]. At this stage, NCC-UB shows the lowest onset temperature may due to the high amount of amorphous materials that still covered the whole NCC surface which weakened the thermal stability of the sample [7]. As bleaching step escalated for sample NCC-B1 and NCC-B2, most of the hemicellulose and lignin are gradually removed and shifted the higher onset degradation temperature.

The lignin removal is reflected in the amount of residue in the range between 300 – 500 °C [7]. As the sample underwent the region between 300 – 500 °C at third stage, the TG graph at Figure 2(a) shows sample NCC-B2 has a lower amount of residue as compared to sample NCC-UB and NCC-B1. The exact values are listed in Table 5. This may due to effect of remaining lignin that coats the NCC surface had protected the NCC from decomposition process thus, improves the thermal performance of the NCC [7]. This statement is supported by Yang et al. [22], claiming that the lignin decomposition can extend over the whole temperature range starting from 200 °C up to 700 °C.

Table 5. Thermal characteristics of NCC from kenaf core

	NCC-UB	**NCC-B1**	**NCC-B2**
Amount of lignin, %	11.5	9.2	2.2
Onset 1st stage (°C)	96.24	95.43	94.36
Onset 2nd stage (°C)	167.25	171.61	183.64
1st weight loss (%)	36.16	35.18	39.03
Onset 3rd stage (°C)	288.86	283.34	298.75
2nd weight loss (%)	33.14	33.57	33.23
Residue up 700 °C (%)	25.04	25.75	21.73

(a)

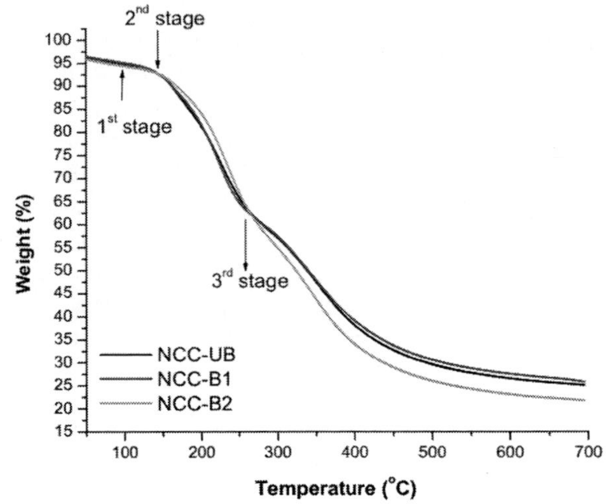

(b)

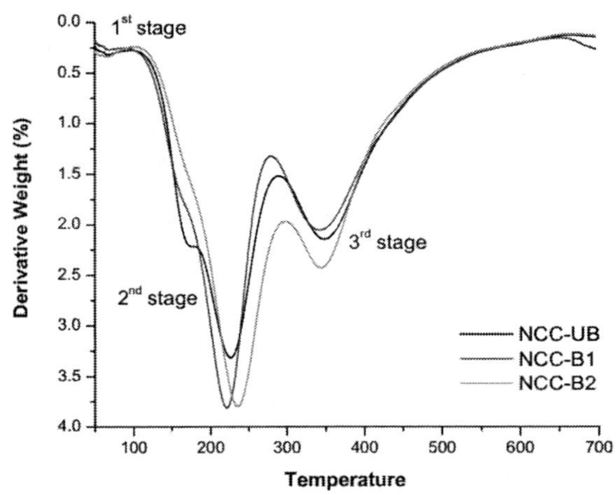

Figure 2. TGA (a) and DTG (b) curves for NCC with different bleaching conditions

As observed from the DTG curves in Figure 2(b), it is clearly shown that those three sample shows two well separated peaks indicated two times of pyrolysis processes which affected by the presence of acid sulfate group in NCCs. First process occurred at a temperature of 200 – 250 °C and the second process occurred at 300 – 400 °C. The first pyrolysis process dominated higher weight loss of the NCC in the range of 35 – 39 wt% as compared to second pyrolysis indicating the overall pyrolysis. A similar observation was shown in a study done by Wang et al. [23]. Sample NCC-UB show lower rate of degradation as compared to samples NCC-B1 and NCC-B2 may indicate the effect of residual lignin that help to delay the degradation process.

4. Conclusion

In this work, the content of lignin was observed to have some influences on the thermal degradation pattern of NCC. NCC having the least amount of lignin (NCC-B2) shows slightly better thermal stability, particularly during the first and second stage of thermal decomposition. However, at the third stage of decomposition (at about 300 °C), NCC that contains higher amount of lignin (NCC-UB and NCC-B1) started to stabilize until a complete decomposition took place, whilst NCC-B2 continued to decompose. The higher amount of residue for both NCC-UB and NCC-B1 was observed, which corresponds well with the amount of residual lignin in the NCC suggesting lignin helps to protect the surface of NCC from heat up to 700 °C. The ability of lignin to protect the surface of NCC especially at high temperature and give it better thermal stability has open the opportunity for it to be applied in the composites and other high thermal end product. The fact that the application of NCC with least lignin need less usage of chemical is believed can help to free the environment from chemical effluent disposal.

Acknowledgement
The authors would like to thank Kementerian Pendidikan Tinggi Malaysia (KPT) for providing the financial support for conducting this research.

References

[1] Brinchi, L., Cotana, F., Fortunati, E. & Kenny, J. M. Production of nanocrystalline cellulose from lignocellulosic biomass: Technology and applications. *Carbohydrate Polymers* **94,** 154–169 (2013).

[2] Hatakeyama, T. & Hatakeyama, H. Lignin Structure, Properties, and Application. *Biopolymers : Lignin, Proteins, Bioactive Nanocomposites* **232,** 1–63 (2010).

[3] Rozman, H. D. *et al.* Effect of lignin as a compatibilizer on the physical properties of coconut fiber-polypropylene composites. *European Polymer Journal* **36,** 1483–1494 (2000)..

[4] Pouteau, C., Dole, P., Cathala, B., Averous, L. & Boquillon, N. Antioxidant properties of lignin in polypropylene. *Polymer Degradation and Stability* **81,** 9–18 (2003).

[5] Gregorová, A., Cibulková, Z., Košíková, B. & Šimon, P. Stabilization effect of lignin in polypropylene and recycled polypropylene. *Polymer Degradation and Stability* **89,** 553–558 (2005).

[6] Morandim-Giannetti, A. A. *et al.* Lignin as additive in polypropylene/coir composites: Thermal, mechanical and morphological properties. *Carbohydrate Polymers* **87,** 2563–2568 (2012).

[7] Rosa, M. F. *et al.* Cellulose nanowhiskers from coconut husk fibers: Effect of preparation conditions on their thermal and morphological behavior. *Carbohydrate Polymers* **81,** 83–92 (2010).

[8] Jonoobi, M., Harun, J., Mathew, A.P., Mohd Zobir, B.H., Oksman, K. (2010). Preparation of Cellulose Nanofibres with Hydrophobic Surface Characteristics. Cellulose. Vol. 17, pp. 299-307.

[9] Surip, S. N., Wan Jaafar, W. N. R., Azmi, N. N. & Anwar, U. M. K. Microscopy Observation on Nanocellulose from Kenaf Fibre. *Advanced Materials Research* **488–489,** 72–75 (2012)..

[10] Chan, C. H., Chia, C. H., Zakaria, S., Ahmad, I. & Dufresne, A. Production and characterisation of cellulose and nano- crystalline cellulose from kenaf core wood. *BioResources* **8,** 785–794 (2013).

[11] Kim, D. Y., Lee, B. M., Koo, D. H., Kang, P. H. & Jeun, J. P. Preparation of nanocellulose from a kenaf core using E-beam irradiation and acid hydrolysis. *Cellulose* **23,** 3039–3049 (2016).

[12] Wise, L. E., Maxine, M. & D'Addieco, A. A. Chlorite holocellulose, its fractionation and bearing on summative wood analysis and on studies on the hemicelluloses. *Technical association of pulp and paper industry***29,** 210–218 (1946).

[13] Lin, S. Y. & Dence, C. W. *Method in lignin chemistry. Methods in Lignin Chemistry* 578 (Springer-Verlag Berlin Heidelberg, 1992). doi:10.1007/978-3-642-74065-7

[14] Jonoobi, M., Harun, J., Shakeri, A., Misra, M. & Oksmand, K. Chemical composition, crystallinity, and thermal degradation of bleached and unbleached kenaf bast (Hibiscus cannabinus) pulp and nanofibers. *BioResources* **4,** 626–639 (2009).

[15] Shi, J., Shi, S. Q., Barnes, H. M. & Pittman, C. U. A chemical process for preparing cellulosic fibers hierarchically from kenaf bast fibers. *BioResources* **6,** 879–890 (2011).

[16] Ng, H. M. *et al.* Extraction of cellulose nanocrystals from plant sources for application as reinforcing agent in polymers. *Composites Part B: Engineering* **75,** 176–200 (2015).

IOP Conf. Series: Materials Science and Engineering **368** (2018) 012039 doi:10.1088/1757-899X/368/1/012039

[17] Al-Dulaimi, A. A. & Wanrosli, W. D. Isolation and Characterization of Nanocrystalline Cellulose from Totally Chlorine Free Oil Palm Empty Fruit Bunch Pulp. *Journal of Polymers and the Environment* **25,** 1–11 (2016).

[18] Kumar, P., Barrett, D. M., Delwiche, M. J. & Stroeve, P. Methods for pretreatment of lignocellulosic biomass for efficient hydrolysis and biofuel production. *Industrial and Engineering Chemistry Research* **48,**3713–3729 (2009).

[19] Mirhosseini, H., Tan, C. P., Hamid, N. S. A. & Yusof, S. Effect of Arabic gum, xanthan gum and orange oil contents on ζ-potential, conductivity, stability, size index and pH of orange beverage emulsion. *Colloids and Surfaces A: Physicochemical and Engineering Aspects* **315,**47–56 (2008).

[20] Travalini, A. P., Prestes, E., Pinheiro, L. A., & Demiate, I. M.. extraction and characterization of nanocrystalline cellulose from cassava bagasse. *Journal of Polymers and the Environment*, 1–9 (2017).

[21] Roman, M. & Winter, W. T. Effect of sulfate groups from sulfuric acid hydrolysis on the thermal degradation behavior of bacterial cellulose. *Biomacromolecules* **5,** 1671–1677 (2004)

[22] Yang, H., Yan, R., Chen, H., Lee, D. H. & Zheng, C. Characteristics of hemicellulose, cellulose and lignin pyrolysis. *Fuel* **86,** 1781–1788 (2007).

[23] Wang, N., Ding, E. & Cheng, R. Thermal degradation behaviors of spherical cellulose nanocrystals with sulfate groups. *Polymer* **48,** 3486–3493 (2007).

Mechanical, thermal and flammability properties of nonwoven kenaf reinforced acrylic based polyester composites: Effect of water glass treatment

M S Salim[1], M F Ahmad Rasyid[1], M A Abdullah[1], R Mat Taib[1] and Z A Mohd Ishak[1,2*]

[1]School of Material and Mineral Resources Engineering, Universiti Sains Malaysia, 14300, Nibong Tebal, Pulau Pinang, Malaysia
[2]Cluster of Polymer Engineering, Science and Engineering Research Centre, Universiti Sains Malaysia, 14300, Nibong Tebal, Pulau Pinang, Malaysia

Email: zarifin.ishak@gmail.com

Abstract. Within this current work, polyester resin on acrylic basis will be modified by the incorporation of environment-friendly sodium silicates (water glass). The modified resin system then was applied on non-woven kenaf fibre (KF) mat via resin impregnation process followed by a compression moulding to prepare the composite samples. The purposes of introducing water glass into the resin system were to improve the fibre-matrix wettability and enhancing the fire resistance in the reinforced composites. Mechanical, thermal and surface morphologies of the composites were examined. The variations in water glass composition in enhancing the flame retardant properties of the composites were evaluated by limiting oxygen index (LOI) and UL94 standard. Significant improvement of fire resistance properties of the composites as amount of water glass increases was observed but reduction in mechanical properties of the composites was recorded.

1. Introduction

The European automotive makers nowadays have extensively use natural fibre within the passenger cars not only for acoustic damping materials but also the interior structural parts in the form of polymer composites. Door trims, spare wheel cover, sun roof, headliners, just to name a few that have been developed commercially by utilizing natural fibre as the reinforcement [1]. The natural fibre reinforced composites have established themselves due to their good mechanical properties, their low production costs, and the good environmental properties [2]. Owing to their low density (approx. 1.3-1.5 g/cm^3), natural fibres have a very good lightweight potential. Other features of natural fibre composites are the very good process related and acoustic properties. Additional advantages like good life cycle assessment and easier processability compared to glass fibre materials can also be taken into account [3]. Despite the availability of various types of natural fibre, kenaf fibre, KF (*Hibiscus cannabinus)* are largely cultivated here in Malaysia due to the climate suitability [4]. Realizing the potential of KF in manufacturing industry, Malaysia government has announced that this crop plant could be the third commodity plant of the nation after rubber and palm oil [5].

Conventional thermoset matrices such as polyester and vinyl ester have the ability to be reinforced by natural fibres. However, polyester and vinyl ester matrices tend to suffer from shortcoming since they emit styrene not only during processing but also from the finished products. The introduction of acrylic based polymer resin in the form of Acrodur resin by BASF has emerged as a new solution in polymer binder technology. This resin is environmentally friendly due to its non-corrosive properties and does not emit styrene during processing and in the final products. The only by product upon the curing process is water. Below its curing temperature, the resin exhibits thermoplastic properties which allow the resin to be processed by means of conventional thermoplastic processing. After being thermally cured, the material shows thermoset properties i.e. resistant to wear and possesses excellent thermal, mechanical and chemical stability [6]. Due to the good features of this resin, it is suitable to be used in automotive industry especially for interior trimmings [7]

One disadvantage of natural fibres is their low chemical compatibility with hydrophobic polymers. This leads to elevated water absorption of natural fibre reinforced composites that can have a significant effect on the mechanical properties of the composites. Another disadvantage of natural fibre reinforced composites is their high flammability that limits their application in many fields including in the automotive industry. Sodium silicate or often referred to as water glass has been considered as a proper modifier for polymeric materials [8]. The incorporation water glass within the polymer resin could provide the solution of both shortcomings i.e. improving the fibre to matrix interaction and the same reduce the flammability. There is similarity in chemical structure between polymerized water glass (polysilicate) and silane. Thus, water glass might act as coupling agent in improving fibre to matrix interaction as shown in the theoretical model by Medina and Schledjewski (2009) (see figure 1). Thongpin et al. (2015) reported that woven mat of jute and abaca fibre treated by sodium silicate significantly increases the tensile strength of the reinforced PLA composites by nearly 25 and 30% respectively. Wang et al. (2010) studied the mechanical properties of sodium silicate treated-moso bamboo particles reinforced polyvinyl chloride composites. They reported an improved tensile strength and Young's modulus of water glass-treated moso bamboo particles. In terms of flame retardant properties, Basak et al. (2014) reported that water glass treatment done to jute fabric by padding method significantly increase the flame retardant properties of the fibre. The LOI value increases from 21 to 43 as a result of 8% water glass treatment onto the fibre.

Based on the literature review, water glass might possess enormous potential as additive to improve the mechanical and flammability properties of the composites. Therefore, the aim of this study is to seek the optimum amount of sodium silicate loading to resin mixture in order to improve fibre matrix interaction and at the same time increase the flame retardant properties of KF reinforced Acrodur composites. The effect of water glass content to mechanical and flammability of the composite will be reported within the current work.

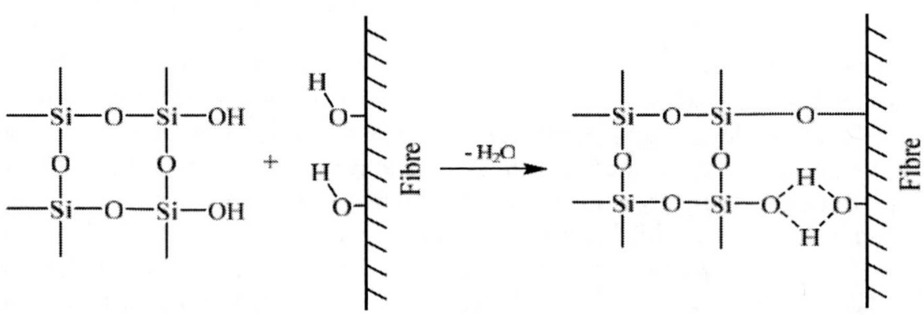

Figure 1. Theoretical model for the surface between water glass and natural fibres.

2. Experimental

2.1 Materials

Kenaf fibre (KF) with an average length of 80 mm was supplied by National Kenaf and Tobacco Board (NKTB), Malaysia. These fibres then underwent a needle punching process to form a non-woven KF mat (NWKF) with an areal density of 1200g/m^2. The preparation of NWKF is given elsewhere [13]. Acrodur resin 950L used in this study was obtained from BASF. This resin is an aqueous acrylic resin based on modified polycarboxylic acid and a polyalcohol (crosslinking agent) that creates a polyester thermoset material upon crosslinking reaction by esterification at a temperature above 130°C (Figure 2).

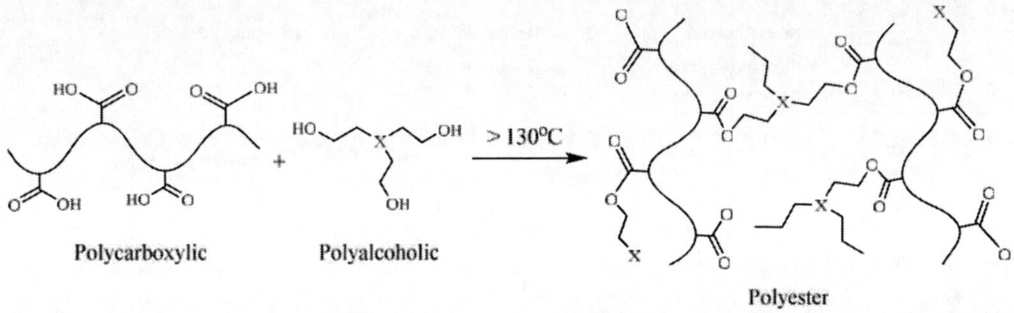

Figure 2. Crosslinking reaction of Acrodur resin.

Sodium silicates belong to the family of soluble silicates, also known as water glasses that represent one of the most versatile inorganic chemicals available. Water glass are manufactured by fusing high purity quartz sand (SiO_2) with sodium or potassium carbonate (Na_2CO_3 or K_2CO_3) in an open hearth furnace at 1100 – 1200°C [8]. The resulting glass is then dissolved using high pressure steam to form liquid silicate or "waterglass" which is clear and slightly viscous. Water glass reacts in aqueous solutions to polysilicates as shown in figure 3 [9]. In this study, sodium silicate solution was obtained from Sigma containing 26 wt.% SiO_2 and 10.6 wt.% Na_2O.

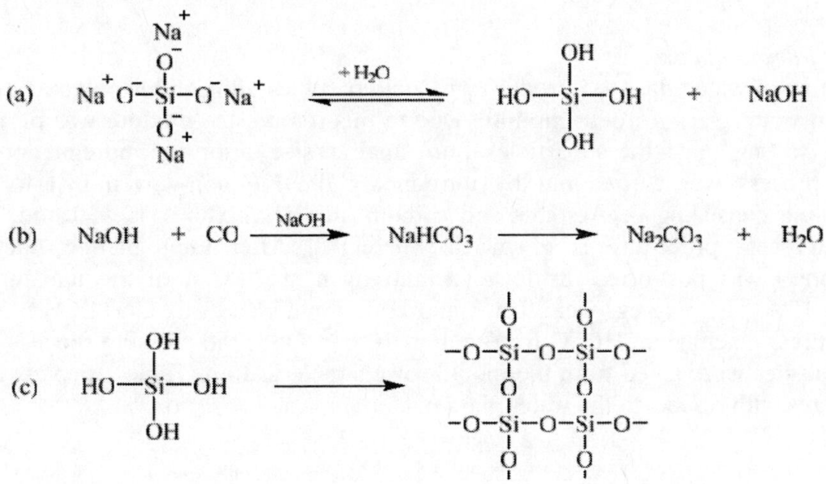

Figure 3. Polycondensation of water glass.

2.2 Viscosity of resin mixture

Different percentage of water glass was added to the Acrodur resin to analyse the compatibility of the resin system with water glass. The stability check of the modified systems was carried out through viscosity measurements and the result is shown in Figure 4. The viscosity of the resin mixture was found to gradually increase from 3900 up to 6000 Pa.s as the water glass content increases from 10 wt.% to 30 wt.%. However, the viscosity increases dramatically to over 14000 Pa.s as 40 wt.% of water glass was added to the resin mixture. High viscosity of resin mixture is not preferable since it might inhibit an adequate impregnation of the NWKF during impregnation process. Therefore, impregnation process was carried out with a maximal 30 wt.% water glass content in the matrix system.

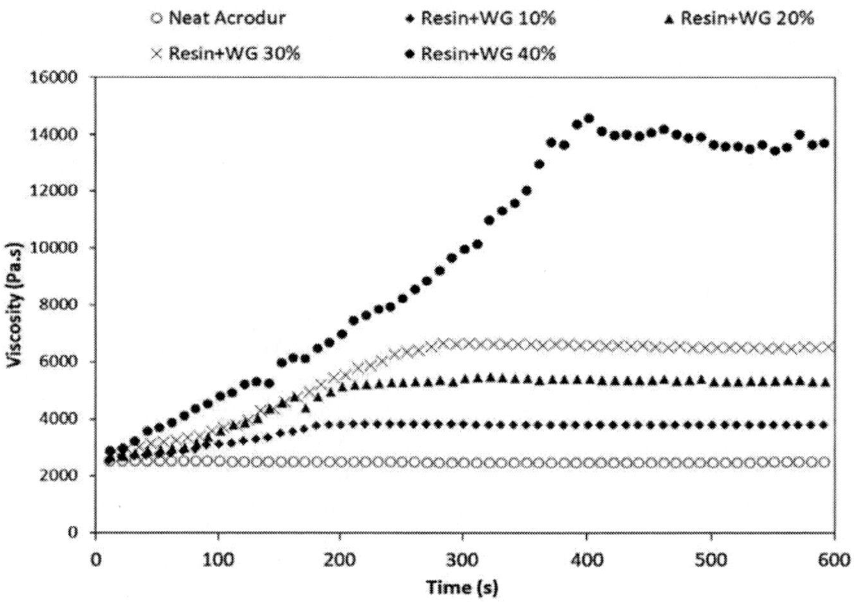

Figure 4. Viscosity of the matrix system mixed with water glass.

2.3 Composite fabrication process

The incorporation of water glass to Acrodur resin yield particles of insoluble polysilicates that increase the viscosity of water glass-Acrodur mixture. Due to this reason, the mixture was properly mixed by electric mixer to segregate these particles into smaller size prior to impregnation process. The impregnation process was carried out by introducing the KF non-woven mat to the fabricated impregnation line consisting of Acrodur and sodium silicate mixture (10, 20 and 30 wt.%). The impregnation process procedure is given elsewhere[14]. After impregnation, the semi-finished material (prepreg) was post-dried inside a vacuum oven at 70°C until the moisture content was approximately 15%. The prepregs were then compression moulded using GOTECH model GT7014-H hydraulic hot press machine at 210 °C for 6 min at 10 bar. Once the pressing process completed, the prepared composites were taken from the mould for characterization. Table 1 depicts the designation of the composites with respect to the water glass content.

Table 1. Designation of KF reinforced Acrodur composites with respect to resin and water glaas content.

Designation	Resin content (wt%)	Water glass content (wt%)
WG 0%	100	0
WG 10%	90	10
WG 20%	80	20
WG 30%	70	30

2.4 Mechanical test

The flexural strength and flexural modulus of the composites were determined using the three-point bending test method as per ASTM D790 standard. The span-to-depth ratio of the test specimens was maintained at 16:1. The flexural test was conducted using Universal Testing Machine Instron 5690 (USA). The experimental was carried out at a crosshead speed of 1 mm/min. The average value of flexural strength and modulus were taken from a total of six composite samples according to the machine direction of non-woven mat. Plain-strain fracture toughness of the KF reinforced composites was determined according to ASTM D5045 by single edge notch bending (SEN-B) approach. The sample geometry [i.e., length (*L*), span length (*S*), width (*W*) and thickness (*B*)] of the specimens satisfies the condition of $2B < W < 4B$, as specified in the standard. The initial crack length (*a*) was generated by tapping at the centre of the notch of the specimen with a new razor blade. The SEN-B specimens were tested using a 5 kN load cell at a crosshead speed of 1 mm/min. The load–displacement curves generated by the test machine were recorded and the maximum load upon fracture was used to calculate the K_C value of the composite as in Equation 1.

$$K_C = Y\frac{3PS\sqrt{a}}{2BW^2} \quad (1)$$

where *P* is the maximum load and Y is the shape factor.

2.5 Flammability test

2.5.1 *UL 94*. The flammability of the different samples was characterized using UL 94-V method. The material classifications are based on the burning behaviour as shown in Table 2.

Table 2. Flammability rating of UL94-V.

	V-0	V-1	V-2
Burning time after flame application (s)	≤ 10	≤ 30	≤ 30
Total burning time (s) for 10 flame applications	≤ 50	≤ 250	≤ 250
Burning and afterglow times of samples after second flame applications	≤ 30	≤ 60	≤ 60
Dripping of burning specimens	no	no	yes
Specimen completely burn	no	no	no

2.5.2 *Limiting oxygen index (LOI)*. The LOI of the composites were determined by FTT Oxygen Index Tester from Fire Testing Technology Ltd. (UK), according to ASTM D 2863. The LOI is the minimum percentage oxygen that is required to maintain flaming combustion of a specimen. The 3mm in thickness of the composites laminates were cut into 6.5 mm wide and 120 mm long specimens for the LOI tests. The test specimen was ignited at the top. Between individual test specimens of a given sample, the oxygen levels were adjusted upward or downward by approximately 0.2% oxygen depending on whether the specimen continue to burn. The LOI was recorded as the lowest percentage oxygen level that allowed 50 mm of the specimen to be burned or continued flaming for 180s duration. The LOI of the composites for each water glass loading was calculated from the average value of 15 specimens as suggested by the aforementioned ASTM standard.

3. Results and discussions

3.1 Determination of crosslinking percentage

DSC analysis was conducted on the composites to determine the crosslinking percentage of the matrix with the respective amount of water glass content. DSC data was used to detect the enthalpy of residual crosslinking capability ($\Delta H_{residual}$) in the composites samples and infer the degree of crosslinking by comparing it to the enthalpy of crosslinking reaction (ΔH_{resin}) of an uncured Acrodur resin. The uncured Acrodur resin was first dried under vacuum at 50°C for 3 hours to eliminate the water content without initiating the crosslinking reaction prior to DSC analysis. The DSC scan of dried Acrodur resin at 10°C/min is shown in figure 5 (a). The peak obtained at 200°C corresponds to the matrix curing with a reaction enthalpy of ΔH_{resin} = 140 J/g. Figure 5 (b-e) show the DSC curves for composites with water glass content of 0, 10, 20 and 30%, respectively. It is important to emphasize that the weight of reinforcement and additive (fibre and sodium silicate) have been excluded in the enthalpy calculation for fair comparison with the uncured resin. The first peak at 100°C is related to the residual moisture within the composite while the second peak at 200°C can be attributed to the post-curing of the Acrodur matrix. The matrix degree of crosslinking was then estimated using Equation 2.

$$Crosslinking\ percentage\ (\%) = \frac{\Delta H_{Resin} - \Delta H_{Residual}}{\Delta H_{resin}}\ x\ 100 \qquad (2)$$

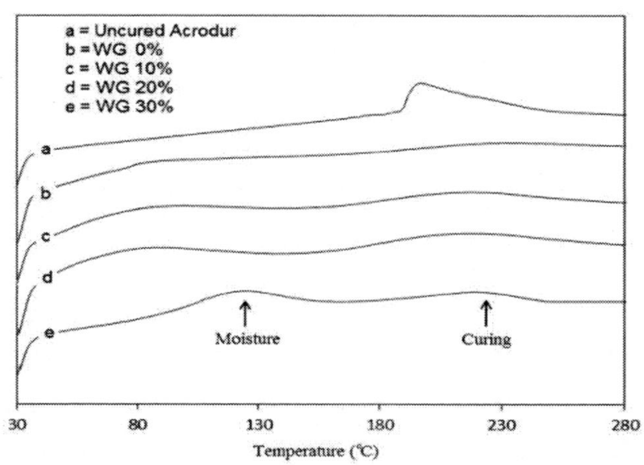

Figure 5. DSC curve for uncured Acrodur resin and composites.

Table 3 summarizes the crosslinking percentage of KF reinforced Acrodur composites with respect to the difference in water glass content. Based on the DSC assessment, incorporation of water glass significantly reduced the degree of the crosslinking. The reduction of degree of crosslinking could significantly affect the mechanical performance of composites [15].

Table 3. Degree of crosslinking of composites.

Composites	$\Delta H_{residual}$ (J/g)	Degree of Crosslinking (%)
WG 0%	22.26	84.10
WG 10%	31.58	77.44
WG 20%	43.05	69.25
WG 30%	52.23	62.69

3.2 Mechanical properties KF reinforced composites

Figure 6 shows the flexural and fracture toughness of the composites with respect to the water glass content. It can be seen that the flexural strength decrease by increasing amount of water glass content. A reduction by 3%, 16% and 44% for flexural strength were recorded by the incorporation of 10% 20% and 30wt% of water glass, respectively. This tendency can be probably explained by an interference of the reaction between prepolymer substance of polycarboxylic acid and the polyalcohol to form polyester with a reaction of one or even both matrix components with water glass inhibiting the curing process of Acrodur resin. As a consequence, less conversion of the resin into rigid thermosetting polyester occurred, therefore reducing the flexural strength of the composites. In terms of flexural modulus, it is interesting to note that the incorporation of 10wt% water glass (WG10%) imparted a slight increase in flexural modulus by 10%. This is due to the introduction of stiff polysilicate within the composite system as a result of polycondensation reaction between water glass and Acrodur resin (see Figure 3) However, the inhibition of crosslink by the addition of higher water glass content (above 20wt%) gives adverse effect to the flexural modulus of composites. A reduction of flexural modulus by 17% and nearly 47% was recorded by WG20% and WG30% composites, respectively. As mentioned by Galvão et al. (2013), uncured functional groups can act as plasticizers, reducing the mechanical properties of composites. In contrast, the fracture toughness of the composites was found to increase as the water glass content increases (see figure 6a). For instance, incorporation of water glass content by 20% and 30% improved the fracture toughness of composites by 34% and 62% respectively. Due to the lower crosslink percentage, some flexibility properties of uncured Acrodur matrix still remained, contributed the ability of absorbing energy, and therefore increased the fracture toughness.

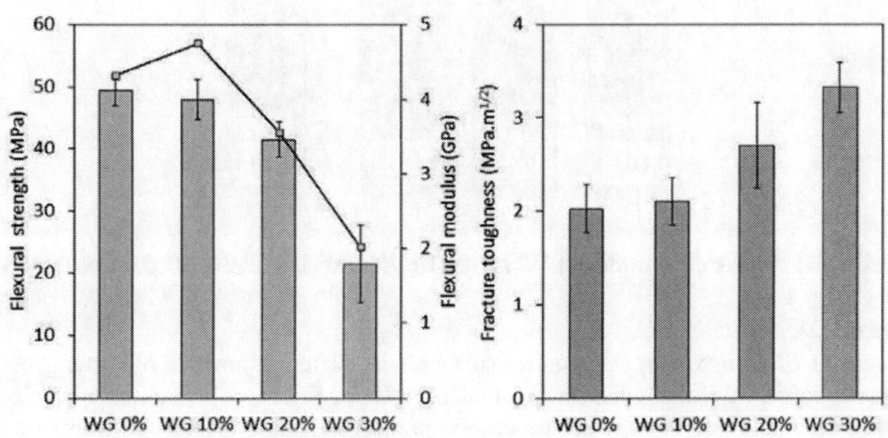

Figure 6. The mechanical properties of Acrodur reinforced KF composites with respect to the water glass content a) flexural properties and b) fracture toughness.

3.3 Flammability properties

Table 4 depicts the UL94 classification and LOI value of the composites with different amount of water glass content. The composites impregnated with solely Acrodur resin and Acrodur with 10% water glass are classified in the category of V-1 due to the burning time exceeding 30s after the application of flame. On contrary, the composites impregnated by Acrodur resin with the additional of 20% and 30% of water glass fall in the best category of V-0 because the flame extinguishes by itself within 10s when the flame is taken away from the specimens. Figure 7 shows that all composites samples containing water glass are not completely burned. It can be deduced that the more water glass content in the resin the better the flame retardant efficiency of the composites. The improvement in the flame retardant of the composite could be evaluated quantitatively by referring to the LOI value which refers to the minimum concentration of oxygen (%) that will support combustion of a material. As expected, the LOI value increase dramatically with the increasing amount of water glass. In the presence of fire, the crystals-like structure of silicates helps to provide an insulating barrier between the product and the flame, and will thus slow down the spread of fire. This result consistent with the study by Cheng and Zhou (2016). They reported an increase in LOI value from 24.5% to 39.5% as the amount of sodium silicate loading increase in the polyurethane.

Table 4.UL94 classification and water absorption value of composite with different amount of water glass content.

Composites	UL94 Classification	LOI
WG 0%	V-1	24
WG 10%	V-1	33
WG 20%	V-0	38
WG 30%	V-0	46

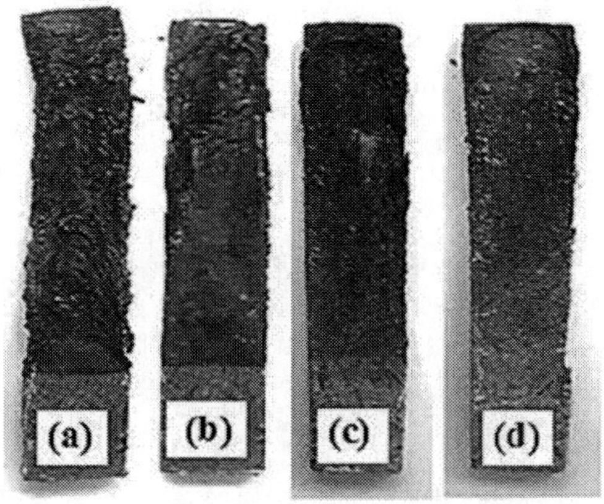

Figure 7. UL 94 tested sample (a) WG 0%, (b) WG 10%, (c) WG 20% and (d) WG 30%

3.4 SEM and EDX analysis

Figure 8 shows the SEM image of composites surface with different amount of water glass content. By increasing the amount of water glass in the matrix system, fewer polymers can be observed in the composite's surface as a very open surface can be perceived. Composites with thermoset resin (WG 0%) polymer shows a smooth, homogeneous surface (Figure 8a), while the compound with 30wt% water glass (WG 30%) exhibits an irregular structure with many air gaps (Figure 5d). This is due to the evaporation of uncured resin during composite fabrication process by compresion moulding.

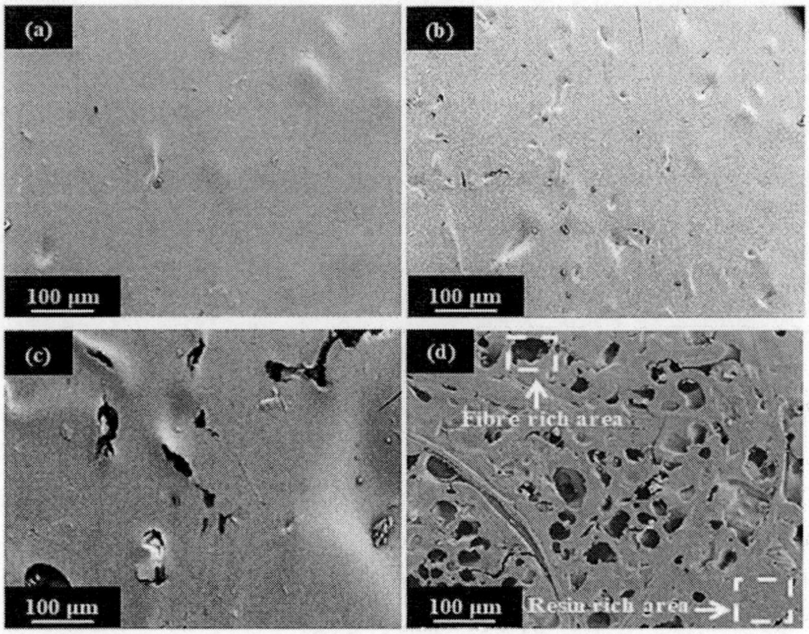

Figure 8. SEM microscope images of composites surfaces (a) WG 0%, (b) WG 10%, (c) WG 20% and (d) WG 30%.

In this research, work Energy Dispersive X-Ray (EDX) analysis was applied on the surfaces of the SEM-samples. The results were shown in figure 9. By using the EDX technique, the composition of the compounds could be detected. Samples impregnated just with thermoset matrix showed only carbon and oxygen signals (Figure 9a), the samples impregnated with sodium water glass showed a small peak for carbon and sodium, a middle signal for oxygen, and a significant signal for silicon (Figure 9b-d).The presence of water glass on the composites surface explains the good behaviour of suppressing flame during the flammability test. It can also be perceived that the decreasing of the carbon signal takes place by increasing the water glass amount on the composite. This could again be a sign of the decreasing of the thermoset polymer amount in the composite due to the interference of water glass on the curing reaction of the polyester system. The EDX analyses on the resin rich area and fibre rich area for WG30% were shown in figure 9(d) and (e), respectively. It is interesting to note that the silicon peak was observed not only on the polymer but also on the fibre surface. The presence of this peak on the fibre rich area might approve the hypothesis where the incorporation of water glass could improve the fibre-matrix bonding by silane coupling mechanism. This is consistent with the study by Thongpin et al. (2015) where sodium silicate treatment on jute and abaca fibres could enhance the fibre-matrix bonding thus improving the tensile strength of the composites. In this study however, the increase in composite strength due to the improved fibre-matrix bonding was constraint by the inhibition of resin curing.

IOP Conf. Series: Materials Science and Engineering **368** (2018) 012040 doi:10.1088/1757-899X/368/1/012040

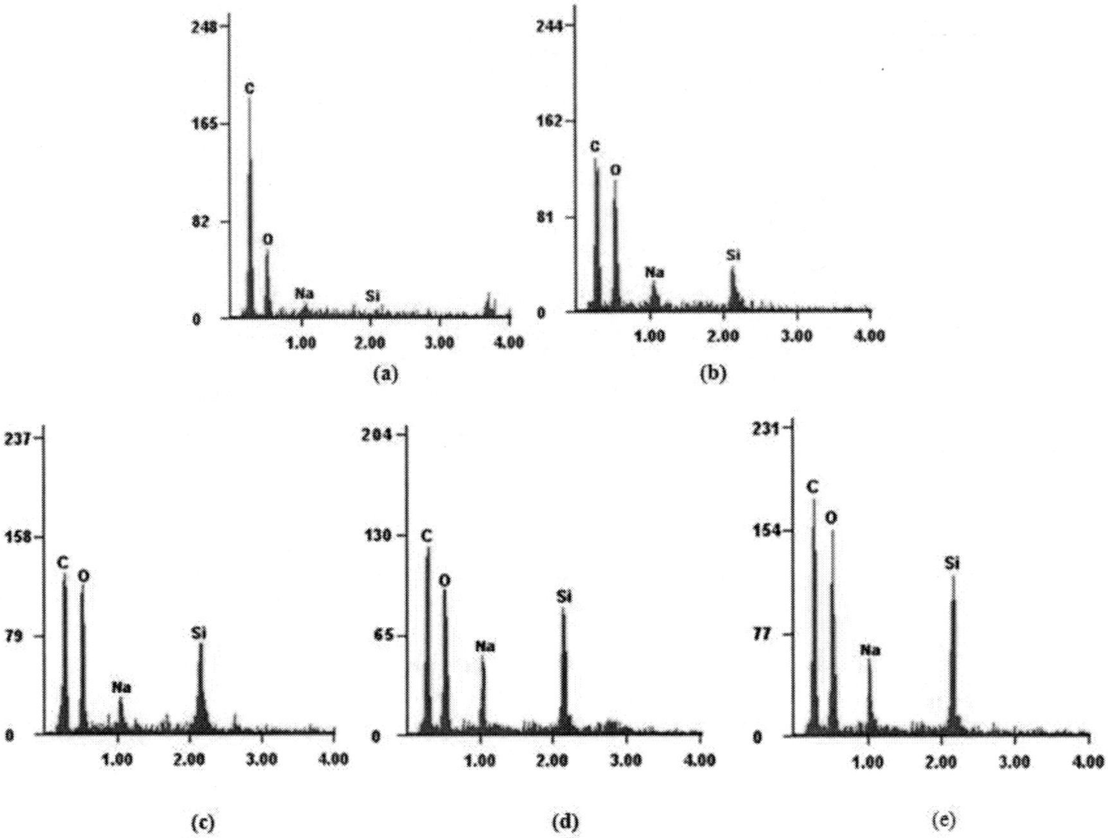

Figure 9. EDX analysis of KF reinforced Acrodur resin with different amount of water glass content a) WG0%%, b) WG10%, c) WG20%, d) WG30% (resin rich area) and e) WG30% (fibre rich area).

4. Conclusions

Within this research work sodium water glass was used as environment friendly inorganic additive in a thermoset matrix system. Water glass should act in the composite as a binder as well as a flame retardant system. However, the improvement of fibre to matrix bonding was constrained by the interference of water glass on the curing reaction of the polymer avoiding the generation of the rigid polyester. The fracture toughness somehow increased as the water glass content increases. The presence of uncured functional group within the resin Acrodur imparts ductility, contributed the ability of absorbing energy thus increased the fracture toughness. Very promising is therefore the flame retardant effectiveness of composite due to the water glass loading. The flame extinguished by itself during the fire testing when the flame was taken away from the sample containing as low as 20% water glass content giving an excellent UL94 classification of V-0. The improvement in flame retardant efficiency of the composites containing water glass content was quantitatively verified by the increasing LOI of the composites as amount of water glass in the composites increase. In order for the developed composites to be used in automotive industry, the amount of water glass content should be appropriately selected to ensure a good compromisation between flame retardant properties and strength of the composites is achieved.

References

[1] Koronis G, Silva A, Fontul M. Green composites: A review of adequate materials for automotive applications. Compos Part B Eng. 2013 Jan;44(1):120–7.

[2] Holbery J, Houston D. Natural-fibre-reinforced polymer composites in automotive applications. Jom. 2006;58(11):80–6.

[3] Karus M, Kaup M. Natural Fibres in the European Automotive Industry. 2002;7(1):119–31.

[4] Hadi M, Basri A, Abdu A, Junejo N, Hamid HA. Journey of kenaf in Malaysia : A Review. Acad J. 2014;9(11):458–70.

[5] BERNAMA. LKTN targets 5,000ha of kenaf by 2015. TheStar Online [Internet]. 2013 May 8; Available from: http://www.thestar.com.my/business/business-news/2013/05/08/lktn-targets-5000ha-of-kenaf-by-2015/

[6] Nordmann G, Kalbe M, Reck B, Gerst M, Norton J. Eco-Friendly Acrylic Copolymers Offering Clean Manufacturing , Reduced VOC Emissions , Excellent Performance. In: 10th Annual Automotive Composites Conference and Exhibition 2010. Troy, Michigan, USA; 2010. p. 1–9.

[7] Karbstein H, Funk J, Norton J. Lightweight Bio-Composites with Acrodur ® Resin Technology. In: 13th Annual Automotive Composites Conference and Exhibition, ACCE 2013. 2013. p. 415–31.

[8] McDonald M, Hamilton J, Thompson J. Recent Developments in Soluble Silicate Based Binders and Coatings. Conf Inst Briquet Agglom [Internet]. 2009;29–40. Available from: http://www.pqcorp.com/Portals/1/lit/Recent Silicate Binder Developments.pdf

[9] Medina L, Schledjewski R. Water Glass As Hydrophobic Additive for Natural Fi- Ber Reinforced Composites. J Nanostructure Polym Nanocomposites. 2009;107–14.

[10] Thongpin C, Srimuk J, Hipkam N, Wachirapong P. Effect of natural fiber types and sodium silicate coated on natural fiber mat / PLA composites : Tensile properties and rate of fire propagation. IOP Conf Ser Mater Sci Eng. 2015;87.

[11] Wang H, Sheng K, Chen J, Mao H, Qian X. Mechanical and thermal properties of sodium silicate treated moso bamboo particles reinforced PVC composites. Sci China Technol Sci. 2010;53(11):2932–5.

[12] Basak S, Samanta KK, Chattopadhyay SK, Das S, Narkar R, Dsouza C, et al. Flame Retardant and Antimicrobial Jute Textile Using Sodium Metasilicate Nonahydrate. Polish J Chem Technol. 2014;16(2):106–13.

[13] Salim MS, Ishak ZAM, Abdul Hamid S. Effect of Stitching Density of Nonwoven Fiber Mat towards Mechanical Properties of Kenaf Reinforced Epoxy Composites Produced by Resin Transfer Moulding (RTM). Key Eng Mater. 2011;471–472:987–92.

[14] Salim MS, Ariawan D, Ahmad Rasyid MF, Ahmad Thirmizir MZ, Mat Taib R, Mohd. Ishak ZA. Effect of fibre surface treatment on interfacial and mechanical properties of non-woven kenaf fibre reinforced acrylic based polyester composites. Polym Compos. 2017;

[15] Asmussen E, Peutzfeldt a. Influence of pulse-delay curing on softening of polymer structures. J Dent Res. 2001;80:1570–3.

[16] Galvão MR, Caldas SGFR, Bagnato VS, Rastelli AN de S, de Andrade MF. Evaluation of degree of conversion and hardness of dental composites photoactivated with different light guide tips. Eur J Dent. 2013;7(1):86–93.

[17] Cheng JJ, Zhou FB. Flame-retardant properties of sodium silicate/polyisocyanate organic–inorganic hybrid material. J Therm Anal Calorim. 2016;125(2):913–8.

IOP Conf. Series: Materials Science and Engineering **368** (2018) 012041 doi:10.1088/1757-899X/368/1/012041

Preparation and characterization of red seaweed/calcium carbonate composite films

EWN Chong[1*], HPS Abdul Khalil [1], T Y Ying [1], HA Tajarudin[1]

[1]*School of Industrial Technology, Universiti Sains Malaysia, 11800 Minden, Penang, Malaysia*

*Corresponding author: ecarborea929@gmail.com

Abstract. This study aims to evaluate the influence of inorganic filler on the properties of seaweed-based film. Seaweed composite films were prepared from raw red seaweed (*Kappaphycus alvarezii*) in the presence of glycerol with the incorporation of calcium carbonate ($CaCO_3$). Solution casting technique was used to form films and dried at 40 °C in the ventilation oven for 24 hours. The effects of different $CaCO_3$ concentrations [0.5%, 1.0%, 1.5%, 2.0% (wt. %)] on the physical, mechanical, and thermal properties were determined. Molecular interaction between the functional groups of red seaweed and $CaCO_3$ was examined through Fourier transform infrared spectroscopy (FTIR). The obtained result demonstrated that the addition of $CaCO_3$ improved the neat seaweed film properties in terms of tensile strength, hydrophobicity, thermal properties and reduced the problem of brittleness encountered in seaweed-based film. However, excessive loading of $CaCO_3$ had led to weaker film properties. Raw red seaweed and $CaCO_3$ has shown compatibility through the analyses of FTIR, thermal, and mechanical of the composite films.

1. Introduction

Seaweed is an essential resource of natural biopolymer that has been used in a vast range of applications worldwide. The main derivatives of seaweed such as alginate and carrageenan have been applied broadly in the field of cosmetic, packaging, pharmaceutical, food and agricultural industry owning to their availability, high organic content, impressive phycocolloids, biodegradability, good gas barrier and mechanical properties [1]. Seaweed phycocolloids are well known to form colloid system either in a gel form or solubilized particles in the presence of water. However, they are hydrophilic in nature. Thus they are usually poor in mechanical properties, water barrier properties and thermal stability [2]. In order to alleviate such drawbacks, biopolymer materials can be reinforced by fillers, blending with other biopolymer materials or through chemical modification to form effective interlocks with the hydroxyl group [3]. Reinforcement fillers in the form of organic and inorganic have been identified in the previous works. Examples of organic reinforcement fillers that had been used to incorporate into sulphated seaweed based matrix were cellulose nanocrystalline (NCC) [5], oil palm shell nanoparticles (OPS) [6], cellulosic pulp fiber [7] while inorganic reinforcement fillers were montmorillonite (MMT) [8-9], silver nanoparticles (AgNPs) [8], zinc oxide (ZnONPs), and copper oxide (CuONPs) nanoparticles [10].

Among the fillers, $CaCO_3$ is one of the most common reinforcing fillers in composite material for paper, rubber, concrete and plastic industries due to its natural abundance, cost-effectiveness, non-toxicity, and non-abrasiveness [11]. Calcium carbonate is normally obtained from carbonatite-lava, stalactities, stalagmites, sketetons, eggshells or shells of marine animals. It can also be obtained from excavating carbonate containing rock [12]. In the recent years, $CaCO_3$ is widely utilized as reinforcing filler to biopolymers such as poly (lactic acid), chitosan, starch, gluten and polycaprolactum to develop nanotablets, hydrogel, microcapsules and composite films with enhanced physical, mechanical and thermal properties [11-13].

Raw seaweed was selected as a matrix in this work since the previous study done by Siah *et al* (2015) had demonstrated the feasibility of using raw seaweed to develop edible film without derivatives extraction [1]. Film properties can be further improved by incorporating filler or other biopolymer materials. This study aimed to fabricate composite films using raw red seaweed as the matrix and $CaCO_3$ as reinforcing filler. It also highlights the influence of using $CaCO_3$ towards physicochemical, mechanical and thermal properties of the red seaweed-based films. Although studies on developing composite films subjected to biopolymer as a matrix have been done extensively, no study has been explored for the fabrication of raw red seaweed/$CaCO_3$ composite films thus far.

2. Experimental

2.1. Materials and film fabrication

Films were prepared using solution casting method. 4 g of red seaweed, *Kappaphycus alvarezii* from Tawau, Sabah, Malaysia was used to form film by dissolving in 200 ml of distilled water in the presence of 2g glycerol, functioned as a plasticizer to improve film's flexibility and reduce film brittleness. Precipitate calcium carbonate with different loadings [0.5, 1, 1.5, 2 (wt. %)] was added to the solution and heated to 90 °C for an hour to allow full gelatinization. The solution was degassed in a vacuum oven and cooled it to room temperature. After that, it was poured unto a casting plate with the diameter of 20 cm and dried in the oven at 40 °C for 24 hours. All dried films were conditioned in a chamber at 50 % RH prior to film characterization.

2.2. Film characterization

2.2.1. Contact angle measurements. The static water contact angles were carried out using a Contact Angle Analyser (KSV CAM 101; KSV Instruments Ltd., Finland) under ambient conditions. A drop of water (5 μL) was deposited on the surface of the solid films through a syringe. Images of the drop profile were recorded immediately after the water touched the film surface. The contact angle (θ) was calculated and expressed in degrees. The mean values of contact angles from 5 replicate measurements for each sample were calculated.

2.2.2. Film thickness. Film thickness was measured at 10 random positions using a precision digital micrometer (Mitutoyo, Kanagawa, Japan) to the nearest 0.0001 mm and the mean value for each film was determined for tensile calculations.

2.2.3. Mechanical properties. Films thickness was determined using a precision digital micrometer (Mitutoyo, Kanagawa, Japan) to the nearest 0.001 mm prior to the mechanical test. Film samples were cut to 10 cm × 1 cm and were conditioned at 50 % RH for a week before all analyses. Tensile strength (TS), modulus of elasticity (YM) and percentage of elongation (%E) were determined using a tensile tester operated according to ASTM D-882-02 standard with initial grips separation of 100 mm and a cross-head speed of 50 mm/min [14]. At least 5 replicates were performed for each sample to obtain the mean values.

2.2.4. Thermogravimetric Analysis (TGA). A TGA apparatus, Mettler-Toledo thermogravimetric analyzer model TGA/DSC 1 was used to go through thermogravimetric analysis (TGA). Approximately 10 mg for each sample was heated under nitrogen purge from 30 °C to 800 °C at the heating rate of 10°C/min. The residue was at 800 °C was expressed as a residue at weight loss (%).

2.2.5. Fourier Transform Infrared (FTIR) analysis. The analysis was carried out with (Perkin-Elmer) Fourier Transform Infrared (PC1600, USA) in attenuated total reflectance (ATR) mode. Films were dried overnight in the oven at 60°C before performing FTIR analysis. The spectra were recorded in between 500 to 4000 cm^{-1}.

3. Results and discussion

3.1. Contact angle

Surface hydrophobicity and wettability of a fluid on a solid surface were determined by static contact angle measurement using sessile drop technique. Contact angles, θ, of the neat red seaweed film and red seaweed/ $CaCO_3$ composite films were illustrated in Figure 1. The contact angle for the neat red seaweed film was 74.37°. A film with the contact angle higher than 65° is normally considered as hydrophobic [15]. In this case, all films including neat red seaweed film were considered hydrophobic despite the hydrophilic nature of seaweed and glycerol as plasticizer. The higher contact angle value achieved by the neat seaweed in the presence of glycerol as a plasticizer could be due to well dispersion of glycerol and the inter- and intra-molecular hydrogen bonding between the hydroxyl groups of red seaweed and glycerol in the matrix [16]. Therefore, this reduced the surface free energy and prevented water penetration through the film. The previous study had shown similar case whereby contact angles increased in the presence of glycerol as a plasticizer [16]. Moreover, it was noticed that contact angle values increased when 0.5 % $CaCO_3$ was added into the matrix. This indication showed that film hydrophobicity improved with the addition of $CaCO_3$ as filler. However, the contact angle of the composite films started to decrease from the addition of 1% $CaCO_3$ onwards. This could be ascribed to excessive addition of $CaCO_3$ into the matrix which probably caused by agglomeration in the composite films. Previous studies revealed that agglomeration occurred when excessive filler content was loaded in cornstarch films and K-carrageenan films [13, 17]. This phenomenon disrupted the interfacial adhesion between the filler and matrix thus, decreased films contact angles.

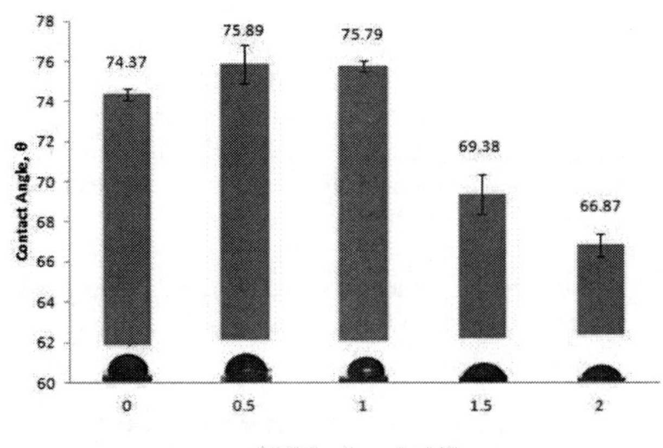

Figure 1. Contact angles of composite films.

3.2 Film thickness

The film thickness is one of the essential parameter used in tensile calculations [6]. The mean value of neat seaweed film thickness was 0.0582 mm while the thickness of the composite films was in the range of 0.0600 to 0.0680 mm as tabulated in 'Table 1'. The increase of film thickness when $CaCO_3$ was added could be due to the increase of film dry mass [18]. It was observed that the trend was not consistent with the addition of $CaCO_3$ increased. This trend was in agreement with the previous study whereby corn starch film reinforced with $CaCO_3$ nanoparticles exhibited an inconsistent trend with no significant changes in film thickness as the addition of $CaCO_3$ increased from 0.02% to 0.5% [13].

3.3 Mechanical properties

The results of tensile strength (TS), Young modulus (YM) and the percentage of elongation (%E) of the composite films were presented in Table 1. Composite films with the addition of 0.5 % $CaCO_3$ achieved the highest TS and YM while the neat red seaweed film showed the lowest TS and YM among all the films. TS and YM increased when 0.5% of $CaCO_3$ was added. The increase of TS and YM could be a good indication of miscibility of filler and matrix which was further identified by FTIR-ATR. It was believed that strong intermolecular hydrogen bonding formed between the hydroxyl groups of filler and the matrix, causing the films exhibited high affinity and high TS. Besides, $CaCO_3$ was able to reduce chain mobility while enhancing macroscopic rigidity of the composite films which resulted in higher TS and YM compared to the neat red seaweed film [13]. However, TS and YM decreased gradually with the addition of 1% $CaCO_3$ onwards. The decrease could be due to the disruption of the molecular interaction between filler and matrix caused by agglomeration whereby more intra-molecular bonds than inter-molecular bonds were formed between the filler and the matrix [13, 19]. TS and YM in this work were higher compared to the TS and YM of cornstarch films reinforced with $CaCO_3$ nanoparticles which were 1.83 to 2.24 MPa and 1.92 to 2.41 MPa respectively [13]. This could be attributed to the nature of gelling properties in seaweed due to the functional compound of 3,6-anhydro-D-galactose and D-galactose-4-sulphate [18].

Table 1. Mechanical properties of composite films.

$CaCO_3$ loadings (%)	Thickness (mm)	TS (N/mm2)	YM (MPa)	Elongation (%)
0.0	0.0582	38.32	182.87	23.26
0.5	0.0648	45.37	198.22	22.84
1.0	0.0680	43.76	191.11	22.96
1.5	0.0600	43.24	189.09	22.82
2.0	0.0650	43.23	187.05	21.40

Aside from TS and YM, the percentage of elongation is another important parameter to evaluate film flexibility and stretchability. In this work, the percentage of elongation decreased gradually due to the increase of film rigidity when $CaCO_3$ was added from 0.5% to 2.0%. Similar cases were found in the study of oil palm shell nanofiller in seaweed-based composite film and nano-$CaCO_3$ in poly(urethane methacrylate) whereby the elongation decreased as the filler loading increased [6, 11].

3.4 Thermal properties

Thermal properties of the neat red seaweed film and the composite films were evaluated by TGA. T_{on} (The initial temperature of decomposition), T_{max} (The maximum temperature of decomposition) and residue after weight loss (%) at 800 °C were determined and tabulated in Table 2. As compared to the neat seaweed film, T_{on} and T_{max} shifted to higher values when 0.5% $CaCO_3$ was added. This could be due to well dispersion of the filler in the red seaweed matrix that led to increase in contact area between the filler and the matrix. Furthermore, there could be the new formation of hydrogen bonds between $CaCO_3$ and the red seaweed which was further confirmed by FTIR. When stronger intermolecular hydrogen bonds formed between the filler and the matrix, more energy, and higher temperature were required to disorganize the intermolecular interactions between the filler and the matrix thereby, led to higher T_{on} and T_{max} [13, 20]. In this work, however, T_{on} and T_{max} started to decrease from the addition of 1% $CaCO_3$ onwards. This could be due to the aggregation of fillers in the matrix that weaken the intermolecular interactions between $CaCO_3$ and red seaweed thereby, decreased the thermal stability [13]. Similar results were reported on thermal stability of starch/nano $CaCO_3$ composite films [21]. On the other hand, the residues after weight loss at 800 °C for the composite films increased as the addition of $CaCO_3$ increased. This finding suggested that the increase of residue was caused by the content of $CaCO_3$. The increase of residues with the increasing of $CaCO_3$ was also observed in starch/nano $CaCO_3$ composite films which were in accordance with the present work [21].

Table 2. Thermal properties of the composite films.

$CaCO_3$ loadings (%)	T_{on}	T_{max}	Residue (%)
0.0	210.17	225.83	14.95
0.5	212.38	226.79	21.90
1.0	211.65	223.50	22.23
1.5	211.33	223.45	22.80
2.0	210.83	222.97	23.36

3.5 FTIR

The interaction between the functional groups of red seaweed and $CaCO_3$ in the presence of glycerol was characterized by FTIR spectroscopy. The miscibility and the effect of molecular interaction between the filler and the matrix can be identified when changes such as shifting, width, and intensity of the spectrum bands were noticed [22].

The FTIR-ATR spectra of the neat red seaweed film and the composite films were presented in 'Figure 2'. Nine characteristics bands were identified and were numbered 1-9. Based on the overall spectra, no obvious differences were observed. This could be due to the small amount of $CaCO_3$ (0.5 to 2 wt%) incorporated into the red seaweed matrix. However, there were slight differences in the shifting of wavenumber (cm^{-1}) for certain functional groups as shown in Table 3. Broad and wide peaks noticed at band 1 (3329 cm^{-1}) for the neat red seaweed film was assigned to the stretching vibration of hydroxyl (–OH) groups due to the formation of inter- and intra- molecular of hydrogen bonds. The stretching vibration of –OH groups for the composite films (bands 1) shifted to lower frequencies as $CaCO_3$ were added ('Table 3'). Previous study showed that stronger hydrogen bonds were formed as the wavenumber shifted to lower frequencies [22]. This suggested that stronger hydrogen bonds were formed between the –OH groups of red seaweed and $CaCO_3$ thus, resulted in stronger interaction between the filler and the matrix. This was confirmed by the increase of tensile strength and thermal stability when 0.5% $CaCO_3$ was added.

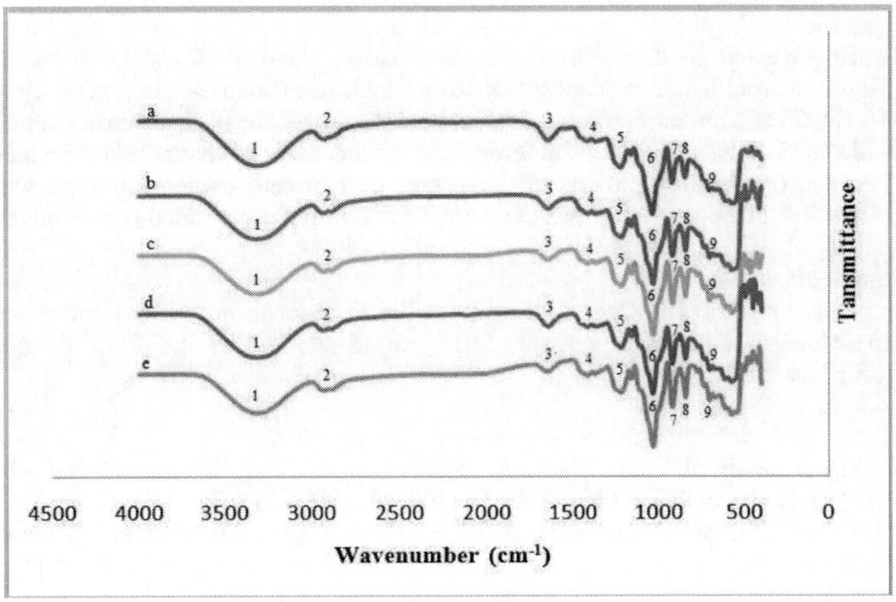

Figure 2. IR-spectra of neat red seaweed film (control film) and composite films : a) neat red seaweed film (control) b) 0.5%, c) 1.0%, d) 1.5%, e) 2.0% of $CaCO_3$ loadings.

Table 3. FTIR-ATR spectra of the neat red seaweed film and the composite films.

$CaCO_3$ loading content (%)	Wavenumber (cm^{-1})								
	Band 1	Band 2	Band 3	Band 4	Band 5	Band 6	Band 7	Band 8	Band 9
0.0	3329	2932	1647	1416	1219	1030	922	845	732
0.5	3319	2937	1645	1416	1217	1034	922	845	700
1.0	3323	2938	1645	1416	1217	1032	922	845	700
1.5	3327	2934	1647	1416	1219	1032	922	845	700
2.0	3327	2934	1647	1416	1219	1032	922	845	700

Stretching vibration of long chain alkyl group was identified at bands 2. There was slight shifting in wavenumber from 2932 cm^{-1} to 2934 cm^{-1}, represents the anti-symmetric stretching of aliphatic C-H and bands 3 at 1647 cm^{-1} and 1645 cm^{-1} were attributed to C=O carbonyl stretching and a combination of hydrogen bonding with COO [20].The shifting of wavenumber indicated that there could be alteration of molecular interaction between red seaweed and $CaCO_3$. Intense absorption peaks from bands 5 to 7 is common to all polysaccharides due to C-C, C-O stretching of the pyranoid ring and C-O-C stretching of glycosidic bonds. Bands 5 (1219 cm^{-1}, 1217 cm^{-1}) appeared in all films were corresponded to stretching vibration of S=O bond, represents the characteristics of sulphate esters which usually found in carrageenan types [23-24]. Bands 6 at 1030 cm^{-1}, 1034 cm^{-1}, and 1032 cm^{-1} were assigned to C-C and C-O stretching vibrations. The anomeric region from 950 cm^{-1} to 700 cm^{-1} particularly the bands 7 at 922 cm^{-1} corresponded to 3,6-anhydro-D-galactose and the bands 8 at 845 cm^{-1} were corresponded to D-galactose-4-sulphate due to the stretching vibration of C-O-C [23-24]. Bands 9 at 732 cm^{-1} and 700 cm^{-1} were assigned to the bending of pyranose ring [23].

4. Conclusion

Red seaweed/$CaCO_3$ composite films were successfully fabricated and characterized based on physicochemical, mechanical and thermal properties. From this work, the composite film with 0.5%

$CaCO_3$ appeared to exhibit the best physical, mechanical and thermal properties among all composite films. FTIR also revealed good miscibility between seaweed and 0.5% $CaCO_3$ as –OH groups had shifted to a lower wavenumber, indicating that strong hydrogen bonds formed between seaweed and 0.5% $CaCO_3$. However, film performance decreased in terms of physical, mechanical and thermal as the addition of $CaCO_3$ exceeded 0.5%. Therefore, it can be concluded that appropriate loading content of $CaCO_3$ into the red seaweed matrix is important to fabricate composite films with improved functional properties. The composite films developed can be used as potential packaging material.

Acknowledgements
The authors would like to acknowledge Ministry of Higher Education for the Fundamental Research Grant Scheme-Malaysia's Rising Star Award 2015 (FRGS-203/PTEKIND/6711531) and the support from all involved departments.

References

[1]Siah W M, Aminah A and Ishak A 2015 Edible films from seaweed (Kappaphycus alvarezii) *International Food Research Journal.* **22** 2230-2236
[2]Martins J T, Bourbon A I, Pinheiro A C, Souza B W, Cerqueira, M A and Vicente A A 2013 Biocomposite films based on κ-carrageenan/locust bean gum blends and clays: physical and antimicrobial properties *Food and Bioprocess Technology.* **6** 2081-2092
[3]Kontopoulou M 2014 Advances in Chemically Modified and Functionalized Polymers *Macromolecular Reaction Engineering.* **8** 67-68
[4]Abdolmohammadi S, Siyamak S, Ibrahim N A, Yunus W M Z W, Rahman M Z A, Azizi S and Fatehi A 2012 Enhancement of mechanical and thermal properties of polycaprolactone/chitosan blend by calcium carbonate nanoparticles *International journal of molecular sciences.* **13** 4508-4522
[5]Huq T, Salmieri S, Khan A, Khan R A, Le Tien, C Riedl B, Fraschini C, Bouchard J, Uribe-Calderon J, Kamal M R and Lacroix M 2012 Nanocrystalline cellulose (NCC) reinforced alginate based biodegradable nanocomposite film *Carbohydrate polymers.* **90** 1757-1763
[6]Abdul Khalil H P S, Tye Y Y, Ismail Z, Leong J Y, Saurabh C K, Lai T K, Chong E W N, Aditiawati P, Tahir P M and Dungani R 2017 Oil Palm Shell Nanofiller in Seaweed-based Composite Film: Mechanical, Physical, and Morphological Properties *BioResources.* **12** 5996-6010
[7]Abdul Khalil H P S, Tye Y Y, Chow S T, Saurabh C K, Paridah M T, Dungani R and Syakir M I, 2016 Cellulosic Pulp Fiber as Reinforcement Materials in Seaweed-Based Film *BioResources.* **12** 29-42
[8]Rhim J W and Wang L F 2014 Preparation and characterization of carrageenan-based nanocomposite films reinforced with clay mineral and silver nanoparticles *Applied Clay Science.* **97** 174-181
[9]Alboofetileh M, Rezaei M, Hosseini H and Abdollahi M 2014 Antimicrobial activity of alginate/clay nanocomposite films enriched with essential oils against three common foodborne pathogens *Food Control.* **36** 1-7
[10]Oun A A and Rhim J W 2017 Carrageenan-based hydrogels and films: Effect of ZnO and CuO nanoparticles on the physical, mechanical, and antimicrobial properties *Food Hydrocolloids.* **67** 45-53
[11]Maurya S D, Purushothaman M, Krishnan P S G and Nayak S K 2014 Effect of nano-calcium carbonate content on the properties of poly (urethane methacrylate) nanocomposites *Journal of Thermoplastic Composite Materials.* **27** 1711-1727
[12]Wang Y, Liu A, Ye R, Wang W and Li X 2015 Transglutaminase-induced crosslinking of gelatin–calcium carbonate composite films *Food chemistry.* **166** 414-422
[13] Sun Q, Xi T, Li Y and Xiong L 2014 Characterization of corn starch films reinforced with $CaCO_3$ nanoparticles *PloS one.* **9**
[14]Standard A S T M, D882-02 Standard test method for tensile properties of thin plastic sheeting

IOP Conf. Series: Materials Science and Engineering **368** (2018) 012041 doi:10.1088/1757-899X/368/1/012041

2002

[15]Shankar S and Rhim J W 2016 Preparation of nanocellulose from micro-crystalline cellulose: the effect on the performance and properties of agar-based composite films *Carbohydrate polymers*. **135** 18-26

[16]Pang J, Liu X, Zhang X, Wu Y and Sun R 2013 Fabrication of cellulose film with enhanced mechanical properties in ionic liquid 1-allyl-3-methylimidaxolium chloride (AmimCl) *Materials*. **6** 1270-1284

[17]Zarina S and Ahmad I 2014 Biodegradable composite films based on κ-carrageenan reinforced by cellulose nanocrystal from kenaf fibers *BioResources*. **10** 256-271

[18]Arham R, Mulyati M T, Metusalach M, and Salengke S 2016 Physical and mechanical properties of agar based edible film with glycerol plasticizer International *Food Research Journal*. **23** 1669-1675

[19]Abdul Khalil H P S, Yap S W, Tye Y Y, Tahir P M, Rizal S and Fazita M N 2017 Effects of Corn Starch and *Kappaphycus alvarezii* Seaweed Blend Concentration on the Optical, Mechanical, and Water Vapor Barrier Properties of Composite Films. *BioResources*. **13** 1157-1173

[20]Marvdashti L M, Koocheki A and Yavarmanesh M 2017 Alyssum homolocarpum seed gum-polyvinyl alcohol biodegradable composite film: Physicochemical, mechanical, thermal and barrier properties *Carbohydrate polymers*. **155** 280-293

[21] Swain S K, Pradhan G C, Dash S, Mohanty F, and Behera L 2017 Preparation and characterization of bionanocomposites based on soluble starch/nano $CaCO_3$. *Polymer Composites*. https://doi.org/10.1002/pc.24326

[22]Wang Y, Liu A, Ye R, Wang W and Li X 2015 Transglutaminase-induced crosslinking of gelatin–calcium carbonate composite films *Food chemistry*. **166** 414-422

[23]Gómez-Ordóñez E and Rupérez P 2011 FTIR-ATR spectroscopy as a tool for polysaccharide identification in edible brown and red seaweeds *Food Hydrocolloids*. **25** 1514-1520

[24]Fernando I S, Sanjeewa K A, Samarakoon K W, Lee W W, Kim H S, Kim E A, Gunasekara U K D S S, Abeytunga D T U, Nanayakkara C, de Silva E D and Lee H S 2017 FTIR characterization and antioxidant activity of water soluble crude polysaccharides of Sri Lankan marine algae *Algae*. **32** 75-86

IOP Conf. Series: Materials Science and Engineering **368** (2018) 012042 doi:10.1088/1757-899X/368/1/012042

Effect of microcrystalline cellulose on the strength of oil palm empty fruit bunch paper

S I Faris [1], Z M A Ainun [1*] and M Jawaid [2*]

[1]Program of Pulp and Paper & Pollution Control, Laboratory of Biopolymer and Derivatives, Institute of Tropical Forestry and Forest Products, Universiti Putra Malaysia, 43400 UPM Serdang, Serdang, Selangor, Malaysia
[2]Laboratory of Biocomposites, Institute of Tropical Forestry and Forest Products, Universiti Putra Malaysia, 43400 UPM Serdang, Serdang, Selangor, Malaysia

*Corresponding authors: ainunzuriyati@upm.edu.my; jawaid@upm.edu.my

Abstract. Oil palm empty fruit bunch (EFB) is one of the most copious residues which are left behind after the oil refining process. It is categorized as short fiber which can be used in the production of pulp and paper. In most cases, EFB fibers need other supporting agent to enhance their wet and dry strength of paper properties. Therefore, the aim of this study is to investigate the effect of paper strength after incorporating different percentage of microcrystalline cellulose (MCC) of 0, 3, 6 and 9% with EFB fibers. The standard method in producing the hand-sheet was applied. Comparison is carried among all samples to determine the optimum percentage of MCC which is suitable to strengthen the paper. Results showed that physical properties of paper in terms of porosity do not have significant effect while the opacity was slightly decreased. The mechanical properties such as tensile strength and tear resistance for MCC papers are better especially paper tensile strength. The scanning electron microscopy (SEM) image showed the morphological structure of paper surface that contains without and with MCC. In conclusion mechanical properties of the hand-sheets are improved with the incorporation of less than 10% MCC into the pulp fibers.

1. Introduction

One of the most modern industries and the oldest crafts is papermaking. Paper is an aqueous deposit of any plant fiber in the sheet form in its technical form. The word paper originated from the Latin world "Papyrus" (Cyperus papyrus), a native plant to the Nile valley, which was made into sheets that could be used to write on by the Egyptians [1]. Extensive use of wood as a raw material for pulp and paper production are affected by the increasing demand of paper in the world due to several reasons; population growth, better literacy, development of communication, and industrialization in developing countries causing global deforestation which gave harmful effect to the environment [2-3]. Tewari [4] estimated that there will be over million tons of paper demand in 2020. Due to the necessity to supply the paper industry with raw materials, new sources of primary pulp fibers had to be identified [5]. There are ranges of non-woody and cellulosic plant materials from which pulp fibers can be extracted. Common species used include rice straw, wheat straw, bamboo, bagasse, kenaf, flax, cotton, sisal, jute, and hemp, to name a few [6]. They can also be divided into three groups based on their availability which are agricultural residues, natural growing plants and non-wood crops grown primarily for their fibers [7]. This has caused a significant growth in the production of pulp for papermaking from the non-wood fibrous materials such as oil palm empty fruit bunch (EFB) [6], [8], straws [9], grass [10] and kenaf [11] and due to their cost-effectiveness and abundance [7].

IOP Conf. Series: Materials Science and Engineering **368** (2018) 012042 doi:10.1088/1757-899X/368/1/012042

EFB is one of the most abundant lignocellulose materials available which are not optimally used and have been left until rotten in the plantation by the miller [8]. Waste generated from oil palm plantation includes empty fruit bunches (EFB), oil palm trunks (OPT), oil palm fronds (OPS), palm shells and palm oil mill effluent palm (POME). The vast quantity of biomass available in the industry can also be transformed into significant additional products, which may be a possible revenue generator for the oil palm-producing countries [12]. EFB is one of the promising potential raw materials to be used for the pulp and paper manufacturing [13]. Lignocellulose material such as EFB is primarily composed of cellulose, hemicellulose, and lignin. Studies done by Yunos [14] have reported that the amount of hemicellulose in EFB was estimated to be around 15–35 % while cellulose and lignin contents were around 30–50 %, and 20–30 %, respectively. Rosli [6] reported that EFB are weaker than the trunk in tensile and burst indices but despite the relatively short fiber (<1mm), they give exceptional tear index.

Chemical pulping involves cooking of raw materials using aqueous chemical solutions at elevated temperature and pressure to extract pulp fibers. Example of chemical pulping include kraft, soda, and sulphite [15]. Kraft pulping involves the fiber cooking in a solution of sodium hydroxide (NaOH) and sodium sulfide (NaS_2). According to Tanaka [16], active alkali (AA) concentration is the most effective factor on the yield and properties, compared to liquor/EFB ratio, cooking temperature and cooking time for kraft pulping.

Addition of additive and filler can improve the physical properties of paper in term of wet and dry strength. Uniformity of the surface, dimension stability and unchanging quality in usage were affected by the dispersion of the additive and filler [17]. Filler addition can give benefits such as cost and energy saving [18], raised in furnish drainage rate, enhancement in optical properties, printability, and appearance of paper produced [19].

Nanocellulosic materials that have been explored in recent decades for using in pulp and paper is microfibrillated cellulose (MFC), which is prepared by mechanical processes that involve very high shear forces to defibrillate cellulose fibers. Cellulose nanocrystals or microcrystalline cellulose (MCC) are the second material, which are prepared by the acid hydrolysis of cellulose fibers, followed by mechanical action, and the third materials are nanocomposites, which may be the combination of first and second types of nanocellulosic materials, along with polymers or other matrices [20]. Incorporating cellulose with the pulp fiber also can help to improve the wet and dry strength of the paper. Many published studies have reported the properties of sheets made by incorporating micro-fibrillated cellulose (MFC) such as [21-24], and nano-fibrillated cellulose (NFC) [25-29] as papermaking additive. Fewer studies have been published on the effect of incorporating microcrystalline cellulose (MCC) with paper pulp fibers on the properties of paper sheets prepared by the standard method for paper sheet making.

The microcrystalline cellulose (MCC) is formed through strong hydrogen bonding among individual cellulose crystals/whiskers and is a type of cellulose that has micro-scaled fibers liberated from larger plant based cellulose fibers and has a broad range of potential applications on tensile strength (wet and dry), tear resistance, burst strength, opacity, and porosity of paper sheets [20, 30]. Ioe Zou [31] reported that addition of 5% MC as additive in the pulp sheet increased the physical properties of pulp sheet. Elongation and TEA increased respectively with concentration of MC at the beginning and reached the highest value on a dose of 8% concentration of MC [17]. Thus, the aim of this work was to study the effect of adding certain amounts of microcrystalline cellulose in kraft paper made from oil palm empty fruit bunches.

2. Materials and Methods

2.1 Raw materials

Oil palm empty fruit bunch fibers were received in bale from a local mill at Pahang. Toluene, acetic acid and acetone were purchased by Bendosen Laboratory Chemical. Ethanol, sodium hydroxide (NaOH) and sulphuric acid were acquired from R&M Chemicals. The MCC was also obtained from R&M Chemicals with degree of polymerization, 210, pH at 5.5, conductivity of more than 75.0 μs/cm and water soluble substances, 0.25.

2.2 Chemical composition of EFB
The chemical composition of EFB was determined according to the following standard methods: ethanol–benzene extractives (TAPPI T204 cm-97), lignin (TAPPI T222 cm-02), holocellulose (Wise et al., 1946) and α-cellulose (TAPPI T 203 cm-99).

2.3 Preparation of pulp and papermaking
The kraft pulping was carried out using Twin Digester (MK Twin Tub Digester). Pulping condition was selected according to Tanaka [16] by adjusting the level of active alkali (AA). The parameters for kraft pulping were fixed at 18% active alkali and 25% sulphidity. The ratio of 1:10 liquor to EFB was selected. The pulping was done for 120 min at 170°C. After completing pulping, the pulps were washed and screened in a Somerville Fractionator. Prior to screening, these pulps were dispersed in a Hydropulper. Finally, unbleached pulps were obtained and applied for papermaking by using Handsheet Machine. Papermaking was carried out based on TAPPI Standard T 205 sp-02. On the other hand, MCC was sonicated using ultrasonic processor (Sonics Vibra-Cell VCX 400) for 5 mins and 10 pulses for 5 secs before adding it into the pulp slurry for papermaking.

In papermaking, there are four samples prepared which contained EFB pulp and EFB pulp added with MCC at 3%, 6% and 9%, w/w MCC over pulp. Firstly, the pulp slurry was disintegrated using Pulp Disintegrator (Regmed DSG 200) at 25,000 rpm to ensure uniform distribution of MCC and EFB fibers. Prior to papermaking, the papers were conditioned at 50% relative humidity and 23°C temperature for 24 hr before testing the paper physically, mechanically and optically. The samples were coded as MP0, MP3, MP6 and MP9.

2.4 Morphological observation of MCC powder and paper
The morphological properties of MCC powder and papers were observed via Scanning Electron Microscopy (SEM) (Hitachi S-3400N and Jeol JXA 840A) under an accelerating voltage of 15 kV. Before scanning, samples were coated with gold using a sputter coater system (Edwards Sputter Coater; BOC Edwards, Sussex, United Kingdom) to obtain excellent image by avoiding any charging effect.

2.5 Determination of paper properties
The paper physical properties comprised grammage, thickness, and bulk. The mechanical properties involved tensile, tear and folding endurance while optical properties consisted of brightness and opacity. The Standard Methods referred are listed as in Table 1.

Table 1. TAPPI Standard used in this study.

Properties	Method
Grammage	TAPPI T 220 sp-01
Thickness	TAPPI T 411 om-97
Bulk Thickness	MS ISO 534:2007
Tensile strength	TAPPI T494 om-01
Tear strength	TAPPI T414 om-98
Brightness	MS ISO 2471:2001
Opacity	MS ISO 2471:2010
Folding endurance	MS ISO 5629:1999

3. Results and Discussions

3.1 Chemical composition of EFB
In order to gain clearer picture on chemical composition of EFB fiber, comparison was done among findings by Sharma [32] and Onuorah [33]. The EFB did not go through any process such as pulping prior to chemical analysis. The content of extractive from this study was found slightly higher than the

other two findings which may be due to the dissimilarities in soil characteristics, climatic surroundings or agricultural arrangement [31]. Onuorah [33] concluded that water and alcohol-benzene solubility can affect the total pulp yield, drainage, paper quality and high alcohol benzene solubility (ABS) can be signed of potential problems which could influence paper formation while Rosli [6] implied that the strands of stem and EFB showed slightly lower solubility in alcohol-benzene as their parenchyma cells had been, more or less, removed.

The lignin content also was considered as high than Onuorah [33] but not much difference than other findings. This shows that EFB is suitable to be pulped using kraft/soda pulping due to percentage amount of lignin as EFB utilize fewer chemicals. Bleaching can be done more easily as lignin content is low, which indicates that this material can undergo bleaching with the minimal usage of chemical [34].

The holocellulose content obtained in this study is lower than Onuorah [33] that can be significantly related to different plant varieties used. However, the chemical composition was at par with findings by Ferrer [35] and α-cellulose content was also almost similar with Onuorah [33]. The cellulose is very important in papermaking process because it affects the strength of paper. Cellulose also will affect the property and economic production of fibers for various uses [34].

Table 2. Chemical composition of EFB fiber.

Analysis	Content, %			
	This study	Ferrer [34]	Sharma [32]	Onuorah [33]
Alcohol benzene extractive	4.80[a]	1.17	4.10	1.6 ± 0.01[b]
Acid insoluble lignin	25.74[a]	24.45	30.99	17.80 ± 0.28[b]
Holocellulose	66.99[a]	66.97	72.00	77.35 ± 0.16[b]
α-cellulose	43.66[a]	47.91	-	43.51 ± 0.30[b]

[a]Average values of duplicate experiments
[b]Average values ± standard deviations of triplicate experiments

3.2 Effect of MCC incorporation on oil palm EFB pulp property

Table 3 displayed the effect of adding MCC with the EFB kraft pulp in the papermaking process. Canadian Standard Freeness (CSF) test showed that there was a decrease in the freeness of the pulp with the increase of MCC percentage. Addition of 3% of MCC resulted in higher freeness compared to the control sample while sample MP9 showed the same amount of freeness with the control. Increased in freeness may be the result of MCC loss with the water during the test. As the percentage of the MCC increased, the bonding between the fibers resulted in decline trend in freeness.

Table 3. Freeness of kraft EFB pulp.

Sample	CSF (ml)
MP0	465
MP3	525
MP6	490
MP9	465

[a]Average values ± standard deviations of triplicate experiments

3.3 Effect of MCC incorporation on oil palm EFB paper: morphological observation

Morphological properties of MCC are shown in Fig 1 which exhibited a disordered, ruptured and aggregated microstructure. The particle size distribution of the commercial MCC indicated that the volume weighted mean diameter is 277.27 μm. The evaluation of hand-sheet structure was determined using SEM. Fig 2 showed the morphology of the reference sample which has not been incorporated with any additive. Addition of MCC to the hand-sheet have a tendency to smooth the surface of the paper while as the concentration of the MCC increases the SEM assessment showed an increase in the bonding of the fiber which tends to cover the pores (Figure 3). Structural properties in addition with MC has been reported previously by Bahar [17] which conclude that the presence of MC, develop greater surface area and binds cellulose fibers with another because of their small size fiber. Increment of surface area will lead to escalation of hydrogen bonding. Study on retention of nanocellulose on paper sheet has not been extensively done as they were chemically alike to the pulp and hard to detect in paper. Fibrils cannot be seen with microscopic methods after drying process of the handsheet [26]. The same goes to the MCC as their properties were similar to the pulp.

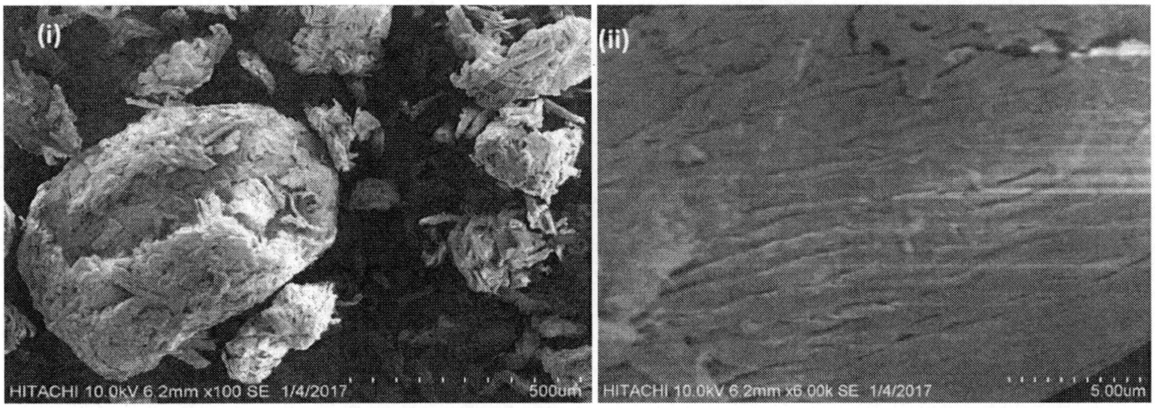

Figure 1. SEM analysis of the commercial MCC at (i) 100x and (ii) 6,000x magnifications

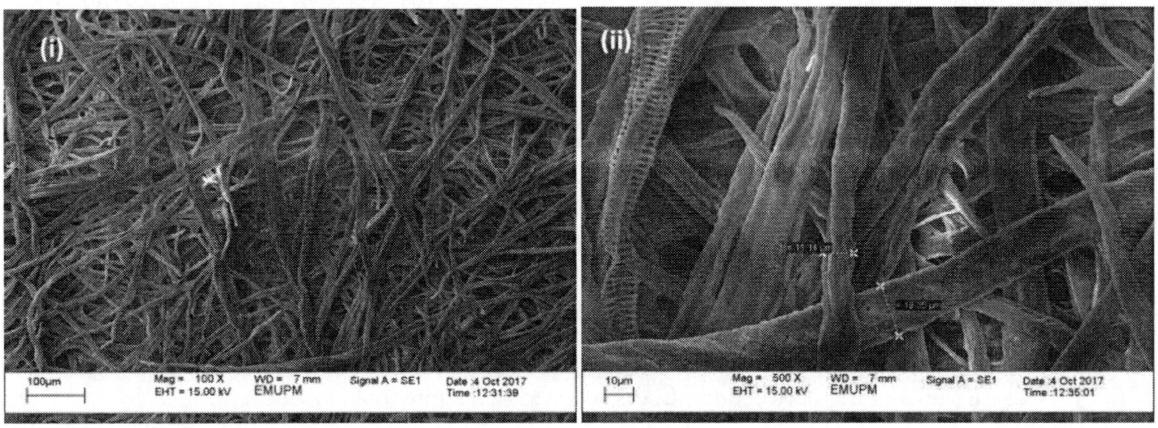

Figure 2. SEM analysis of the sheet structure composed of kraft EFB paper, MP0 at (i) 100x and (ii) 500x magnifications

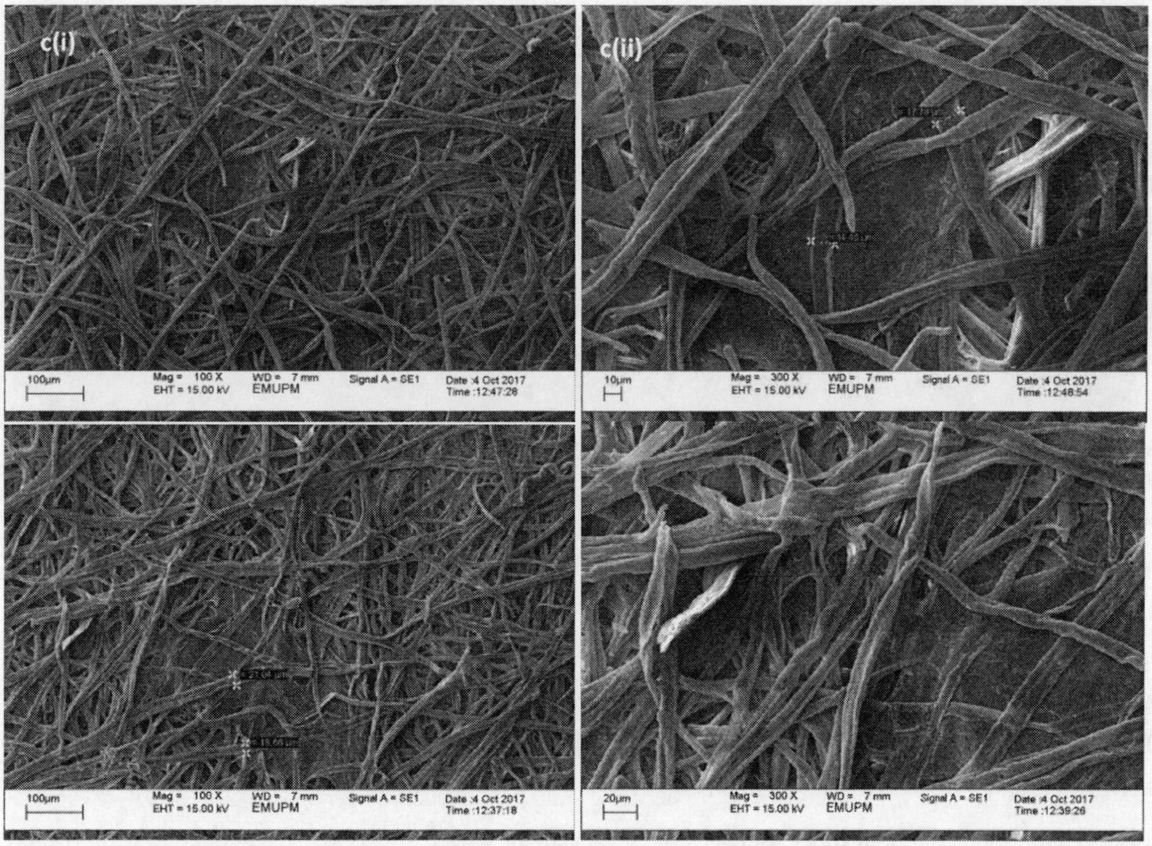

Figure 3. SEM analysis of the sheet structure composed kraft EFB and EFB-MCC papers (3%, 6%, and 9%), (a) MP3, (b) MP6, (c) MP9 .The corresponding image acquired at 100x and 300x for (a), (b), and (c).

3.4 Effect of MCC incorporation on oil palm EFB paper: physical properties

An EFB paper is produced as reference to compare the potential benefits of incorporating MCC with the kraft EFB pulp. This is presented in Table 4. The thickness of the paper produced after incorporating with MCC is lower than the reference sheet. Petroudy [28] also reported that decrement in thickness happened with addition of 5% MFC caused by combination of the pulp with MFC which both are anionic in neutral water thus they repel each other [22]. In fact, thickness, grammage and bulk thickness of the paper decreased with the increased of percentage of the MCC. Addition of 3% MCC give about 5% reduction in thickness. There is no significant difference between the grammage of the reference hand-sheet compared to the hand-sheet that was added with MCC.

Table 4. Physical properties of EFB paper (MP0) and EFB-MCC paper (MP3, MP6 and MP9).

Sample	MP0	MP3	MP6	MP9
Grammage (g m^{-2})	32.68	33.08	32.03	33.38
Thickness (µm)	81.65 ± 4.4[a]	77.61 ± 4.0[a]	71.4 ± 2.9[a]	63.23 ± 2.5[a]
Bulk Thickness (µm/g m^{-2})	2.50	2.35	2.23	1.89

[a]Average values ± standard deviations of 10 replications

3.5 Effect of MCC incorporation on oil palm EFB paper: mechanical properties

The mechanical properties of paper are controlled by fiber strength, the bonding degree of fiber network and strength of the bonds [22]. Fiber-to-fiber joint strength as well as on bonded area highly affect the bonding strength. The plateau in tensile strength may represent the amount of bonded area when the sheet is so strongly bonded that bonds are not broken when the paper is subjected to tensile forces but the fibers themselves are broken [27]. As expected, the result indicated that incorporation of MCC help to increase the tensile strength of the paper (Figure 4). As tensile strength of paper depends on the fiber strength and the bonding strength, this showed that MCC promoted higher amount of bonded areas (Figure 4) as they have larger surface areas. The positive effect of MCC addition in the papermaking was also reported by Ioe Zou and Hsieh, (2007).

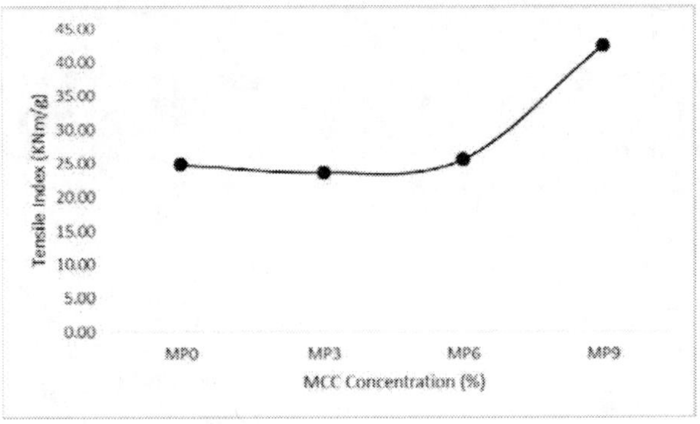

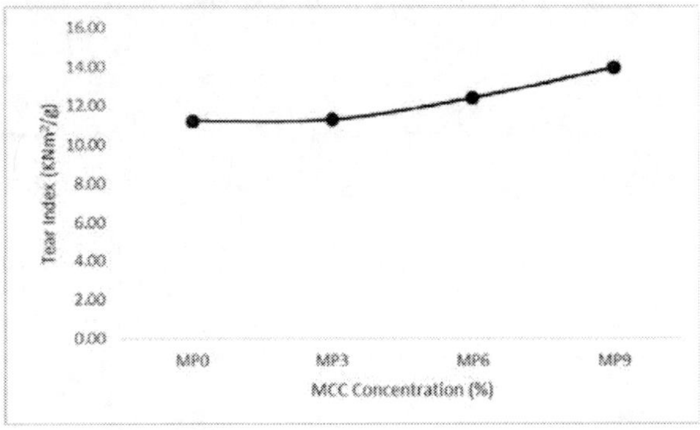

Figure 4. The effect of different percentage of MCC on a) tensile index and b) tear index of EFB paper (MP0) and EFB-MCC paper (MP3, MP6 and MP9).

Both tensile index and tear index showed linear increment with respect the MCC concentration (Figure 4). Addition of 9% MCC during the papermaking process increased both tensile index and tear index by 71.4% and 24.5% respectively. Increasing the number of bonds or the mechanical entanglement resulted in the enhancement of strength of the fiber network. The increased bonding hold the fiber and contribute to a greater network strength [22]. Petroudy [27] reported that tear resistance of the sheet negatively affected by addition of MFC but result obtained shows positive effect on the tear index of the paper sheet which supported Hassan [21] discovery as tear resistance did not deteriorated the strength of paper sheet with addition of 30% MFC to bagasse pulp.

Elongation and toughness energy absorption (TEA) show negative effect at the beginning for 3% addition of MCC but gradually increase for 6% and reaching highest value on dose of 9% concentration of MCC which is higher than the reference pulp (Figure 5). Positive result for the TEA and elongation by adding cellulose was previously reported by several studies. Bahar [17] reported that addition of 2% nanocellulose (70 nm) raised the TEA by 50% and elongation by 24%. An increase occurred by as much as 3 times for the hand-sheet which was added with 10% of nanofibrillated cellulose (NFC) to the unbeaten pulp [27]. By increasing the dose of MCC concentration improved fiber-fiber adhesion which is probably contribute to this effect.

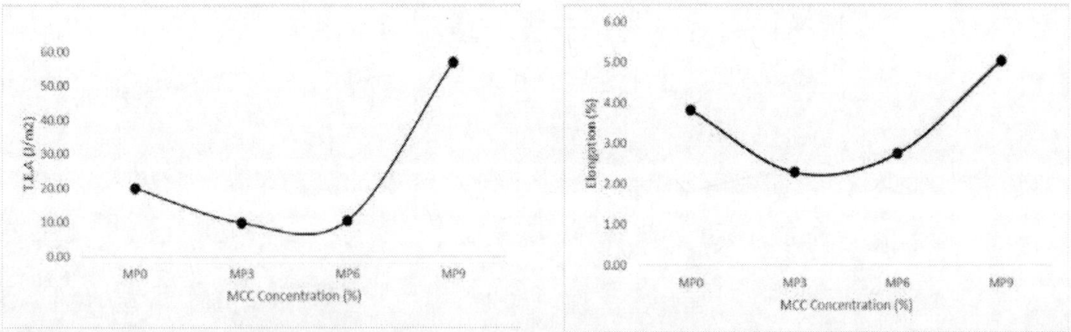

Figure 5. Effect of different percentage of MCC addition on a) T.E.A and b) elongation of EFB paper (MP0) and EFB-MCC paper (MP3, MP6 and MP9).

Increase of bonding can support folding endurance paper property. Fig. 6 exhibited that increase of folding endurance are linear with the increase of MCC concentration. Reference pulp showed lower folding endurance than Tanaka [16]. As EFB are short fiber, they tend to produce close sheet due to lower air permeability and bulk [35]. Folding endurance reached the highest with 9% addition of MCC which is almost 3 times compared to the reference paper.

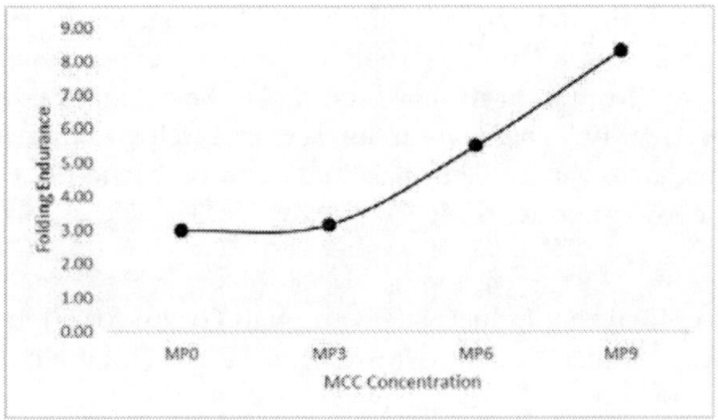

Figure 6. Effect of different percentage of MCC addition on folding endurance of EFB paper (MP0) and EFB-MCC paper (MP3, MP6 and MP9).

3.6 Effect of MCC incorporation on oil palm EFB paper optical properties

Both results indicated that opacity and brightness decreased linearly (Figure 7) with respect to MCC concentration. Brightness showed significant difference between the EFB paper and the EFB-MCC paper while opacity showed a little difference as the opacity ranging from between 99.99 to 99.93. The concentration of 9 % MCC concentration gave 0.06 % and 29.00 % decrement in terms of opacity and brightness respectively. Study done by Petroudy [27] reported that replacing 5 % soda bagasse pulp with MFC resulted in significant decreased of opacity. Kajanto [26] also reported that optical properties of paper worsened with addition of nanocellulose as 2-3 % points lower opacity is measured at constant weight basis. These slight decreases of brightness and opacity resulted from improvement of interfiber bonding with the MCC causing reduction on the light-scattering surface. The produced sheet was more compact which consequently reduce the brightness and opacity.

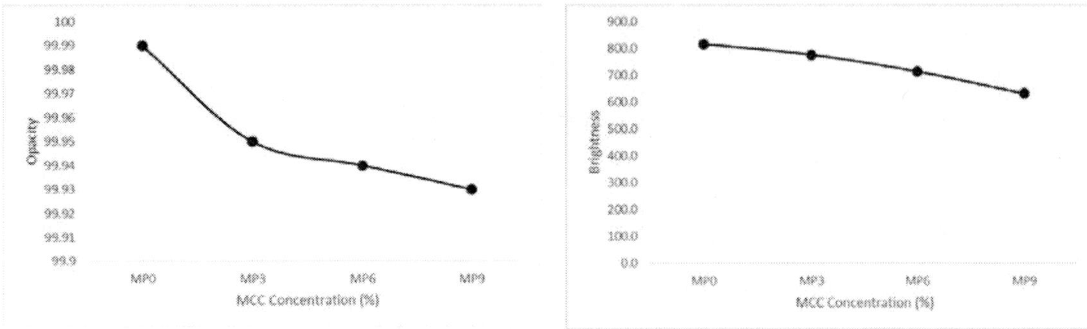

Figure 7. The effect of different percentage of MCC on a) opacity and b) brightness of EFB paper (MP0) and EFB-MCC paper (MP3, MP6 and MP9).

4. Conclusions

In conclusion, all results demonstrated excellent properties especially paper mechanical properties due to the incorporation of MCC. The results indicated that this type of MCC has great potential as an additive and reinforcing agent in papermaking by increasing the strength of paper even at low concentration. Looking at beneficial effect on tensile and tear indices allows high potential for high grade and specialty usages of writing and printing. Combination of low porosity but high strength is crucial characteristics required for such niche application of specialty paper. The MCC also has extra advantage as they are derived from pulp itself which make them highly compatible with the fiber and are more environmentally friendly. Their larger surface area helps to increase the binding and help to reduce the frequency of paper to rupture. Further study can be carried out such as drainage time and beating of pulp prior to incorporation with MCC.

Acknowledgements

The authors are grateful to Malaysian Industry-Government Group for High Technology(MIGHT) for financial support of this work Under Newton-Ungku Omar Fund Grant No: 6300873.

References

[1] Ekhuemelo D O, Oluwalana S A, Adetogun A C, "Potentials of agricultural waste and grasses in pulp and papermaking," *J. Res. For. Wildl. Environ.*, **4**, no. 2, pp. 79–91, 2012.

[2] Abrantes S, Amaral M E, Costa A P, Duarte A P, "Cynara cardunculus L. alkaline pulps: Alternatives fibres for paper and paperboard production," *Bioresour. Technol.*, **98**, no. 15, pp. 2873–2878, 2007.

[3] Enayati A A, Hamzeh Y, Mirshokraie S A, Molaii M, "Papermaking potential of canola stalks," *BioResources*, **4**, no. 1, pp. 245–256, 2009.

[4] Tewari P K, Batra V S, Balakrishnan M, "Efficient water use in industries: Cases from the Indian agro-based pulp and paper mills," *J. Environ. Manage.*, **90**, no. 1, pp. 265–273, 2009.

[5] Plazonic I, Barbaric-Mikocevic Z, Antonovic A, Plazonić I, Barbarić-Mikočević Ž, Antonović A, "Chemical Composition of Straw as an Alternative Material to Wood Raw Material in Fibre Isolation," *Drv. Ind.*, **67**, no. 2, pp. 119–125, 2016.

[6] Rosli W, Daud W, Law K, "Oil palm fibers as papermaking material: potentials and challenges," **6**, pp. 901–917, 2011.

[7] Mahale S, Goswami-giri A S, "Environmental Friendly Approach in Papermaking using Natural Organic Waste," **4**, no. 14, pp. 489–493, 2015.

[8] Harsono H, Putra A S, Maryana R, Rizaluddin A T, H'ng Y Y, Nakagawa-izumi A, and Ohi H, "Preparation of dissolving pulp from oil palm empty fruit bunch by prehydrolysis soda-anthraquinone cooking method," *J. Wood Sci.*, **62**, no. 1, pp. 65–73, 2016.

[9] Vargas F, González Z, Sánchez R, Jiménez L, Rodríguez A, "Cellulosic pulps of cereal straws as raw material for the manufacture of ecological packaging," *BioResources*, **7**, no. 3, pp. 4161–4170, 2012.

[10] Madakadze I C, Masamvu T M, Radiotis T, Li J, Smith D L, "Evaluation of pulp and papermaking characteristics of elephant grass (Pennisetum purpureum Schum) and switchgrass (Panicum virgatum L.)," *African J. Environ. Sci. Technol.*, **4**, no. July, pp. 465–470, 2010.

[11] Mossello A A, Harun J, Md Tahir F, et al., "A review of literatures related of using kenaf for pulp production (beating, fractionation, and recycled fiber)," *J. Mod. Appl. Sci.*, **4**, no. 9, pp. 21–29, 2010.

[12] Hashim R, Nadhari W N A W, Sulaiman O, et al., "Characterization of raw materials and manufactured binderless particleboard from oil palm biomass," *Mater. Des.*, **32**, no. 1, pp. 246–254, 2011.

[13] Rushdan I, Latifah J, Hoi W K, Mohd Nor M Y, "Commercial-Scale Production of Soda Pulp and Medium Paper From Oil Palm Empty Fruit Bunches," *J. Trop. For. Sci.*, **19,** no. 3, pp. 121–126, 2007.

[14] Yunos N S H M, Baharuddin A S, Md Yunos K F, et al, "The physicochemical characteristics of residual oil and fibers from oil palm empty fruit bunches," *BioResources*, **10**, no. 1, pp. 14–29, 2015.

[15] "Basic Overview of Pulp and Paper," pp. 11–39, 2015.

[16] Tanaka R, Wan R W D, Magara K, Ikeda T, and Hosoya S, "Chlorine-Free Bleaching of Kraft Pulp from Oil Palm Empty Fruit Bunches," *Japan Agric. Res. Q. JARQ*, **38**, no. 4, pp. 275–279, 2004.

[17] Bahar N, Hidayat T, Elyani N, and Rostika I, "The Potential of Nanocellulose from Acacia mangium Pulp for Specialty Papermaking," *J. Selulosa*, **3**, no. 1, pp. 1–8, 2013.

[18] Dong C, Song D, Patterson T, Ragauskas A, and Deng Y, "Energy Saving in Papermaking through Filler Addition," pp. 8430–8435, 2008.

[19] Shen J, Song Z, Qian X, and Ni Y, "Carbohydrate-based fillers and pigments for papermaking: A review," *Carbohydr. Polym.*, **85**, no. 1, pp. 17–22, 2011.

[20] Chauhan V S and Chakrabarti S K, "Use Of Nanotechnology For High Performance Cellulosic And Papermaking Products," **46**, pp. 389–400, 2012.

[21] Hassan E A, Hassan M L, and Oksman K, "Improving bagasse pulp paper sheet properties with microfibrillated cellulose isolated from xylanase-treated bagasse," *Wood Fiber Sci.*, **43**, no. 1, pp. 76–82, 2011.

[22] Taipale T, Österberg M, Nykänen A, Ruokolainen J, and Laine J, "Effect of microfibrillated cellulose and fines on the drainage of kraft pulp suspension and paper strength," *Cellulose*, **17**, no. 5, pp. 1005–1020, 2010.

[23] CharaniP R, Dehghani-Firouzabadi M, Afra E, Blademo Å, Naderi A, and Lindström T, "Production of microfibrillated cellulose from unbleached kraft pulp of Kenaf and Scotch Pine and its effect on the properties of hardwood kraft: Microfibrillated cellulose paper," *Cellulose*, **20**, no. 5, pp. 2559–2567, 2013.

[24] Gao Y, Li Q, Shi Y, and Cha R, "Preparation and Application of Cationic Modified Cellulose Fibrils as a Papermaking Additive," *Int. J. Polym. Sci.*, **2016**, 2016.

[25] Ahola S, Österberg M, and Laine J, "Cellulose nanofibrils - Adsorption with poly(amideamine) epichlorohydrin studied by QCM-D and application as a paper strength additive," *Cellulose*, **15**, no. 2, pp. 303–314, 2008.

[26] Kajanto I and Kosonen M, "The potential Use of micro- and nano fibrillated Cellulose as a reinforcing Element in Paper and Board based Packaging," *TAPPI Int. Conf. Nanotechnol. Renew. Mater. 2012*, **2**, no. 6, pp. 393–408, 2012.

[27] Sehaqui H, Zhou Q, and Berglund L A, "Nanofibrillated cellulose for enhancement of strength in high-density paper structures," *Nord. Pulp Pap. Res. J.*, **28**, no. 2, pp. 182–189, 2013.

[28] Petroudy S R D, Syverud K, Chinga-Carrasco G, Ghasemain A, and Resalati H, "Effects of bagasse microfibrillated cellulose and cationic polyacrylamide on key properties of bagasse paper," *Carbohydr. Polym.*, **99**, pp. 311–318, 2014.

[29] Bharimalla A K, Deshmukh S P, Patil P G, and Vigneshwaran N, "Micro/nano-fibrillated cellulose from cotton linters as strength additive in unbleached kraft paper: Experimental, semi-empirical, and mechanistic studies," *BioResources*, **12**, no. 3, pp. 5682–5696, 2017.

[30] Adel A M, El-Gendy A A, Diab M A, Abou-Zeid R E, El-Zawawy W K, and Dufresne A, "Microfibrillated cellulose from agricultural residues. Part I: Papermaking application," *Ind. Crops Prod.*, **93**, pp. 161–174, 2016.

[31] Ioe Zou Y and Hsieh, J S 2007 Microcrystalline Cellulose (MCC) for Papermaking and Surface Sizing. International Conference for the forest Product Industry. Knoxville.

[32] Sharma A K, Anupam K, Swaroop V, Lal P S, and Bist V, "Pilot scale soda-anthraquinone pulping of palm oil empty fruit bunches and elemental chlorine free bleaching of resulting pulp," *J. Clean. Prod.*, **106**, pp. 422–429, 2014.

[33] Khalil H P S A, Alwani M S, Ridzuan R, Kamarudin H, and Khairul A, "Chemical composition, morphological characteristics, and cell wall structure of Malaysian oil palm fibers," *Polym. - Plast. Technol. Eng.*, **47**, no. 3, pp. 273–280, 2008.

[34] Ferrer A, Vega A, Ligero P, and Rodríguez A, "Pulping of empty fruit bunches (EFB) from the palm oil industry by formic acid," *BioResources*, **6**, no. 4, pp. 4282–4301, 2011.

[35] Rodríguez A, Sánchez R, Eugenio M E, Yáñez R, and Jiménez L, "Soda-anthraquinone pulping of residues from oil palm industry," *Cellul. Chem. Technol.*, **44**, no. 7–8, pp. 239–248, 2010.

IOP Conf. Series: Materials Science and Engineering **368** (2018) 012043 doi:10.1088/1757-899X/368/1/012043

Potential of oxidative polymerization coating of cellulose nano-fiber by dopamine

H Nishida and F Koga

Department of Biological Functions and Engineering, Graduate School of Life Science and System Engineering, Kyushu Institute of Technology, 2-4 Hibikino, Wakamatsu, Fukuoka 808-0916, Japan

*Corresponding author: nishida@lsse.kyutech.ac.jp

Abstract. In order to develop functional natural fiber reinforcements derived from cellulose nanofibers (CNF), surface modification of CNF by oxidative polymerization of dopamine was carried out. Resulting polydopamine (PDA)-coated CNF was blended with polypropylene (PP) to prepare the surface-modified CNF/PP nanocomposites. Obtained PDA-coated CNF and its nanocomposites were analyzed as to their morphologies, structures, and various properties. From analyses by Fourier transform infrared spectroscopy, scanning electron microscopy equipped with energy dispersive spectroscopy, X-ray photoelectron spectroscopy, and thermogravimetry, it was clarified that the treated CNF surface was homogeneously coated by PDA without re-aggregation of CNF. Moreover, it was confirmed that the PDA-coated CNF/PP nanocomposites had excellent mechanical performances and good antistatic properties.

1. Introduction

Cellulose as a typical natural fiber is the most abundant renewable biomaterial. The effective utilization of natural fibers as indispensable component in composites with polymers for developing novel eco-friendly composites with acceptable properties is one of the most required research fields [1]. The blended cellulosic fibers in matrix polymers have valuable effects on various composite properties [2]. Many studies concerning the composites have been focusing on effects on mechanical strength of chemical treatments [3], fiber size [4] when the cellulosic fibers were used as fillers and reinforcements, and on the development of appropriate compatibilizers [5].

Many researchers have been extensively studying the preparation method of nanofibers from wood and other plant fibers. The development of cellulose nanofibers (CNF) has been attracting significant interest in the last few decades due to the unique characteristics of CNF that endows such as high surface area-to-volume ratio, high Young's modulus, high tensile strength and low coefficient of thermal expansion [6]. CNFs have shown great potentials in nanocomposites as well as other several applications including biomedical, gas barrier films and optically transparent functional materials.

Cellulose/polymer nanocomposites also have been attracting much interest of researchers for their mechanical performance. However, it is well known that the performance of biomass/polymer composites is strongly depending on the interface adhesion between cellulose surface and matrix polymers [7]. Therefore, the appropriate compatibilizer between biomass and polymer is the key point to demonstrate the ability of biomass.

Dopamine and its catecholic derivatives are able to undergo oxidative polymerization in the presence of oxygen as an oxidant. During the oxidative polymerization of dopamine, a tightly adherent

IOP Conf. Series: Materials Science and Engineering **368** (2018) 012043 doi:10.1088/1757-899X/368/1/012043

polydopamine (PDA) layer is created on the surface of a substrate [8] based on interactions between the PDA layer and the substrate surface including covalent and noncovalent bonds [9]. Actually, the PDA coating on polymer surfaces has good stability and durability in various environments [10,11].

Previously, there are some reports as to the PDA coating of cellulose fibers for developing flame retardant composites [12], Fe^{3+}-pollutant detector [13], etc. In this study, in order to develop functional CNF-reinforced nanocomposites, surface modifying of CNF by oxidative polymerization of dopamine was carried out. Obtained PDA-coated CNF was analyzed with Fourier transform infrared (FTIR) spectroscopy, scanning electron microscopy equipped with energy dispersive spectroscopy (SEM-EDS), X-ray photoelectron spectroscopy (XPS), and thermogravimeter (TG). These analyses clarified that CNF surfaces were homogeneously coated by polydopamine without re-aggregation. Moreover, the PDA-coated CNF was blended with commodity plastic polypropylene (PP) to prepare PDA-coated CNF/PP nanocomposites. The nanocomposites showed excellent mechanical strength and good antistatic properties.

2. Experimental

2.1 Materials
Cellulose nanofiber (CNF, 1 wt% aqueous solution) was kindly provided from Chuetsu Pulp & Paper Co., Ltd., Takaoka, Japan). The CNF was freeze-dried to obtain a solid-type CNF. Buffer: tris(hydroxymethyl)-aminomethane (Tris-HCl, pH8.5, 0.1M) was purchased from SIGMA-Aldrich (St. Louis, USA) and dopamine hydrochloride (LKT Laboratories Inc., St. Paul, USA) was purchased from Wako Pure Chemical Industries, Ltd. (Wako, Japan). Polypropylene (PP): extrusion grade NOVATEC-PP FY6, density 0.90 g/cm^3, melt flow index 2.5 g/10 min (Japan Polypropylene Corporation, Tokyo, Japan) was used as a matrix polymer.

2.2 PDA coating of CNF
Solid-type CNF 5 g was added into a Tris-HCl buffer solution 250 mL, and then dopamine hydrochloride (dopamine) 500 mg was added. The solution was stirred at 1200 rpm on a magnetic stirrer at room temperature at around 20 °C under atmospheric conditions to start the oxidative polymerization of dopamine. During the polymerization, the solution became gradually black. After 24 and 48 h, solid products were filtered to isolate polydopamine (PDA)-coated CNF.

As a reference, the dopamine-HCl was polymerized in the same manner in the Tris-HCl buffer solution without CNF. The solution also became gradually black, and after 24 h solid products were filtered to isolate PDA.

2.3 Preparation of PDA-coated BLCNF/PP composites
PDA-coated CNF was blended with PP in a twin-screw extruder IMC-160B (Imoto Machinery Co., Ltd., Japan, screw diameter 20 mm, L/D 25) equipped with an air vent was used for the blending under the cylinder temperature profiles of 160, 180, 180, and 180°C for four zones from hopper to die, respectively, with a screw rotational speed of 25 rpm. After retention time of 5-7 min in the cylinder, nanocomposite strands were extruded from die, followed by pelletization by cutting the strands with a pelletizer.

2.4 Characterization
Fourier transform infrared (FTIR) spectra were recorded on a Perkin Elmer GX2000R infrared spectrophotometer (Waltham, USA) in a range of 500–4000 cm^{-1} at a resolution of 4 cm^{-1}. Reflection spectra were measured by the single-reflection ATR method on a Golden Gate Diamond attenuated total reflectance (ATR) (10500) module with a germanium crystal.

SEM-EDS images were observed using a scanning electron microscopy equipped with energy dispersive spectroscopy (SEM-EDS) (JCM 6000, JEOL Ltd. Tokyo, Japan). The samples were firstly coated with platinum using a vacuum sputter coater prior to observation by SEM-EDS.

X-ray photoelectron spectroscopy (XPS) measurements were made on an AXIS-HSi surface analysis instrument (Kratos Analytical Ltd., Manchester, UK) with a Mg-Kα X-ray (photon energy 1253.6 eV).

Thermogravimetric (TG) and differential TG (DTG) measurements were conducted on a SEIKO Instruments Inc. EXSTAR 6200 TG system (Chiba, Japan) in aluminum pans (5 mm in diameter) at a prescribed heating rates, φ, of 9 °C min^{-1} in a temperature range of 50 to 550 °C under a constant nitrogen flow (100 mL min^{-1}) using about 5 mg of sample. A blank aluminum pan was used as a reference. The pyrolysis data were collected at regular intervals (about 20 times ºC^{-1}) by an EXSTAR 6000 data platform, and recorded into an analytical computer system.

Antistatic properties of CNF/PP nanocomposites were evaluated by measurement of surface electric resistivity of samples. The electric resistivity of CNF/PP nanocomposite sheets was measured on a Hiresta UP (Mitsubishi Chemical Analytech Co., Ltd., Chigasaki, Japan) equipped a probe UR-100 (measurement range: 10^{10}–10^{18} Ω, main electrode: ϕ 50 mm, inner diameter of guard: ϕ 53.2 mm). After charging at an applied voltage 1000 V for 60 s, the surface electric resistivity of samples was measured at 23 °C.

2.5 Estimation of PDA deposition on CNF from FTIR spectrum

In order to estimate the deposition amount of PDA on CNF surface, a calibration curve was prepared using mixtures of intact CNF and PDA that was synthesized without CNF. FTIR spectra of mixtures in a range of CNF:PDA=100:0-60:40 (wt/wt) were measured and an isolate and PDA-specific absorption peak at 1542 cm^{-1} was used to prepare the calibration curve. The peak in an absorbance plot was simulated using Gauss function in a PeakFit software version 4.12 to estimate the intensity of the peak area, resulting in the calibration curve shown in figure A1 in Appendices.

2.6 Mechanical tests of composites

Samples for mechanical tests were fabricated using a hot press machine IMC-180C (Imoto Machinery Co., Ltd, Kyoto, Japan) in templates: 100 × 100 × 0.3 and 1 mm^3 for tensile and bending tests, respectively. The hot press was conducted at 180 ºC for 5 and 10 min for melting and pressing, respectively, at 50 MPa. Obtained film and sheet samples were cut to prepare the tensile (40 × 40 × 0.3 mm^3) and bending (40 × 40 × 1 mm^3) tests specimens, respectively.

Tensile and bending tests were conducted using an IMC-18E0 model machine (Imoto Machinery Co. Ltd, Kyoto, Japan) at a rate of 5 mm min^{-1} at 23 ºC. Five replicates were employed for both mechanical tests.

3. Results and Discussion

3.1 Preparation of PDA-coated CNF

Figure 1 shows the color change of CNF after the oxidative polymerization of dopamine. The surface of CNF became gradually black, which suggests the deposition or coating of PDA on CNF surface.

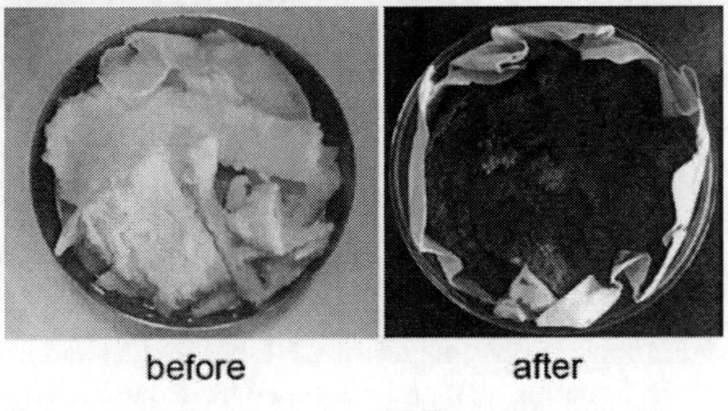

Figure 1. Color change of CNF before and after PDA-coating.

IOP Conf. Series: Materials Science and Engineering **368** (2018) 012043 doi:10.1088/1757-899X/368/1/012043

So far, a serious problem on the nanocomposite preparation was the re-aggregation of CNF to microscale fibers during the dispersion in polymer matrix. In order to confirm the re-aggregation of CNF during the oxidative polymerization, SEM observation of the treated CNF was conducted. Results are shown in figure 2. The diameter of original CNF (20-50 nm) was maintained in the same size of 20-50 nm after the PDA-coating, while reference sample, which was treated in the same manner without dopamine, was observed to increase in the diameter by around 100 nm. These results suggest that the treatment with dopamine suppressed the CNF re-aggregation during the oxidative polymerization.

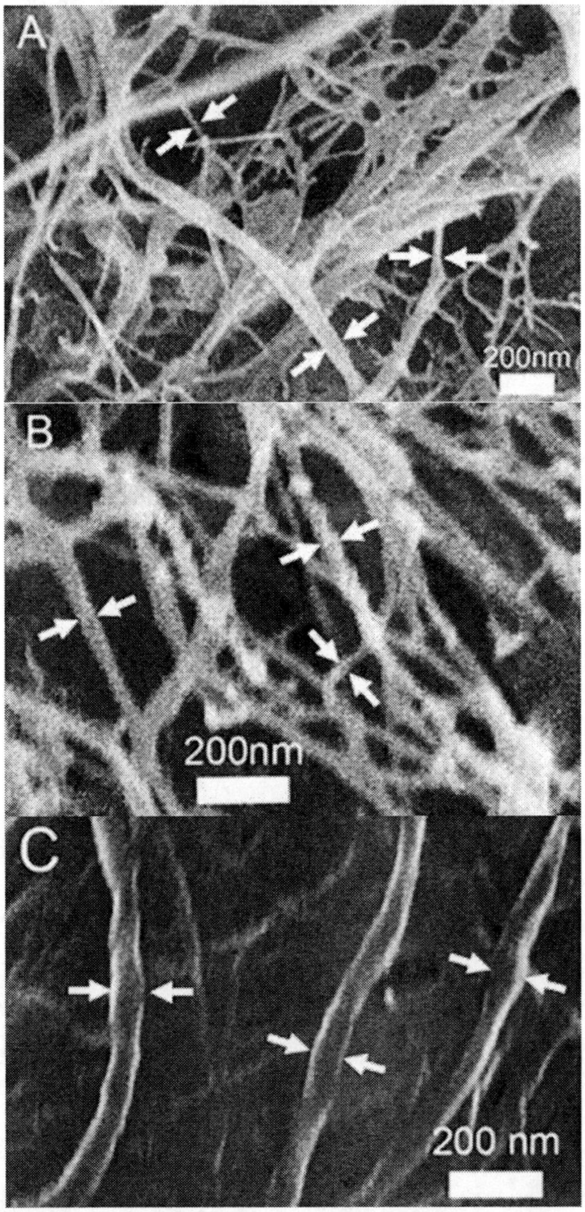

Figure 2. SEM images of CNF before (A) and after PDA-coating (B), and reference CNF sample (C). Arrows show typical diameters of fibers.

3.2 Confirmation of PDA-coating on CNF surface

To confirm the progress of oxidative polymerization on the CNF surface, the treated CNF surfaces were analyzed with FTIR, TG/DTG, SEM-EDS, and XPS.

Changes in FTIR spectrum of CNF before and after the oxidative polymerization of dopamine are shown in figure 3. The dopamine-treated CNF showed new absorption peaks at 1610 and 1540-1550 cm^{-1} assigned to $\nu_{C=O,quinone}$ [14] and $\nu_{C=C,\ aromatic\ ring}$ [15], which are specific absorption bands derived from catechol group. These absorption peaks were detected from the black product prepared from dopamine only as the reference without CNF, which are PDA [16]. The referential PDA showed an additional weak peak around 1500 cm^{-1} assignable to $\delta_{N\text{-}H}$ of secondary amine [8]. These results reveal that the black coating material appeared on CNF surface is PDA as the oxidative polymerization of dopamine.

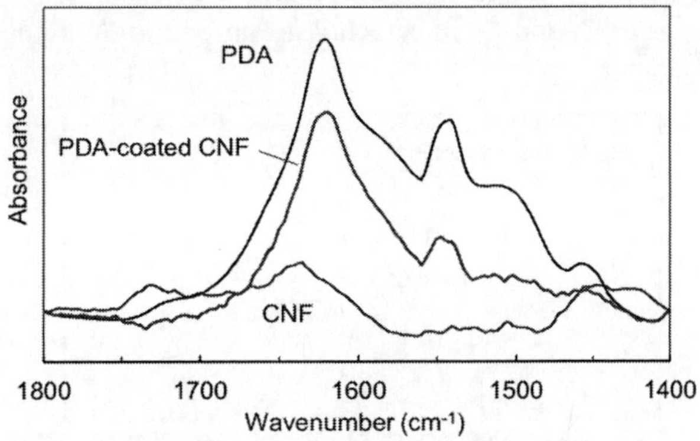

Figure 3. FTIR spectra of CNF surfaces before and after PDA-coating.

In order to determine the CNF-surface coating by PDA, SEM-EDS analysis was employed. Figure 4 show the SEM-EDS images of CNF before and after the dopamine treatment. Morphologies of both SEM images are nearly the same, however, the N-*K*α X-ray mapping images were obviously different between the treated CNF and intact CNF. The treated CNF surfaces were covered by N-element, meaning the PDA coating of CNF surface.

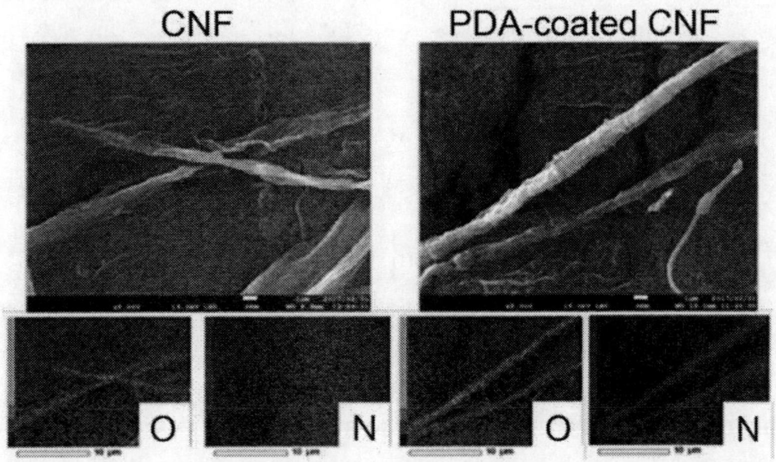

Figure 4. SEM-EDS images of CNF before and after PDA-coating.

Figure 5 depicts the N1s XPS core-level spectra of CNF surfaces before and after PDA-coating. Both the PDA-coated CNF and the referential PDA indicated broad peaks in a wide binding energy range of 395-402 eV. Interestingly, the PDA-coated CNF indicated a main peak around 398 eV and a minor peak around 400 eV, while the referential PDA appeared the main and minor peaks oppositely. In previous reports, these peaks around 398 and 400 eV have been assigned to tertiary/aromatic amine (C=N–R) and secondary amine (R–NH–R) functionalities [15,17]. These XPS results clearly showed the oxidation reaction of amino group to form N-C bonds (scheme 1). Above FTIR results also indicated the difference in the additional absorption peak around 1500 cm^{-1} assignable to secondary amine (R–NH–R) functionality of the referential PDA.

These XPS, SEM-EDS, and FTIR results indicate that the oxidative polymerization of dopamine progresses according to an expected reaction mechanism shown in scheme 1. Here, it is considered that from the XPS and FTIR results CNF affects the mechanism to promote the formation of tertiary/aromatic amine (C=N–R), suggesting the relative increase in reactivity of C_6-H bond compared to hydroxyl groups at C_1 and C_2 in catechol group due to the interaction between cellulose units and thereby steric hindrance.

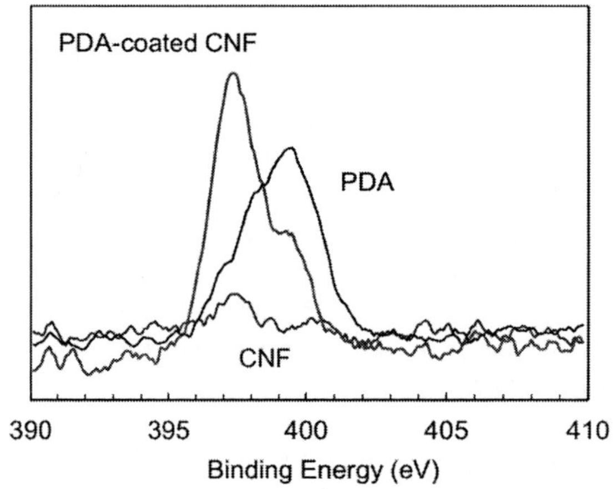

Figure 5. N1s XPS core-level spectra of CNF before and after PDA-coating.

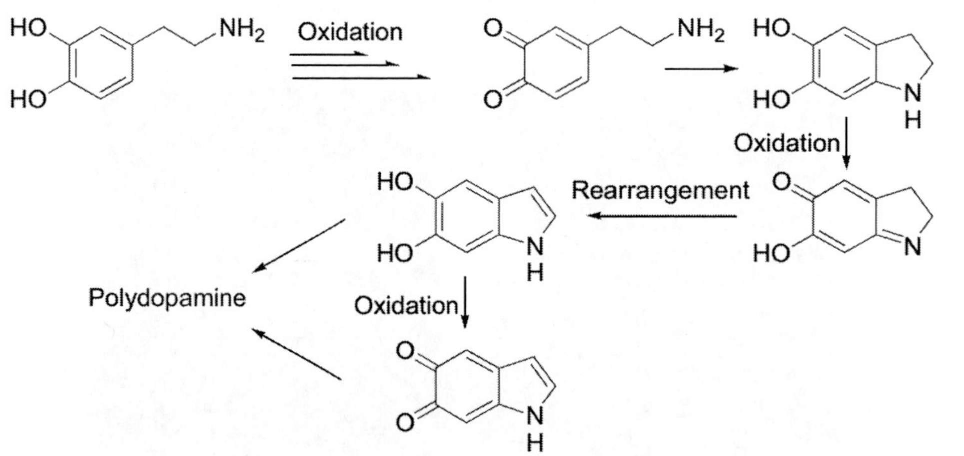

Scheme 1. Expected mechanism of oxidative polymerization of dopamine.

Deposition amount of PDA on CNF surfaces was calculated from a calibration curve (figure A1 in Appendices). As results, the amounts of deposited PDA was 2 and 10 wt% for products after 24 and

48 h of the oxidative polymerization, respectively, resulting in a significant increase in deposition after 48 h compared to 24 h.

In figure 6, TG and DTG profiles of samples before and after PDA-coating are shown. TG/DTG profiles of CNF showed a one-step degradation process in a temperature range of 280-380 °C, while PDA degraded in a three-step process in a wide temperature range of 130-470 °C. The PDA-coated CNF showed an intermediate broad one-step-like degradation process, and, interestingly, the large amount of char was produced in comparison with CNF and PDA. This indicates that the PDA-coating layer makes interact with CNF surface, resulting in changes in the thermal degradation behaviors. Previously, it has been reported that the interactions between the PDA layer and the various polymer substrates include covalent and noncovalent interactions such as the hydrogen bonding interaction, π-π interaction, and electrostatic interaction [9].

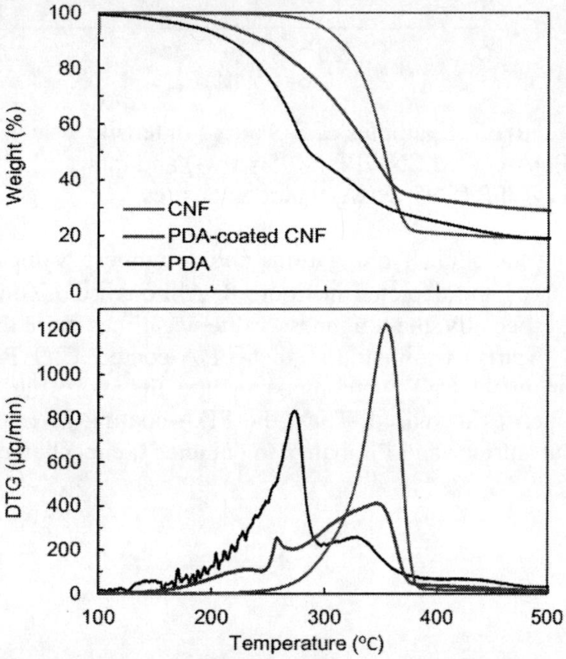

Figure 6. TG and DTG profiles of CNF before and after PDA-coating.

3.3 Mechanical properties of PDA-coated CNF/PP composites

In general, typical effect of CNF is reinforcement of composites. Thus, to evaluate the reinforcing effect the PDA-coated CNF was blended with PP in a composition ratio of PDA-coated CNF: PP = 5:95 (wt/wt), in which the amounts of PDA-coating were 2 and 10 wt% vs. CNF. Tensile tests of obtained PDA-coated CNF/PP and intact CNF:PP (5:95 wt/wt) composite films were carried out. Resulting strain-stress (S-S) curves are shown in figure 7.

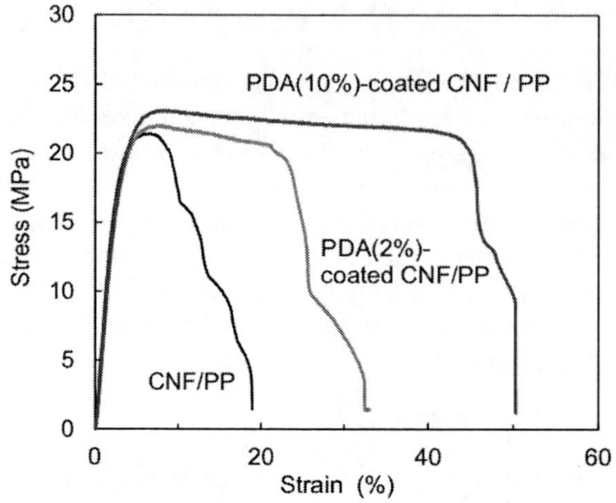

Figure 7. Examples of S-S curve in tensile tests of PDA-coated CNF/PP (5/95 wt/wt) and intact CNF/PP (5/95 wt/wt) nanocomposites.

Tensile strength, Young's modulus, elongation, and toughness values of the composites were calculated from the S-S curves and depicted in figure 8. Although the Young's modulus values were nearly equal, other values, especially, the toughness value was times more than these of intact CNF/PP composite. These results clearly demonstrated that PDA-coated CNF/PP composites had higher mechanical properties than intact CNF/PP composite, and the more the PDA-coating amount, the higher the mechanical properties increased. Thus, the PDA-coating layer effectively functioned as a compatibilizer between CNF surface and PP matrix to enhance the mechanical performance.

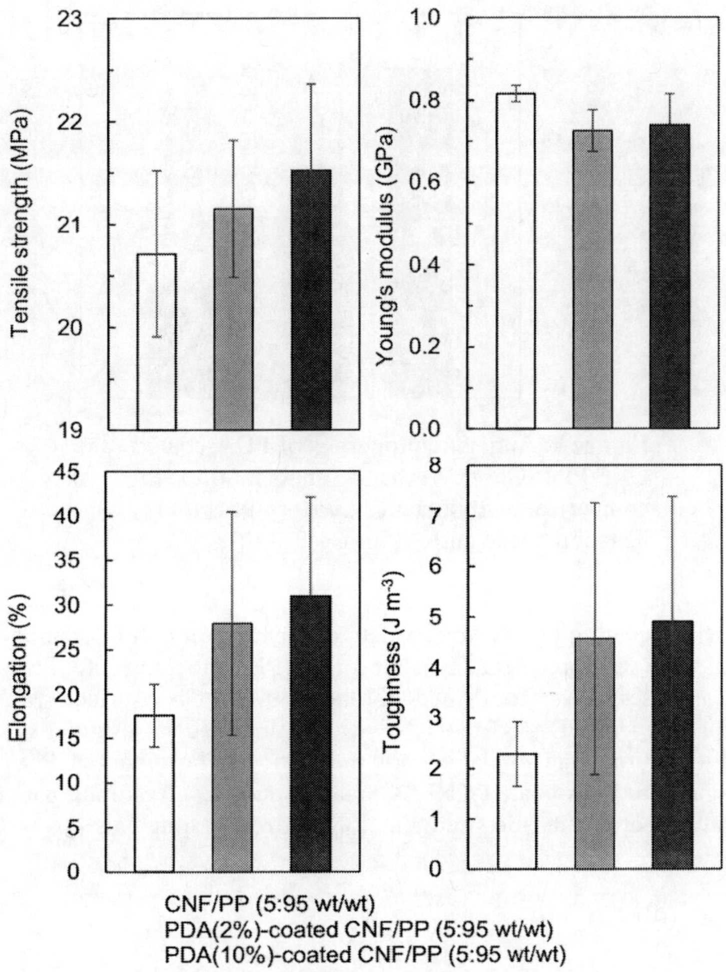

Figure 8. Mechanical properties of PDA-coated CNF/PP (5/95 wt/wt) and intact CNF/PP (5/95 wt/wt) nanocomposites.

3.4 Antistatic properties of PDA-coated CNF

The PDA-coated CNF not only acts as a reinforcement of composites, but also is expected to function as an effective antistatic agent of hydrophobic plastics, because of the conjugated structure of PDA. In order to evaluate the function as an antistatic agent, PDA-coated CNF: PP = 5:95 (wt/wt) composite films were evaluated in comparison with PP and intact CNF/PP = 5:95 (wt/wt) composite films. Obtained surface resistivity values are illustrated in figure 9, showing the lowest resistivity values compared to PP and intact CNF/PP composite films. Thus, it was revealed that the PDA-coated CNF is able to act as the excellent antistatic agent.

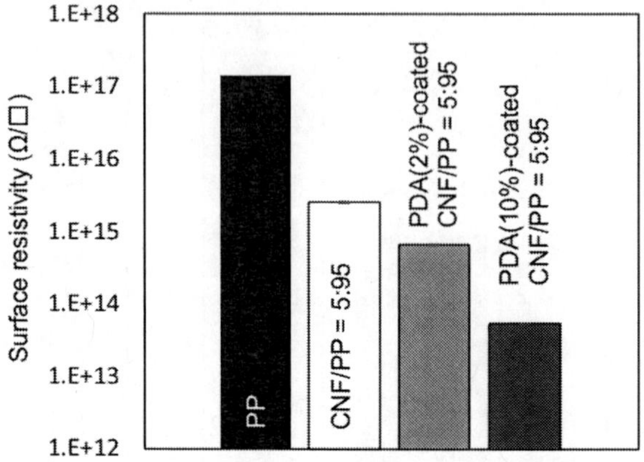

Figure 9. Antistatic properties of PDA-coated CNF/PP (5/95 wt/wt) nanocomposite films in comparison with intact CNF/PP (5/95 wt/wt) nanocomposite and PP films.

4. Conclusion

In this study, the surface coating of CNF by oxidative polymerization of dopamine was carried out to develop a functional CNF reinforcement for natural fiber/PP nanocomposite. From FTIR, SEM-EDS, XPS, and TG/DTG analyses, it was confirmed that the modified CNF surface was successfully coated by PDA. The morphology of PDA-coated CNF was maintained its original diameter in nano-size without re-aggregation during the oxidative polymerization. The obtained PDA-coated CNF were blended with PP to prepare PDA-coated CNF/PP nanocomposites. Resulting nanocomposites showed excellent mechanical properties: tensile strength, elongation, and toughness, as well as good antistatic properties.

5. Appendices

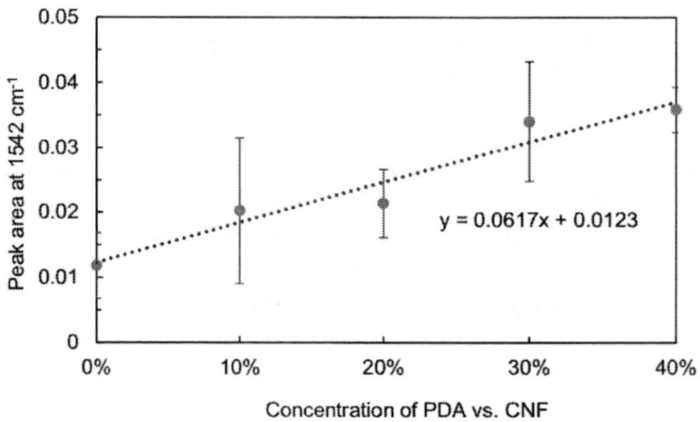

Figure A1. Calibration curve of PDA-coating on CNF surface based on PDA-specific absorption peak at 1542 cm^{-1}.

.

References

[1] Thakur V K, Thakur M K and Gupta R K 2014 Raw natural fiber–based polymer composites *Int. J. Polym. Anal. Ch.* **19** 256-71

[2] Thakur V K and Thakur M K 2014 Processing and characterization of natural cellulose fibers/thermoset polymer composites *Carbohyd. Polym.* **109** 102-17

[3] Rao H R, Rajulu A V, Reddy G R and Reddy K H 2011 Tensile properties and SEM analysis of bamboo and glass fiber reinforced epoxy hybrid composites *Int. J. Appl. Eng. Res.* **6** 27-32
[4] Takagi H and Ichihara Y 2004 Effect of fiber length on mechanical properties of green composites using a starch-based resin and short bamboo fibers *JSME Int. J. Ser.* **A 47** 551-5
[5] Bonse B C, Caranti L A, Magnabosco R, Delijaicov S, Josefovich M P P M and Da Costa R A 2011 Effect of fiber size, compatibilizer and bamboo fiber content on physical properties of PP-g-MA compatibilized polypropylene/bamboo fiber composites in *Proc. Polym. Processing Soc. 27th Annual Meeting PPS-27* (Marrakech, Morocco) P. 1939
[6] Chirayil C J, Mathew L and Thomas S 2014 Review of recent research in nano cellulose preparation from different lignocellulosic fibers *Rev. Adv. Mater. Sci.* **37** 20-8
[7] Yamashiro K and Nishida H 2015 Structural and compositional changes of bamboo fibers during super-heated steam treatment and thereby composite preparation *Int. J. Biomass & Renewable* **4** 8-16
[8] Jiang J, Zhu L, Zhu L, Zhu B and Xu Y 2011 Surface characteristics of a self-polymerized dopamine coating deposited on hydrophobic polymer films *Langmuir* **27** 14180–7
[9] Lee H, Scherer N F and Messersmith P B 2006 Single-molecule mechanics of mussel adhesion *Proc. Natl. Acad. Sci. U.S.A.* **103** 12999–3003
[10] Lee H, Dellatore S M, Mille W M and Messersmith P B 2007 Mussel-inspired surface chemistry for multifunctional coating *Science* **318** 426–30
[11] Xi Z Y, Xu Y Y, Zhu L P, Wang Y and Zhu B K 2009 A facile method of surface modification for hydrophobic polymer membranes based on the adhesive of poly(DOPA) and poly(dopamine) *J. Membr. Sci.* **327** 244–53
[12] Li Y, Wang B, Sui X, Xu H, Zhang L, Zhong Y and Mao Z 2017 Facile synthesis of microfibrillated cellulose/organosilicon/polydopamine composite sponges with flame retardant properties *Cellulose* **24** 3815–23
[13] Han Y, Wu X, Zhang X, Zhou Z and Lu C 2016 Dual functional biocomposites based on polydopamine modified cellulose nanocrystal for Fe^{3+}-pollutant detecting and autoblocking *ACS Sustainable Chem. Eng.* **4** 5667–73
[14] Leng C, Liu Y, Jenkins C, Meredith H, Wilker J J and Chen Z 2013 Interfacial structure of a DOPA-inspired adhesive polymer studied by sum frequency generation vibrational spectroscopy *Langmuir* **29** 6659–64
[15] Zangmeister R A, Morris T A and Tarlov M J 2013 Characterization of polydopamine thin films deposited at short times by autoxidation of dopamine *Langmuir* **29** 8619-28
[16] Dreyer D R, Miller D J, Freeman B D, Paul D R and Bielawski C W 2012 Elucidating the structure of poly(dopamine) *Langmuir* **28** 6428–35
[17] Ding Y, Weng L-T, Yang M, Yang Z, Lu X, Huang N and Leng Y 2014 Insights into the aggregation/deposition and structure of a polydopamine film *Langmuir* **30** 12258–69

IOP Conf. Series: Materials Science and Engineering **368** (2018) 012044 doi:10.1088/1757-899X/368/1/012044

Effect of chemical pretreatment on pulp and paper characteristics of bamboo *gigantochloa scorthechinii* kraft fibers

Z M A Ainun[1*], K I Muhammad[1], H Rasmina[1], H A Hazwani[2], A Sharmiza[3], A K Naziratulasikin[1] and J Latifah[3]

[1]Pulp and Paper & Pollution Control Program, Laboratory of Biopolymer and Derivatives, Institute of Tropical Forestry and Forest Products (INTROP), Universiti Putra Malaysia, 43400 UPM Serdang, Selangor, Malaysia.

[2]Malaysian Timber Industry Board (MTIB), Menara PGRM, 56100 Cheras, Kuala Lumpur, Malaysia.

[3]Forest Products Division, Forest Research Institute Malaysia (FRIM), 52109 Kepong, Selangor, Malaysia.

*Corresponding author: ainunzuriyati@upm.edu.my

Abstract. In chemical pulping process, lignin is targeted to be dissolved and eliminated as much as possible in order to obtain better fiber separation. A primary study was carried out to investigate the effect of sodium hydroxide pretreatment prior to bamboo chemical pulping. The selected bamboo species was *Gigantochloa scortechinii* which is a well-known bamboo in Malaysia. The bamboo chips were soaked in 2 % sodium hydroxide solution for 6 hours at 60 °C. These treated chips were cooked at 14 % and 18 % of alkali active via kraft pulping process. The unbleached pulp was obtained and made into paper. The pulp and paper were tested according to pulp yield, pulp freeness, paper optical, paper physical and paper mechanical properties. The results exhibited that by using treated bamboo chips, the value of pulp yield, freeness and paper brightness increasing. Slight decrement was observed for mechanical paper properties such as tensile, tear and burst strength. Chemical pretreatment prior to pulping process was investigated to offer significant effects to the pulp and paper properties of bamboo *Gigantochloa scortechinii*. Hence, mild kraft pulping can be carried out to produce bamboo unbleached pulps. In addition, dependency on bleaching sequences can also be declined.

1. Introduction

Various chemical pretreatment methods have been developed and introduced such as acid, alkaline, lime, steam explosion and ionic liquid in order to pretreat lignocellulosic biomass. Lignin and hemicellulose can be partially removed by conducting pretreatment process. In addition, pretreatment also may decrease cellulose crystallinity and increase porosity of accessible surface area of biomass [1-2]. Pretreatment methods can be categorized as physical, physic-chemical, chemical, biological, electrical or combination [3-4]. Among these pretreatment methods, chemical pretreatment was examined and verified to be a promising one [5]. Therefore, it has becoming a promising method in order to remove lignin and/or hemicellulose, decrease degree of polymerization and reduce crystallinity of cellulosic components [6]. Hence, pretreatment process has been widely applied in pulp and paper industry specifically on the delignification of cellulosic materials [7].

The purpose of chemical pulping is to dissolve as much as possible lignin which binds the cellulose fibers. Hence, better fiber separation in terms of quantity or quality can be obtained. In pulp and paper technology, the choice of pulping process is identified based on final product, type of fiber and economic factors as well. Generally, there are 4 processes usually being used in pulping process, kraft, soda, sulfite and neutral sulfite semi-chemical process. Kraft pulping process is a process carried out in an elevated temperature and pressure in a solution of sodium sulfide and sodium hydroxide which dissolves lignin that contains in lignocellulosic materials. In producing chemical pulp of bamboo, kraft pulping is generally chosen compared to soda pulping [8]. The process offers acceptable delignification in terms of yield and viscosity. In addition, the reason kraft is suitable to bamboo is because of its fiber dimensions and chemical content resembles near to woods [8]. The kraft process provides satisfactory delignification as well as high yield and viscosity, but it also needs more bleaching agent due to darker pulp produced. The cooking condition is advised to be carried out at 16 % active alkali, 22 % sulfidity, 2.5 hours of cooking time at maximum 165 °C and liquor to wood ratio of 1:4.5 [9]. Thus, it is important to apply or combine several methods in order to reduce lignin content in biomass more economically and produce better quality of pulp [10].

Cell wall of plant is composed of three components namely as cellulose, lignin and hemicellulose. The most complexity structure and difficult to degrade is lignin. This is due to its high molecular weight and insolubility [11]. Hence, in the process of enhancing hydrolysis of cellulose, pretreatment of lignocellulose materials can be conducted in order to remove lignin and also hemicellulose content [12]. The pulp and paper miller applied both mechanical and chemical pulping processes to produce the desired pulps. Mechanical pulping process uses intensive electrical energy and produce paper of higher colour reversion rate due to the high content of lignin while chemical pulping needs excessive chemicals to degrade and dissolve the lignin which at the end leaving high strength cellulose fibers from cell wall. Nevertheless, chemical pulping produces lower yield [13].

Bamboo is a type of grass plant and most used as non-wood raw materials in the manufacturing of food, furniture, handicraft, construction, paper, board, textile, biomass energy [14], reinforcement fiber and more. In comparison to wood, bamboo may produce cellulose 2-6 times more [15] while its biomass increased by 10-30 % daily compared to wood biomass which increased by 2.5 % daily. Bamboo can be harvested after 4 years but wood trees need to wait 8-20 years before harvesting [15]. Bamboo species named *Gigantochloa scortechinii* which is well-known bamboo in Malaysia were used throughout this study.

Based on literature review, chemical pretreatment was introduced to woody materials. Looking at this matter, therefore a study has been conducted in order to determine whether the chemical pretreatment gives remarkable effects to the pulping and papermaking of bamboo.

2. Methodology

2.1 Materials

The bamboo stalks were obtained from Bentong, Pahang. These stalks were splitted into six strips prior to chipping process using Mini Chipper. These chips were then screened using Chip Classifier to a size of 2-3 cm in terms of width and length. The accepted size of bamboo chips was dried until 10 % moisture content in order to avoid any fungal attack. The chemicals used throughout the study were natrium hydroxide and sodium sulfite which were purchased from R&M Chemicals.

2.2 Pre-treatment of bamboo chips, kraft pulping and pulp testing

The bamboo chips were soaked in 2% sodium hydroxide solution for 6 hr at 60 °C. The pre-treated bamboo chips were then washed in order to remove soapy substance of natrium hydroxide. The pre-treated bamboo chips were placed into pulp digester and cooked for 120 min at 160 °C. The selected sulfidity was 25 % while the alkali active was at 14 % and 18 %. Two parameters of active alkali were selected in order to investigate the effect of alkali active concentration on pulp and paper properties of untreated and treated bamboo chips. Once cooking accomplished, the cooked pulp was washed and disintegrated using a hydropulper. The pulp at this stage is called unbleached bamboo pulp. The unbleached pulp was then screened to separate any unwanted fibers in terms of length by using Somerville Screener. Finally, the accepted and rejected yields were calculated.

2.3 Beating of unbleached bamboo pulp and papermaking process
The beating was conducted using a PFI Mill which involved 24 g oven dried of unbleached pulp each time proceeding beating process. Before that, the pulp was dispersed in 2 L of distilled water. The dispersion was done in a British disintegrator until 3,000 rev for about 25 min. The dispersed pulp was later placed into the PFI Mill until 5000 rev. After completing the beating process, the beaten pulp was diluted in a pulp Stock Divider. The following steps were done according to TAPPI Standard T 205.

2.4 Paper Testing
Paper testing consists of physical, optical and mechanical paper properties. The physical paper test involved paper grammage and thickness. The optical paper properties comprise of paper opacity and brightness while mechanical paper properties include tensile, burst, tear and folding endurance. Standard methods referred throughout the experiments are listed as in Table 1.

Table 1. The standards applied for physical, optical and mechanical paper properties.

Paper properties	Name of test	Standard Method
Physical	Grammage	ISO 536-2012
	Thickness	ISO 534-1988
Optical	Opacity	T 425 om-01
	Brightness	T 425 om-02
Mechanical	Tensile	T 494 om-01
	Burst	T 403 om-97
	Tear	T 414 om-98
	Folding endurance	T 511 om-02

3. Results and discussion

3.1 Effect of chemical concentrations on pulp properties
The accepted pulp yield of chemical pulping process usually achieved about 40 to 50 %. This is different compared to mechanical pulping which pulp yield may reach up to 80 to 90 %. The results in Table 2 exhibited that by treating the bamboo chips prior to pulping resulted in significant affects to the accepted pulp yield. The pulp yield was improved for both 14% and 18% sodium hydroxide concentrations of 12.39 % and 31.05 % increment respectively. These can be allied to the function of sodium hydroxide which swollen the fibers directed to easier refining process. Due to this condition, the internal surface of cellulose will increase. The degree of polymerization and crystallinity of bamboo polysaccharides may decrease too [6]. These phenomena accumulated and promoted the cleavage of structural linkages between lignin and carbohydrate thus causes simultaneous disruption of lignin structure [16]. However, the pulp freeness showed increasing values which are expected to be lower than the freeness of untreated samples. In fact, the pulp freeness for 18% samples had no changes. This condition can be explained by referring to the condition of pulps. The pulps might be swollen but have high possibilities of flocking together which made these pulp flocs to pass through the refining part. On the other hand, this condition occurred might be due to high dose of chemicals had been introduced to pulps either during pretreatment or pulping processes. The ability of holding water by the pulp fibers are getting weaker which more towards drainage the water out from pulp slurry systems.

Table 2. The pulp yields and pulp freeness for untreated and treated samples of 14 % and 18 % soda pulping concentration.

Sodium hydroxide concentration	14 %		18 %	
Type of samples	**Untreated**	**Treated**	**Untreated**	**Treated**
Pulp screened yield, %	41.24	46.35	34.85	45.67
Pulp freeness, ml	410	535	560	560

3.2 Effect of chemical pre-treatment concentrations on paper optical properties

The opacity of the papers produced from both 14 % and 18 % pulping conditions were determined to be decreasing 4.64 % and 0.75 % accordingly. In contrary, the brightness of both treated samples increased 77.60 % and 2.30 % respectively. Huge increment of treated pulps at 14 % pulping condition can be associated to the disappearance of lignin content during pretreatment and pulping processes. Pulping process which involved sodium hydroxide as one of the pulping agent reduces amorphous hydroxyl groups from the fibers. This means that hydroxyl groups present in amorphous region act as hemicellulose and lignin were removed [17]. The decrement of hemicellulose and lignin may help to reduce the dependency on bleaching process in order to brightening the pulps for various products.

3.3 Effect of chemical pre-treatment concentrations on paper physical properties

Based on Table 3, grammage for 4 samples ranged from 67.34 g/m^2 to 69.96 g/m^2. These values showed weight of pulps or fibers that bonding together on 1 square meter of paper. The treated paper of 14% pulping condition provided highest grammage value compared to other samples. This result can be connected with thickness value which same sample showed highest value of thickness. Both thickness and grammage for paper produced from treated chips and undergone 14% pulping condition were recorded 190.43 μm and 69.96 g/m^2 accordingly.

3.4 Effect of chemical pre-treatment concentrations on paper mechanical properties

Based on mechanical properties showed in Table 3, both treated samples exhibited slight differences of result between each other. At the same time, both treated samples declined from untreated samples for tensile, burst and tear indices. The paper folding property presented extreme drop 53.68 % and 45.65 % for treated samples of 14 % and 18 % pulping condition respectively. This revealed that pretreatment was trivially done to the bamboo chips.

Table 3. The paper properties for untreated and treated samples of 14 % and 18 % soda pulping concentration.

Pulping concentration	14 %		18 %	
Type of samples	**Untreated**	**Treated**	**Untreated**	**Treated**
Physical				
Grammage, g/m^2	67.34	69.96	68.35	67.74
Thickness, μm	164.11	190.43	172.13	167.25
Optical				
Opacity, %	97.66	93.13	93.03	92.33
Brightness, %	22.23	39.48	37.73	38.60
Mechanical				
Tensile index, kN/m,	27.56	25.43	27.50	27.48
Burst index, kPa.m2/g	2.95	2.78	2.76	2.80
Tear index, kN/m	2.63	2.59	2.77	2.60
Folding endurance	136	63	138	75

According to Lehto and Alen [16], kraft pulping involves three stages namely as extraction phase, bulk delignification stage and residual delignification. To relate, the bamboo chips was most probably have been over treated which involves all the three stages of pulping. The last stage that is called residual delignification proceeding delignification slowly and loss of carbohydrate is increasing. It is estimated that prolong pretreatment with sodium hydroxide and proceeding with kraft pulping accelerated the dissolution of small amount of carbohydrate thus resulted in the slight decrement of paper strength. In other words, the pretreatment is entailed to break down the lignin structure and disrupt the crystalline structure of cellulose [17]. Therefore, the chemicals of pulping can easily access the cellulose to hydrolyze into monomers [18]. These conditions influenced the performance of fiber bonding that is proportionally relationship to paper strength. Looking at this matter, the fixed time for pretreatment should be revised to be less than 6 hours while the concentrations of sodium hydroxide as low as 1 or 1.5 % only.

4. Conclusions

Chemical pretreatment is found out to be one of the methods which can be introduced prior to pulping process. Based on the findings, it affected pulp and paper properties in all aspects. The point that can be highlighted was the increment of pulp brightness that was recorded as 77.60%. This showed that pretreatment method can be applied to reduce the dependency of bleaching sequences. In order to obtain better performance of pulp and paper properties, time of pretreatment and concentration of sodium hydroxide during pretreatment process should be perused and adjusted lesser.

Acknowledgements

The authors would like to thank Laboratory of Pulp and Paper, Institute of Tropical Forestry and Forest Products (INTROP) and Forest Research Institute Malaysia (FRIM) for their technical advice and guidance.

References

[1] Kumar P, Barrett D M, Delwiche M J, Stroeve P. 2009 Methods for pretreatment of lignocellulosic biomass for efficient hydrolysis and biofuel production. *Ind. Eng Chem Res.* 48:3713-29.

[2] Hongzhang C, Junying Z. 2013 Clean production technology of integrated pre-treatment of lignocelluloses. *Afr J Agric Res.* 8:339-348.

[3] Alvira P E, Tomas-PejoE, Ballesteros M, Negro M J. 2010 Pretreatment technologies for an efficient bioethanol production process based on enzymatic hydrolysis: a review. *Bioresource Technol.* 101: 4851-61.

[4] Chiaramonti D, Prussi M, Ferrero S, Oriani I, Ottonello P, Torre P, Cherchi F. 2012 Review of pretreatment processes for lignocellulosic ethanol production, and development of an innovative method. *Biomass Bioenergy.* 46: 25-35.

[5] Yang B, Wymen C E. 2008 Pretreatment: the key to unlocking low-cost cellulosic ethanol. *Biofuels Bioprod Bioref.* 2: 26-40.

[6] Mtui Y S. 2009 Recent advances in pretreatment of lignocellulose wastes and production of value added products. Afr J Biotechnol. 8: 1398-415.

[7] Zheng Y, Pan Z, Zhang R. 2009 Overview of biomass pretreatment for cellulosic ethanol production. *Int J Agric Biol Eng.* 2:51-68.

[8] Vu T.H.M., Pakkanen H, Alen R 2004 Delignification of bamboo (Bambusa procera acher) Part 1. Kraft pulping and the subsequent oxygen delignification to pulp with a low kappa number *Indust. Crops and Prod.* 19:49-50.

[9] Krisdianto, Sumarni G, Ismanto A 2007 Sari Hasil Penelitian Bambu. [11 Desember 2008]

[10] Ermawar R A, Yanto D H Y, Fitria, Hermiati E 2006 Lignin degradation content in rice straw pre-treated by white-rot fungi *J Widya Riset* 9(3):197-202

[11] Perez J, Dorado JM, De la Rubia T, Martinez J 2002 Biodegradation and biological treatments of cellulose, hemicellulose and lignin: an overview *Int. Microbiol.* 5:53-63

[12] Sun Y & Cheng J 2002 Hydrolysis of lignocellulosic materials for ethanol production: a review *Bioresource Technol.* 83:1-11

[13] Akhtar M, Scott G M, Swaney R E, Kirk T K. 1998 Overview of Biomechanichal and Biochemichal Pulping Research In: Eriksson K, Paulo ECA, editors. Enzyme Applications in Fiber Processing. Washington DC: American Chemical Society

[14] Kobayashi F, Take H, Asada C, Nakamura Y 2004 Methane production from steam-exploded bamboo *J Biosci. and Bioeng* 97:426-428

[15] Herliyana E N, Noverita, Lisdar I S 2005 Fungi pada bambu kuning (Bambusa vulgaris schard var. vitata) dan bambu hijau (Bambusa vulgaris schard var vulgaris) serta tingkat degradasi yang diakibatkannnya *J Teknol. Hasil Hutan* 18:2-10.

[16] Lehto J T & Alén R J 2015 Chemical pretreatments of wood chips prior to alkaline pulping - A review of pretreatment alternatives, chemical aspects of the resulting liquors, and pulping outcomes *BioResources*, 10(4), 8604–8656

[17] Kabir M M, Wang H, Lau K. T & Cardona F 2013 Effects of chemical treatments on hemp fibre structure *Applied Surface Science*, 276, 13–23.

[18] Harmsen P, Huijgen W, Bermudez-Lopez L, Bekker R 2010. Literature review of physical and chemical pretreatment processes for lignocellulosics biomass. 1st ed. Wegeningen: Food & Biobased Research.

[19] Brodeur G, Yau E, Badal K, Collier J, Ramachandran K B, Ramakrishnan R 2011 Chemical and physiocochemical pretreatment of lignocellulosic biomass: a review. *Enz Res* 2011.

IOP Conf. Series: Materials Science and Engineering **368** (2018) 012045 doi:10.1088/1757-899X/368/1/012045

Tensile shear strength of natural wood decorative plywood from lesser known Sungkai (*Peronema canescens*) tree

N A Farizan [1, 2], Catherine B[1], and M Hamami Sahri[1]

[1]Forestry Complex, Faculty of Science and Natural Resource, Universiti Malaysia Sabah, 88450 Kota Kinabalu, Sabah, Malaysia

[2]Civil Engineering Department, Politeknik Kota Kinabalu, Jalan politeknik, 88450 Kota Kinabalu, Sabah, Malaysia

E-mail: opnike79ers@gmail.com

Abstract. Sungkai (*Peronema canescens*) small logs has been harvested from a private sungkai plantation (Sabahpuri Nursery Sdn Bhd) in Telipok, Sabah. The logs were then rotary peeled to 0.6mm thick and made into 4' x 4', 3 plies Urea Formaldehyde (UF) plywood with Acacia Hybrid wood used as the core veneer (1.2 mm thick) at a local plywood facility; Cymao Plywood Sdn Bhd in Sandakan. Remaining dried veneer were brought back to University Malaysia Sabah for further study. Laboratory scale 3 plies plywood (Sungkai-Acacia, Sungkai-Seraya) were made using the same source of UF resin and veneer, and used for comparison studies. Commercial 3 plies Nyatoh plywood were also used as control samples. The samples were then tested according to Japanese Agricultural Standard for Plywood 2003 (amended: 2008) *Tensile Shear Strength of Type II Natural Wood Decorative Plywood*. All of the samples pass the Tensile Shear Strength (TSS) minimum standard of 0.7 N/mm^2 in untreated condition or even after hot and cold water immersion treatment. The untreated mill samples of Sungkai-Acacia plywood perform the best with mean TSS = 1.74 N/mm^2, SE = 0.05 while Nyatoh shows the lowest mean tensile shear strength with 1.06 N/mm^2, SE = 0.07. Further statistical analysis shows there was no significance difference between Sungkai mixed plywood group samples. However, Sungkai mixed plywood samples shows a significant difference ($p < 0.05$) with superior strength when compared to commercial Nyatoh plywood samples. If these criteria met the final requirement for the end use of the veneer, then the potential of Sungkai as an alternative or supplemental raw material is something worth given due consideration.

1. Introduction

Malaysian timber industry is one of the top 3 largest commodity based nation export earner with total exports value amount of RM 21.86 billion in year 2016. Veneer and plywood seconded the largest timber product export list with 23 %. Japan was the largest importer of Malaysian plywood to date with United states comes in second [1].

The competition on Malaysian commercial tropical tree for the use as timber industry raw material has long been acknowledged. The depleting resources from the primary forest to produce high quality veneer for the production of plywood has been the effects from high demand for other uses of timber based products. The nation is in its efforts to be more systematic and finds sustainable way to maintain timber resources. Forest plantation has widely been implemented throughout the country and the searches on suitable tree species for plantation purpose are still continuing until today. There are plenty species that has been planted more than 15 years and already reached maturity in Malaysia. Two dominant species were Rubberwood (*Hevea brasiliensis*) and *Acacia mangium*. Other species that has been considered under the plantation programme includes Teak (*Tectona grandis*); Sentang (*Azadirachta excelsa*); Khaya (*Khaya ivorensis/ Khaya senegalensis*); Kelempayan/Laran (*Neolamarckia cadamba*); Batai (*Paraserianthes falcataria*); Binuang (*Octomeles sumatrana*); and *Eucalyptus spp*.

Looking at the two primary plantation species. *Acacia mangium* species were found unable to produce high quality face veneer. The high volume of knots on its wood prevent it from being selected to be used as face veneer [2]. The current commercial use of mangium species is more towards the production for general purpose plywood and pulp production. Meanwhile Rubber tree has long been acknowledged as the prime source of solid wood for furniture production. More suitable or appropriate plantation tree species has to be studied to address the issue so that it can add more alternatives or supplement the current raw material shortage in the timber industry.

Sungkai tree is one of the lesser known tree species that has yet been fully explored and utilised by the Malaysian timber industry at the current state. The high survival character and the ability to grow far from the usual habitat are the strong trait of Sungkai *(Peronema canescens)* which is similar to other species from the Verbenaceae family. Highly significant differences in height, DBH growth and survival were observed among growth sites and between tree clones [3]. Sungkai tree are very much suitable to be planted for agroforestry and also to rejuvenate low productivity land such as shifting cultivation area and scrubland. A harvest cycle of 15 years is said to be economic [4]. Young leaf observed to have been sprouting out from the dislocated trunk in several days/weak after it has been cut from its stump. This shows a magnificent adaptability and survivability of the wood to grow in severe condition.

Matured Sungkai trees has the ability to survive and endure several wild fire during the dry season and quickly rejuvenate and sprout its leaves when the season back to normal or raining. It is in that condition which the tree adapting to dry and rainy season creates a beautiful ring porous structure within the wood. Sungkai is considered one among a few tropical tree with a distinct growth ring feature.

The wood is cream-coloured with bits of light yellow, light pink or light brown colour. The timber closely resembles teak wood with brighter shades of colour. Because of the bright colour, it sometimes being referred as white teak in the commercial trade. The tree is also known as Jati londo, Jati sabrang, Kurus and Sungkai (Indonesia); and Khoeilai, Sakae and Sangkae (Thailand).

The wood radial and surface texture is attractive and beautiful. Wood texture is moderately fine to moderately coarse and uneven. Grain is straight to wavy. It has a certain character that can be manipulate to be used as face veneer in decorative plywood production and further to be used for the furniture industry. Sungkai wood is moderately hard and medium in weight with a density of 520-730 kg/m^3 air dry [5].

Difference in veneer grade quality from a single species does not significantly affects density, bending strength and elasticity of a plywood [6]. However there are difference in wood characteristic between different species. So there might be a chance where there are differences for plywood properties of combined species. The effect of ageing process, wood species, glue type and drying temperature on bonding strength were said to be statistically significant [7].

The objectives of the study were to determine whether sungkai as face veneer can pass Article 7 in Japanese Agricultural Standard for Plywood minimum tensile shear strength requirement.

IOP Conf. Series: Materials Science and Engineering **368** (2018) 012045 doi:10.1088/1757-899X/368/1/012045

2. Materials and Method

2.1 Materials

The Sungkai wood used in this research was received from Sabahpuri Nursery Sdn Bhd which has an individual plantation plot at Telipok, Kota Kinabalu, Sabah. Raw materials used for the study was 13 years old Sungkai (*Peronema canescens*). 11 small logs from 6 different 13 years old trees with an average length of 137 cm each with average DBH of 22 cm were collected from the plot. It was felled and transferred with the permission of Sabah Forestry Department to the veneer peeling facilities at Cymao Plywood Sdn Bhd mill in Sandakan, Sabah and processed into 0.6 mm veneer. 12 Acacia hybrid small logs were collected from Sabah Forestry Development Authority (SAFODA) plantation plot at Ulu Kukut, kota Belud a day before being transported together with Sungkai logs to Sandakan. Acacia logs were peeled to 1.2 mm thick to be used as core veneer for the 3 plies plywood production process. Meanwhile the actual Cymao Plywood mill production samples of natural forest Seraya 1.2 mm veneer were requested and collected from the manufacturing facility to be used for comparison study. Commercial Nyatoh 3 plies plywood were also acquired as a control sample for the study. UF resin used throughout the study were compliments from Cymao Plywood Sdn Bhd.

2.2 Plywood production.

There were 12 piece Sungkai-Acacia plywood 1200 mm (width) x 1200 mm (length) x 2.4 mm (thick) which were made using the mill facilities. Sungkai 0.6mm veneer was used as face and back meanwhile 1.2 mm acacia veneer were used as the core veneer to produce the samples. At the University, 10 samples of 300mm (width) x 300mm (length) x 2.4 mm (thick) Sungkai-Acacia and Sungkai-Seraya plywoods were made from the same veneer and resin source acquired at Cymao Plywood Sdn Bhd earlier. This time the university facilities are being used to produce the laboratory scale samples for comparison study. The average adhesive consumption was 250 g/m^2 in double glue line. Assembly time was 25 min. The boards were pre-pressed at room temperature for 5 min. The hot press conditions were: Temperature, 135 °C; pressure, 1.2 MPa and time, 5 min.

2.3 Samples preparation

Samples from each group of plywood were cut according to the Japanese Agricultural Standard for Plywood 2003 (amended: 2008) *Tensile Shear Strength of Type II Natural Wood Decorative Plywood.* [8] Each plywood type was divided into two sub group which are the normal untreated group and the 3hr hot water immersion treatment group. At least 30 replicates for each group for every plywood types tested towards the JAS standard requirements.

2.4 Sample testing

Sample testing were done at UMS forestry complex, using Gotech universal testing machine (Model AI-7000 L10). The dimension of each sample shown in Figure 1 are in millimetres (mm). Normal untreated replicates will go directly for testing according to their respective types of plywood. Treated samples will undergo 3-hour water immersion in water bath with 60 ± 3 °C and later on immersed in water of room temperature until the samples cool down before being test for tensile shear strength.

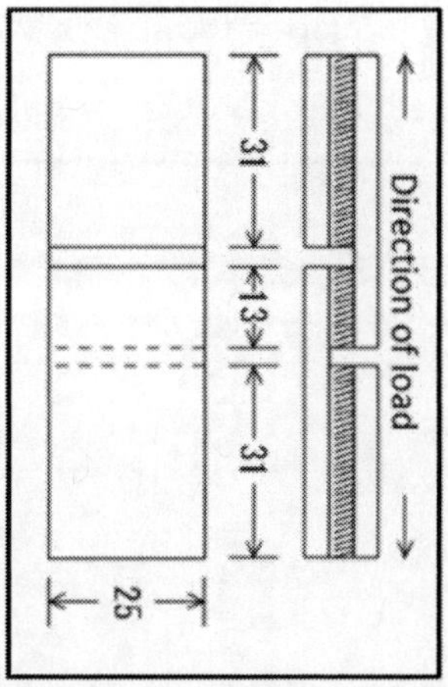

Figure 1. Test specimens dimension (mm)

Figure 2. test specimen grip position

The test specimens will be gripped vertically on the upper and lower part of the test specimens as shown in figure 2. Tensile force of 5mm/min will be loaded according to the direction load shown in figure 1.

3. Results and Discussion

All data collected were recorded and transferred into IBM SPSS 20 Statistical Analysis Program. One-way ANOVA were used followed by Duncan Mean Range Test (DMRT) to check the significance of the mean tensile shear strength statistically.

Table 1. Mean tensile shear strength of sample plywood.

Samples	Mean ± Standard error	
	Normal	3hr Hot Water Immersion
Sungkai Acacia B mill	1.74 ± 0.05^{a}	1.44 ± 0.07^{a}
Sungkai Acacia B lab	1.64 ± 0.09^{a}	1.59 ± 0.09^{a}
Sungkai Seraya B lab	1.59 ± 0.11^{a}	1.56 ± 0.09^{a}
Sungkai Acacia A mill	1.05 ± 0.03^{b}	1.19 ± 0.04^{b}
Nyatoh Commercial	1.06 ± 0.07^{b}	0.53 ± 0.05^{c}

Note: Data is expressed as means ± standard error; the means value is in N/mm2 unit, a different set of samples used for each treatment condition. Values with the same letter are not significantly different after one-way ANOVA followed by the Duncan post hoc test ($p < 0.05$).

As the result shown in Table 1. There was no significance difference between samples made at the mill and samples made at the laboratory for Sungkai-Acacia B and Sungkai-Seraya plywood. Highest mean tensile shear strength value was shown by mill sample Sungkai-Acacia B for normal condition samples with 1.74 ± 0.05 N/mm^2. 3-hours hot water immersion was done as an accelerated stress/ageing to the bonds between veneers. Laboratory sample of Sungkai-Acacia and Sungkai-Seraya plywood showed better mean tensile shear strength of compare to other samples with 1.59 ± 0.09 N/mm^2 and 1.59 ± 0.09 N/mm^2, respectively.

Sungkai-Acacia A mill and shows a significant difference compare to previous Sungkai-Acacia B and Sungkai-seraya B with 1.05 ± 0.03 N/mm^2 in normal condition and 1.19 ± 0.04 N/mm^2 after 3-hour Hot water immersion treatment. This may the result of type A sample has longer length (25mm) in the centre of the test specimens. Nyatoh commercial samples also exhibit no significant difference with Sungkai-Acacia A for normal condition samples but showed a significant difference with lower mean tensile shear strength value of 0.53 ± 0.05 N/mm^2. A further investigation is proposed and new

Type B replicates of Nyatoh Commercial should be done to clarify the reason there is a significant drop on mean tensile shear strength.

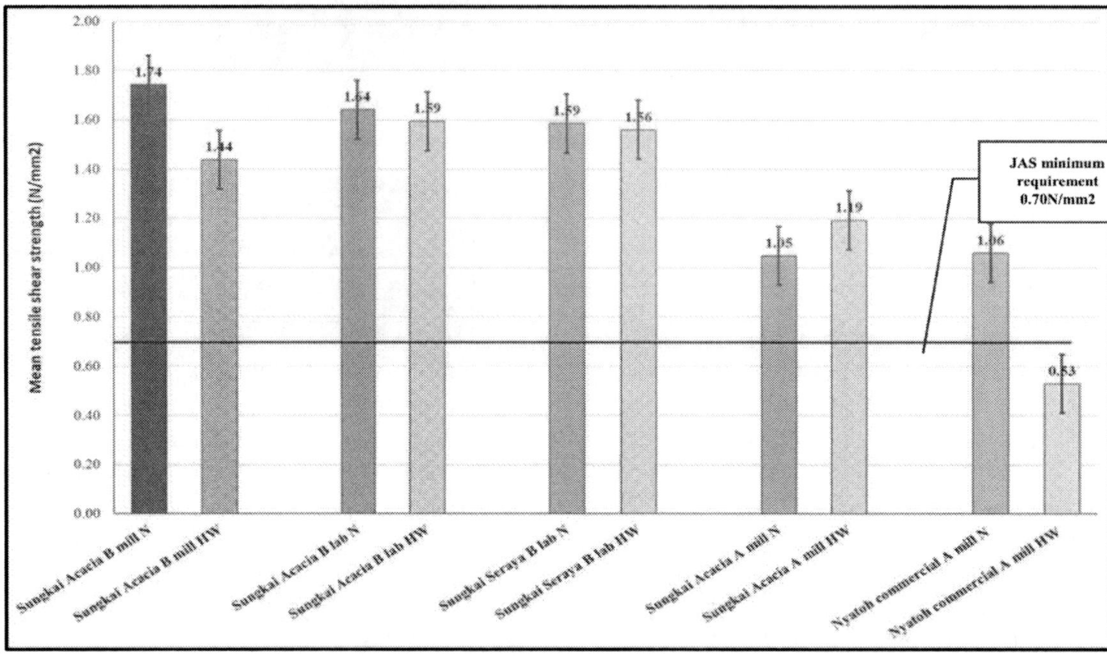

Figure 3. Mean tensile shear strength and standard error.

Overall, normal untreated condition samples show no significant difference when compared with treated 3-hour hot water immersion. It is interesting that Sungkai plywood were able to withstand the Type II accelerated stress/ageing test and come up with mean tensile shear strength with no significant difference from the normal condition samples and surpass the minimum JAS standard requirements of 0.7N/mm2 for tropical hardwood plywood. On top of that Sungkai plywood were also significantly better than the control Nyatoh commercial sample in normal condition and even in treated condition.

4. Conclusion

Based on the experimental results, it can be concluded that sungkai wood veneer offers attractive potential as an alternative or supplement raw material to produce good face veneer. It shows great result when used together with another potential plantation species and also with a current commercial natural forest species. It even surpasses the strength of a sample of current commercial species significantly. This finding should add more knowledge on the wood itself. If these criteria met the final requirement for the end use of the veneer, then the potential of Sungkai as an alternative or supplemental raw material is something worth given due consideration.

Acknowledgement

The author would like to express his appreciation towards the research team and also Sabahpuri Nursery Sdn Bhd, Cymao Plywood Sdn Bhd, Sabah Forestry Development Authority and Sabah Forestry Department for their continuous support and assistance.

References

[1] Anonymous 2015 Malaysia Timber Council Annual Report 2015. Malaysia Timber Council, Kuala Lumpur.

[2] H. Hamdan 2011. Properties of acacia mangium planted in peninsular malaysia itto project on improving utilization and value adding of plantation timbers from sustainable sources in malaysia project no. Pd 306/04(1). Forest Research Institute Malaysia, Kepong, Selangor Darul Ehsan. 69

[3] Kyu-Suk Kang, Mu-Seok Han, In-Sik Kim, Song-Hee Nam 2013 *Plant Breed. Biotech.* **1(3)** 245~252

[4] Wahyudi, A. R Mojiol, Z. Muttaqin 2016 *Borneo Science* **37 (1)** 72

[5] De Graaf, N.R. Hidelbrand, R.R. Van der zwan, and J.M. Fundter 1994. *Peronema canescens Jack. In: Soerianegara, I. and R.M.J. Lemmens (eds.) Plant Resources of South East Asia No. 5 (1);Timber* Tress: Major Commercial Timber. Bogor: Plant Resources of South East Asia (Prosea).

[6] Vassil Jivkov, Ralitsa Simeonova, Assia Marinova 2013 *Innovation in Woodworking Industry and Engineering Design; Science journal* **2** 86

[7] C. Demirkira, G. Colakoglua, S. Colaka, I. Aydina and Z. Candan 2016 *Acta Physica Polonica A* **129 (6)** 1232

[8] Anonymous 2003 *Japanese Agricultural Standard for Plywood, Notification No. 233, Amendment Notification No 1751(2008)*, Ministry of Agriculture, Forestry and Fisheries.

Distribution of zinc oxide nanoparticles on unbleached and bleached bamboo paper via in-situ approaches

Z Sobri[1], Z M A Ainun [1*] and E S Zainudin[2*]

[1]Pulp and Paper & Pollution Control Program, Laboratory of Biopolymer and Derivatives, Institute of Tropical Forestry and Forest Products, Universiti Putra Malaysia, 43400 UPM Serdang, Selangor, Malaysia.
[2]Laboratory of Biocomposites, Institute of Tropical Forestry and Forest Products, Universiti Putra Malaysia, 43400 UPM Serdang, Selangor, Malaysia.

*Corresponding authors: ainunzuriyati@upm.edu.my; edisyam@upm.edu.my

Abstract. Metal oxide nanoparticles such as zinc oxide are increasingly being recognized for potential use in research and health-related application. It is usually incorporated in the modern healthcare products functioning as antimicrobial properties. These inorganic compounds could be applied for antimicrobial paper. Therefore, the objective of this study is to investigate the growth distribution of zinc oxide nanoparticles prepared via in-situ approaches on unbleached and bleached pulps. In order to produce these pulps, hydrothermal method is used throughout the study. The micrograph obtained from the scanning of Field Emission Scanning Electron Microscope exhibited that zinc oxide nanoparticles appeared in the range of 64.1 nm to 87.5 nm supported by the analysis of Field Emission Scanning Electron Microscopy with Energy Dispersive X-Ray Analysis, as high as 27% zinc and 39% oxygen. As expected, zinc oxide nanoparticles decreased the mechanical properties of paper due to the interference of fiber bonding. Antimicrobial test demonstrated acceptable results which required further test.

1. Introduction

Zinc oxide is applied as traditional medicinal ointments for boils and carbuncles since 2000 B.C. [1]. In modern technologies, zinc oxide has gone far in electronic, optic and biomedical applications [2]. The slurry form containing zinc oxide has been used in calamine lotion which is suitable for rashes skinner [1]. Moreover, zinc oxide has served effectively to reduce the microbial growth and has been used as biomimetic membrane in bioscience area [3]. To the extent that, zinc oxide can be nano-synthesized via hydrothermal method, sol-gel method, wet chemical reaction and biological technique [4-6].

In terms of papermaking, zinc oxide has been used as prior source in optoelectronic printing which allow detection and controlling of light. Tetrapodal zinc oxide microstructures are applied to produce such photosensor [7]. Paper also is treated with nanorod zinc oxide in order to function as effective protein concentrator for biomarker in heart disease [8]. Ma et al., [9] and Martins et al., [10] produced antimicrobial paper by coating the paper surface with zinc oxide-starch nanocomposites and nanofibrillated cellulose composites with zinc oxide nanoparticles (NFC/ZnO nanocomposites) respectively. Previously, a study has been conducted by Li et al., [11] which showed resistance of antibacterial activity by immersing nano-zinc oxide cotton fabrics against salt and alkaline artificial sweat solution.

Unbleached paper is widely used as food packaging or paper bag materials in most developed countries. Some countries like Japan used it as writing printing paper that is more environmental friendly material, as it escapes bleaching process. However, bleached papers are more preferred for some official correspondence like writing and printing; has better mechanical properties such as tensile, elongation and tear indices by lignin removal[12]. Brightness of paper that available in market is related to the application of acid, chlorine dioxide and hydrogen peroxide during bleaching process [12-14]. In order to improve the pulp and paper properties, combination of mechanical pulping and chemical bleaching are carried out which further influence the size and distribution of pore [15].

Chen et al., [16] stated that size of pore can influence the ability of fibres to absorb chemical additives. Therefore, a study was conducted in order to investigate the distribution of zinc oxide nanoparticles via hydrothermal method on both unbleached and bleached paper using in-situ techniques. According to Guo et al., [17] , the technique was simple and effective especially for small-scale laboratory application.

2. Materials and methods

2.1. Materials and chemicals

Bamboo culms of *Gigantochloa scortechicinii* (Semantan) were purchased from Sungai Siput, Perak, Malaysia. The bamboo was received 2 m in length prior to chipping process. The chemicals involved in bleaching process were sodium hydroxide (NaOH), sodium chlorite ($NaClO_2$), acetic acid (CH_3COOH) and hydrogen peroxide (H_2O_2). Other chemicals listed as zinc acetate dihydrate ($Zn(CH_3COO)_2.2H_2O$), zinc nitrate hexahydrate ($Zn(NO_3)_2.6H_2O$) and hexamine ($C_6H_{12}N_4$) were used in synthesizing zinc oxide nanoparticles. All chemicals were purchased from R&M Chemicals.

2.2. Preparation of unbleached and bleached pulps of bamboo

Bamboo culms were grinded into chips until 2-3 cm width and length. These chips were cooked using soda pulping technique, 17% sodium hydroxide in twin digester at 170 °C for 90 mins. After pulping process, unbleached pulp was washed and screened using Somerville Fractionator. At this stage, unbleached pulp was obtained. In order to produce bleached pulp, bleaching need to be carried out on unbleached pulp. However, prior to bleaching, the unbleached pulp needs to be beaten until 6,300 revolutions by using PFI-mill according to TAPPI T248 Standard Method. The next step was bleaching process which involved delignification stage (D) and extraction stage (Ep) following D_1EpD_2 bleaching sequences. During D_1 stage, unbleached pulp was put in the transparent plastic bag containing 2 % of $NaClO_2$ and 3 % of CH_3COOH. This pulp was placed in water bath at 70 °C for 180 mins. In order to continue bleaching process, a mixture of NaOH at 1.5 % and H_2O_2 at 1 % was applied to the pulps for 90 mins. Finally, D_2 stage involved the immersion of pulps was carried out for another 90 mins by adding $NaClO_2$ and CH_3COOH at 1.25 % and 3.00 % concentrations accordingly.

2.3. Synthesis of zinc oxide nanoparticles

Nanoparticles of zinc oxide was synthesized using hydrothermal method[4]. Firstly, zinc acetate dihydrate was stirred in ethanol until the mixture was completely dissolved. Then, this mixture was poured into another mixture of sodium hydroxide and ethanol. The stirring continued for 2 hrs at 70 °C. The latter solution promptly changed into cloudy after adding the mixture of sodium hydroxide, zinc nitrate hexahydrate and hexamine. The solution was stirred at 90 °C for another 3 hrs in order to allow optimum growth of zinc oxide. The solution was then filtered and dried in the oven at 100 °C. Finally, whitish solid powder was attained and sent for characterization purpose. This experiment was conducted in order to prove that the produced powder was obtained in nano-size.

2.4. Preparation of zinc oxide unbleached and bleached paper via in-situ approach

The in-situ approach was conducted for both unbleached and bleached pulps by generating zinc oxide nanoparticles in the pulp slurry. The solution of sodium hydroxide and zinc acetate dihydrate was heated and stirred for 2 hrs. An amount of 24 g unbleached or bleached pulps was soaked for 30 mins and dried for another 30 mins at room temperature. The soaking process was repeated for three times. The pulps were then stirred in the solutions of hexamine and zinc nitrate hexahydrate at 90 °C for 3 hrs. This solution was sonicated at 450 Hz for 30 mins and disintegrated before preceding the papermaking process. Papermaking was done using laboratory handsheet machine.

2.5. Characterization of zinc oxide nanoparticles, zinc oxide unbleached and bleached paper
The zinc oxide nanoparticles which were hydrothermally synthesized were observed using Field Emission Scanning Electron Microscope (Model: FESEM JEOL JSM 7600F). Besides, another 4 samples were prepared which comprises of unbleached paper, bleached paper, zinc oxide unbleached paper and zinc oxide bleached paper. The formation and distribution of zinc oxide for these papers were observed using FESEM. In addition, the Energy Dispersive X-ray (EDX) was also applied in order to detect the elements that attached on the fibers that formed a sheet of paper.

2.6. Characterization of optical, physical and mechanical properties of paper
The optical properties of paper involved paper brightness and opacity. Physical properties of paper consisted of grammage and bulk thickness while mechanical properties contained paper tensile strength, tear strength and folding endurance. All test were referred to TAPPI Standard Methods namely as following : TAPPI Standard T 452 om-02 for brightness [18], TAPPI Standard T425 om-01 for opacity [19], TAPPI Standard T411 om-97 for thickness [20], TAPPI Standard T414 om-98 for tear resistance [21], TAPPI Standard T511 om-02 for folding endurance [22], TAPPI Standard T404 cm-02 for tensile [23].

2.7. Antimicrobial test
The test was carried out by using method from Cavalieri et al., [24], by placing 6 mm diameter of paper disc onto a plate of agar with microbes. The microbe culture was standardized to 0.5 McFarland standards which was approximately 10^8 cells [25]. An amount of 100 mg/ml of streptomycin standard were used for each bacteria with three replicates for each plates; *Staphylococcus aureus* (ATCC 43300), *Salmonella choleraesuis* (ATCC 10708) and *Escherichia coli* (ATCC 25922). These plates were inverted and incubated at 30-37 °C for 18-24 hrs, until the sufficient growth has occurred. After incubation, each plate was examined. The diameters of the zones of complete inhibition were measured, including the diameter of the disc. Zones were measured to the nearest whole millimeter, using sliding calipers or a ruler, which was held on the back of the inverted petri plate.

3. Results and discussion

3.1. Morphological observation of zinc oxide nanoparticles and zinc oxide papers
The morphological observation of control paper and zinc oxide nanoparticles which observed via FESEM were exhibited as in Figure 1 and 2. There were two type of shapes observed in the powder which namely as rod-shape and spherical-shape. The spherical-shape is also known as nano-seeds of zinc oxide nanoparticles. The micrograph showed the size distribution was 356 nm in diameter and 1.52 μm in length of nanorod. On the other hand, the nano-seeds of zinc oxide achieved nano-sized with range of 64.1 nm to 87.5 nm. The distribution of particles size of zinc oxide–starch nanocomposites studied by Ma et al., [19] was from 35 nm to 45 nm of spherical shapes while Xu & Cai, [8] obtained nanorod shapes of zinc oxide with diameter 40 nm to 60 nm with 300 nm to 450 nm in length.

The occurrence of these two shapes might be due to duration of heating during synthesis process. The same conditions were obtained by Baruah et al., [26]. In the processes of producing photocatalytic paper using zinc oxide, the paper sheets are dipped in pretreatment solutions containing zinc acetate dihydrate and sodium hydroxide. The dipping was done until three times and was dried at 90 °C for 15 mins which temperature was higher. Besides that, the nanoseed zinc oxide paper is dipped into solutions of zinc nitrate hexahydrate and hexamine for 10 hrs at 90 °C and dried at 70 °C for 6 hrs. Hence, this explains the reason of the presence of two shapes of nano-zinc oxide.

IOP Conf. Series: Materials Science and Engineering **368** (2018) 012046 doi:10.1088/1757-899X/368/1/012046

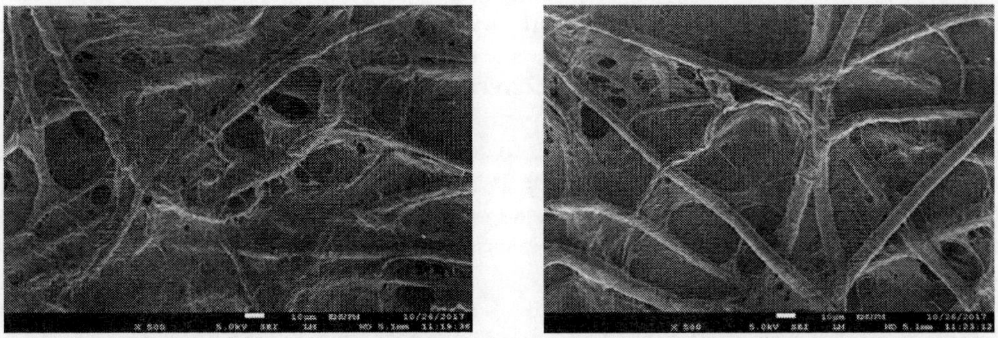

(a) (b)

Figure 1. Surface of control paper is observed via FESEM at 500X magnification showing (a) bleached control paper, and (b) unbleached control paper.

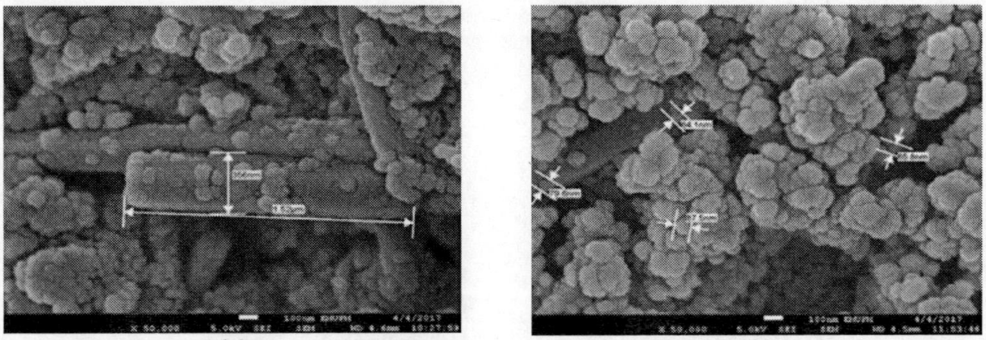

(a) (b)

Figure 2. Flocs of zinc oxide particles observed via FESEM at 50kX magnification showing nano-rod and nano-spherical shapes.

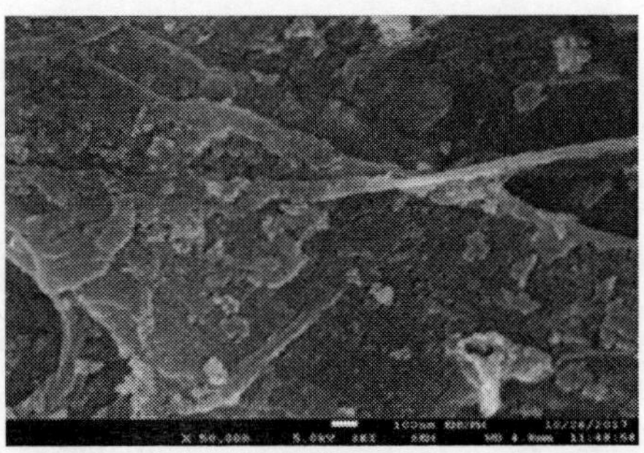

Figure 3. FESEM image of unbleached paper which contained zinc oxide nanoparticles at 50kX magnification. Majority of zinc oxide was observed in lumps.

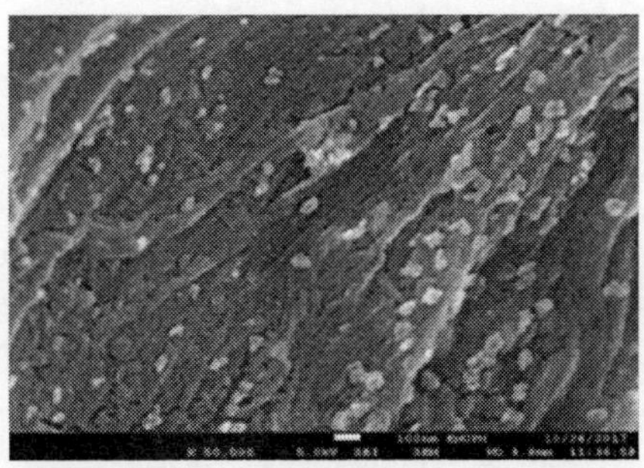

Figure 4. FESEM images of bleached paper which contained zinc oxide nanoparticles at 50kX magnification. Majority of zinc oxide was observed in spherical form.

Figure 3 and 4 exhibited the micrographs of zinc oxide unbleached and bleached paper accordingly. The distribution of zinc oxide was found more in bleached paper which most part of paper were located with spherical nano-particles of zinc oxide. On the other hand, majority of zinc oxide also existed on unbleached paper in lump form but fewer than on unbleached paper surfaces. Although the distribution types were different but zinc nanoparticles on both unbleached and bleached paper were ranged similarly as 57 nm until 143 nm. In order to confirm the existence of zinc oxide, EDX scanning was carried out which results are shown in Graph 1.

The deposition of nanoparticles on paper is confirmed as zinc oxide which occurs as zinc and oxygen on both unbleached and bleached papers. The element composition of zinc and oxygen on bleached paper was found higher at 27.25 % (zinc) and 38.72 % (oxygen) than unbleached paper at 25.27 % (zinc) and oxygen 33.72 % (oxygen). This may be due to the surface properties of both pulp as explained by Andreasson et al., [15]. The Total Chlorine Free bleached pulps were analyzed using Nuclear Magnetic Resonance (NMR) having larger average pore radius up to 16.5 nm compared to unbleached pulps with 15.0 nm. The pores were reduced to 14.8 nm after treated with high pH and salt concentration. Porosity test measures void of fraction in a material and Kamel et al., [27] stated that increment of paper porosity can consequently enhance the retention of any polymer solution due to the adsorption on surfaces and high potential penetration to fill in the voids.

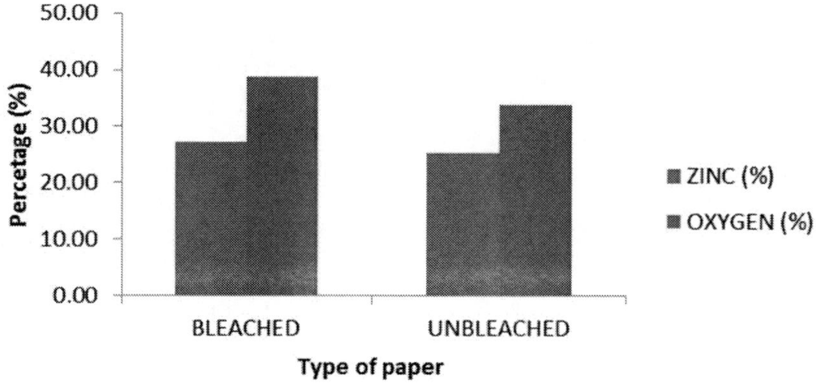

Graph 1. Percentage of zinc and oxygen on unbleached and bleached paper scanned via EDX.

3.2. Physical, optical and mechanical properties of zinc oxide paper

There are three category of paper testing were carried out namely as physical, optical and mechanical paper properties. All test results were performed in Table 1. Paper physical test involved grammage and thickness of each paper sample. The grammage achieved by zinc oxide paper either unbleached or bleached are higher compared to the control papers. The addition about 13.26% and 15.13% of grammage for unbleached and bleached respectively most probably were due to the existence of zinc oxide nanoparticles. These results clearly proved that the in-situ synthesis was successfully done. The thickness showed a slight increment after incorporating the zinc oxide nanoparticles. This can be related to the presence of these particles which affect the arrangement of fibers that formed the papers.

Table 1. The table show physical, optical and mechanical properties of control and zinc oxide paper.

Paper testing	Control paper		Zinc oxide paper	
	Bleached	**Unbleached**	**Bleached**	**Unbleached**
Grammage (g/m^2)	55.30	56.20	65.16	64.79
Thickness (μm)	125.68	128.60	130.00	136.77
Opacity	81.80	96.60	81.81	98.99

Brightness	78.90	19.30	84.30	19.69
Tensile index (Nm/g)	29.60	50.20	26.42	39.04
Tear index (mN.m²/g)	18.90	22.30	11.84	11.99
Folding endurance	37	326	32	123

Optical paper properties involved paper opacity and brightness. The results showed that the presence of zinc oxide nanoparticles enhance the brightness of the paper by 0.40 % and 5.40 % for unbleached and bleached paper respectively. The opacity percentages of bleached and unbleached papers were slightly increased about 0.01 % and 2.39 % accordingly. The nanoparticles are observed to improve the brightness and opacity of papers which reacted as fillers. Mineral fillers such as precipitated calcium carbonate, clay and other functional filler materials in papermaking improve paper opacity and printability as well. Producing bright and opaque paper with good white colour has been generally achieved by applying zinc pigments and zinc oxide [28]. In instance, starch-zinc oxide and alginic acid-zinc oxide may modify fibre sheets up to 89.9 % and 93.1 % of opacity respectively compare to control fibre sheets which has 88.3 % of opacity [29]. Bulk zinc oxide coated paper is thicker and more opaque than base paper by having intensity of XRD spectra peaks from bulk-zinc oxide that is reduced to 60 % of intensity from the base paper [30].

Mechanical properties of paper consist of tensile, tear and folding test. Huge decrement of 22.2 % in tensile properties was detected by unbleached paper. However, the decrement of tensile properties for bleached paper was lower, about 10.7 %. Bleached fibres are more fibrillated and flexible compared to unbleached fibres due to the bleaching process that provide chemical action to the fibre surfaces. However, unbleached reading was higher than bleached papers due to the chemical treatments that have been carried out. According to Fišerová et al., [31], chemical treatments could reduce fibre strength and length because fibre bonding is condensed, and according to Hossain et al., [32], changes of interfibrillar region caused by chemical treatment contribute to rearrangement of fibrils along the tensile loading direction as this consequently applied higher stressed in the fibre which certainly reduce tensile strength. Tear properties of both unbleached and bleached paper presented same trend as tensile properties. The reductions percentages of both unbleached and bleached papers are 46.23 % and 37.35 %. Folding endurance showed 62.27 % and 13.51 % of decrement for unbleached and bleached papers. This degradation in terms of mechanical properties is expected because any presence of fillers on the surface of fibers which later formed into paper will affect greatly the fibre bonding. Interruption of fibre bonding creates crucial effects to paper strength mechanically.

3.3. Antimicrobial test

Referred to Table 2, it was found out that no inhibition zones are observed from the three bacteria tested. This happened might be because of low concentration of zinc oxide in paper sample as stated by Cavalieri et al., (2005)[24]. He showed that the concentration of the ampicillin interpretive' categories for the break-point panel provide resistant interpretive of 32.0 mcg/ml (0.032 mg/ml) via Minimal Inhibitory Concentration (MIC) testing. The diameter of zinc oxide particles also could be the factor of percentage of inhibition growth such as study done by Baruah et al., (2010)[26], showed that particles size at 8 nm and 30 nm to 10 nm caused inhibition growth of *S. aureus* (95%) and *E. coli* (100%) respectively, compare to particles size at 50 nm to 70 nm for only 50% inhibition growth. Conversely, the paper sample gives the zinc oxide paper a mild antimicrobial properties by not spreading new bacteria or fungal on agar media as can be seen in Figure 5.

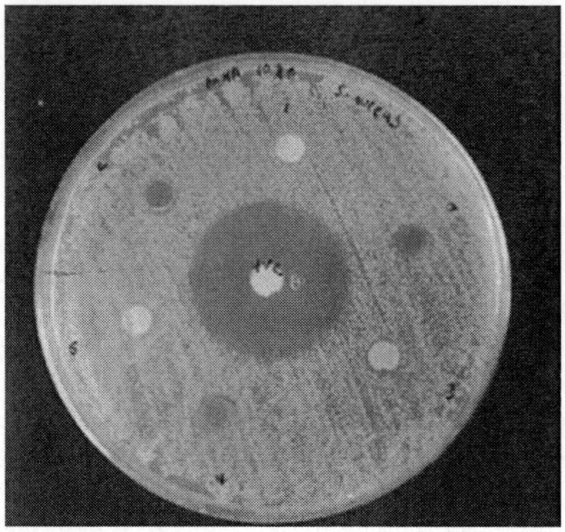

Figure 5. The inhibition zone and paper samples for standard with Streptomycin using *S. aerus.*

Table 2. Antimicrobial screening results

Types of bacteria	Screening results		Observation
Escherichia coli (gram negative)	CB	-	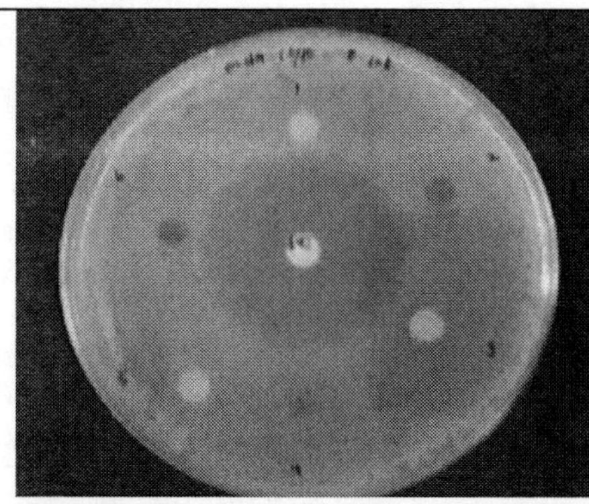
	CUB	-	
	T1-B	-	
	T1-UB	-	
Salmonella choleraesuis (gram positive)	CB	-	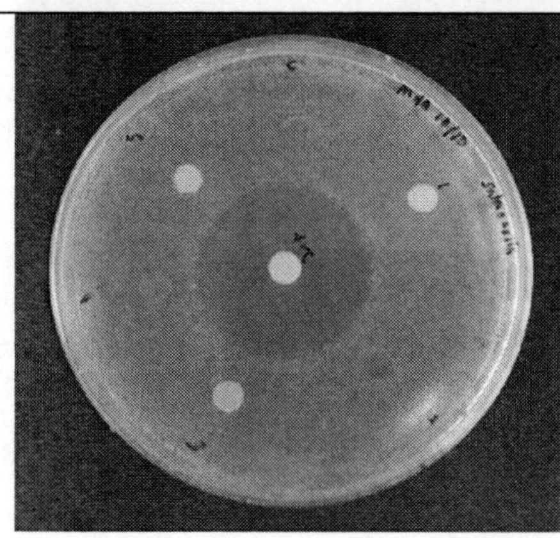
	CUB	-	
	T1-B	-	
	T1-UB	-	

Staphylococcus aerus (gram positive)	CB	-	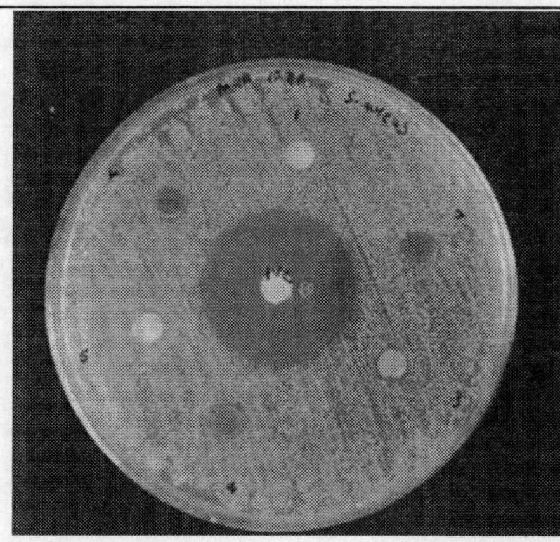
	CUB	-	
	T1-B	-	
	T1-UB	-	

*CB = control bleached, CUB = control unbleached, T1-B = hydrothermal bleached, T1-UB = hydrothermal unbleached

4. Conclusion
This study showed nano-ZnO synthesized using hydrothermal method is highly distributed on bleached paper as for its pore size of paper is more dispersed compare to unbleached paper prepared via in-situ approach. As expected the physical, optical and mechanical properties of paper were affected due to the presence of zinc nanoparticles. Antimicrobial test demonstrated promising results however more testing need to be carried out in next study.

Acknowledgements
The authors would like to acknowledge Malaysia Ministry of Higher Education for awarding the project grant 5524855 under Fundamental Research Grant Scheme.

References
[1] Moezzi A, McDonagh AM, Cortie MB. Zinc oxide particles: Synthesis, properties and applications. *Chem Eng J*. 2012;**185–186**:1–22.
[2] Dhobale S, Thite T, Laware SL, Rode C V., Koppikar SJ, Ghanekar R-K, et al. Zinc oxide nanoparticles as novel alpha-amylase inhibitors. J Appl Phys. 2008;(July 2014).
[3] Varaprasad K, Raghavendra GM, Jayaramudu T, Seo J. Nano zinc oxide-sodium alginate antibacterial cellulose fibres. *Carbohydr Polym*. 2016;**135**:349–55.
[4] Xu B, Cai Z. Fabrication of a superhydrophobic ZnO nanorod array film on cotton fabrics via a wet chemical route and hydrophobic modification. *Appl Surf Sci*. 2008;**254(18)**:5899–904.
[5] Alwan RM, Kadhim QA, Sahan KM, Ali RA, Mahdi RJ, Kassim NA, et al. Synthesis of Zinc Oxide Nanoparticles via Sol – Gel Route and Their Characterization. *Nanosci Nanotechnol*. 2015;**5(1)**:1–6.
[6] Azizi S, Namvar F, Mahdavi M, Ahmad M Bin, Mohamad R. Biosynthesis of silver nanoparticles using brown marine macroalga, Sargassum muticum aqueous extract. *Materials (Basel)*. 2013;**6(12)**:5942–50.
[7] Sandberg M, Tordera D, Granberg H, Sawatdee A, Dedic D, Berggren M, et al. Photoconductive zinc oxide-composite paper by pilot paper machine manufacturing. *Flex Print Electron*. 2016;**1**:1–10.
[8] Tiwari S, Vinchurkar M, Rao VR, Garnier G. Zinc oxide nanorods functionalized paper for protein preconcentration in biodiagnostics. *Nat Publ Gr*. 2017;**(January)**:1–10.
[9] Ma J, Zhu W, Tian Y, Wang Z. Preparation of Zinc Oxide-Starch Nanocomposite and Its Application on Coating. *Nanoscale Res Lett*. 2016;**11(1)**:200.
[10] Martins NCT, Freire CSR, Neto CP, Silvestre AJD, Causio J, Baldi G, et al. Antibacterial paper based on composite coatings of nanofibrillated cellulose and ZnO. *Colloids Surfaces A Physicochem Eng Asp*. 2013;**417**:111–9.

[11] Li Q, Chen S-L, Jiang W-C. Durability of Nano ZnO Antibacterial Cotton Fabric to Sweat. *J Appl Polym Sci*. 2007;**103**:412–6.
[12] Jani SM, Rushdan I. Effect of bleaching on coir fibre pulp and paper properties. *J TropAgric FdSc*. 2014;**42(1)**:51–61.
[13] Seo J-H, Kim H-J. Effect of H2O2 bleaching with ultrasonication on the properties of thermomechanical pulp and unbleached kraft pulp. *Ultrason Sonochemistry*. 2015;**23**:347–53.
[14] Barros DP, Silva VL, Hamalainen H, Colodette JL. Effect of last stage bleaching with peracetic acid on brightness development and properties of eucalyptus pulp. *BioResources*. 2010;**5(2)**:881–98.
[15] Andreasson B, Forsstro J, Wagberg L. Determination of fibre pore structure : influence of salt , pH and conventional wet strength resins. *Cellulose*. 2005;**12**:253–65.
[16] Chen YM, Wan JQ, Huang MZ, Ma YW, Wang Y, Lv HL, et al. Influence of drying temperature and duration on fiber properties of unbleached wheat straw pulp. *Carbohydr Polym*. 2011;**85(4)**:759–64.
[17] Guo Q, Ghadiri R, Weigel T, Aumann A, Gurevich EL, Esen C, et al. Comparison of in situ and ex situ methods for synthesis of two-photon polymerization polymer nanocomposites. *Polymers (Basel)*. 2014;**6(7)**:2037–50.
[18] TAPPI. Standard Test Method for Brightness of Pulp , Paper , and Paperboard (Directional Reflectance at 457 nm). *TAPPI T425 om-02*. 2002;15(Reapproved):1–6.
[19] TAPPI. Opacity of Paper. *TAPPI T425 om-01*. 2001;9–13.
[20] TAPPI. Thickness (caliper) of paper , paperboard , and combined board. *TAPPI T411 om-97*. 1997;1–4.
[21] TAPPI. Internal tearing resistance of paper (Elmendorf-type method). *TAPPI T414 om-98*. 1998;(July):1–7.
[22] TAPPI. Folding endurance of paper (MIT tester) (Revision of T 511 om-08). *TAPPI T511 om-02*. 2006;1–4.
[23] TAPPI. Tensile breaking strength and elongation of paper and paperboard (using pendulum-type tester). *TAPPI T404 C*. 1992;3–7.
[24] Cavalieri SJ, Harbeck RJ, Carter YSM, Ortez JH, Rankin ID, Sautter RL, et al. *Manual of antimicrobial susceptibility testing*. Coyle MB, editor. *Manual of antimicrobial susceptibility testing*. Seattle, Washington: American Society For Microbiology; 2005. 3-151 p.
[25] Murray, Baron, Pfaller. McFarland Standard. *Dalynn Biol*. 2002;
[26] Baruah S, Jaisai M, Imani R, Nazhad MM, Dutta J. Photocatalytic paper using zinc oxide nanorods. *Sci Technol Adv Mater*. 2010;**11(5)**:55002.
[27] Kamel S, El-Sakhawy M, Nada AMA. Mechanical properties of the paper sheets treated with different polymers. *Thermochim Acta*. 2004;**421**:81–5.
[28] Higham RRA. *A Handbook of Papermaking: The Technology of Pulp, Paper and Board Manufacture*. 2nd ed. London: Business Books Limited; 1968. 133-139 p.
[29] Khatri V, Halasz K, Trandafilovic L V., Dimitrijevic-Brankovic S, Mohanty P, Djokovic V, et al. ZnO-modified cellulose fiber sheets for antibody immobilization. *Carbohydr Polym*. 2014;**109**:139–47.
[30] Prasad V, Shaikh AJ, Kathe AA, Bisoyi DK, Verma AK, Vigneshwaran N. Functional behaviour of paper coated with zinc oxide-soluble starch nanocomposites. *J Mater Process Technol*. 2010;**210(14)**:1962–7.
[31] Fišerová M, Gigac J, Ák JBČ. Relationship between fibre characteristics and tensile strength of hardwood and softwood kraft pulps. *Cellul Chem Technol*. 2010;**44(7–8)**:249–53.
[32] Hossain SI, Hasan M, Hasan N, Hassan A. Effect of Chemical Treatment on Physical , Mechanical and Thermal Properties of Ladies Finger Natural Fiber. *Adv Mater Sci Eng*. 2013;**2013**:1–6.

Physical properties of silane-treated sugar palm fiber reinforced thermoplastic polyurethane composites

A Atiqah[1], M Jawaid[1]*, S M Sapuan[2] & M R Ishak[3]

[1]Laboratory of Biocomposite Technology, Institute of Tropical Forestry and Forest Products (INTROP), Universiti Putra Malaysia, 43400 UPM Serdang, Selangor, Malaysia. [2]Department of Mechanical and Manufacturing Engineering, Faculty of Engineering, Universiti Putra Malaysia, 43400 UPM Serdang, Selangor, Malaysia.
[3]Department of Aerospace Engineering, Faculty of Engineering, Universiti Putra Malaysia, 43400 UPM Serdang, Selangor, Malaysia.

*Corresponding author: jawaid_md@yahoo.co.in

Abstract. Fiber modifications are widely utilized to enhance the interface of fiber and matrix adhesion based on eco-friendly fibers such as sugar palm fibers. The sugar palm (SP) fibers in particle size of 150-250μm were immersed in 2 wt. % silane for 3hrs in this research. Composites of untreated and treated sugar palm fibers reinforced thermoplastic polyurethane (TPU) with different fiber loadings, ranging from 0 to 50 wt. % were then prepared by melt mixing compounding followed by hot pressing moulding. The physical property such as density, water absorption and thickness swelling of the composites were evaluated. Ten replicates of SP/TPU composites with the standard dimension were immersed in distilled water for 168hrs. The chemical changes for untreated and treated through Fourier-transform infrared spectroscopy (FTIR) was done to distinguish the untreated and silane treated of the SP/TPU composites. Increasing the fiber content resulted in higher water uptake and thickness swelling. Moreover, the silane treated SP/TPU composites was shown reducing water absorption and thickness swelling properties. This study is a part of exploration potential application of the composites for automotive part applications.

1. Introduction

Natural fibers are taken from natural plants such as sugar palm, kenaf, flax, and bast which are extracted from its plants, stems, leaves or any part thereof which are classified as an eco-friendly material due its biodegradability and abundance. Sugar palm is one of the best examples that exhibited noteworthy potential to be used as reinforcement in polymer for the development of composite materials [1-2]. On the potential notes, sugar palm has an acceptable standard of strength in its mechanical properties as well as low density, good tolerance in its mechanical and thermal properties [3]. However, on the drawback notes, sugar palm has poor adhesion with polymer matrix. This drawback has extended to be investigated further on the distribution load in reinforcement fibers in order to obtain its maximum strength. Many researchers have tackled this drawback issue by using a certain method such as chemical treatment onto fibers in order to enhance its adhesion to get ideal bonding strength.

The reinforcement between fiber and matrix plays the main role in transferring the stresses performing on the matrix to the fibers. In order to have a good properties of the composites, the fiber and matrix interface must be strongly attached. Nevertheless, if the interface is too durable, the composite will have the brittle and low stiffness properties. There are numerous reported works to enhance the interfacial bonding between the polymer matrix and natural fibers through fiber treatment modification such as sisal/PLA [4], ijuk/polypropylene [5], pineapple/kenaf/phenolic [6].

One of the common methods to modify the fiber surface properties is by silane treatment. Silane treatment has been discovered to modify the glass fiber surface to be reinforced in polymer matrices. Currently, researches employed the same technique to natural fibers in order to improve the adhesion bonding of fiber/matrix. The fiber treatment enhances the adhesion between the polymer matrix and the surface of fiber by changing the polarity of the fiber surface [7-8].

The previous study conducted by Atiqah, Jawaid, Ishak and Sapuan [9] have investigated the bonding strength of sugar palm reinforced thermoplastic polyurethane. Various treatment such as alkaline, silane and combined alkaline-silane were selected to undergo single fiber test and found that silane treatment showed a good compatible for sugar palm fiber. Silane treatment was employed in modifying natural fiber-polymer matrix interface and improving the interfacial strength. The hydrophobic chemical derivatives of SiH_4 called silane can be hydrolysed and bonded to the hydroxyl groups of cellulose chains [10]. It had been reported that, the cross-linked network between silane treated fibers and the matrix showed non-swelling behaviour and chemical resistance [11].

This paper reports a study of the influence of fiber treatment (silane) on the physical and structure of sugar palm fiber. In order to analyse the physical property of SP/TPU composites, the variation of fiber loading consisted of 0, 10, 20, 30, 40 and 50 wt. % and 2 wt. % of silane treatment were selected in this investigation.

2. Experimental

2.1. Materials
Estane® 58311 TPU was supplied in pellet form with the density of 1.13 g/cm^3 by Pultrusion Sdn. Bhd. and was used as the polymer matrix. The sugar palm fiber (SPF) was collected from sugar palm tree at Jempol, Negeri Sembilan, Malaysia.

2.2. Preparation of Sugar Palm
Firstly, sugar palm fiber was first washed with tap water for several times to get rid of any impurities and dirt that attached to the SPF. The SPF was kept in an open air for 24hrs and dried in an air circulating oven at 60°C for 48hrs. The dry SPF grounded to get the size of 10mm-15mm using plastic crusher machine then followed by using pulverize machine. Next, the particle SPF were sieved to obtain 125-250 μm.

2.3. Silane Treatment of Sugar Palm
The SPFs in particulate size of 125-250 μm were immersed in 2 wt. % of silane for 3hrs. For silane treatment, 3-aminopropyl-triethoxysilane was used and dissolved in a mixture of methanol-water (90/10 w/w) for their hydrolysis. The pH of the solution was adjusted to 3.5 with acetic acid and stirred continuously for 10 minutes. Then, the fibers were immersed in the solution and left for 3hrs under agitation. The SPF were thoroughly rinsed with distilled water and then oven-dried at 60 °C for 72hrs to completely eliminate any moisture effect from the fibers.

2.4. Fabrication of SP reinforced TPU composites
The SP/TPU composites were prepared using melt compounding technique followed by hot moulding process. Sugar palm with particles size of 125-250 μm and thermoplastic polyurethanes in pellet form were dried in an electric oven at 80°C for 48hrs. Six set of neat TPU and SP/TPU composites (0, 10, 20, 30, 40 and 50) wt. % were fabricated. SP/TPU composites were prepared using melt-mixing compounding, followed by hot pressing moulding process to achieve uniform distribution. Haake polydrive R600 was used in the mixing process at the optimum processing parameters temperature,

time and rotating speed; 190 °C, 11 minutes and 40 rpm, respectively. Vechno Vation 40 ton compression molding machine was used in the hot pressing moulding. The samples were pre-heated for 7 minutes at 190°C. Then they were fully pressed for another 10 minutes at 190 °C. Finally, the sample were cold-pressed for 5 minutes at 25 °C. Table 1 shows the composition of the formulation of untreated and treated sugar palm fiber reinforced thermoplastic polyurethane.

Table 1. Designation and composition of the materials.

Materials	TPU (wt %)	SP (wt %)
Neat TPU	100	0
TPU + Sugar Palm Fiber	90	10
TPU + Sugar Palm Fiber	80	20
TPU + Sugar Palm Fiber	70	30
TPU + Sugar Palm Fiber	60	40
TPU + Sugar Palm Fiber	50	50

3. Characterizations

3.1. Physical Properties Testing

Density was measured in air using a digital weighing scale and in water using densimeter, MD-200S Mirage. The difference in weight of the sample in two media gives the weight in gram was converted to volume in cm^3.

Water absorption specimens of the untreated and treated SP/TPU composites were immersed in distilled water in accordance with ASTM D 570. The SP/TPU composites samples were cut into the size of 20 mm long, 20 mm wide and 3 mm thickness. All specimens were conditioned in and oven at 60° C for 24hrs then put into the sealed plastic bag and cooled in desiccator over granulated silica gel. The samples were then weighed and immersed in a distilled water at room temperature for 72, 150 and 168hrs. Then the specimen was removed from the water, wiped with tissue paper, weighed to measure the weight gain, and put back in the water for continued soaking.

Ten specimens of 20 x 20 x 3 mm samples of each different type of composites were prepared according to ASTM D570 for the testing of thickness swelling.

3.2. FTIR

Fourier transform infrared (FTIR) spectra of the neat TPU, untreated SP/TPU and silane treated SP/TPU composites were tested by using a FTIR machine (SHIMADZU81001. Japan) to evaluate the changes in functional groups on the composites. All spectra were recorded in the range from 4000 cm^{-1} to 350 cm^{-1}.

4. Result and Discussion

4.1. Physical Properties

4.1.1. Density Properties. The results of density measurement of neat TPU, untreated and silane treated SP/TPU composites subjected to different fiber loading were presented as in Figure 1. The density of SP/TPU composites decreased when the sugar palm fibers were treated with silane treatments. It can be seen that neat TPU has lower density (1.07 g/cm^3) neat TPU and SP/TPU composites (10, 20, 30, 40, 50 wt. %) are 1.07, 1.08, 1.14, 1.16 and 1.20 g/cm^3 , whereas the values of density seems to reduce after reinforcing for silane treated fibers which are 1.07, 1.07, 1.09, 1.13 and 1.17 g/cm^3. On the contrary, the density should be higher after fiber treatment, it was suggested that

IOP Conf. Series: Materials Science and Engineering **368** (2018) 012047 doi:10.1088/1757-899X/368/1/012047

due to treatment employed on the fiber surface, the fiber cell wall become more compact that increased the density of the fiber [12, 13].

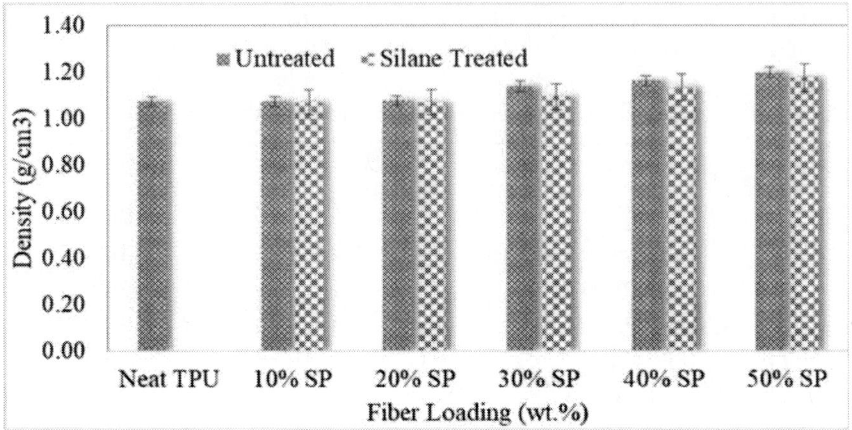

Figure 1. Density of untreated and silane treated SP/TPU composites.

4.1.2. Moisture Absorption Properties. Figure 2 shows the values of the water absorption for the untreated and silane treated of SP/TPU composites, which vary depending upon the fiber loading. As shown in Figure 2, the water absorption of the composites increased with increasing fiber loading, but was lower for neat TPU composites due to the matrix is hydrophobic, whereas the sugar palm fiber composites are hydrophilic. The 2 wt. % of silane treated of SP/TPU composites are significantly lower than those of composites as shown in Figure 2. The gradient of the water absorption graphs for the neat TPU and SP/TPU composites (10, 20, 30, 40, 50 wt. %) are 1.17, 4.23, 4.44, 6.24, 7.47, and 8.21 whereas the values of water absorption for silane treated are 3.24, 3.82, 4.74, 5.73 and 6.83. The neat TPU and 10 wt. % of treated SP composites exhibited less water absorption than those containing 50 wt. % untreated and treated composites. The silane treatment was prone to effective in enhancing the fiber-matrix interface properties [14, 15]. The surface modification of silane after hydrolysis undergo bond formation stage and condensation that induced polysiloxane structures form the reaction hydroxyl groups on the fiber. Hence, the reduction of water absorption of treated SP/TPU composites is due to the improved interfacial adhesion between the TPU chains and the lignocellulosic fiber.

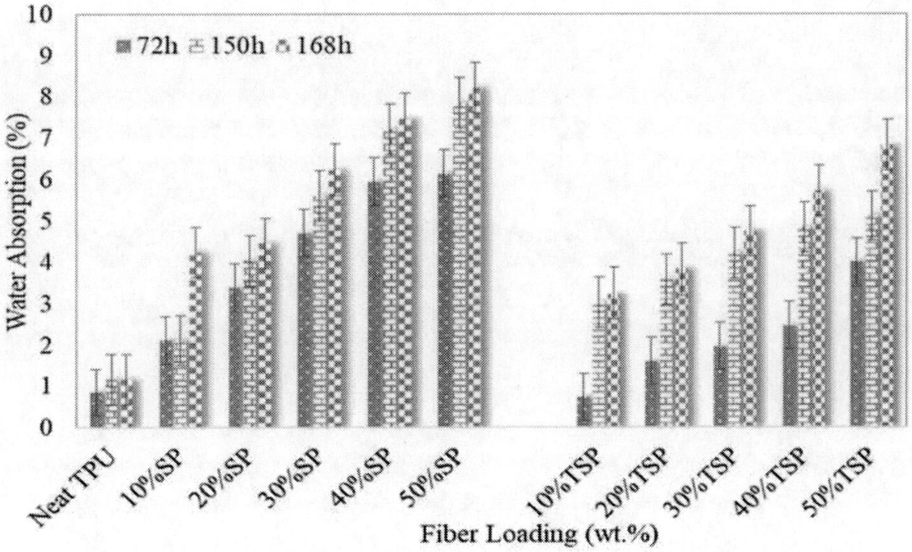

Figure 2. Water Absorption of untreated and silane treated SP/TPU composites.

4.1.3. Thickness Swelling Test. As shown in Figure 3, the thickness swelling of neat TPU, untreated and silane treated of SP/TPU composites were varied according to fiber loading (10, 20, 30, 40 and 50 wt. %). The silane treated of thickness welling are lower as compared than untreated of SP/TPU composites. The gradient of the thickness swelling graphs for the neat TPU and SP/TPU composites (10, 20, 30, 40, 50 wt. %) are 0.99, 3.95, 4.64, 6.18, 7.14 and 7.62 whereas the values of water absorption for silane treated are 3.11, 5.49, 6.28, 5.62, and 6.84. In this case, the higher loading of sugar palm fiber exhibited the higher thickness swelling. However, the treated fiber loading of 40 wt. % showed the less thickness swelling as compared to other formulations. According to Agrawal, Saxena, Sharma, Thomas and Sreekala [16], silane treatments may reduce the number of cellulose hydroxyl group in the surface of the fiber and polymer matrix. The silanols group were formed due to the presence of the hydrolyzable alkolyl group from the moisture that existed on the fiber surface. Then, the silanols react with the hydroxyl group of the fiber to form the stable covalent bonds to the outer cell wall so that the chemical easily absorbed onto the fiber surface. Thus, the hydrocarbon chain that induced by the employing of silane treatment detains the swelling of the fiber by creating a cross-linked network because of covalent bonding between the fiber and matrix.

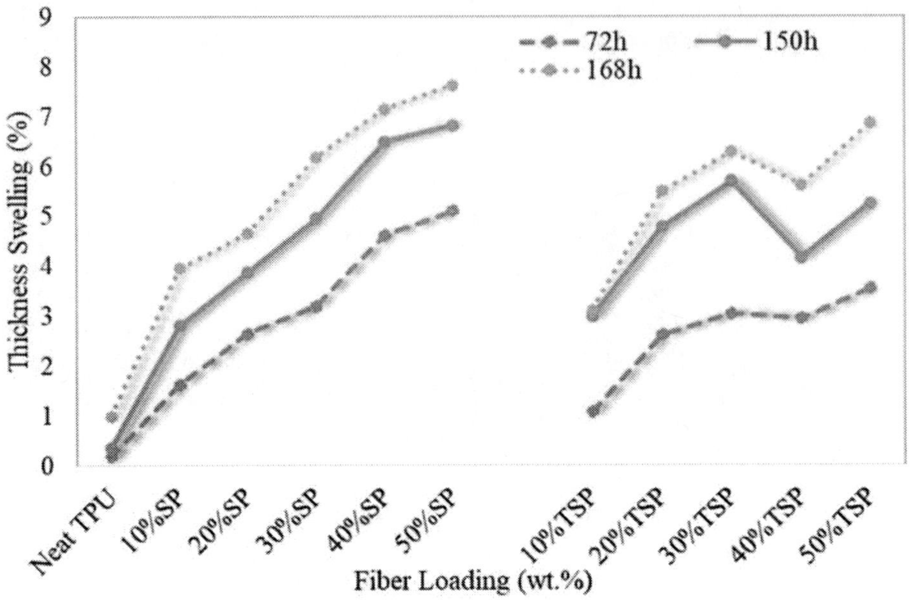

Figure 3. Thickness swelling of untreated and silane treated SP/TPU composites.

4.2. Chemical and Morphological Analysis

4.2.1. FTIR. Figure 4 exhibits the FTIR spectra of the neat TPU, untreated and silane treated sugar palm fibers reinforced TPU (SP/TPU) composites. To attain good physical properties of the composites requires a good interface between fiber and matrix. The sugar palm fiber was treated with silane treatment to improve the interfacial adhesion between sugar palm fibers and TPU matrix in this study. Figure 4 shows the untreated SP, which showed similar attributions nearly the neat TPU. The chemical structure of absorption peak at 921 cm^{-1} was due to the Si-O vibration [17], which agreed on the confirmation of 3-aminoporpyltriethoxysilane of treated SP/TPU composites. The presence on chemical structure as observed in Figure 4 conformed that the silane chemical structure are existed on the sugar palm fibers. The possible chemical structure that existed on the neat TPU, untreated and silane treated were tabulated in Table 1.

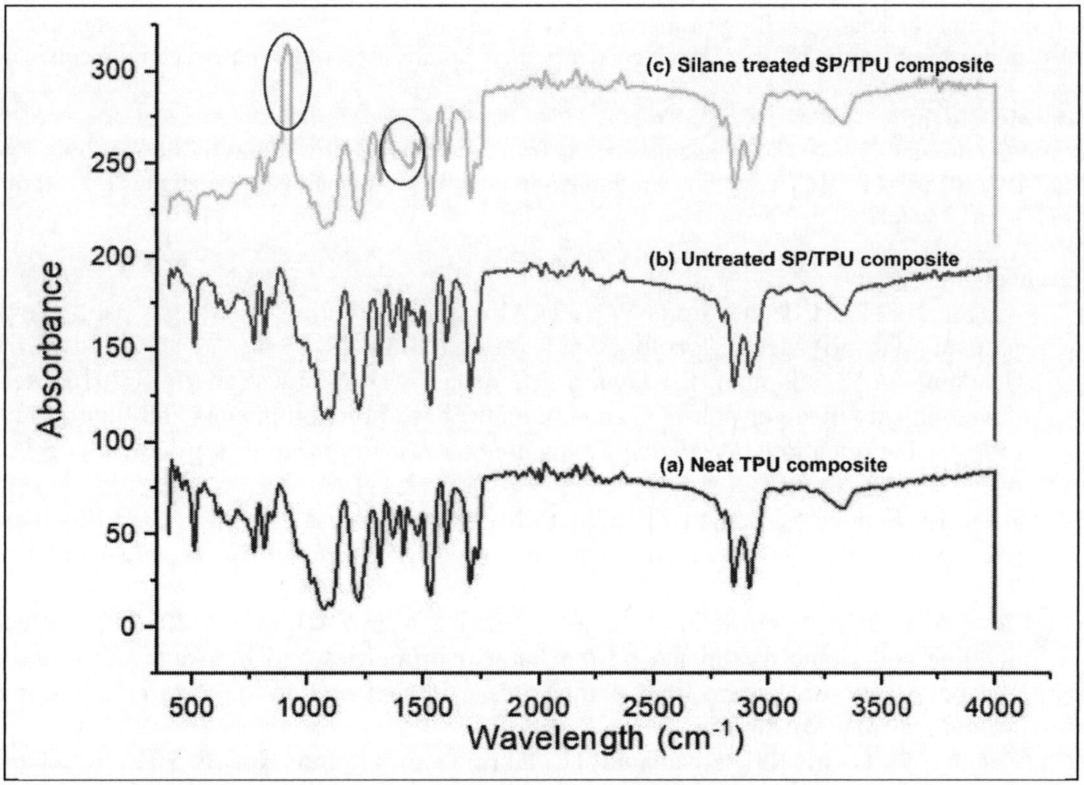

Figure 4. Spectrography from FTIR on neat TPU and untreated fibers (a) neat TPU, (b) untreated SP/TPU composites and (c) silane treated SP/TPU composites.

Table 2. The wave numbers of peaks used for FTIR analysis and corresponding functional groups and vibrational type of neat TPU, untreated and silane treated SP/TPU composites [18-19].

Peak Location (cm^{-1})	Chemical Structure	Motion	Neat TPU	Untreated SP/TPU composites	Silane Treated SP/TPU composites
1510-1550	H-N-CO	Combined motion	1531.228	1535.085	1531.228
1690	C=O	Associated urethane	1700.936	1700.936	1700.936
1740	C=O	Non-bonded urethane Stretching	1731.792	1729.863	1727.935
2800-3000	CH_2 and CH_3	Stretching	2917.818	2919.747	2919.747
1590-1650	N-H	Bending	1627.653	1598.728	1629.851
3200-3420	N-H	Stretching	3330.517	3328.588	3332.445

5. Conclusion

In this study, the SP/TPU were successfully prepared with fiber treatment of silane. The experimental results showed that the physical of TPU based were improved with the incorporation of 50 wt. % of SP and 2 wt. % silane treatment of fibers. The silane treated of SP/TPU reduce the density, water absorption, and thickness swelling properties was the minimum for 40 wt. % of SP compared with those of untreated sugar palm fiber which enhances their suitability for automotive components.

Acknowledgment
The authors are grateful for the financial support from Universiti Putra Malaysia through Putra grant no. GP-IPS/2015/9441501. The author would also like to thank the Ministry of Higher Education for the MyBrain15 scholarship.

References

[1] Mogea J, Seibert B and Smits W 1991 Multipurpose palms: the sugar palm (Arenga pinnata (Wurmb) Merr.) *Agroforestry Systems* **13** 111-29

[2] Sanyang M L, Sapuan S, Jawaid M, Ishak M R and Sahari J 2016 Recent developments in sugar palm (Arenga pinnata) based biocomposites and their potential industrial applications: A review *Renewable and Sustainable Energy Reviews* **54** 533-49

[3] Ishak M, Sapuan S, Leman Z, Rahman M, Anwar U and Siregar J 2013 Sugar palm (Arenga pinnata): Its fibres, polymers and composites *Carbohydrate polymers* **91** 699-710

[4] Orue A, Jauregi A, Unsuain U, Labidi J, Eceiza A and Arbelaiz A 2016 The effect of alkaline and silane treatments on mechanical properties and breakage of sisal fibers and poly(lactic acid)/sisal fiber composites *Composites Part A: Applied Science and Manufacturing* **84** 186-95

[5] Zahari W, Badri R, Ardyananta H, Kurniawan D and Nor F 2015 Mechanical properties and water absorption behavior of polypropylene/ijuk fiber composite by using silane treatment *Procedia Manufacturing* **2** 573-8

[6] Asim M, Jawaid M, Abdan K and Ishak M R 2016 Effect of Alkali and Silane Treatments on Mechanical and Fibre-matrix Bond Strength of Kenaf and Pineapple Leaf Fibres *Journal of Bionic Engineering* **13** 426-35

[7] Kushwaha P K and Kumar R 2010 Studies on water absorption of bamboo-epoxy composites: Effect of silane treatment of mercerized bamboo *Journal of applied polymer science* **115** 1846-52

[8] Goriparthi B K, Suman K and Rao N M 2012 Effect of fiber surface treatments on mechanical and abrasive wear performance of polylactide/jute composites *Composites Part A: Applied Science and Manufacturing* **43** 1800-8

[9] Atiqah A, Jawaid M, Ishak M R and Sapuan S M 2017 Effect of Alkali and Silane Treatments on Mechanical and Interfacial Bonding Strength of Sugar Palm Fibers with Thermoplastic Polyurethane *Journal of Natural Fibers* 1-11

[10] Saha P, Chowdhury S, Roy D, Adhikari B, Kim J K and Thomas S 2016 A brief review on the chemical modifications of lignocellulosic fibers for durable engineering composites *Polymer Bulletin* **73** 587-620

[11] Thakur M K, Gupta R K and Thakur V K 2014 Surface modification of cellulose using silane coupling agent *Carbohydrate polymers* **111** 849-55

[12] Ticoalu A, Aravinthan T and Cardona F 2014 A study into the characteristics of gomuti (Arenga pinnata) fibre for usage as natural fibre composites *Journal of Reinforced Plastics and Composites* **33** 179-92

[13] Sawpan M A 2010 Mechanical performance of industrial hemp fibre reinforced polylactide and unsaturated polyester composites. The University of Waikato)

[14] Abdelmouleh M, Boufi S, Belgacem M N and Dufresne A 2007 Short natural-fibre reinforced polyethylene and natural rubber composites: effect of silane coupling agents and fibres loading *Composites science and technology* **67** 1627-39

[15] Zhou F, Cheng G and Jiang B 2014 Effect of silane treatment on microstructure of sisal fibers *Applied Surface Science* **292** 806-12

[16] Agrawal R, Saxena N, Sharma K, Thomas S and Sreekala M 2000 Activation energy and crystallization kinetics of untreated and treated oil palm fibre reinforced phenol formaldehyde composites *Materials Science and Engineering: A* **277** 77-82

[17] Satyanarayana N, Xie X and Rambabu B 2000 Sol–gel synthesis and characterization of the Ag 2 O–SiO 2 system *Materials Science and Engineering: B* **72** 7-12

[18] Aji I, Sapuan S, Zainudin E and Abdan K 2009 Kenaf fibres as reinforcement for polymeric composites: a review *International Journal of Mechanical and Materials Engineering* **4** 239-48

[19] Kabir M, Wang H, Lau K and Cardona F 2012 Chemical treatments on plant-based natural fibre reinforced polymer composites: An overview *Composites Part B: Engineering* **43** 2883-92

IOP Conf. Series: Materials Science and Engineering **368** (2018) 012048 doi:10.1088/1757-899X/368/1/012048

The effect of alkali treatment on tensile properties of coir/polypropylene biocomposite

Y R Ng[1], S N A M Shahid[1] and N I A A Nordin[1*]

[1]Faculty of Chemical & Natural Resources Engineering, Universiti Malaysia Pahang, 26300 Gambang, Pahang, Malaysia

*Corresponding author: idamalina@ump.edu.my

Abstract. The increase demand for use of natural fiber in biocomposites is due to its inexpensive material, abundantly available and it has reasonable strength. In this study, coir fiber were treated with NaOH solution (3 wt% and 5 wt%) for 2, 4 and 6 hours at room temperature. The treated and untreated coir fiber with size of 160 – 250 μm were extruded with polypropylene (PP) at fiber content of 10 and 30 wt%. The effect of alkali treatment was analyzed for its chemical composition, thermal stability and morphology. The biocomposites were tested for its tensile properties. Results showed that alkali treatment at 5 wt% for 6 hours had higher cellulose composition which was 29% compared to untreated coir. Thermal stability of treated sample increased and had high percentage of residue which showed the improved hydrophobicity of the sample. The surface of treated sample were rough due to removal of impurities and silica bodies. The tensile properties of biocomposite prepared from 5 wt% NaOH treated fiber/PP were higher by 28% compared to untreated coir/PP. This study shows that alkali treated coir fiber had better characteristics compared to untreated coir fiber and lead to the higher tensile properties of the PP-based biocomposite.

1. Introduction

In recent years, there are great interest toward development of natural fiber as reinforcements for polymeric biocomposite. This is because natural fiber reinforced biocomposite have the advantages of low cost, widespread availability, biodegradability, renewability and reasonable mechanical properties. There are various types of natural fibers such as jute, sisal, palm oil biomass, abaca and coir have been studied as reinforcement in biocomposite. Coir fiber is used widely as rope, mat, compost and filler in biocomposite [1]. Coir fiber can be obtained from coconut husk which routinely disposed from coconut plantation.

However, the main drawback of coir fiber is the low thermal stability and compatibility with hydrophobic polymer. The main constituents of coir fibers are 21.5-23.0 wt% hemicellulose, 31-55 wt% cellulose and 15-21 wt% lignin [2]. Coconut fibers has hydrophilic properties due to present of hydroxyl group (-OH) and carboxyl group (-COOH) in hemicellulose and cellulose resulted to incompatibility of coir fibers with hydrophobic polymer. Alkali treatment can be an alternative treatment method for reducing hemicellulose content to make the fibers less hydrophilic and enhance the compatibility of treated fibers with matrix such as polypropylene (PP) [3]. Morphology of alkali

IOP Conf. Series: Materials Science and Engineering **368** (2018) 012048 doi:10.1088/1757-899X/368/1/012048

treated fibers is rougher than untreated fibers viewed under SEM which will enhance the physical interaction of fibers and polymer [4].

In this present study, the concentration and treatment time of sodium hydroxide (NaOH) to treat coir fiber were studied. The characteristics of treated fiber were analyzed. Then, the treated fiber were mixed with PP at 10 and 30 wt% fiber content and the tensile strength of the biocomposites were determined. This research aims to utilize the abundant biomass from agricultural industry and to propose a simple treatment method to improve the characteristics of coir fibers.

2. Materials and Methods

2.1. Materials

Coir fiber was collected from coconut factory located in Sabak Bernam, Selangor. NaOH and potassium hydroxide (KOH) were supplied by Merck, Darmstadt, Germany. Sodium chlorite ($NaClO_2$) was supplied by Acros Organics, Geel, Belgium. Polypropylene is purchased in pellets form from Polypropylene (Malaysia) with code of 606251.

2.2. Alkali treatment

Coir fibers were cut into range of 160 to 250 mm and treated with NaOH solution in different concentration namely 3 wt% and 5 wt% at the room temperature. The treatment time were set at 2, 4, and 6 hours with coir fiber to NaOH solution at ratio of 1:15. After treatment, the fibers were wash with distilled water until neutralize and oven-dried overnight.

2.3. Fabrication of Coir / Polypropylene Biocomposite

The treated coir fibers with size of 160 – 250 µm and PP is dried at 60 °C prior to use. Biocomposites was prepared by twin screw extruder machine (Model: Prism Eurolab 16) at temperature range of 200 – 160 °C and 50 rpm rotor speed. The samples were mixed at 10 wt% and 30 wt% fiber content. Then, the extruded samples were pressed using hot press and cold molding machine (Model: Lotus Scientific-2205) with a pressure of 10 KPa, 160 °C for 10 min to produce 1 mm biocomposite sheet.

2.4 Characterization of Coir Fiber

2.4.1 Chemical Composition Analysis. The analysis was determined according to the procedure described by Iwamoto *et al.* [5]. Firstly, lignin were removed by soaking into 5 wt% of $NaClO_2$ for 1 hour at 70 °C and wash until pH neutral. The dried sample was weighed. Then, the bleached samples were soaked into 6 wt% of KOH for 24 hours at room temperature. The samples were rinsed and dried. The weight of dried sample was recorded. The final residue of the analysis contain only cellulose component.

2.4.2 Thermogravimetric Analysis (TGA). TGA was conducted on Hitachi STA 7200 Thermal Analysis. The coir weighed 6 to 10 mg was placed on aluminum pan. The sample was heated from 50 to 600 °C at heating rate of 10 °C/min under nitrogen flow of 100 mL/min.

2.4.3 Scanning Electron Microscope (SEM). The surface morphology of the untreated and treated fibers was studied using scanning electron microscopy machine (JEOL-JSM 840). The treated and untreated fibers are mounted on carbon tape and viewed with acceleration voltage of 10 kV [6].

2.5 Tensile Test

Tensile test were determined according to ASTM D638 on dumbbell shape specimens by using a universal tensile testing machine model A6-4. The testing condition used were crosshead speed 10 mm/min and load cell 1 kN. The recorded value are the average of 5 results.

IOP Conf. Series: Materials Science and Engineering **368** (2018) 012048 doi:10.1088/1757-899X/368/1/012048

3. Results and Discussion

3.1 Chemical Composition

The chemical composition of treated and untreated fiber are summarize in table 1. Result has shown that the composition of untreated fiber consist of 22.71% lignin, 39.15% hemicellulose and 38.14% cellulose. It was reported the untreated coir has about 21.5-23% hemicellulose, 31-55% cellulose and 15-21% lignin [2]. At longer treatment time the composition of cellulose and lignin were higher while the amount of hemicellulose decreased as the soaking time and NaOH concentration increased.

Table 1. The chemical composition of treated and untreated coir fiber.

Concentration of NaOH (wt%)	Sample	Soaking Time (hours)	Lignin (%)	Hemicellulose (%)	Cellulose (%)
Untreated	1	-	22.71	39.15	38.14
3 wt%	2	2	22.51	31.51	45.98
	3	4	24.29	27.49	48.22
	4	6	24.07	25.47	50.46
5 wt%	5	2	28.72	26.39	44.89
	6	4	28.65	25.12	46.23
	7	6	28.43	22.23	49.34

From table 1, the treatment with 3 wt% and 5 wt% NaOH for 6 hours had increased cellulose content by 32% and 29% respectively. Higher content of cellulose can increased the strength of fiber. As for lignin content, it was increased by 5.9% for 3 wt% and 25% for 5 wt% NaOH at 6 hours treatment. The highest removal of hemicellulose content was found in sample 7 with 43% removal. This result was supported by other studies which had shown that NaOH treatment able to reduce the hemicellulose content in the fiber [7].

Hemicellulose consist of branched and amorphous structure where the hydroxyl group (-OH) is easily broken by alkali treatment, which then react with water molecules (H-OH) and discharged out from the fiber structure. Due to the reaction, the hemicellulose content in the fiber was reduced. The cellulose component has crystalline structure and not easily hydrolyze by chemical treatment, thus the remaining cellulose molecules will formed a chain of fiber–O–Na group between cellulose molecules chain. According to the research conducted by Sun *et al.*, it was found that the treated fiber has a better packing of cellulose chain due to removal of amorphous components and may led to better properties of its biocomposite [8].

3.2 Surface Morphology of Samples (SEM)

The effect of the different concentration treatments on the surface fiber was investigated using the SEM. Figure 1a (i,ii) shows the SEM micrograph of untreated coir fiber. The untreated fiber is covered with a layer of impurities such as wax, pectin, oil and lignin, thus it provide a smooth surface as observed under SEM. The same observation was reported by Carvalho *et al.* whereby it was shown that the surface of coir fiber was covered by a layer of substances such as waxes, lignin and pectin [9]. It can be observed the silica bodies were deposited on the fiber surface.

After alkaline treatment as shown in figure 1b (i,ii) and 1c (i,ii), it can be seen that the fiber had rougher surface due to removal of impurities. Besides that, it can be observed that higher concentration of NaOH can caused more removal of silica bodies. For the 3 wt% NaOH treated fiber, crater can be seen due to removal of silica. While for the 5 wt% NaOH treated fiber, the roughness surface was increased and more crater or pore was revealed indicating more silica was removed. This result is in agreement with Wong *et al.* which demonstrated that the higher concentration of NaOH will increased the surface roughness of the fiber [10].

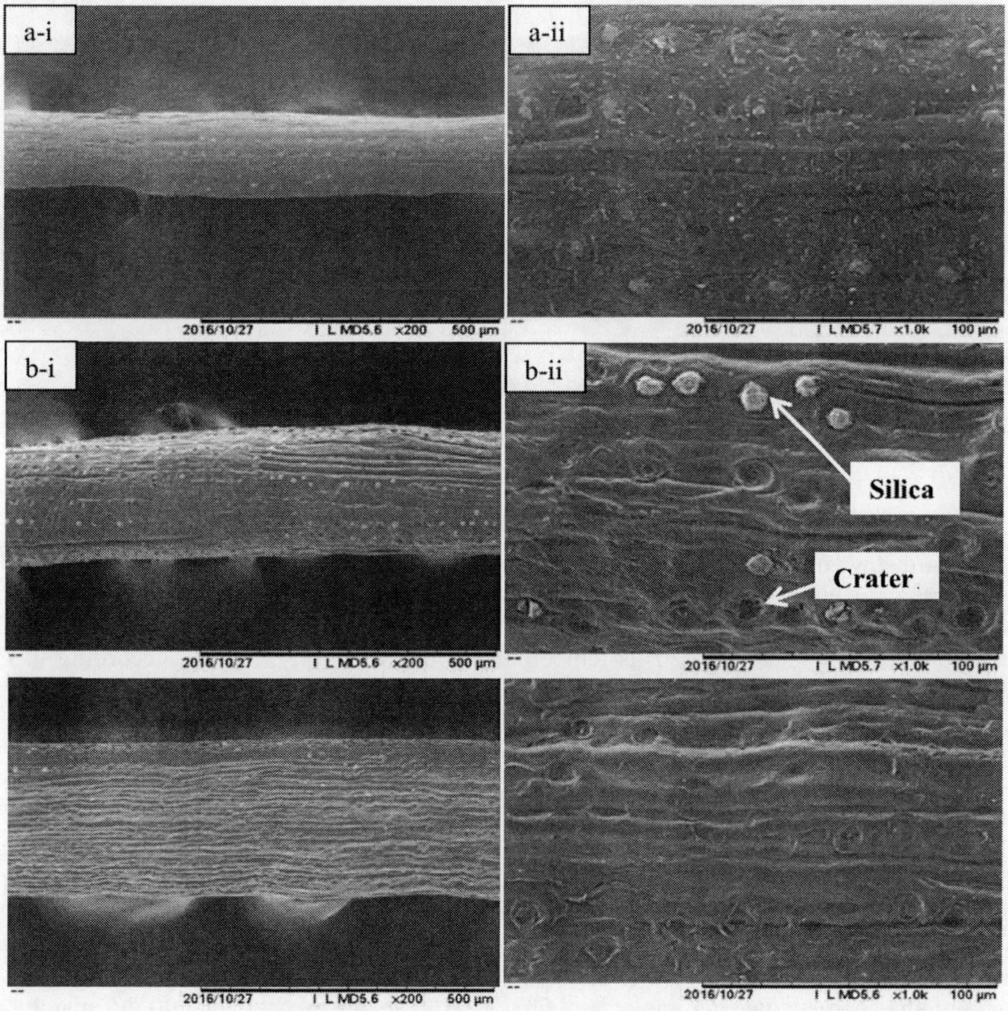

Figure 1. Scanning electron micrographs of (a) untreated fiber, treated fiber with (b) 3wt%, and (c) 5wt% of NaOH treatment for 6 hours.

In addition, the rough surface and dislodging of silica bodies will provide a good anchor for PP which will subsequently improve the physical interaction between fiber and polymer [11].

3.3 Thermal Stability

Figure 2 shows the result of TGA analyses for untreated and coir fiber treated with 3 wt% and 5 wt% NaOH for 6 hours. The thermal composition of coir consists of three decomposition stage. The first stage is the decomposition of moisture content which occurs between 80 and 100 °C. As the fiber is heated, the moisture content from the fiber will evaporated resulting the weight of the material decreased. The moisture content of the untreated coir is the highest and lower for the treated sample as shown in table 2.

The second stage is decomposition of hemicellulose which occurs at 200 to 300 °C. The untreated fibers had higher mass loss compared to the treated fiber. The results is in agreement with study done by Yang *et al.*, reported that hemicellulose degrade within temperature range of 180 to 350 °C [12].

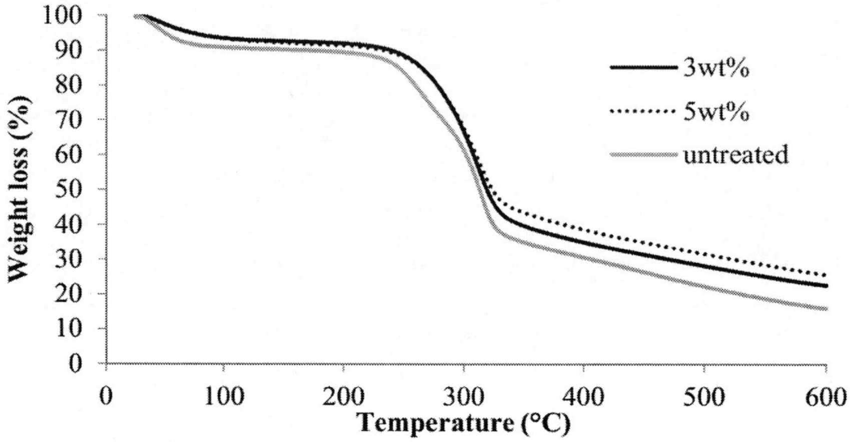

Figure 2. TG analysis for untreated coir fibers.

The third stage of degradation was contributed to lignin degradation. Lignin decomposed at slower rate compare to other components where the degradation temperature start from 190 to 600 °C. According to Yang *et al.*, although the decomposition of lignin could start as early as 160 °C, it decomposed slowly and extended its temperature as high as 900 °C [12]. According to Quajai and Shanks, the decomposition of lignin was occurring in a wider temperature range compared to hemicellulose and cellulose content on the coir [4]. Lignin presents a broad peak throughout the range, degrading between 280 and 500 °C.

Table 2 shows the weight lost and thermal stability of untreated and treated coir fiber. Temperature of fiber at 5% weight loss was increased for treated fiber. At higher concentration of NaOH the fiber had better thermal stability where the 5% and 50% weight loss occurred at higher temperature. This is due to the removal of hemicellulose that have low thermal stability. Based on table 1, untreated coir fiber had the highest composition of hemicellulose, while coir fiber treated with 5 wt% NaOH for 6 hours had the lowest hemicellulose content. The thermal stability result was in agreement with the results shows in table 1. Therefore, the reduction of hemicellulose content in the treated fiber will increased the thermal degradation of the fiber [13]. The residue left at 600 °C mainly containing lignin. Lignin contains polar hydroxyl groups and non-polar hydrocarbon and benzene rings making it less hydrophilic [14]. Thus, the increased residue of alkali treated fiber show the treated fiber had better hydrophobicity compared to untreated fiber. Thus, the compatibility of treated fiber and hydrophobic polymer will increased.

Table 2. Thermal stability of untreated and NaOH treated coir fiber for 6 hours of treatment.

Sample	Weight loss at 100 °C (%)	$T_{5\%}$[a] (°C)	$T_{50\%}$[b] (°C)	Residue at 600 °C (%)
Untreated coir	9.11	49.08	313.17	16.00
Treated fiber (3wt%)	6.47	72.50	319.25	22.62
Treated fiber (5wt%)	6.46	74.83	323.84	25.68

[a] Temperature at 5% weight loss.
[b] Temperature at 50% weight loss.

3.4 *Tensile properties*

The fiber content and fiber matrix interfacial adhesion is main factor that can affect the mechanical properties of biocomposite. The tensile properties of untreated coir fiber and alkali treated fiber biocomposite are shown in table 3. It can be observed the tensile strength increased with the fiber loading up to 30% for 3 wt% and 5 wt% NaOH treated fibers. The tensile modulus also increase as the fiber loading increased. The untreated coir/PP biocomposite has the lowest properties while the 5 wt% NaOH (6 hours) treated coir/PP biocomposite had the highest tensile strength. The tensile strength of this biocomposite sample was increased by 28% compared to untreated biocomposite at 10 wt% fiber content. At 30 wt% fiber content, the tensile strength of treated fiber biocomposite also higher compared to untreated biocomposite.

Table 3. Mechanical properties of coir/PP biocomposite.

Sample	10 wt% fiber loading		30 wt% fiber loading	
	Tensile strength (MPa)	Tensile modulus (MPa)	Tensile strength (MPa)	Tensile modulus (MPa)
Untreated coir/PP	21.66	922.27	22.33	1114.88
3 wt% NaOH[a] treated coir/PP	22.48	1122.39	23.79	1226.39
5 wt% NaOH[a] treated coir/PP	27.75	1328.76	28.74	1338.07

[a] For 6 hours treatment.

The tensile strength of fiber treated with 5 wt% NaOH was high due to increase cellulose content after alkali treatment. Alkali treatment able to reduce hemicellulose, waxes and silica bodies on the surface of the fiber led to improve the physical interaction between fiber and matrix. According to the study done by Mohanty *et al.*, alkali treatment provides the great fiber matrix adhesion through removal of hydrophilic components in fiber and subsequently enhances the tensile properties of the biocomposite [15].

4. Conclusion

In the present study, it is shown that alkali treatment successfully improved the characteristics of coir fiber to be used as reinforced material in biocomposite production. Coir fiber treated with 5 wt% NaOH for 6 hours gave the significant improvement to the fiber. Alkali treated fiber had low hemicellulose content while having higher percentage of cellulose and lignin. The surface of the fiber was clean and rough which beneficial for physical adhesion with polymer. The removal of low thermal stability components such as hemicellulose had increased the thermal stability of treated fiber. Higher residue at 600 °C of alkali treated coir shows that the treated fiber had better hydrophobicity compared to untreated fiber. Tensile strength of biocomposite prepared from alkali treated fiber is higher than untreated coir biocomposite.

Acknowledgements

The authors acknowledge Universiti Malaysia Pahang for the financial support through grant number RDU1703177.

References

[1] Zaman H U and Beg M D H 2014 Preparation, structure and properties of the coir fiber/polypropylene composites *J. Compos. Mater.* **48** 3293–301

[2] Li X, Tabil L G and Panigrahi S 2007 Chemical treatments of natural fiber for use in natural fiber-reinforced composites: A review *J. of Polym. Environ.* **15** 25–33

IOP Conf. Series: Materials Science and Engineering **368** (2018) 012048 doi:10.1088/1757-899X/368/1/012048

[3] Pickering K L, Efendy M A and Le T M 2016 A review of recent developments in natural fibre composites and their mechanical performance *Compos. Part A* **83** 98–112
[4] Quajai S and Shanks R A 2016 Composition , structure and thermal degradation of hemp cellulose after chemical treatments *Polym. Degrad. Stab.* **89** 327–35
[5] Iwamoto S, Abe K and Yano H 2008 The effect of hemicellulose on wood pulp nanofibrillation and nanofiber network characteristics *Biomacromolecules* **9** 1022–26
[6] Yunos N S H M, Baharuddin A S, Yunos K F M, Naim M N and Nishida H 2012 Physicochemical property changes of oil palm mesocarp fibers treated with high-pressure steam *BioResources* **7** 5983–94
[7] Gomes A, Matsuo T, Goda K and Ohgi J 2007 Development and effect of alkali treatment on tensile properties of curaua fiber green composites *Compos. Part A* **38** 1811–20
[8] Sun J X, Sun X F, Zhao H and Sun R C 2004 Isolation and characterization of cellulose from sugarcane bagasse *Polym. Degrad. Stab.* **84** 331–9
[9] Carvalho K, Mulinari D, Voorwald H and Cioffi M 2010 Chemical modification effect on the mechanical properties of hips/coconut fiber composites *BioResources* **5** 1143–55
[10] Wong K J, Yousif, B F and Low K O 2016 The effects of alkali treatment on the interfacial adhesion of bamboo fibres. Proceedings of the Institution of Mechanical Engineers, Part L: *J. of Mater. Des. Appl.* **224** 139–48
[11] Shinoj S, Visvanathan R, Panigrahi S and Kochubabu M 2011 Oil palm fiber (OPF) and its composites: *A review Ind. Crops Prod.* **33** 7–22
[12] Yang H, Yan R, Chen H, Lee D H and Zheng C 2007 Characteristics of hemicellulose, cellulose and lignin pyrolysis *Fuel* **86** 1781–88
[13] Mulinari D R, Baptista C A R P, Souza J V C and Voorwald H J C 2011 Mechanical properties of coconut fibers reinforced polyester composites *Procedia Eng.* **10** 2074–79
[14] Rozman H D, Tan K W, Kumar R N and Abubakar A 2001 The effect of hexamethylene diisocyanate modified ALCELL lignin as a coupling agent on the flexural properties of oil palm empty fruit bunch-polypropylene composites *Polym. Int.* **50** 561–7
[15] Mohanty A K, Misra M and Drzal L T 2001 Surface modifications of natural fibers and performance of the resulting biocomposites : An overview *Compos. Interfaces* **8** 313–43

IOP Conf. Series: Materials Science and Engineering **368** (2018) 012049 doi:10.1088/1757-899X/368/1/012049

Properties of Cellulose Extract from Different Types of Oil Palm Biomass

L N Megashah[1], H Ariffin[1, 2]*, M R Zakaria[1] and M A Hassan[1]

[1] Department of Bioprocess Technology, Faculty of Biotechnology and Biomolecular Sciences, Universiti Putra Malaysia, 43400 UPM Serdang, Selangor, Malaysia
[2] Laboratory of Biopolymer and Derivatives, Institute of Tropical Forestry and Forest Products (INTROP), Universiti Putra Malaysia, Serdang 43400 UPM, Selangor, Malaysia.

Corresponding author: hidayah@upm.edu.my

Abstract. This study focuses on the characteristics of cellulose extracts from different oil palm biomass namely oil palm mesocarp fiber (OPMF), oil palm empty fruit bunch (OPEFB) and oil palm frond (OPF). Cellulose was obtained from these biomass through soda pulping process followed by a totally chlorine free (TCF) bleaching step using peracetic acid solution. It was found that the cellulose extract for OPMF, OPEFB and OPF had 88.8%, 93.0% and 94.6% purity. Further analyses on the properties of cellulose in terms of morphology, thermal stability and crystallinity were conducted. The OPF cellulose obtained after pretreatment exhibited the highest crystallinity at 84.5%, followed by OPEFB (82.4%) and OPMF (69.2%). This directly affected the thermal stability of the celluloses, in which OPF cellulose was more thermally stable compared to celluloses from OPEFB and OPMF, as being recorded by a thermogravimetry analyzer (TGA). Results from this study provide new insight not only for fundamental understanding, but also to forecast the potential applications of cellulose extracts from different type of oil palm biomass, based on the performance of the celluloses reported herein.

1. Introduction

Oil palm or namely as *Elaeis guineensis* was early introduced in central of West Africa and its production was expanded worldwide in many countries due to the acceptance in global market. Malaysia and Indonesia have the largest area for oil palm plantation and are the biggest producer of palm oil in the world, mainly due to the favorable agro ecological zone which contributes to the high yield of oil palm [1-2]. Currently, the uses of oil palm can be seen in food and non-food applications. The large amount of palm oil being produced is in parallel with the high production of biomass, either from the palm oil mill or plantation. About 80 million tons of oil palm biomass is being generated in Malaysia annually [3].

IOP Conf. Series: Materials Science and Engineering **368** (2018) 012049 doi:10.1088/1757-899X/368/1/012049

Various types of biomass can be obtained from the oil palm plantation and palm oil mill, such as oil palm mesocarp fiber (OPMF), oil palm empty fruit bunches (OPEFB), oil palm fronds (OPF), oil palm trunks (OPT), leaves and roots. For example, the OPF will be cut during pruning seasons for harvesting oil palm fresh fruit bunches. It has been reported that the OPF tend to increase its production due to frond pruning before the age of 13 years [4]. OPF represents the major oil palm biomass from plantation area [5]. At present, oil palm biomass are mainly used for fuel generation, biocomposting and mulching.

It has been reported that cellulose represents about 1.5 x 10^{12} tons of the annual biomass production [6]. In biomass, cellulose is accompanied by other components such as hemicellulose, lignin and small amount of extractives. This structure is known as lignocellulosic materials which exhibit by its architectural of biological origin. Complex carbohydrate in cellulose composed by chain of polymer with unbranched β (1, 4) linked D glucopyranosyl units. There are intramolecular hydrogen bonds which are responsible for rigidity and stiffness of cellulose polymer, and bring the strong interaction between cellulose chains. This layer with anhydroglucopyranose unit causes the cellulose chain to held together by weak van der Waals forces [7]. The cellulose can form in both crystalline and amorphous structure by it placement at plant cell wall. For hemicellulose, it tends to have amorphous structure due to its long branches of polymer chain. Various monomeric sugars are found in hemicellulose such as glucose, galactose, mannose, xylose, arabinose and rhamnose. Hemicellulose represents about 25-30% in plant cell walls and comprises a mixture of a polysaccharide chain with low molecular weight as compared to cellulose. Whereas, lignin is a complex polymer in aromatic ring or refer as phenolic monomers. Phenol is very stable organic molecule benzene with an alcohol functional group. The main building alcohol of lignin are ρ-coumaryl alcohol, coniferyl alcohol and sinapyl alcohol [8].

Cellulose has become of interest in product development for commercialization purpose such as in paper making, composite production, textiles, pharmaceutical products, cosmetics and other advanced materials. Nevertheless, cellulose from different bioresources may have different characteristics, and the potential applications of cellulose from various bioresources could actually be expanded if the characteristics are well studied. To the best of our knowledge, there is lack of detailed study on the characteristics of cellulose obtained from oil palm biomass. Since there is abundant of oil palm biomass in Malaysia and they are readily available, it is important to have a set of data on the characteristics of cellulose obtained from different oil palm biomass in order to forecast the actual potential applications of each biomass. In this study, cellulose obtained from pre-treated OPMF, OPEFB and OPF were characterized for their chemical, morphological, thermal stability and crystallinity properties. Results obtained showed variation, suggesting the potential applications of these materials could be varied.

2. Experimental

2.1 Materials

Three types of biomass were used namely oil palm empty fruit bunch (OPEFB), oil palm mesocarp fiber (OPMF) and oil palm frond (OPF). OPEFB and OPMF were collected from Seri Ulu Langat Palm Oil Mill, Dengkil Selangor, Malaysia. OPF was obtained from Taman Pertanian Universiti (TPU), Universiti Putra Malaysia. OPF was shredded and pressed in order to remove its juice. Afterwards, all samples were thoroughly washed with tap water and sun dried to lower its moisture content in order to avoid fungal growth in samples. The dry samples were kept in sealed plastic bag and stored at room temperature prior to further treatment.

2.2 Chemical pulping

Prior to pulping process, the size of the biomass samples was standardized by cutting it into 1-2 cm in length. Pulping was conducted in a 15L batch reactor of lab-scale twin digester (model GTD-15L, GIST Co Ltd., Korea). Sodium hydroxide solution (14% w/v) was used. Fiber samples were mixed with the alkaline solution at ratio of 1:4 and the pulping process took place at 160°C with pressure at about 0.5 MPa. The condition for soda pulping was modified from [9]. Pulping process was conducted for 1 – 1.5 hour. After cooling, the obtained cellulose pulp was washed several times using tap water until the pH became neutral. The pulp was spun to remove water and dried at 60°C overnight.

2.3 Pulp bleaching

Dried cellulose pulp was subjected to bleaching process using a totally chlorine free (TCF) method. Peracetic acid (PA) solution was prepared by mixing 50% acetic acid and 30% hydrogen peroxide at equal volume ratio (1:1 v/v) with addition of catalyst, *i.e.* 98% sulphuric acid (H_2SO_4) at concentration of 1.1% (w/v) (modified from [10]). The pure PA solution was allowed to stand for 3 days prior to bleaching process. For bleaching, dried cellulose pulp samples were mixed with distilled water and PA solution at ratio 1:5:5 and vigorously stirred for 24 hours at 70 °C. Afterwards, the solid residues were collected and washed several times with tap water until the pH became neutral. The obtained cellulose samples were dried at 60 °C overnight. After the samples were cooled, they were kept in an airtight container until further use.

2.4 Chemical composition analysis

Samples were analyzed for their chemical composition using a Standard Test Method of TAPPI (Technical Association of Pulp & Paper Industry). The analysis was conducted to characterize the composition of untreated and treated fibers based on following TAPPI standard of cellulose (T-203cm-99), hollocellulose (Tappi useful Method 249-75), lignin (T-222om-88) and ethanol-toluene extractives (T-204cm-97). The chemical analysis was done in replicates to obtain the mean and standard deviation values.

2.5 Morphology

Surface morphology of cellulose was observed using a scanning electron microscope (SEM), JCM-6000PLUS NeoScope Benchtop SEM (JEOL Ltd., Japan). Firstly, the samples were coated with Pt using Hitachi Ion Sputter E-1030 and the magnifying view was at 500x -1000x magnification with 15kV of beam voltage. Diameter size of cellulose was analyzed using a polarized microscope.

2.6 Thermal stability

Thermal stability of samples was analyzed using a thermogravimetry analyzer (TGA) model EXSTAR6000 (Hitachi High-Tech Science Corporation, Tokyo, Japan). TG and DTG thermograms were obtained by conducting the TG analysis at temperature range in between 50 °C to 550 °C with heating rate of 10 °C/min. The analysis required pure nitrogen as the inert gas which was purged at flow rate of 100ml/min.

2.7 Crystallinity

Samples crystallinity was determined using a benchtop X-ray Diffractometer (XRD) model Rigaku MiniFlex. Operating conditions set were 40kV and 15mA. The peak measurement was conducted by taking 2θ value between 5 to 50° at a scan speed at 2.00°/min. The crystallinity index (CrI) was calculated based on calculation by Segal method [11] as following:

$$CrI = \frac{\text{I002} - \text{Iam}}{\text{I002}} \times 100 \quad (1)$$

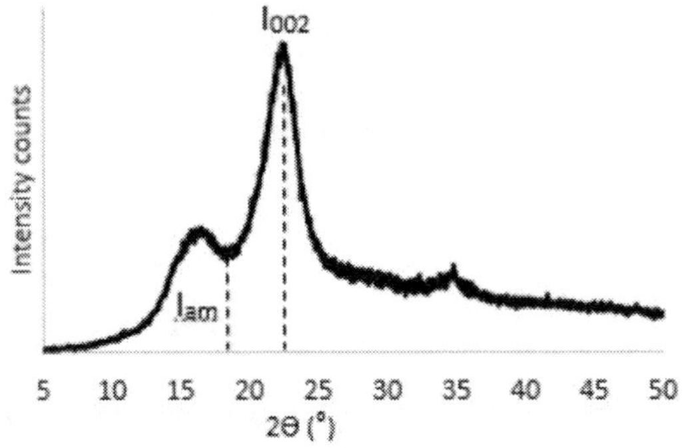

Where, I_{002} 2θ was from highest intensity at 22° and I_{am} 2θ was from minimum intensity at 18° which known as amorphous scattering.

3. Results and Discussion

Three different types of oil palm biomass were used to investigate the effect of pretreatment process on various anatomical parts of oil palm. The success of pretreatment process in extracting cellulose from oil palm biomass can be determined physically by observing the color changes after the pretreatment. The original color of the fibers was brown. After pretreatment, the color changed to white indicates the major composition of the fiber which is cellulose. The change in color could be contributed by solubilization of lignin at high temperature in alkaline charge solution [12]. The cellulose is known as represent white in color. In this study, all samples turned to white color after pulping and bleaching. The composition of the samples before and after pretreatments is shown in Table 1. OPF had the highest composition of cellulose after the pretreatment with cellulose composition of 94.6%, followed by OPEFB and OPMF with 93.0% and 88.8%.

It is seen that the chemical composition varied greatly depending on the part of the plant. For instance, OPMF exhibited the highest composition of lignin at 32.4%, as compared to OPEFB and OPF at 18.6% and 16.9%, respectively. Cellulose on the other hand is the least in OPMF, followed by OPEFB and OPF. This observation is similar to that reported by [13] in which different chemical composition was reported for oil palm fiber from different anatomical part. This could be explained by the function of each fiber. For instance in palm fruits, oil is stored in mesocarp. Since oil is hydrophobic, lignin content in OPMF is hence high, because lignin is the hydrophobic part which could function to allow the fruit to store the oil and protect it from wet condition, for example during raining.

After pretreatment, all samples exhibited non-detectable amount of lignin which indicates that the lignin component was being removed almost completely after the pretreatment. Despite of lignin removal, some hemicellulose retained after pretreatment as shown by the % of hemicellulose remained in the samples. OPMF showed the highest hemicellulose content as compared to OPEFB and OPF. This is expected to be contributed by the content of recalcitrant hemicellulose in the sample. It has been shown that some portion of hemicellulose is trapped within cellulosic microfibrils. Also, covalent linkages between arabinose and lignin have been proposed [14]. These two conditions lead to a condition which is called recalcitrant hemicellulose. Since OPMF contained the largest amount of lignin initially, it is expected that more hemicellulose fraction is bound covalently to lignin in OPMF compared to the OPEFB and OPF. It is hence the highest amount of residual hemicellulose was found in OPMF. This is in agreement with earlier report which exhibited that hemicellulose hydrolysis is limited by high lignin content [14].

Table 1. Chemical composition (%) of different oil palm biomass pretreated with pulping and TCF bleaching.

Oil palm biomass	Cellulose (%)	Hemicellulose (%)	Lignin (%)	Extractives (%)
OPMF				
Untreated OPMF	28.2 ± 0.8	32.7 ± 4.8	32.4 ± 4.0	6.5 ± 0.1
OPMF cellulose extract	88.8 ± 1.7	11.2 ± 1.7	ND	ND
OPEFB				
Untreated OPEFB	37.1 ± 4.4	39.9 ± 0.75	18.6 ± 1.3	3.1 ± 3.4
OPEFB cellulose extract	93.0 ± 0.5	7.0 ± 0.5	ND	ND
OPF				
Untreated OPF	45.0 ± 0.6	32.0 ± 1.4	16.9 ± 0.4	2.3 ± 1.0
OPF cellulose extract	94.6 ± 3.2	5.4 ± 3.2	ND	ND

*Note: Chemical composition (%) expressed on oven dry basis (OD). Data represent in replicates (mean values). ND indicates not detectable.

Higher amount of extractives was found in OPMF sample, which was contributed by the residual oil content in the OPMF. OPMF stores oil, and some oil is still present after pressing of the oil palm fruits [15].

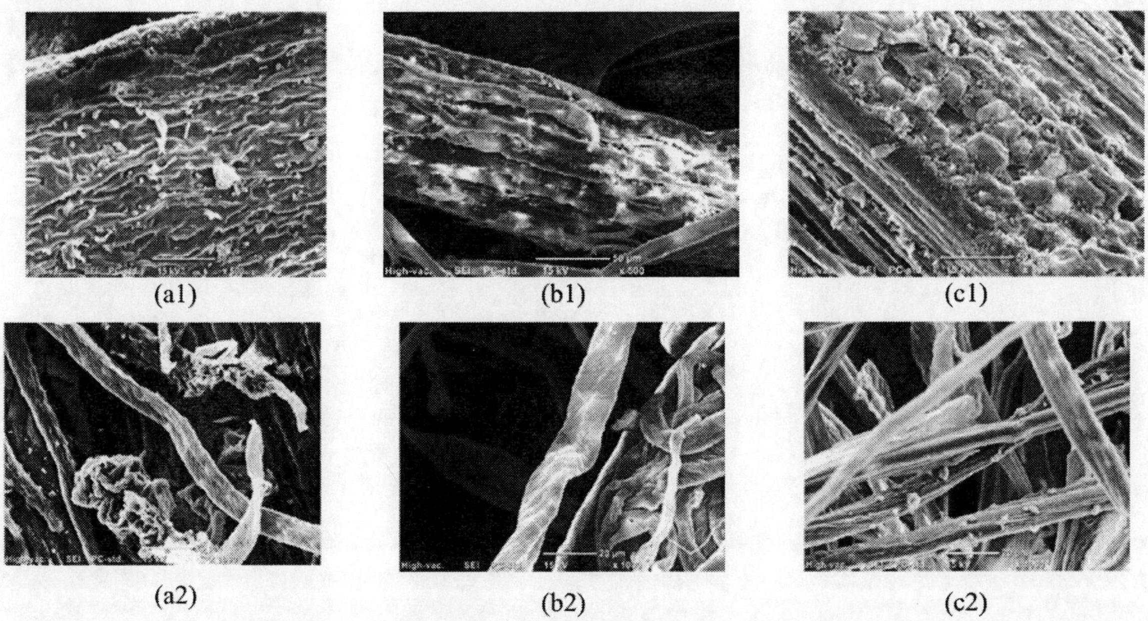

Figure 1. SEM images of untreated OPMF, OPEFB and OPF (a1-c1) and cellulose extracts from OPMF, OPEFB and OPF (a2-c2)

SEM images of cellulose obtained from pretreated fibers are shown in Figure 1 (a2-c2). Several microfibrils are seen from all images, indicating disruption of lignocellulosic structure as compared to the untreated oil palm biomass in Figure 1 (a1-c1). This observation is mainly due to the removal of lignin and hemicellulose [16]. OPMF shows incomplete microfibrillation as compared to OPEFB and OPF, as seen by the presence of non-fibrillated structure. This may indicate the presence of other components, which support the chemical composition results shown in Table 1. The diameter size of fibers was less than 10μm as estimated from SEM images.

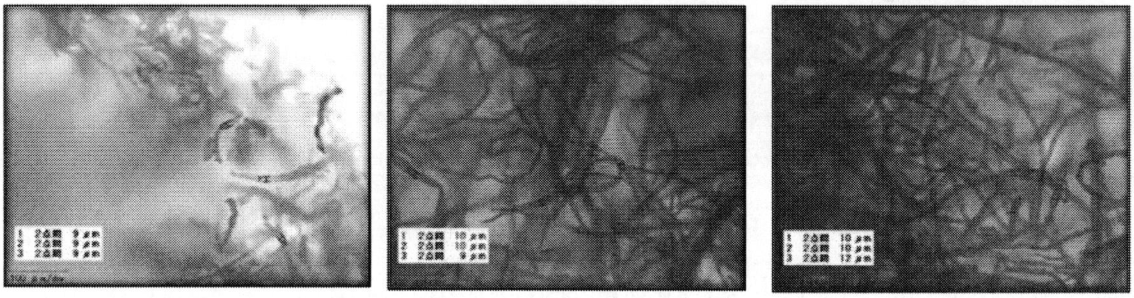

Figure 2. Polarized images showing cellulose fibers from (L-R) OPMF, OPEFB and OPF

Polarized microscope images (Figure 2) further exhibit the diameter size of the microfibrils cellulose. All the biomass had almost similar diameter size, in the range of 9 - 11μm. However there was no image shown for untreated biomass due to large micro bundles fibrils to be seen by polarized microscope.

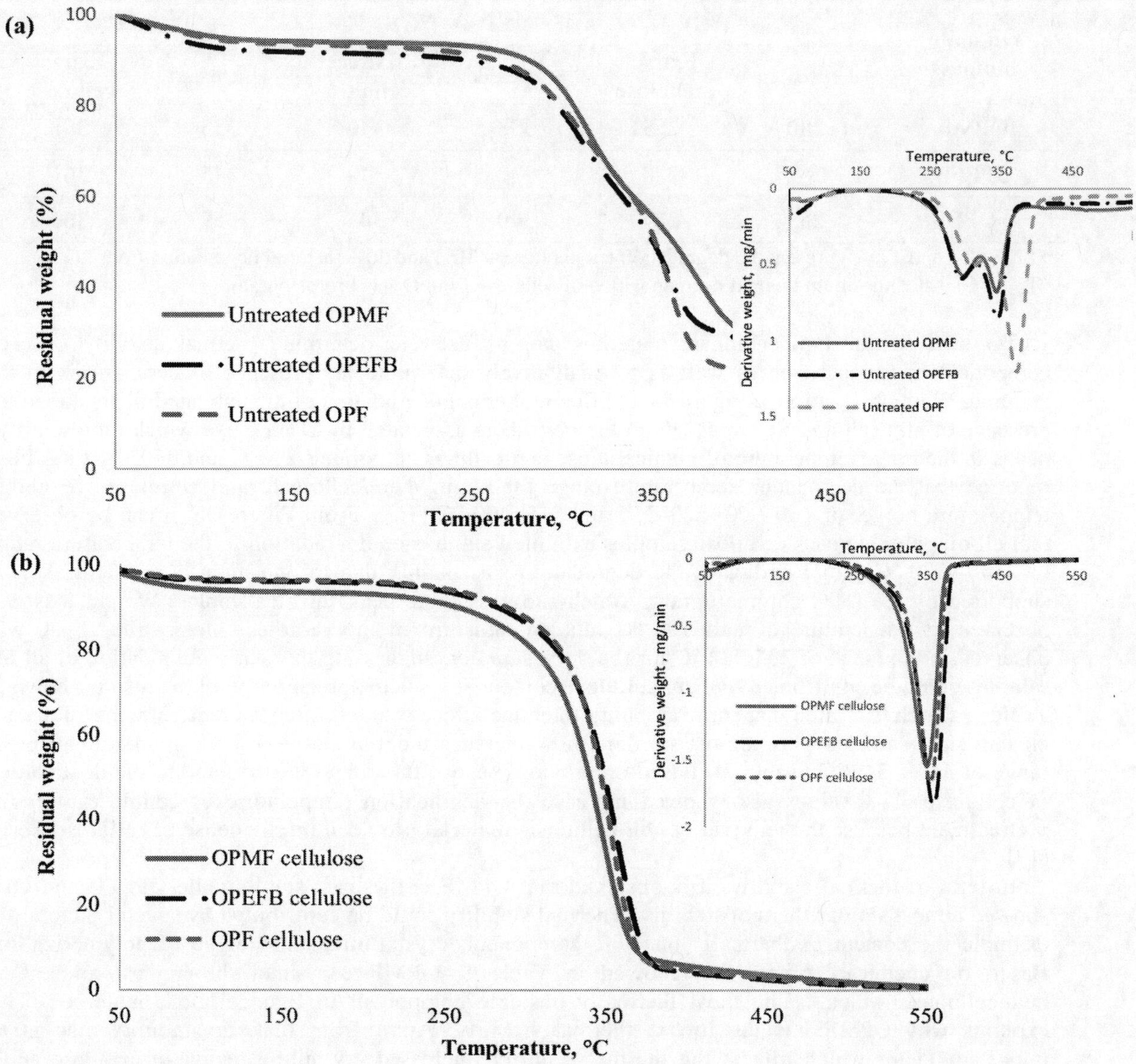

Figure 3. TG and DTG chromatograms of (a) untreated and (b) cellulose extracts from OPMF, OPEFB and OPF.

IOP Conf. Series: Materials Science and Engineering **368** (2018) 012049 doi:10.1088/1757-899X/368/1/012049

Table 2. Data evaluated from TG and DTG curve

Oil palm biomass	$T_{d10\%}$ (°C)		$T_{d50\%}$ (°C)		T_{max} (°C)	
	Untreated	Extract cellulose	Untreated	Extract cellulose	Untreated	Extract cellulose
OPMF	280	255	338	340	335	351
OPEFB	247	280	332	349	345	361
OPF	253	284	340	347	359	360

*$T_{d10\%}$ (°C) and $T_{d50\%}$ represent the degradation temperature at 10% and 50% material degradation from TG data.
*T_{max} (°C) is the maximum thermal decomposition of cellulose from DTG chromatogram.

TG analysis for lignocellulosic materials can be used to determine thermal stability of each component in the material, as well as to qualitatively determine the presence of components in the material. From observation in Figure 3a, a different thermal degradation of the untreated fibers due to the presence of non-cellulosic components in the raw fibers associated by DTG curve which represent two peaks in the range temperature of hemicellulose and cellulose at around 290°C and 370°C. It has been reported that the degradation temperature range for lignin, hemicellulose, and cellulose are within temperature ranges of 160–900, 220–315 and 315–400 °C [16]. From Figure 3b, it can be observed that all oil palm biomass cellulose samples exhibited single step degradation in the temperature range of 230 – 380 °C, which indicates the degradation was mainly due to one main component. This is supported by the DTG chromatograms which showed single peak for all samples. Weight loss was observed at temperature around 100 °C, due to moisture in the samples. Meanwhile, $T_{d10\%}$ was observed in the range of 255-284 °C for the three samples, suggesting that early degradation of all the samples could be contributed by hemicellulose content. This is in agreement with the results shown in Table 1, which exhibited that some amount of hemicellulose was left after the pretreatments. $T_{d50\%}$ for all untreated samples were almost similar, show increase thermal stability after pretreatment to the range of 340 – 349 °C (Table 2). Based on previous report, this range is attributed to the degradation of cellulose. T_{max} values also showed increase the degradation temperature of cellulose after the pretreatment because the removal of non-cellulosic material provided highly dense of cellulose fibrils [17].

In term of thermal stability, it is observed that OPMF is the least stable while OPEFB and OPF showed almost similar thermal stability. Thermal stability could be contributed by several factors, for example the content of thermally unstable component, crystallinity and degree of polymerization. Based on chemical composition showed in Table 1, OPMF contained the highest amount of hemicellulose, which is the most thermally unstable component in lignocellulose materials. This explains why OPMF has the lowest thermal stability. Apart from that, crystallinity also is an important factor which affects the thermal stability. Increased crystalline region in cellulose could increase its thermal stability [10].

In corresponding to the crystallinity, XRD data was analyzed and the results are shown in Table 3. It is apparent that the highest crystallinity was recorded for OPF cellulose extract (84.5%) followed by OPEFB (82.4%). Distinct observation can be seen for OPMF cellulose extract, in which the crystallinity of the sample was only 69.2%. This observation could be contributed by hemicellulose content in OPMF cellulose extract. It has been widely known that hemicellulose is amorphous [18]. The higher content of hemicellulose in OPMF cellulose extract was hence influencing the overall crystallinity of the sample. In comparison to other reports, the crystallinity of oil palm biomass before pretreatment was almost similar for all biomass. There were variations in crystallinity of samples after pretreatment, which is expected due to the pretreatment methods used.

Table 3. Crystallinity of cellulose from untreated and treated oil palm biomass fibers in comparison with other studies.

Oil palm biomass	Crystallinity (%)		Reference
	Before pretreatment	After pretreatment (Cellulose extract)	
OPMF	34.3	69.2	This study
	15 - 18	51 - 55	[19]
	38.5	56.8	[20]
OPEFB	45.0	82.4	This study
	43	-	[21]
	50	62	[22]
OPF	54.5	84.5	This study
	35	-	[23]
	-	57	[24]

4. Conclusions

Characteristics of cellulose from different bioresources may vary despite of its similar chemical structure. The characteristics could be vary due to the nature of the bioresources which may influence the physical characteristics of the cellulose and hence, its performance for product development. In this study, we demonstrated the characteristics of different oil palm biomass: OPMF, OPEFB and OPF. OPF and OPEFB exhibited superior performance in terms of cellulose purity, thermal stability and crystallinity compared to OPMF. Variation in properties of cellulose from oil palm biomass may suggest its suitability in different applications. For instance, pretreated cellulose from OPF and OPEFB which showed better crystallinity and thermal properties may be suitable as reinforcement material for biocomposites. Cellulose from OPMF on the other hand can be used as starting material for cellulose nanofiber production since it has been reported that some amount of hemicellulose in cellulose sample may assist in nanofibrillation. The results from this paper are important to provide fundamental understanding on oil palm biomass cellulose as well as to estimate its actual potential for product development. The variation in the results obtained suggesting that different applications of material could be proposed and developed.

Acknowledgement
The authors would like to acknowledge Universiti Putra Malaysia for financial support through Putra Berimpak (vote no: 9523000) and Putra-IPS (vote no: 9519300). Appreciation also goes to Taman Pertanian Universiti (TPU), Universiti Putra Malaysia and Seri Ulu Langat Palm Oil Mill for providing the oil palm biomass.

References

[1] Fairhurst T and Härdter R 2003 *The Oil Palm. Management for Large and Sustainable Yields* (Canada) International Potash Institute. (IPI), 191-230.

[2] Abdullah N and Sulaiman F 2013 Oil palm biomass as a sustainable energy source: A Malaysian case study *Biomass and Bioenergy* **3** 97–103

[3] Aljuboori A H R 2013 Oil Palm Biomass Residue in Malaysia : Availability and Sustainability *Int. J. Biomass Renewables* **2** 13–8

[4] Henson I A N E, Betitis T, Tomda Y and Chase L D C 2012 The Estimation Of Frond Base Biomass (FBB) Of Oil Palm (FBB) **24** 1473–9

[5] Safwan M, Ong M, Lin H and Tieng W 2016 Characterization of nanocellulose recovery from Elaeis guineensis frond for sustainable development *Clean Technol. Environ. Policy*

[6] Khalil H P S A, Jawaid M, Hassan A, Paridah M T and Zaidon A 2012 Oil Palm Biomass Fibres and Recent Advancement in Oil Palm Biomass Fibres Based Hybrid Biocomposites *Composites and Their Applications* ed N Hu (Rijeka: InTech) p. 187-220 Ch. 8

[7] Fahma F and Iwamoto S 2010 Isolation , preparation , and characterization of nanofibers from oil palm empty-fruit-bunch (OPEFB) *Cellulose* **17** 977–85

[8] Harmsen P, Huijgen W, López L and Bakker R 2010 Literature Review of Physical and Chemical Pretreatment Processes for Lignocellulosic Biomass *Food Biobased Res.* 1–49

[9] Hasanjanzadeh H, Hedjazi S, Ashori A, Mahdavi S and Yousefi H 2014 Effects of hemicellulose pre-extraction and cellulose nanofiber on the properties of rice straw pulp *Int. J. Biol. Macromol.* **68** 198–204

[10] Nascimento P, Marim R, Carvalho G, Celso R, Cid G, Celso R and Cid G 2015 Nanocellulose Produced from Rice Hulls and its Effect on the Properties of Biodegradable Starch Film *Materials Research* **19** 1–8

[11] Segal L, Creely J J, Martin A E and Conrad C M 1959 An Empirical Method for Estimating the Degree of Crystallinity of Native Cellulose Using the X-Ray Diffractometer *Text. Res. J.* **29** 786– 794

[12] Md Yunos N S H, Baharuddin A S, Md Yunos K F, Naim N and Nishida H 2012 Physicochemical property changes of oil palm mesocarp fibers treated with high-pressure steam *BioResources* **7** 5983–94

[13] Saka S and Bae H-J 2016 Chapter 11 - Secondary Xylem for Bioconversion *Secondary Xylem Biology* ed Y S Kim, R Funada and A P Singh (Boston: Academic Press) pp 213–31

[14] Kapu N and L. Trajano H 2014 Review of hemicellulose hydrolysis in softwoods and bamboo *Biofuels, Bioprod. Biorefining* **8**

[15] Neoh B K, Thang Y M, Zain M Z M and Junaidi A 2011 Palm pressed fibre oil: A new opportunity for premium hardstock *Int. Food Res. J.* **18** 769–73

[16] Zeng Y, Zhao S, Yang S and Ding S ScienceDirect Lignin plays a negative role in the biochemical process for producing lignocellulosic biofuels *Curr. Opin. Biotechnol.* **27** 38–45

[17] S J C C, George N and Narayanankutty S K 2016 Isolation and characterization of cellulose nanofibrils from arecanut husk fibre *Carbohydr. Polym.* **142** 158–66

[18] Tian D, Chandra R P, Lee J S, Lu C and Saddler J N 2017 Biotechnology for Biofuels A comparison of various lignin - extraction methods to enhance the accessibility and ease of enzymatic hydrolysis of the cellulosic component of steam - pretreated poplar *Biotechnol. Biofuels* 1–10

[19] Yasim-Anuar T A T, Ariffin H, Norrrahim M N F and Hassan M A 2017 Factors Affecting Spinnability of Oil Palm Mesocarp Fiber Cellulose Solution for the Production of Microfiber *Bioresources* **12** 715–34

[20] Souza N F, Pinheiro J A, Silva P, Paulo J and Morais S 2015 Development of Chlorine-Free Pulping Method to Extract Cellulose Nanocrystals from Pressed Oil Palm Mesocarp Fibers *J. Biobased Mater. Bioenergy* **21** issue 2 pp: 532-540

[21] Haji Badri K 2016 Thermal behaviors of oil palm empty fruit bunch fiber upon exposure to acid-base aqueous solutions *Malaysian J. Anal. Sci.* **20** 1095–1103

[22] Ramli R, Junadi N, Beg H and Yunus R 2015 Microcrystalline Cellulose (MCC) from oil palm empty fruit bunch (EFB) fiber via simultaneous ultrasonic and alkali treatment *Int. J. Chem. Nucl. Mater. Metall. Eng.* **9** 8–11

[23] Nordin N A, Sulaiman O, Hashim R, Haafiz M and Kassim M 2016 Characterization of Different Parts of Oil Palm Fronds (Elaeis Guineensis) and Its Properties Intl. *J. Advan. Sci. Eng. Infor. Technol.* **6**, 74–76.

[24] Syamani F A and Suryani A 2015 Changes in Oil Palm Frond Fiber Morphology , Cellulose Crystallinity and Chemical Functional Groups during Cellulose Extraction Phases *Chemistry and Materials Research* **7** 105–114

Mechanical properties of laminated kenaf woven fabric composites for below-knee prosthesis socket application

M H Nurhanisah[1*], F Hashemi[1], M T Paridah[1], M Jawaid[1*], J Naveen[2].

[1]Laboratory of Biocomposite Technology, Institute of Tropical Forest and Forest Products, Universiti Putra Malaysia, 43400 UPM Serdang, Selangor, Malaysia.
[2]Department of Mechanical and Manufacturing engineering, Faculty of Engineering, Universiti Putra Malaysia, 43400 UPM Serdang, Selangor, Malaysia.

*Corresponding authors: mhnurhanisah@gmail.com, jawaid@upm.edu.my

Abstract. Prosthesis is a process of creating artificial human body part. The primary function of below-knee prosthesis is controlling the leg during static and dynamic conditions. The aim of this study is to develop a prosthetic socket made from kenaf-glass fiber composite. The development of prosthetic socket with kenaf-glass fiber was carried out as an attempt to substitute prosthetic socket made from fiberglass polyester composites. Lamination method was adopted to fabricate the prosthetic socket. The laminates contain woven kenaf, glass silk knitted fabric and nylon knitted fabric. The effect of kenaf fabric layering sequence on the volumetric and mechanical properties were evaluated. The flexural and impact strength of 2 layers of kenaf fiber-based laminates are 7.11 MPa and 16.7 kJ/m^2 respectively which is higher than the single layer kenaf fabric-based composites. From the results woven kenaf fabric composites has the potential to replace the existing glass fabric-based polymer composites. It is a great initiative to fabricate a prosthetic socket which biodegradable, environmental friendly, locally available, lightweight, comfortable and psychosocially acceptable.

1. Introduction

Natural fibers as a reinforcement in composite materials gaining more attention among researchers and industries due to good mechanical strength, biodegradability, renewability, environmental friendly and cost reduction than synthetic fibers [1]. Natural fiber has a potential to be developed as an alternative t o replace the synthetic fiber composites. Natural fiber composites nowadays are getting acceptance in many medical device applications. For instance, Irawan et al. has developed the prosthetic socket mad e of ramie fiber reinforced epoxy composite [2]. Also the studies found bamboo fiber reinforced comp osites have the potential to replace with cotton and nylon composites in orthotic and prosthetic laminat ions [3]. In the other study [4], the authors revealed the opportunity to further development in rattan fi ber as above knee prosthetic socket.

Kenaf or Hibiscus cannabinus L. is plant that traditionally is used for twine, rope and coarse cloth [5]. Kenaf fibers are new type of economic crop and has gained a world attention in many application of eco-friendly assets such as automotive, furniture, sports industries and food packaging [6]. It has excellent fiber strength and great potential to be used as industrial materials and competing with hemp and flax in global market [7]. Since kenaf fiber has superior mechanical properties it could be a potential reinforcement in the polymer matrix instead of Glass. Also it is locally available, renewable, eco-friendly and low cost material [8].

The prosthesis is an artificial extension that replaces a missing body part, such as an upper or

IOP Conf. Series: Materials Science and Engineering **368** (2018) 012050 doi:10.1088/1757-899X/368/1/012050

lower body extremity [9]. Therefore, the prosthesis must be comfortable to wear, easy to don and doff, cosmetically pleasing, light in weight, durable, function well and offer easy maintenance. Limb prosthesis characteristically has three main parts, which are interface, components and cover. As for the prosthetic legs, they consist of a socket, pylon and foot, which make up the main parts. The prosthetic leg socket is the uppermost part of a prosthesis that makes contact with the residual limb and transfers the forces of walking to the ground. Thus, the material of a prosthetic socket is the most important in determining the quality of fitting. Prosthetic socket basically made from glass fiber composite material. Fiberglass is a strong and lightweight reinforcement used in composites. However, few drawbacks have found in fiberglass due to disadvantages issues such as non-environmental friendly, expensive, and high hardness [10-11].

Previous study has mentioned that the incorporation of woven kenaf fibres in the lamination socket could improve the mechanical properties of prosthetic socket [12]. In this study, prosthesis socket has been made by laminated process with the different numbers kenaf layers in the composite. Therefore, the study was conducted by comparing two samples represent as composite A which has one layer of kenaf and composite B which composed of two layers of kenaf to determine the mechanical strength of woven kenaf fabric composite. The flexural and impact tests were conducted for this research. In determining the material's ability to resist deformation under load, the flexural tests were required due to the reason that prosthesis should be functioning exactly or at least emulate closely the missing limbs or body parts [13]. The flexural modulus is a measure of the resistance to deformation of the composite in bending [4]. The flexural strength analysis done for prosthetic socket to identify the strength needed to support the body weight and extreme movement when the patient walks with it [4].

2. Material and method

2.1 Materials

The kenaf woven fiber, with a linear density of 150 tex for its yarns, were supplied by Sri Jentayu Sdn. Bhd. (Kuala Lumpur). Glass-silk, helanca and polyvinyl alcohol (PVA) knitted fabrics were obtained from Endolite Sdn. Bhd. The acrylic resin and härterpulver UN 3106 hardener which also obtained from Endolite Sdn. Bhd. The hexagon mould was used as positive cast and the materials such as PVA film, helanca stockinette and glass silk stockinette were used in layering as same as laminated fabrication of prosthetic socket.

2.2 Fabrication of laminated composites

Kenaf woven composite laminates were produced using vacuum infusion in room temperature (25 – 30°C). The laminated composite is shown in Figure 1. Two PVA film are used as the first layer to form a separator between liner and mould, and the second is to enclose the whole layup and to hold the resin together in reinforcing materials [14]. The glass-silk knitted fabric provides strength to the socket and the helanca knitted fabric improves the elasticity of the composites. Two types of composites namely; composites A and B, with different kenaf fiber content were fabricated. The composite A was produced with 6 layers of different materials while the composite B was produced with 7 layers different materials included woven kenaf. The Figure 2 shows the arrangement of composites A and B. The vacuum processor with 60 kPa is used to remove the air trapped during the lamination process.

Figure 1. The laminated of woven kenaf composite in hexagon mould using vacuum infusion.

Composite A

Polyvinyl Alcohol (PVA)
Helanca Stockinette
Kenaf Woven
Glass Silk Stockinette
Helanca Stockinette
Polyvinyl Alcohol (PVA)

Composite B

Polyvinyl Alcohol (PVA)
Helanca Stockinette
Kenaf Woven
Kenaf Woven
Glass Silk Stockinette
Helanca Stockinette
Polyvinyl Alcohol (PVA)

Figure 2. The sequence of material layer in Composite A and B.

3. Characterization of the composites

3.1 Volumetric composition

Prior to analysis of the mechanical properties of the composite laminates, the precise volume determination of the reinforcement and matrix is important for further analysis of the mechanical properties. To this end, the weight fraction and accurate density of the compositions need to be measured. First of all, the aerial weight (i.e, mass) of the composites and compositions were measured.

The composites and compositions were cut in to the size of 23×4 cm^2 and weighed precisely (±0.0001 g). The aerial weight of the kenaf woven fabric was obtained from the data sheet of the supplier. In order to measure the aerial weight of the knitted reinforcements accurately, they have been stretched and cut into the required dimensions. The average aerial weight of the composite and constituents are given in Table 1. The mass of the matrix for each composite was calculated by subtraction of the total mass of the reinforcements from the mass of the composite.

Table 1. The aerial weight and density of the composites and constituents.

	Composite A	Composite B	Kenaf woven fabric	Glass silk stockinette	Helanca stockinette	PVA film	Matrix for composite A	Matrix for composite B
Aerial weight (g)	21.48	26.57	1.18	1.52	0.5848	0.9607	15.70	19.60
Density (g/cm^3)	1.23 (0.02)	1.25 (0.01)	1.51 (0.04)	2.88	1.12 (0.01)	1.18 (0.00)	1.18	1.18

By knowing the number of laminates, mass of the composites, and composite compositions, the weight fraction of the compositions can be calculated by dividing the mass of the constituents to the mass of the composite. The density of the composites, kenaf woven fabric, glass silk stockinette, helanca stockinete, PVA film, and cured matrix were measured by using the Mettler Toledo (XS205) density kit. It is important to note that glass silk stockinette is a spun yarn that made of glass yarn and silk yarn. During the sampling for density measurement the care should be taken into account in a way that the sample contains both glass yarn and silk yarn. Distilled water was used as an immersion liquid and its temperature was controlled before each measurement for correction of the distilled water density. For determination of the cured matrix density, first the acrylic resin mixed with hardener in the petri dish with the same ratio that used for fabrication of the composites, and then the mixture was put under the vacuum for 1 hr to remove the air bubbles inside the matrix. Minimum six samples with the size of 2×2 cm were cut for measurement of the density. Finally, the volume fraction of the compositions and void were calculated by using the following equations 1 and 2.

$$V_i = \frac{\rho_c}{\rho_i} W_i \tag{1}$$

$$V_v = 1 - \sum V_i \tag{2}$$

where, V_i is the volume fraction of the composition, ρ_c is the density of the composite, ρ_i is the density of the composition, W_i is the weight fraction of the composition, and V_v is the void volume fraction.

3.2 *Flexural properties*

Figure 3 shows the set up for the flexural tests. Two types of composites were tested using the three-point bending test according to the the EN ISO 14125 standard by using the 5kN Instron 4201 Universal Testing Machine. The tests were done under the laboratory condition with temperature 23 ± 2 °C and relative humidity of 50 ± 10%. The load was applied midway between the supports with a crosshead speed of 2.2 mm/min and each sample was loaded to failure. Based on the standard the dimensions of the specimens for flexural testing were cut into the size of 40 mm x 7 mm (length x width). Figure 4 shows the total of specimen for flexural tests. The average values were reported including standard deviations.

Figure 3. Flexural specimen tests

3.3 Impact properties

Impact specimens were tested using Instron CEAST 9050 Pendulum Systems (IZOD IMPACT TESTER) with 2.75 joules pendulum. Test samples were prepared based on ISO 180:2000 (E) which are 80 x 10 x 4 mm (length x width x thickness). The energy required to break the sample was divided by unit area of residual cross section of sample to obtain impact resistance value [15].Impact resistance was calculated by using the following equation 3:

$$^{a}\text{iN} = \frac{\text{Ec}}{h.b} x10^{3} \qquad (3)$$

Where,

Ec = corrected energy, in joules, absorbed by breaking the test specimen;

h = thickness of the specimen (mm)

b = width of the specimen (mm)

3.3 Walking ability testing

Each prosthetic socket prototype was tested to determine the comfort level. Respondent was asked to use the prosthetic socket and evaluate the comfortability of each prosthetic socket. Test results were compared between one layer and two layer of laminated kenaf woven fabric composites.

4. Results and discussion

4.1 Volumetric composition

Table 2 shows the weight fractions and volumetric compositions of the composites A and B. The void content for composite A and B is 0.00 and 0.02 respectively. It indicates that the composites have good quality. In general, the composites with void content in the range of 0-2% considered as a high quality composites [16]. In addition, the 2% void content in composite A is due to its lower volume fraction of matrix compared to composite B. As can be seen in Table 2, the volumetric composition values for glass silk knitted fabric, Helanca knitted fabric and PVA film of composite B is less than composite A, eventhough same number of reinforcement layer were used for both composites. This is attributed to the higher total volume of the composites B due to using of high kenaf fiber content.

Table 2. The determined fibre weight fractions, density and volumetric composition of the manufactured composite A and composite B.

Composite	Weight fraction					Volume fraction					
	Kenaf woven fabric	Glass silk knitted fabric	Helanca knitted fabric	PVA film	Matrix	Kenaf woven fabric	Glass Silk knitted fabric	Helanca knitted fabric	PVA film	Matrix	Void
A	0.06	0.07	0.05	0.09	0.73	0.04	0.03	0.06	0.09	0.76	0.02
B	0.09	0.06	0.04	0.07	0.74	0.07	0.02	0.05	0.07	0.78	0.00

4.2 Flexural Properties
The failure of the samples was observed under the flexural loading. The composites samples failed due to the failure in the tension side of the samples. Figure 4 shows the flexural stress-strain trend for composites A and B. Both composites showed ductile behavior. The considerable improvement in maximum stress and failure at ultimate strain can be seen in composites type B. Addition of 2 layers of kenaf layers in the composite B increases the bending resistance of the composite B.

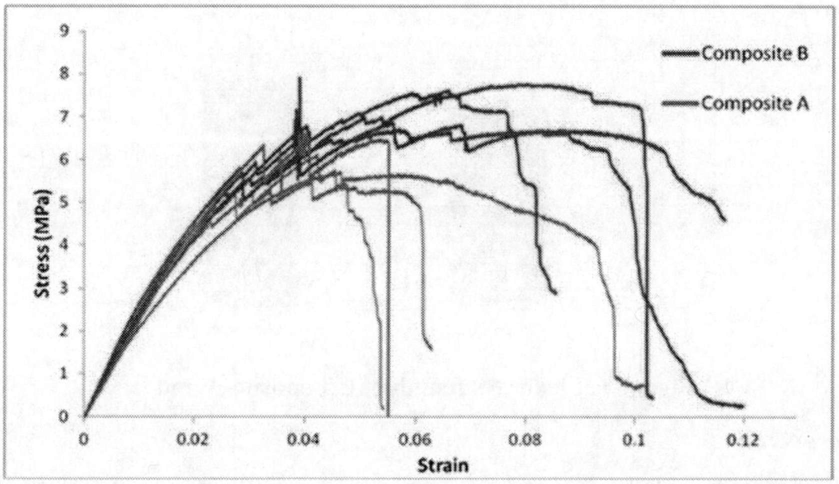

Figure 4. The flexural stress versus strain curves for composites A and B.

Figure 5 shows the flexural modulus for composites A and B. The flexural modulus for composite B (234.6 MPa) is slightly higher than composite A (229.6 MPa). It means that flexural modulus improved 2% by addition of 1 layer of kenaf fiber in composite B.

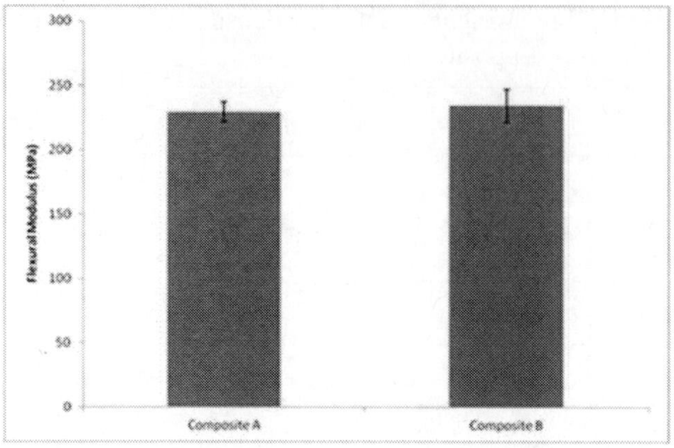

Figure 5. Flexural modulus of Composite A and B

Figure 6 shows the average values of the flexural strength of composite A and B. As can be seen the flexural strength of composite B (7.11 MPa) is 13.64 % higher than composite type A (6.14 MPa). This may be due to the fact that cellulose molecules in the kenaf fabric made a strong hydrogen bond with the matrix.

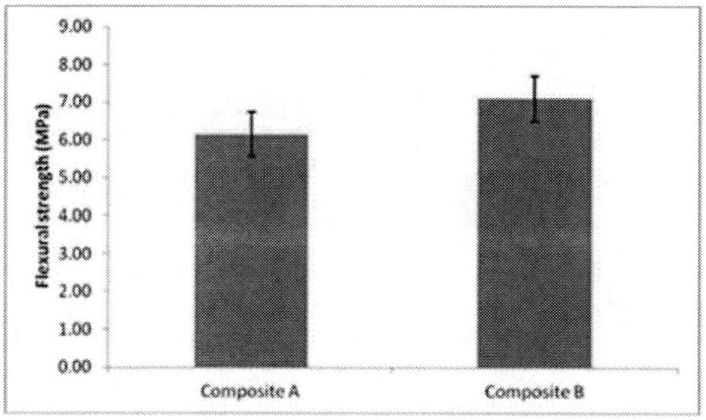

Figure 6. Flexural strength of Composite A and B

4.3 Impact properties

The impact strength of kenaf fiber woven composite is presented in Figure 7. The impact strength of composite A is 15.39kJ/m^2 whereas for composites B it is 16.7 kJ/m^2. Therefore, the results showed that two layers of kenaf woven composite provide better impact for prosthetic socket material. Addition of 2 kenaf layers in the composite B enhances the dissipation of kinetic energy. The prosthesis socket material needs to have an excellent ability on impact force which provides safety and unbreakable material while having impact load from the residual limb[4].

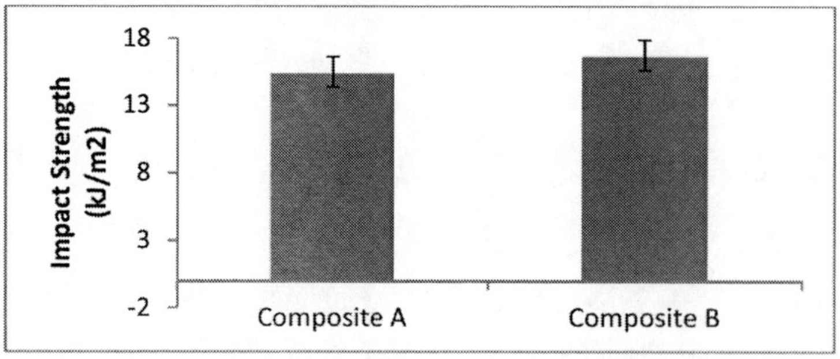

Figure 7. Impact strength of Composite A and B

4.4 The prosthetic socket ability testing
Figure 8 shows the prosthetic socket made from kenaf woven fabric composite has been attached with a human leg. The Ability of the artificial prosthetic socket has been evaluated during static and dynamic conditions. The patient felt that this prosthetic socket is more comfortable and easy to remove and refit. The feedback from the patient showed that prosthetic socket has great ability to donning by patient.

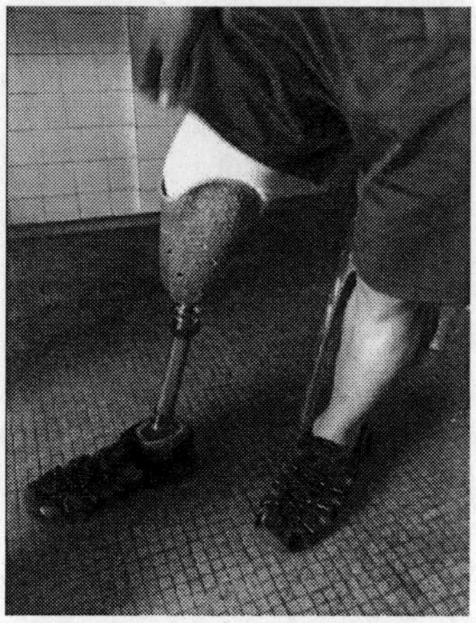

Figure 8. Patient donning the prosthetic socket made from kenaf woven fabric composite

4.5 The significance of incorporating single and double layer of kenaf in laminated composite for prosthetic leg socket
Based on result of volumetric, flexural and impact testing, the incorporation of double layer of woven kenaf fiber has shown better mechanical properties than single kenaf layer based composites. The mechanical properties are required in determining the material for prosthetic socket fabrication which should to be works as following the missing body parts [13] which expected to replace the function of lost leg.

Flexural strength is used to determine the mechanical properties of restorative materials [17]. It is important to determine the material design and flexural properties [18], due to the ability of the composite to withstand bending forces and rapture in surface of composite. The suitable material composite needs to provide an ability to endure bending forces from body weight of wearer which some activities might causes the breaking part of prosthetic socket. Therefore, the result for this study shows good properties by incorporating two layers and it could increase the ability on resist breaking by energy load at the prosthetic socket surface.

The durability of material is highly concerned in prosthetic leg socket because the physical capability of wearer might be limited due to their condition and a better impact strength could provide a benefit to active wearer as well. Thus, in composite B, which provide 2 layers of kenaf shown the better impact properties which provide resist fracture under stress applied in any stress and resist breaking under a shock load [19].

5. Conclusion

Kenaf fiber based prosthetic socket showed better mechanical properties and comfort. Therefore, kenaf woven fabric composite has the potential to develop below knee prosthetic socket as an alternative material to the existing pure glass fiber based prosthetic socket. The kenaf woven fabric composite offer good strength, environmental friendly and improves the aesthetic value of the product.

Acknowledgement

The authors gratefully thankful to Ministry of Higher Education for providing Exploratory Research Grant Scheme (ERGS) - 5527184.

References

[1] Mohammed AA, Bachtiar D, Siregar JP. Rejab M R B M and Hasany S F 2016 Physicochemical Study of Eco-Friendly Sugar Palm Fiber Thermoplastic Polyurethane Composites *BioResources* .11 9438–9454.

[2] Irawan A,Soemardi T, Widjajalaksmi K and Reksoprodjo 2011 Tensile and flexural strength of ramie fiber reinforced epoxy composites for socket prosthesis application *International Journal of Mechanical and Materials Engineering.* 6 46–50.

[3] Hanspal R, Fisher K and Nieveen R 2003 Prosthetic socket fit comfort score *Disability and Rehabilitation.* 25 1278–1280.

[4] Irawan A P and Sukania I W 2011 Mechanical Characteristics Rattan Fiber Reinforced Epoxy Composites Reco As Above Knee Socket Prosthesis Materials *The International Conference on Innovation in Polymer Science and Technology* (Bali, Indonesia: Indonesia Polymer Association.)

[5] Kramer A, Sardo K and Slocumb W 2015 Analysis of Bamboo Reinforced Composites for use in Orthotic and Prosthetic Application. *American Academy of Orthotists & Prosthetists*

[6] Saba N, Paridah M T and Jawaid M 2015 Mechanical properties of kenaf fibre reinforced polymer composite: A review *Construction and Building Materials.* 76 87–96.

[7] Chen Y, Sun L, Chiparus O, Negulescu I, Yachmenev V and Warnock M 2005 Kenaf/ramie composite for automotive headliner *Journal of Polymers and the Environment.* 13 107–114

[8] Mohd Y, Phongsakorn P, Haeryip S, Jeefferie A, Puvanasvaran P, Kamarul A and Kannan R 2011 Mechanical Properties of Kenaf/Polyester Composites *International Journal of Engineering & Technology.* 11 127–131

[9] Strait E, Mc Gimpsey G and Bradford T 2006 Limb Prosthetics Services and Devices *White Paper.* (January) 1–35.

[10] Joshi S V, Drzal, L, Mohanty A and Arora S 2004 Are natural fiber composites environmentally superior to glass fiber reinforced composites? *Composites Part A: Applied Science and Manufacturing*, 35 371–376.

[11] Etcheverry M and Barbosa S E 2012 Glass fiber reinforced polypropylene mechanical properties enhancement by adhesion improvement *Materials.* 5 1084–1113

[12] Rosalam Che Me, Rahinah Ibrahim P M T 2012 Natural Based Biocomposite Material for Prosthetic Socket Fabrication *Alam Cipta* 5 27–34.

[13] Arvela E, Soderstrom M, Alback A, Aho P S ,Venermo M and Lepantalo M 2010 Arm vein conduit vs prosthetic graft in infrainguinal revascularization for critical leg ischemia *Journal of Vascular Surgery* 52 616–623

[14] Fleer B and A Bennett Wilson J 1962 Construction of the Patellar-Tendon-Bearing Below-Knee Prosthesis *Artif. Limbs* 6 25–73.

[15] Cairns N, Murray K, Corney J and McFadyen A 2014 Satisfaction with cosmesis and priorities for cosmesis design reported by lower limb amputees in the United Kingdom: Instrument development and result *Prosthetics and orthotics international*. 38 467–73

[16] Fraisse A, Beauson J, Brøndsted P and Madsen B 2016 Thermal recycling and re-manufacturing of glass fibre thermosetting composites *IOP Conf. Ser. Mater. Sci. Eng.* 139 12020

[17] Aramide F O, Oladele I O and Folorunso D O 2009 Evaluation of the Effect of Fiber Volume Fraction on the Mechanical Properties of a Polymer Matrix Composite *Leonardo*14134–141.

[18] Shibata S, Cao Y and Fukumoto I 2006 Lightweight laminate composites made from kenaf and polypropylene fibres *Polymer Testing*.25 142–148

[19] Khan G A, Terano M, Gafur M and Alam M S 2013 Studies on the mechanical properties of woven jute fabric reinforced poly(l-lactic acid) composites *Journal of King Saud University - Engineering Sciences* 281–6.

Institute of Physics
Dirac House, Temple Back
Bristol BS1 6BE UK

ISSN: 1757-8981
ISBN 978-1-5108-6483-2
9 781510 864832